A Specialist Periodical Report

Nuclear Magnetic Resonance
Volume 1

A Review of Recent Published Literature up to June 1971

Senior Reporter
R. K. Harris, *School of Chemical Sciences, University of East Anglia*

Reporters
N. Boden, *University of Leeds*
P. Diehl, *University of Basel, Switzerland*
E. G. Finer, *Unilever Research Ltd., Welwyn*
M. I. Foreman, *University of Strathclyde*
D. G. Gillies, *Royal Holloway College, University of London*
R. Grinter, *University of East Anglia*
P. M. Henrichs, *University of Basel, Switzerland*
R. G. Jones, *University of Essex*
W. T. Raynes, *Sheffield University*
D. Shaw, *Varian Associates Ltd., Walton-on-Thames*

© Copyright 1972

The Chemical Society
Burlington House, London, W1V 0BN

ISBN: 0 85186 252 7

Library of Congress Catalog Card No. 72–78527

Printed in Great Britain by
Alden & Mowbray Ltd
at the Alden Press, Oxford

Foreword

This volume of *Specialist Periodical Reports* is the first in a series covering the topic of Nuclear Magnetic Resonance. It largely follows the practice set by other *Specialist Periodical Reports*, but it should be obvious from the list of contents that on the whole this volume is intended to be oriented towards a discussion of the phenomenon of n.m.r. itself rather than towards a summary of applications of n.m.r. in given areas of chemistry. This is more readily done for some aspects of n.m.r. than for others; thus Chapter 3, Relaxation, is heavily phenomenon-orientated, whereas Chapter 8, Macromolecules and Solids, necessarily has some degree of chemical orientation. However, a deliberate attempt has been made to exclude chapter-headings based on compound-type or on a particular n.m.r.-active nucleus (*e.g.* '^{13}C Resonance'). Clearly, the division of n.m.r. into the areas given by the chapter-headings is arbitrary. In some instances there is overlap of material; it is hoped that such overlap is helpful and not excessive. Some cross-referencing has been included, but it is not feasible or desirable to cross-reference every time overlap occurs.

The extent of coverage of the chapters varies. This is deliberate, and follows from the phenomenon-orientation of the volume. Thus, since the aim is to cover fully the *phenomenon* of n.m.r., many papers which are judged (subjectively!) to represent trivial applications of the technique are not discussed. This is most apparent in those areas of n.m.r. which are most widely used in everyday chemical research. Thus Chapter 1, Chemical Shifts, contains no discussion of many research papers which mention chemical shift data (these probably constitute the majority of all published papers referring to n.m.r.!). On the other hand, a much higher proportion of the published papers mentioning n.m.r. bandshapes is included in Chapter 6 because the total output of research in this area is lower and, at the present time, contains more novelty (even in Chapter 6, however, nearly all references merely giving coalescence temperatures are omitted).

Some other aspects of n.m.r. besides the 'trivial applications' category have also been omitted. This may be for one of two reasons:

(*a*) Omissions occur if the work is judged to be of little importance to chemistry. Most aspects in this category are of studies which would normally be classified as physics, *e.g.* most n.m.r. studies of metals and alloys.

(*b*) Some areas of n.m.r. are omitted from this volume but further advances will be treated in later volumes. The Fourier Transformation technique has not, for instance, been treated fully here, because it is known that a comprehensive review is due to appear elsewhere at about the same time as this publication. A similar comment applies to Chemically Induced Dynamic Nuclear Polarization and to INDOR (something of these topics will appear in the forthcoming *Specialist Periodical Report* on Electron Spin Resonance). Broad-line n.m.r. studies of solids have in general not been treated in the present volume, but they will be discussed in Volume 2.

It is anticipated that *Specialist Periodical Reports* volumes on n.m.r. will appear annually, and it is intended that most of the chapter-headings (and the corresponding Reporters) will be the same from year to year. However, some chapters may be biennial and occasional re-grouping of some topics may be necessary.

The literature of n.m.r. is increasing at an alarming pace (making annual Reports a necessity), as the work mentioned in this volume demonstrates. It might be objected that there are already a number of review series on n.m.r. and that the Chemical Society series is duplicating other efforts. We believe, however, that this series is of particular value for several reasons:

(*a*) It complements other volumes of the Chemical Society's *Specialist Periodical Reports*.

(*b*) Its annual survey of the literature is of great assistance to the practising chemist.

(*c*) Its phenomenon-orientation differentiates it from some other publications. In this context the *Annual Reports* — formerly *Annual Reviews* — *of N.M.R.* series, edited by E.F. Mooney, contains complementary articles, *inter alia*, under such headings as '^{19}F N.m.r. Spectroscopy' on a regular basis. Other Chemical Society articles, such as the chapter on N.m.r. Spectroscopy in the *Specialist Periodical Reports* volumes entitled 'Spectroscopic Properties of Inorganic and Organometallic Compounds', are also complementary because they are largely compound-orientated.

It is hoped that the articles in this issue represent *critical* surveys of the literature, *i.e.* that they attempt to evaluate published results as well as to mention or reproduce them. The choice of Reporters, as experts in their fields, is intended to ensure this. The critical approach is, of course, easier to request than to fulfil, particularly in view of the volume of research literature which has to be covered. However, a corollary of this approach needs to be stated, namely that it makes some comments controversial. In this context it should be made clear that each chapter is written independently and that the comments and attitudes are those of the Reporter or Reporters concerned and do not necessarily represent a consensus by all the Reporters. As an example, I quote Dr. Raynes' closely reasoned but controversial views on the notation for the direction of chemical shifts (views with which I personally am in some disagreement.)

The period of coverage of this volume may be considered abnormal, since

Foreword

it is the first of the series. The nominal coverage is from July 1970 to June 1971 inclusive, but in many instances material from the first half of 1970 is included in total (as in Chapter 6) or in part. In addition, most chapters contain some material which is introductory to the topic under consideration, and thus they refer to earlier important papers. Chapter 10 specifically covers a wider time-span (May 1969 to May 1971 inclusive) in order to make it follow consecutively on an earlier review of the subject. Naturally, a clear end to the coverage is impossible to achieve, as some journals arrive late or are mislaid, but omissions from the relevant period will be rectified in the next volume of the series.

The topicality and value of a volume of this type demands rapid publication, which bears heavily on all concerned. The problems of a first volume in a series make for special difficulties in this direction. Tribute should be paid to the Editorial Staff of the Chemical Society, and to the printers for coping admirably with this problem.

R.K. Harris
December 1971

UNITS AND NOTATION

Some explanation of policy on the matter of units and notation is particularly desirable in a volume of this type at this point in history, owing to the current debates and changes concerning SI units, and, to a lesser extent, concerning n.m.r. conventions. The problems of introducing SI do not, fortunately, impinge very greatly on n.m.r. since all the important measurable parameters of a spectrum — chemical shifts, coupling constants, and relaxation times — do not differ in the old and the new systems. However, attempts to express those parameters in terms of more fundamental quantities (*i.e.* to evaluate them theoretically) do lead to difficulties and, in particular, to changes in equations. The research literature is still largely non-SI, and the decision was taken for the current volume that each Reporter should be free to use the system of his choice provided there is self-consistency *within each chapter*. Thus Chapter 1 uses mainly SI equations (except where explicitly stated otherwise) whereas Chapter 2 basically uses non-SI expressions.

Such individual freedom seems appropriate at the present time; more uniformity may be justifiable for the future. Some particular points may, perhaps, be noted here:

(*i*) As has been mentioned in the Foreword, Dr. Raynes discusses in Chapter 1 the current controversy regarding the appropriate signs to be used for chemical shifts.

(*ii*) Magnetic susceptibilities differ by a factor of 4π in SI and c.g.s. systems because of the effects of rationalization. Thus

$$\chi_{SI} = 4\pi\chi_{c.g.s.}$$

This point is of importance when the effect of magnetic anisotropy on chemical shifts is discussed. Further, the literature contains values of $\chi_{c.g.s.}$ expressed in two ways — as cm^3 molecule^{-1} and as cm^3 mole^{-1}, differing by Avogadro's Constant.

(*iii*) One item of uniformity in the present volume is that the recommended symbol B is used for magnetic induction field rather than the symbol H, which should be reserved for magnetic field intensity. The distinction is of little significance in c.g.s. since B and H are then of the same dimensions and approximately equal. However, in SI the units of B and H differ by Ξ, the permeability of the medium, which is usually of the order of $4\pi \times 10^{-7}$ V A^{-1} s m^{-1}. Unfortunately, many n.m.r. spectroscopists still use the symbol H (and compound the confusion by quoting gauss units rather than oersteds).

(*iv*) As with other areas of chemistry, energy and entropy units are of importance for n.m.r. (*e.g.* for measurements of barriers to rate processes using n.m.r. bandshapes). Standardization on the calorie or the joule is only made in this volume within individual chapters. It may be noted that

$$1 \text{ cal}_{th} \equiv 4.184 \text{ J}$$

(*v*) There appears to be some controversy over the definition (and hence units) of the reduced coupling constant, K, in SI. The original definition by Pople and Santry[1] (which won acceptance over the proposals of Lynden-Bell and Sheppard[2]) was expressed by them in two ways, as in equations (1) and (2):

$$K_{AB} = (2\pi/\hbar\gamma_A\gamma_B) J_{AB} \qquad (1)$$

$$E_{AB} = (K_{AB})_{\alpha\beta}\mu_{A\alpha}\mu_{B\beta} \qquad (2)$$

where E_{AB} is the energy perturbation caused by the interaction of magnetic moments $\mu_{A\alpha}$ and $\mu_{B\beta}$ (summation over α and β subscripts is assumed, in the normal convention for tensors). The units are usually cm^{-3} in the c.g.s. system (the values are normally of the order of 10^{20} cm^{-3}. For SI, McGlashan[3] (or rather Whiffen) has chosen to re-define K by:

$$K_{AB} = 8\pi^2 J_{AB}/\Xi_0\gamma_A\gamma_B\hbar \qquad (3)$$

where Ξ_0 is the permeability constant ($4\pi \times 10^{-7}$ V A^{-1}s m^{-1}). Such a re-definition leaves K with the same units as reciprocal volume, as before. Earlier, however, Lynden-Bell and Harris[4] suggested the retention of the definitions (1) and (2) in SI, and quoted a number of reduced coupling constants on this basis. They then have units* of N A^{-2} m^{-3} ($\equiv$ kg m^{-2}s^{-2}A^{-2}). The relationship between the c.g.s. and SI units becomes:

$$1 \text{ N A}^{-2} \text{ m}^{-3} \equiv 10 \text{ cm}^{-3}$$

This latter method has the advantage that the definition (1) is a relationship in simple physical terms, and does not have the constant $\Xi_0/4\pi$ explicitly written (this constant will then only appear when expressions for K are derived in terms of VB or MO theory). The change of units may be felt to be a disadvantage, but it should be noted that a similar situation obtains for polarizability: in SI the equations have been retained and the 'dimensions' of the units therefore differ from those in the c.g.s. (e.s.u.) system. In Chapter 2 of this volume, Grinter has adhered to the choice of Lynden-Bell and Harris.

---

* The first printing of this book erroneously mentions the units as N A^2 m^{-3}.

[1] J.A. Pople and D.P. Santry, *Mol.Phys.*, 1964, **8**, 1.
[2] R.M. Lynden-Bell and N. Sheppard, *Proc.Roy.Soc.*, 1962, **A269**, 385.
[3] M.L. McGlashan, 'Physico-Chemical Quantities and Units', Royal Institute of Chemistry, 1971, (second Edition).
[4] R.M. Lynden-Bell and R.K. Harris, 'Nuclear Magnetic Resonance Spectroscopy', Nelson, 1969.

Contents

Chapter 1 Nuclear Shielding
By W. T. Raynes

    1 Introduction   1

    2 Chemical Shift Scales   2

    3 Basic Aspects of Nuclear Shielding   5
        A Magnetic Field-dependent Chemical Shifts   5
        B Developments in Theory   6

    4 Calculations of Nuclear Shielding   10
        A *Ab Initio* Calculations   10
        B Semi-empirical Calculations   14

    5 Transmissions of Shielding Effects within Molecules   16
        A Introduction   16
        B Inductive and Resonance Effects   17
        C Magnetic Anisotropy Effects   21
        D The Ring Current Effects   25
        E Electric Field Effects   34
        F Intramolecular Dispersion Forces   38
        G Steric Effects   40
        H Intramolecular Hydrogen-bonding   41
        I Isotope Shifts   42
        J Shielding Anisotropies   43

    6 Shieldings of Particular Nuclear Species   45
        A Introduction   45
        B Proton Chemical Shifts   45
        C Carbon Chemical Shifts   46
        D Fluorine Chemical Shifts   48
        E Nitrogen Chemical Shifts   49
        F Chemical Shifts of Nuclei Other Than H, C, F, and N   49

Chapter 2 Nuclear Spin–Spin Coupling
By R. Grinter

    1 Introduction   51

2 Basic Theory — 51

- A Ramsey's Equations — 51
- B The Average Energy Approximation — 53
- C Useful General Expressions — 53
- D The Finite Perturbation Theory (FPT) — 54
- E Coupling by π-Electrons — 55
- F The Variation Method — 55
- G SI Units — 55

3 Recent Theoretical Work — 56

- A Calculations on H_2 or HD — 56
- B *Ab Initio* Calculations of Coupling in Small Molecules — 59
- C Semi-empirical Calculations — 59
- D Theoretical Work on π-Electron Coupling — 61
- E The Anisotropy of Electron-coupled Nuclear Spin Interactions — 63
- F Solvent Effects on Coupling — 64

4 Coupling of Directly Bonded Nuclei — 65

- A Coupling between ^{13}C and ^1H — 65
- B Coupling between ^{15}N and ^1H — 67
- C Coupling between ^{31}P and ^1H — 68
- D Coupling between ^1H and Other Nuclei — 69
- E $^1J(^{13}C^{13}C)$ and $^1J(^{13}C^{14/15}N)$ — 70
- F Coupling between ^{13}C and ^{31}P — 70
- G Coupling between ^{13}C and ^{19}F — 71
- H $^1J(^{13}CM)$ where M is a Metal — 72
- I Couplings to ^{19}F, except $^1J(^{19}F^{31}P)$ — 72
- J Couplings between ^{31}P and ^{19}F — 73
- K Other One-bond Couplings involving ^{31}P — 74

5 Coupling between Atoms Separated by Two Chemical Bonds, ^2J — 77

- A Geminal Proton–Proton Coupling across Carbon — 78
- B Geminal Proton-Proton Coupling across Other Elements — 81
- C $^2J(^{13}CCH)$ — 82
- D Coupling of Other Elements to Hydrogen through Carbon — 82
- E $^2J(^{19}FP^{19}F)$ and $^2J(^{19}FPH)$ — 85
- F $^2J(^{19}FM^{19}F)$, where M is a Metal — 85
- G $^2J(^{19}FC^{19}F)$ — 87
- H $^2J(^{13}CO^{31}P)$ and $^2J(^{13}CC^{31}P)$ — 87

	I $^2J(^{31}PO^{31}P)$ and $^2J(^{31}PS^{31}P)$	87
	J $^2J(^{31}PM^{31}P)$, where M is a Metal	87
	K Other Two-bond Couplings	88
6	**Coupling between Atoms Separated by Three Chemical Bonds, ^3J**	88
	A Coupling of Hydrogen Atoms through Two Carbon Atoms	88
	B ^3J(HNCH) and ^3J(HOCH)	94
	C $^3J(^{13}CCCH)$	95
	D Coupling to Fluorine through Two Carbon Atoms. ^3J(HCC^{19}F), $^3J(^{13}CCC^{19}F)$, and $^3J(^{19}FCC^{19}F)$	95
	E Three-bond Couplings involving ^{31}P and ^{1}H	97
	F Miscellaneous Three-bond Couplings	99
7	**Coupling between Nuclei Separated by Four or More Chemical Bonds**	99
	A Long-range Coupling between Protons	100
	Long-range H–H Coupling in Saturated Systems	100
	Long-range H–H Coupling in Unsaturated Systems	101
	B Long-range Couplings involving at least One Nucleus other than H	106
	Saturated Systems—Coupling between ^{1}H and ^{19}F	106
	Saturated Systems—Coupling between Other Nuclei	106
	Unsaturated Systems	106
	C 'Through-space' Coupling	108
8	**Experimental Work of Significance for the Measurement or Interpretation of Spin–Spin Coupling Constants**	111
	A Multiple-resonance Experiments	111
	B Bandshape Analyses	111
	C Effects of Paramagnetic Materials	112
	D Other Experimental Techniques	112

Chapter 3 Nuclear Spin Relaxation
By N. Boden

1	**Introduction**	115
2	**Nuclear Spin Relaxation in Gases**	116
	A Introduction	116
	B Theoretical Developments	117

C	Diatomic Gases	118
D	Polyatomic Gases	119
	Pure Gases	119
	Stepped T_1/ρ vs. ρ Plot	119
	CH_4–O_2 and CH_4–NO Mixtures	120
E	Self-diffusion Measurements	121

3 Spin Relaxation in Liquids — 121

A	Pure Liquids	121
	Molecular Rotation	121
	Studies of Anisotropic Rotation	125
	Relaxation by internal rotation of functional groups	126
	Relaxation in water	127
	Determination of nuclear quadrupole coupling constants	128
	Studies of Scalar Relaxation	129
	Relaxation in glasses and viscous liquids	129
B	Molecular Motion and Association in Mixtures of Liquids	131
C	Electrolyte Solutions	134
	Diamagnetic Solutions	134
	Paramagnetic Solutions	135
D	Studies of Critical Phenomena in Liquids	137
E	^{13}C Relaxation Studies	138
F	Methods for Selective Measurement of T_1 and T_2	140

4 Spin-echo Experiments in Liquids — 142

A	Self-diffusion Measurements	142
B	Spin-echo Studies of Fluid Flow	143
C	Carr–Purcell Spin-echo Experiments	143

5 Nuclear Spin Relaxation and Diffusion in Liquid Crystals — 144

A	Thermotropic Liquid Crystals	144
B	Lyotropic Liquid Crystals	146

6 Spin Relaxation in Solids — 147

A	N.M.R. Saturation in Solids	147
B	Nuclear Dipolar Relaxation	148
C	Relaxation by Rotational Motion	149
D	Spin–Lattice Relaxation by Spin–Rotational and Anisotropic Chemical Shielding Interactions	150

  E Spin–Lattice Relaxation due to Paramagnetic Impurities 151
  F Spin–Lattice Relaxation by Nuclear Quadrupole Interaction 152
  G Spin–Lattice Relaxation Time Measurements in Miscellaneous Solids 153
    Molecular Solids 153
    Ionic Solids 155
    Antiferromagnetic Solids 156
    Metallic Hydrides 156
    Hexagonal Ice 156
    Molecules Bound to Surfaces 156
  H Self-diffusion in Molecular Crystals 157
  I Ferroelectric Phase Transitions 159

7 Pulsed N.M.R. in Solids 161

  A Application of Multiple-pulse Experiments to Line-narrowing in Solids 161
  B Application of Multiple-pulse Experiments to the Study of Molecular Motion 164
  C Pulsed Double Nuclear Magnetic Resonance Experiments 165

Chapter 4 Experimental Techniques
By D. G. Gillies

1 Introduction 169

2 The Magnet 170

  A Permanent Magnets 170
  B Electromagnets 170
  C Superconducting Magnets 170
  D Control of Homogeneity 170
  E Stability 171

3 Sensitivity 171

4 Frequency Generation 171

  A Single Resonance 171
  B Double Resonance 174
    Homonuclear 174
    Heteronuclear 175
    Noise 175
    Universal Shift Reference 176

5 The Probe		176
A	Single-coil Systems	176
B	Crossed-coil Systems	177
C	Double Resonance	178
D	Field–Frequency Lock	178
E	General Construction	178
F	Field Modulation	179
G	Variable-temperature Operation	179
H	Rapid Sample Spinning	180

6 The Preamplifier — 180

7 The Receiver — 180

8 Modulation Schemes — 182

9 Computer Techniques — 184

10 Pulsed Spectrometers — 185

11 Measurement of Relaxation Times		188
A	Values of T_1	188
	C.W. Methods	188
	Pulse Methods	189
B	Values of T_2	190
	C.W. Methods	190
	Pulse Methods	190

12 Miscellaneous Topics		190
A	Stochastic Resonance	190
B	Fourier Difference Spectroscopy	190

Chapter 5 Spectral Analysis
By R. G. Jones

1 New Spin Systems Studied — 191

Analysis of N.M.R. Spectra of the Type $[AMX]_2$ with $J(XX') = 0$ — 191

2 New Methods of Tackling Known Spin Systems — 195

A Superoperator Direct Method Least-squares Analysis of N.M.R. Spectra — 195

B A Systematic Approach to the $[AB]_2$, $[AB]_2MX$, and Derived Systems — 196

C Use of Double Quantum Transitions	198
Double Resonance: Perturbation of Double Quantum Transitions	198
Effect of Double Irradiation (Tickling) of Single Quantum Transitions or Double Quantum Transitions in Systems Simplified by Double Resonance	199
3 Studies of Known Spin Systems	**200**
A Systems involving Protons Only	200
Aliphatic Systems	200
Olefinic Systems	200
Heterocyclic Systems	202
Aromatic Systems	202
B Compounds containing Fluorine	203
2,2'-, 3,3'-, and 4,4'-Dihalogenobiphenyl Compounds ABCD, ABCDX	203
Pyridine Compounds	204
Pentafluorophenol, $(AM)_2X$	204
Benzene, Naphthalene, Phenanthrene, and Biphenyl Compounds	204
Tetrafluorodichlorocyclopropanes, $[AX]_2$, A_4	204
1,1,2,2-Tetrafluoroethane $[AMX]_2$	205
C Compounds containing Phosphorus	205
Bis(dialkylphosphines)	205
The $[AMX_{18}]_2$ System	206
Tri-(3-furyl)phosphine, ABCX	206
1-Phospha-2,6-dioxacyclohexanes	206
1-Phospha-2,5-dioxa(diaza, dithia)cyclopentanes	207
Bis-(5,5-dimethyl-1-phospha-2-oxo-6-thionocyclohexane)	208
D Organometallic Compounds	208
Triphenyl Group IV Element Lithium Compounds, $[AB]_2C$	208
Allenic Derivatives of Tin, Lead, and Mercury AB_2X	208
Cycloheptatrienyl Manganese Tricarbonyl Compounds, $A[BCXY]_2$, ABCDEXYZ	208
Aryl Derivatives of Mercury, $[AB]_2C$, $[AB]_2X$, $[AB]_2$	209
4 General Comments	**209**
Books and Reviews	209
Computer Programs	210

Chapter 6 Bandshape Phenomena for Fluids
By R. K. Harris

1 General Introduction		211
A	Current Developments	211
B	Coverage of the Report	212
C	Nomenclature, Notation, and Units	213
D	Reviews and Symposia	214
2 Experimental Features		215
	Miscellaneous	218
3 Exchange of Magnetic Sites		218
A	Theoretical Work	218
B	Examples of Intramolecular Exchange	226
	Hindered Internal Rotation	226
	Nitrogen Inversion	230
	Ring Inversion	231
	Phosphorus-containing Compounds	233
	Metal Complexes	236
C	Examples of Intermolecular Exchange	237
	Studies of Ligand Exchange (The Swift–Connick Approach)	238
	Other Studies of Intermolecular Exchange	240
4 Effects of Quadrupolar Nuclei		242
A	Theoretical Work	242
B	Resonance of Spin-$\frac{1}{2}$ nuclei Coupled to Quadrupolar Nuclei	245
C	N.M.R. Spectra of Quadrupolar Nuclei	246
	Nitrogen	246
	Other Nuclei	247
D	Effects of Magnetic Site Exchange on N.M.R. Spectra for Spin Systems containing Quadrupolar Nuclei	248
5 Relaxation Effects — Linewidths		250
A	Theoretical Work	250
B	Paramagnetic Effects	251
C	Gas-phase Studies	252
D	Miscellaneous Width Studies	253
6 Relaxation Effects—Saturation		253
7 Bandshapes in Multiple Resonance		254
	Cases including Exchange	255

Contents xvii

Chapter 7 Multiple Resonance
By D. Shaw

1	Introduction	257
2	Theory of Multiple Resonance	257
3	Double Resonance in the Presence of Chemical Exchange	260
4	Chemical Applications of Multiple-resonance Techniques	262
	Spin Decoupling	262
	Generalized Overhauser Effect	264
	INDOR	264
	Nuclear Overhauser Effect (NOE)	265
	Transient Phenomena	266
5	Relaxation Effects in Multiple Resonance	267
	Relaxation in the Presence of Incoherent Decoupling	267
	Overhauser Studies	268
6	Conclusion	271

Chapter 8 Macromolecules and Solids
By E. G. Finer

1	Introduction	273
2	N.M.R. Studies of Macromolecules	273
	A Lipid–Water and Soap–Water Systems	273
	Pulse and Wide-line Methods	273
	High-resolution Methods	276
	B Polyamino-acids, Peptides, and Proteins	278
	Proton Studies	278
	^{13}C Studies	282
	^{19}F Studies	283
	C Synthetic Polymers	283
	D Other Macromolecules	285
3	N.M.R. Studies of Solids	286
4	N.M.R. Studies of Small Molecules Interacting with Macromolecules or Solids	289
	A Bound Water	289
	B Bound Ions	292
	C Other Small Molecules	293

Chapter 9 Medium Effects
By M. I. Foreman

1 Introduction	295
2 Hydrogen-bonding Effects	295
A Proton Shifts	295
B Heteronuclear Shifts	297
3 Ions in Solution	300
4 Aromatic Solvent-induced Shifts (ASIS Effects)	303
5 N.M.R. Shift Reagents	310
A Cobalt(II) and Nickel(III) Complexes	310
B Lanthanide Metal Complexes	315

Chapter 10 Oriented Molecules
By P. Diehl and P. M. Henrichs

1 Introduction	321
2 Structure Determination	321
3 Orientation Parameters	324
4 Anisotropies in Indirect Couplings	326
5 Anisotropies of Chemical Shifts	327
6 Relaxation	329
7 Determination of Nuclear Quadrupole Coupling Constants	330
8 The Structure of the Liquid Crystal Phase	331
9 Miscellaneous	332

Author Index	333

1
Nuclear Shielding

BY W. T. RAYNES

1 Introduction

Three alternative procedures suggest themselves for a review on the subject of nuclear magnetic shielding in molecules. First, one could consider the various chemical and physical influences (*e.g.* the inductive effect, conjugation, magnetic anisotropy) which are believed to be responsible for the differences in shielding between different compounds. These could be discussed in turn, each one being illustrated by examples drawn from the shieldings of all species of magnetic nuclei in a wide range of compounds. Secondly, one could consider the shieldings of a particular nuclear species (*e.g.* the proton), illustrate the shielding changes that occur along several series of compounds containing that species of nucleus, and interpret these changes in terms of varying contributions from one or more of the above-mentioned phenomena. This would then be repeated in turn for other species of nucleus. The third approach involves selecting a particular compound (*e.g.* $C_6H_5NO_2$) and showing how each of the chemical and physical influences makes its own contribution to the shielding of all the nuclei in the compound. This would then be repeated in turn for other compounds.

Of these procedures, it is traditionally the second that has been adopted for dealing with observed shieldings in a wide range of compounds. The third procedure is seldom used, partly because much of the information required (*e.g.* carbon, nitrogen, and oxygen shieldings) has been unavailable, at least until very recently, and partly because it is obviously unrealistic except when dealing with a group of closely related compounds. Since the present review is 'phenomenon-oriented' rather than 'compound-oriented' the first procedure will, for the most part, be adopted. However, because many workers tend to study the shieldings of only one nuclear species — presumably because of personal interest or limited experimental facilities — a brief survey of recent work in terms of individual species of nuclei has also been included.

This review covers work that appeared in the literature between July 1st 1970 and June 30th 1971. However, since this is the first in the present review series, some important papers appearing in the first half of 1970 will also be discussed. Because of the vast abundance of publications on nuclear shielding, some restrictions have had to be imposed. For instance, no space has been given to experimental aspects, and the details of wave-mechanical calculations have, for the most part, been excluded. The ways in which nuclear shieldings can be changed by intermolecular effects (*e.g.* dispersion forces, reaction fields) are omitted here but are discussed in Chapter 9. The main concern in this chapter is the interpretation of recent work on observed shieldings of isolated molecules in terms of the various physical and chemical influences which chemists have found to be useful for their understanding. A further omission, requiring the Reporter's apologies, is the discussion of papers in foreign language journals.

Two conventions that are used in this chapter will be stated at this point. For the sake of economy, most numerical values of shieldings and chemical shifts are given without the appellation p.p.m. (parts per million). Secondly, the convention has been adopted that the chemical shift is positive if the nucleus under consideration has a more positive shielding constant than the reference nucleus. The reason for this choice will be clarified in Section 2.

2 Chemical Shift Scales

It is appropriate to commence with a brief discussion of chemical shift scales — a topic which has aroused some controversy. The aim here is not to approve particular choices of reference for particular nuclei but to deal with two general points which need to be stressed. A more detailed review of the definition of chemical shifts has been given by Rummens.[1]

In defining chemical shifts there are two possible starting points, which may be termed 'theoretical' and 'experimental'. The theoretical approach starts with a molecule in a uniform magnetic field B. The field B_{local} at a selected point in the vicinity of a molecule is different from B because of the small field B' arising from the induced motions of the electrons. Thus:

$$B_{local} = B - B' \qquad (1)$$

The negative sign is placed here explicitly so that one may refer to the phenomenon as 'magnetic shielding', as any point magnetic dipole (of very small magnitude) placed at the selected point would be shielded from the full effect of B by the influence of the induced electronic motions. We are implying that B' is positive but, of course, this does not exclude the possibility of a negative value of B', in which case the phenomenon is termed 'antishielding'. Provided that the field B is not too large, B' is proportional to B, so that:

$$B' = \sigma B \qquad (2)$$

where σ is the magnetic shielding constant. The value of σ obviously depends on the location of the selected point. (It also depends, in general, on the direction of B relative to axes fixed in the molecule. However, no loss of significance for the arguments given below occurs if we imagine the molecule to be tumbling freely in the field with the point-dipole held in a fixed position relative to molecule-fixed axes.) With the negative sign given explicitly in

[1] F. H. A. Rummens, *Org. Magn. Resonance*, 1970, **2**, 209.

equation (1), it follows that σ is positive when shielding occurs and negative when antishielding occurs.

In practice, one is almost always interested in the value of σ at the site of a particular magnetic nucleus in the molecule. For these special positions σ is called the nuclear (magnetic) shielding constant. These nuclear shielding constants are, of course, fundamental physical properties of a molecule and are, therefore, the quantities which are obtained from quantum-mechanical calculations. From the standpoint of chemical theory, however, it is often the difference between the nuclear shielding constant in the compound of interest and that in some suitable reference compound (σ_{ref}) with which one is concerned. This quantity δ_{th} is the chemical shift and is defined by equation (3). There are three points to be noted here. First, a definition of δ_{th} as being $\sigma_{\text{ref}} - \sigma$ would be unsatisfactory: as one would say in everyday language, it

$$\delta_{\text{th}} = \sigma - \sigma_{\text{ref}} \qquad (3)$$

would be 'illogical'. Secondly, the choice of reference depends on the particular nucleus in the particular compound of interest. For instance, the proton chemical shifts of the halogen derivatives of methane would be referred to the protons of methane. Again, the nitrogen chemical shifts of the substituted derivatives of pyridine would be referred to the nitrogen of pyridine. The chemical shifts of the protons of the methyl halides relative to tetramethylsilane in the first example, and the chemical shifts of the nitrogen nuclei of the substituted pyridines relative to the nitrogen of, say, Me_4N^+ in the second example, would be quantities devoid of theoretical interest. Thirdly, our definition of δ_{th} is in no way dependent upon the way in which chemical shifts are measured or, indeed, upon the very existence of nuclear magnetic resonance.

The experimental approach to defining the chemical shift starts with the observation of signals on an n.m.r. spectrometer. Two experimental procedures are used; 'frequency-sweep', in which the field is held at a fixed value and the frequency is varied through resonance, and 'field-sweep', in which the frequency is held fixed and the field is varied through resonance. For the frequency-sweep experiment one defines the chemical shift δ_v of the nucleus of interest from the reference nucleus by equation (4), where v and v_{ref}

$$\delta_v = (v - v_{\text{ref}})/v_{\text{ref}} \qquad (4)$$

are the frequencies required for resonance of the nucleus of interest and of the reference nucleus respectively. From the resonance condition of equation (5),

$$v = \gamma B(1-\sigma)/2\pi \qquad (5)$$

where γ is the magnetogyric ratio, we see that the definition δ_v becomes that given by equation (6), where we have assumed that the nucleus of interest

$$\delta_v = (\sigma_{\text{ref}} - \sigma)/(1 - \sigma_{\text{ref}}) \qquad (6)$$

and the reference nucleus are of the same species. If these two nuclei are different then equation (6) is replaced by one involving γ and γ_{ref}.

For the field-sweep experiment, the chemical shift δ_B is defined by equation (7), where B and B_{ref} are the fields required for resonance of the nucleus of

$$\delta_B = (B - B_{\text{ref}})/B_{\text{ref}} \qquad (7)$$

interest and the reference nucleus respectively. Using the resonance condition of equation (5) this definition becomes equation (8), where, again, it has been

$$\delta_B = (\sigma - \sigma_{\text{ref}})/(1 - \sigma) \qquad (8)$$

assumed that the nucleus of interest and the reference nucleus are of the same species.

The difference between the two experimental definitions, which is a difference of sign, will be discussed first. Which definition is preferable? In the early days of n.m.r. the definition δ_B was used exclusively. More recently, authors have been using δ_ν. Since the concern here is with a definition, the answer to the question is a matter of opinion, based on physical intuition. I argue here in favour of the use of δ_B over δ_ν. On the δ_B scale a nucleus that is highly shielded has a high chemical shift; whereas on the δ_ν scale it has a low chemical shift. It runs counter to physical sense, for instance, that the protons of methyl fluoride be given a higher chemical shift than those of methane when, in fact, it is the protons of methane that are the more highly shielded. (One could define a temperature scale in which the boiling point of water was 0° and the freezing point 100°. However, this has never been done.) Another advantage of δ_B over δ_ν is the near identity of the definitions of δ_B and δ_{th} on account of the smallness of σ_{ref}. However, here it should be pointed out that with increasing accuracy in chemical shift measurement,[2] δ_B and δ_{th} will cease to be equivalent in practice. For the important field of proton resonance, in which tetramethylsilane (TMS) is the most commonly used reference, our arguments favour the 'τ-scale' to the 'δ-scale'.

In a very recent paper[3] the definition δ_ν has been suggested on the grounds that 'most of the newer instruments for multinuclear studies use a frequency sweep, so that a scale with chemical shifts positive to increasing frequency seems logical'. Still more recently, Brey[4] has supported δ_ν on essentially the same grounds with the words 'such a convention seems absolutely necessary to maintain the sanity of anyone who tries to operate a modern spectrometer using a heteronuclear lock or applying internuclear double resonance'. To the present author there seems no real difficulty in using a chemical shift scale in the laboratory that is suitable for the purposes of measurement but reporting results on a scale most suitable for theoretical discussion. After all, molecular spectroscopists have for years measured their absorption or emis-

[2] K. W. Gray and I. Ozier, *Phys. Rev. Letters*, 1971, **26**, 161.
[3] E. D. Becker, *J. Magn. Resonance*, 1971, **4**, 142.
[4] W. S. Brey, *J. Magn. Resonance*, 1971, **5**, 159.

sion lines on a wavelength scale but reported their results on a frequency scale. (A frequency scale is necessary here since one is directly interested in energies and energy differences whereas in n.m.r. spectroscopy one is interested in shieldings and shielding differences.) Thus in reply to Becker and Brey one may pose the argument that if a defined scale of chemical shifts is to be set up, then the requirement that increased chemical shift should parallel increased shielding should take precedence over one based upon the particular fashion in which chemical shifts are measured. The former is absolute whereas the latter could conceivably change with time, as fashions sometimes do.

The second point to be considered is to question the need, in the long term, for any kind of chemical shift scale based on a chosen reference nucleus. The nuclear shielding constant is the molecular parameter in which one is interested. The experimental determination of the chemical shift using either definition δ_v or δ_B is only a step on the way to determining σ. What is required with some urgency is the direct determination by experiment of σ for some suitable chemical compound which would then serve as the primary reference standard. The best available value at present[5] is for water at 25 °C, which has a proton shielding constant (corrected for the bulk susceptibility of the water):

$$\sigma(H_2O, 25\,°C) = 25 \cdot 97(\pm 0 \cdot 30)\,\text{p.p.m.} \qquad (9)$$

Obviously, considerable improvement in precision is called for. However, once a precise value for water or another chosen compound has been obtained it will be possible for accurate and precise values of the shielding constants of the secondary reference standards used in experiments (*e.g.* TMS at a given concentration in carbon tetrachloride) to be found. (Another quantity of particular importance, so far experimentally uninvestigated, is the temperature dependence of the shielding constant for any substance.) What is being argued for here is what one might call the 'σ-scale' *i.e.* the reference nucleus is any nucleus stripped of its electrons, in which case, of course, σ_{ref} is zero.

3 Basic Aspects of Nuclear Shielding

A. Magnetic Field-dependent Chemical Shifts.—In an important paper Ramsey[6] has examined the possibility of observing magnetic field-dependent nuclear shielding. For very large applied fields equation (2) will no longer adequately express the dependence of B' on B, since σ is itself field-dependent. For a particular component $\sigma_{\alpha\beta}$ of the nuclear shielding tensor we would have to write (in standard tensor notation):

$$\sigma_{\alpha\beta} = \sigma_{\alpha\beta}^{(0)} + \tfrac{1}{2}\sigma_{\alpha\beta\gamma\delta}^{(2)} B_\gamma B_\delta + \ldots \qquad (10)$$

where $\sigma_{\alpha\beta}^{(0)}$ is the value of $\sigma_{\alpha\beta}$ for low values of B (*i.e.* the usual shielding constant) and $\hat{\sigma}^{(2)}$ is a tensor which allows for the initial change in shielding

---

[5] T. Myint, D. Kleppner, N. F. Ramsey, and H. G. Robinson, *Phys. Rev. Letters*, 1966, **17**, 405.
[6] N. F. Ramsey, *Phys. Rev. (A)*, 1970, **1**, 1320.

when the field is large. The tensor $\hat{\sigma}^{(1)}$ has been omitted from equation (10) since all of its components are zero, as is required by the independence of the shielding to the sign of B.

Let it be assumed that the molecule is tumbling rapidly in the external field. Only averages over all orientations need be considered, so that equation (10) becomes equation (11), where $\sigma^{(0)}$ and $\sigma^{(2)}$ are obtainable from the

$$\sigma = \sigma^{(0)} + \sigma^{(2)} B^2 + \ldots \qquad (11)$$

components of $\hat{\sigma}^{(0)}$ and $\hat{\sigma}^{(2)}$ respectively. For the observation of the chemical shift of a given nucleus from that of a reference nucleus, equation (11) gives equation (12), where $\delta^{(0)}$ is the chemical shift at low field. It can be seen that if

$$\delta = \delta^{(0)} + (\sigma^{(2)} - \sigma^{(2)}_{\text{ref}}) B^2 \qquad (12)$$

$\sigma^{(2)} - \sigma^{(2)}_{\text{ref}}$ is large enough, or if the precision of the method of determining δ is sufficiently high, it will be possible to obtain different values of δ at, say, 100 MHz and 300 MHz.

To obtain expressions for the components of $\hat{\sigma}^{(2)}$ it is necessary to go to the fourth order of perturbation theory. Ramsey considers a typical component of $\hat{\sigma}^{(2)}$ and, in an order of magnitude calculation, shows that $\sigma^{(2)} B^2 / \sigma^{(0)}$ is of the order of 10^{-9} for typical values of $\sigma^{(0)}$. However, for nuclei with shielding constants of 10^{-2} (as found, for instance, for cobalt in some of its complexes) this ratio becomes large enough for the detection of field-dependent chemical shifts. An initial search for such shifts has not found them[7] but with the dramatic improvement in the precision of chemical shift measurement recently reported by Gray and Ozier[2] the prospects of observing such shifts have been considerably enhanced.

Two other phenomena have to be considered when measuring chemical shifts with large magnetic fields (or at very high precision). If an external reference is employed then the magnetic susceptibilities of the sample and reference may be changed by different amounts by the field so that the bulk susceptibility correction will be changed. Secondly, for substances in which there is an anisotropy in the molecular susceptibility certain orientations will be preferred, so that the observed shielding will not be fully averaged over all directions of the molecule in the magnetic field. Hence the shielding will be different from that at lower fields, when no significant molecular orientation occurs.

B. Developments in Theory.—The familiar expression of Ramsey consists of two terms. In deriving this expression it is customary to take the origin of the vector potential of the external magnetic field at the nucleus of interest. However, there is no particular need to make this choice. Some time ago it was pointed out[8] that calculations could be simplified by other gauge choices. Let us assume that the origin of co-ordinates is at the nucleus of interest and

[7] S. I. Chan, personal communication.
[8] S. I. Chan and T. P. Das, *J. Chem. Phys.*, 1962, 37, 1527.

that the origin of the vector potential is at a point with a position vector $\mathbf{r}_0$ from the co-ordinate origin. Then we obtain, for the component $\sigma_{\alpha\beta}$, Ramsey's equation in its most general form. Thus:

$$\sigma_{\alpha\beta} = \sigma_{\alpha\beta}^{d} + \sigma_{\alpha\beta}^{dg} + \sigma_{\alpha\beta}^{p} + \sigma_{\alpha\beta}^{pg} \qquad (13)$$

where, in SI units*, $\sigma_{\alpha\beta}^{d}$, $\sigma_{\alpha\beta}^{dg}$, $\sigma_{\alpha\beta}^{p}$, and $\sigma_{\alpha\beta}^{pg}$ are defined as in equations (14)—(17).

$$\sigma_{\alpha\beta}^{d} = \frac{\Xi_0 e^2}{2m} \langle 0 | \sum_k r_k^{-3}(r_k^2 \delta_{\alpha\beta} - r_{k\alpha} r_{k\beta}) | 0 \rangle \qquad (14)$$

$$\sigma_{\alpha\beta}^{dg} = \frac{\Xi_0 e^2}{2m} \langle 0 | \sum_k r_k^{-3}(r_{0\alpha} r_{k\beta} - r_{0\gamma} r_{k\gamma} \delta_{\alpha\beta}) | 0 \rangle \qquad (15)$$

$$\sigma_{\alpha\beta}^{p} = \frac{\Xi_0 e^2}{2m^2} \sum_n{}' \frac{\langle 0|\sum_k r_k^{-3} l_{k\alpha}|n\rangle \langle n|\sum_k l_{k\beta}|0\rangle + \langle 0|\sum_k l_{k\beta}|n\rangle \langle n|\sum_k r_k^{-3} l_{k\alpha}|0\rangle}{W_0 - W_n} \qquad (16)$$

$$\sigma_{\alpha\beta}^{pg} = \frac{\Xi_0 e^2}{2m^2} \varepsilon_{\beta\gamma\delta} r_{0\gamma} \sum_n{}' \frac{\langle 0|\sum_k r_k^{-3} l_{k\alpha}|n\rangle \langle n|\sum_k p_{k\delta}|0\rangle + \langle 0|\sum_k p_{k\delta}|n\rangle \langle n|\sum_k r_k^{-3} l_{k\alpha}|0\rangle}{W_0 - W_n}$$

(17)

In equations (14)—(17) the symbols Ξ_0, e, and m denote respectively the permeability of free space, the electronic charge, and the electronic mass. $\mathbf{r}_k$ is the position vector of the k'th electron from the nucleus of interest, W_n represents the energy of the n'th excited state, $\mathbf{p}_k$ denotes the linear momentum operator of the k'th electron, and $\mathbf{l}_k (= \mathbf{r}_k \times \mathbf{p}_k)$ is the orbital angular momentum operator of this electron. The prime on the summations in equations (16) and (17) denotes a summation over all values of n except $n = 0$, including the continuum of excited states. $\delta_{\alpha\beta}$ is the substitution tensor ($= 1$ if $\alpha = \beta$; $= 0$ if $\alpha \neq \beta$) and $\varepsilon_{\beta\gamma\delta}$ is the alternating tensor [$= 1$ if $(\beta\gamma\delta)$ is an even permutation of (xyz); $= -1$ if $(\beta\gamma\delta)$ is an odd permutation of (xyz); $= 0$ if any two of $(\beta\gamma\delta)$ are identical]. It is obvious that if $\mathbf{r}_0 = 0$, equation (13) reduces to the familiar two-term expression $\sigma_{\alpha\beta}^{d} + \sigma_{\alpha\beta}^{p}$, with the 'conventional' diamagnetic and paramagnetic terms respectively. The terms $\sigma_{\alpha\beta}^{dg}$ and $\sigma_{\alpha\beta}^{pg}$ are both gauge-dependent — hence the superscript g. Here we are using the words 'diamagnetic' and 'paramagnetic' to denote terms that need respectively the ground-state wavefunction only and the eigenfunctions of the molecule in all its electronic states for their calculation.

During the review period there has been no major development in the general theory of nuclear shielding. Here we comment on a short communication by Sadlej.[9] The most common method of dealing with the intractable

---

* For an introduction to the use of SI units in n.m.r. the reader is referred to the book 'Nuclear Magnetic Resonance Spectroscopy', by R. M. Lynden-Bell and R. K. Harris, Nelson, London, 1969. Equations (14)—(17) may be written in c.g.s. units by deleting Ξ_0 and introducing the factor c^2 into the denominator of each equation.

[9] A. J. Sadlej, *Mol. Phys.*, 1970, **19**, 749.

paramagnetic terms is by use of the average-energy approximation. For a linear molecule with the bond axis as the z-axis and the choice $r_k = 0$, we obtain for the nucleus of interest with this approximation:

$$\sigma_{xx} = \frac{\Xi_0 e^2}{2m} \langle 0| \sum_k \frac{y_k^2 + z_k^2}{r_k^3} |0\rangle - \frac{\Xi_0 e^2}{m^2 \Delta E} \langle 0| \sum_{k,k'} r_k^{-3} l_{kx} l_{k'x} |0\rangle, \quad (18)$$

where ΔE is the mean excitation energy. For a gauge change along the z-axis to the point z_0, we now obtain, using equations (15) and (17), with the average energy approximation in the latter, the new contributions σ_{xx}^{dg} and σ_{xx}^{pg}, as defined in equations (19) and (20).

$$\sigma_{xx}^{dg} = -\frac{\Xi_0 e^2}{2m} z_0 \langle 0| \sum_k z_k r_k^{-3} |0\rangle \quad (19)$$

$$\sigma_{xx}^{pg} = \frac{\Xi_0 e^2}{m^2 \Delta E} z_0 \langle 0| \sum_{k,k'} l_{kx} p_{k'y} r_k^{-3} |0\rangle \quad (20)$$

Now in order that σ_{xx} be gauge-invariant, we follow Sadlej by imposing the equality of σ_{xx}^{dg} and $-\sigma_{xx}^{pg}$. This determines ΔE, and so equation (18) becomes equation (21). This expression for σ_{xx} is entirely dependent on the

$$\sigma_{xx} = \frac{\Xi_0 e^2}{2m} \langle 0| \sum_k r_k^{-3}(y_k^2 + z_k^2)|0\rangle - \frac{\Xi_0 e^2}{2m^2} \frac{\langle 0| \sum_{k,k'} l_{kx} l_{k'x} r_k^{-3} |0\rangle \langle 0| \sum_k z_k r_k^{-3} |0\rangle}{\langle 0| \sum_{k,k'} l_{kx} p_{k'y} r_k^{-3} |0\rangle} \quad (21)$$

ground-state wavefunction and is also gauge-invariant. In effect, Sadlej's result is equivalent to an earlier result of Kern and Lipscomb[10] in which one imposes the identity of equations (16) and (17), thereby selecting the gauge which causes σ_{xx}^p and σ_{xx}^{pg} to cancel one another. However, Sadlej's derivation relies on a smaller number of assumptions. The numerical validity of equation (21) has been extensively studied and it does appear to give good results for the shielding of nuclei bonded to heavier nuclei and, in particular, for proton shielding.

A series of important papers dealing with the π-electron contribution of aromatic molecules to proton magnetic shielding has recently appeared.[11] This treatment employs the current-density concept in contrast to the test-dipole concept of McWeeny.[12] An expression for the contribution $J'_\pi(r)$ of the π-electrons to the total current density is obtained and shown to be gauge-invariant.[11a] The main concern of these papers, however, is with the so-called 'London Approximation'. With the current-density approach

[10] C. W. Kern and W. N. Lipscomb, *Phys. Rev. Letters*, 1961, **7**, 19; *J. Chem. Phys.*, 1962, **37**, 260.
[11] (a) A. T. Amos and H. G. Ff. Roberts, *Mol. Phys.*, 1971, **20**, 1073; (b) *ibid.*, p. 1081; (c) *ibid.*, p. 1089.
[12] R. McWeeny, *Mol. Phys.*, 1958, **1**, 311.

(and one application of this approximation) Amos and Roberts obtain for the secondary field (F') perpendicular to the plane of the ring, induced by an external field B, the expression:

$$F' = -\frac{\Xi_0 e^2 B}{2m} \sum_{s,t=1}^{M} C_{st} \int \frac{r \times (\omega_s \nabla \omega_t - \omega_t \nabla \omega_s)}{r^3} dr \quad (22)$$

where C_{st} is a parameter obtainable from the zero- and first-order bond matrices, r denotes the position vector whose origin is taken at the proton for which the shielding is desired, and ω_s and ω_t are two of the $M\,2p_z$-type orbitals on the carbon atoms s and t ($s \neq t$). The London Approximation is:

$$r\,\omega_s\omega_t \simeq \tfrac{1}{2}(R_s+R_t)\omega_s\omega_t \quad (23)$$

where R_s and R_t are the position vectors of the origins of the s and t $2p_z$ orbitals respectively from the proton of interest. It is based on the fact that ω_s and ω_t are localized so that it is only where they overlap, i.e. about the point $r = \tfrac{1}{2}(R_s+R_t)$, that the product $r\omega_s\omega_t$ will have any significant magnitude. This approximation was used by Pople[13] in his well-known MO treatment of molecular magnetic properties. There is no difficulty with the London Approximation (LA) for the numerator of equation (22). However, for the denominator Amos and Roberts consider the two possibilities:

$$\text{LA I} \quad r^{-3} \simeq \left|\tfrac{1}{2}(R_s+R_t)\right|^{-3} \quad (24)$$

$$\text{LA II} \quad r^{-3} \simeq \tfrac{1}{2}(R_s^{-3}+R_t^{-3}) \quad (25)$$

Calculations with these approximations yield absolute contributions to the proton shielding in benzene of $-1\cdot05$ and $-2\cdot40$ respectively. Since the observed non-local contribution to the shielding is about $-1\cdot5$ (but possibly smaller[14]), it seems that the choice LA I is to be preferred. Further support for LA I is provided by a calculation of the chemical shifts of the α and β protons in naphthalene relative to benzene, in which both LA I and LA II overestimate the shifts but LA I by considerably less than LA II.

Nevertheless, a more exact calculation[11b] proves that the fair agreement with experiment that LA I gives is rather fortuitous. Avoiding the London Approximation altogether and calculating the various integrals by numerical methods, Amos and Roberts have demonstrated that the apparent superiority of LA I arises from the cancellation of errors. The more exact calculation gave $-1\cdot07$ for the non-local contribution to the proton shielding in benzene. No real improvement occurs for a scaled version of LA II. In the third paper[11c] the test-dipole approach is used to calculate the proton shieldings in a number of aromatic molecules. This work shows that LA I is as good as LA II and the scaled LA II (usually used in test-dipole calculations). However, none of the three methods gives good agreement with the experimental shifts of these aromatic hydrocarbons from benzene.

[13] J. A. Pople, *J. Chem. Phys.*, 1962, **37**, 53.
[14] J. A. Pople, *J. Chem. Phys.*, 1964, **41**, 2559.

McWeeny[15] has discussed the question of gauge invariance in the calculation of magnetic properties and he has proved that the required matrix elements are invariant against translation through the uniform magnetic field as well as against change in the origin of the vector potential. With regard to the gauge-invariant atomic orbitals χ_n frequently used in the calculations of magnetic properties, McWeeny raises doubts as to the validity of using a

$$A = \tfrac{1}{2}B \times (r - r_0) + (\mu \times r)/4\pi\Xi_0 r^3 \qquad (26)$$

vector potential A, as defined in equation (26), in the determination of χ_n from the field-free orbitals ϕ_n, viz.

$$\chi_n = \phi_n \exp(-i\alpha A_n \cdot r) \qquad (27)$$

The dipole part of A should not be included in equation (27) since it is already gauge-invariant and the solutions obtained without its use are already gauge-invariant.

4 Calculations of Nuclear Shielding

A. Ab Initio Calculations.—The most extensive series of calculations on nuclear shielding during the review period has been reported by Ditchfield et al.[16] using the method of finite perturbation theory.[17] The shielding constants of selected nuclei (carbon, nitrogen, oxygen, and fluorine) in a large number of simple compounds containing one or more of these elements together, in most cases, with hydrogen were evaluated. The basic atomic orbitals chosen were combined into five sets with varying degrees of restriction on the parameters involved. Here we briefly describe their results for just three of these sets:

(i) STO-5G. This set consisted of a single function for hydrogen ($1s$) and five functions for the other atoms ($1s$, $2s$, $2p$). Each function ϕ_μ was defined by:

$$\phi_\mu(\zeta, r) = \zeta^{3/2} \phi'_\mu(\zeta r) \qquad (28)$$

and the valence shell ζ values were determined by minimizing the calculated total molecular energy. The ϕ'_μ were defined in terms of a sum of five Gaussian functions. Thus for $\phi'_{1s}(r)$:

$$\phi'_{1s}(r) = \sum_{k=1}^{5} d_{1s,k} g_{1s}(\alpha_{1k}, r) \qquad (29)$$

where g_{1s} is a normalized Gaussian orbital with specified coefficients d and α. Similar expressions exist for $\phi'_{2s}(r)$ and $\phi'_{2p}(r)$ with a common value of α being used for these latter two functions.

---

[15] R. McWeeny, *Chem. Phys. Letters*, 1971, **9**, 341.
[16] R. Ditchfield, D. P. Miller, and J. A. Pople, *J. Chem. Phys.*, 1970, **53**, 613; 1971, **54**, 4186.
[17] J. A. Pople, J. W. McIver, and N. S. Ostlund, *J. Chem. Phys.*, 1968, **49**, 2960.

(ii) LEMAO-5G. This set consisted of a single function for hydrogen ($1s$) and five functions for the other atoms ($1s$, $2s$, and $2p$) with equations (28) and (29) as before. However, the coefficients d and α were now found by minimizing the calculated energy of the isolated atoms (hence, least energy minimal atomic orbitals) and then the parameters ζ were found by variation in the molecular calculations. As before, however, a common α is used for $2s$ and $2p$ functions.

(iii) 4-31G. In this set the inner $1s$ shell of the heavy atom was represented by a single function, as before, but the remaining valence orbitals ($1s$ in hydrogen and $2s$ and $2p$ in the other atoms) were described by inner and outer parts which are respectively the sums of three and one Gaussian functions. Otherwise the parameters were determined as for the LEMAO set. However, unlike (i) and (ii), the calculations were not performed in terms of the experimentally obtained molecular geometry of each molecule but in terms of a 'standard geometrical model'.[18] By splitting the orbitals with an inner and outer weighting, flexibility of effective atomic size can be introduced. Some anisotropy is also introduced by allowing different inner and outer weightings in different directions.

Table 1 gives the calculated values of σ^d, σ^p, σ, and $\Delta\sigma$ (from a suitable reference) for a selection of the results of Ditchfield et al. on carbon and nitrogen shieldings together with the experimentally observed values of $\Delta\sigma$. For a given compound it can be seen that σ^d varies only a little with the choice of wavefunction, most of the change involving σ^p. For different compounds σ^d changes by amounts considerably smaller than does σ^p, but changes in the former are by no means negligible. From the tables of Ditchfield et al. it is clear that the STO-5G basis gives poor agreement for carbon shieldings although it is somewhat better for the other nuclei. Much improvement is found with the LEMAO-5G and 4-31G sets. The large deshielding of the carbon atoms of C_2H_4 and C_2H_2 observed relative to methane is predicted with each of these basis sets but is much underestimated by STO-5G. In CH_3CN the low shielding of the nitrile carbon is not indicated at all in STO-5G but is given quite well by the other two sets. In performing these calculations the origin of the vector potential was taken at the molecular centre of mass.

It may be remarked, not only in connection with the results of Ditchfield et al. but also in connection with all theoretical calculations of chemical shifts, that exact agreement between calculated and observed shifts is not to be expected. In the first instance, experimental values may be in substantial error. Thus in Table 1 three more values of $\Delta\sigma$ for nitrogen shieldings have been included (from references e and f) which differ from those quoted by Ditchfield et al. by amounts ranging from 5—18 p.p.m. In addition, most experimental values are obtained with liquids and there may be solvent effects which have different extents of influence on different solutes. Even for isolated molecules, the observed shielding constant is a weighted mean over the occupied rotational and vibrational states, which means that, in most cases,

[18] J. A. Pople and M. Gordon, J. Amer. Chem. Soc., 1967, **89**, 4253.

Table 1 Calculated shieldings of carbon and nitrogen nuclei in some simple compounds and comparison of calculated chemical shifts from methane and ammonia respectively with those obtained experimentally

	STO-5G				LEMAO-5G				4-31G				
	σ^d	σ^p	σ	$\Delta\sigma$	σ^d	σ^p	σ	$\Delta\sigma$	σ^d	σ^p	σ	$\Delta\sigma$	$\Delta\sigma$ (exptl.)
Carbon shieldings													
Molecule													
CH$_4$	295·4	−131·0	164·4	—	296·2	−132·0	164·2	—	296·2	−75·1	221·1	—	—
C$_2$H$_6$	319·9	−145·2	174·7	+10·3	321·4	−155·8	165·6	+1·4	320·8	−90·7	230·1	+9·0	−8·0[a]
C$_2$H$_4$	311·0	−225·1	85·9	−78·5	312·3	−285·8	26·5	−137·7	311·7	−216·9	94·8	−126·3	−126[b]
C$_2$H$_2$	—	—	158·1	−6·3	302·3	−197·9	104·5	−59·7	301·9	−159·0	142·9	−78·2	−76[b]
CH$_3$F	321·2	−159·3	161·9	−2·5	320·3	−205·4	114·9	−49·3	320·2	−140·4	179·8	−41·3	−77·5[a]
*CH$_3$CN	313·9	−152·3	161·6	−2·8	314·6	−169·4	145·2	−19·0	314·7	−86·3	228·4	+7·3	−2·6[c]
CH$_3$*CN	372·3	−206·6	165·7	+1·3	372·1	−285·1	87·0	−77·2	371·8	−265·6	106·2	−114·9	−120·0[c]
Nitrogen shieldings													
Molecule													
NH$_3$	353·6	−177·4	176·2	—	354·1	−157·3	196·8	—	353·7	−109·5	244·2	—	—
CH$_3$NH$_2$	379·4	−209·0	170·4	−5·8	380·3	−184·8	195·5	−1·3	379·1	−131·4	247·7	+3·5	−23[d], −5[e]
CH$_3$CN	346·7	−442·9	−96·2	−272·4	346·7	−583·9	−237·2	−434·0	342·9	−429·5	−86·6	−330·8	−241[d], −246[f]
HCONH$_2$	382·3	−238·9	143·4	−32·8	383·6	−248·3	135·3	−61·5	382·6	−189·6	193·0	−51·2	−108[d]
(CH$_3$)$_2$NH	422·9	−206·6	194·2	+22·0	424·5	−236·3	188·2	−8·6	422·2	−157·0	265·2	+21·0	−12[e]

[a] H. Spiesecke and W. G. Schneider, *J. Chem. Phys.*, 1961, **35**, 722; [b] estimated from experimental data; [c] J. B. Stothers and P. C. Lauterbur, *Canad. J. Chem.*, 1964, **42**, 1563; [d] D. Herbison-Evans and R. E. Richards, *Mol. Phys.*, 1964, **6**, 191; [e] M. Witanowski and H. Januszewski, *Canad. J. Chem.*, 1969, **47**, 1321; [f] M. Witanowski, *Tetrahedron*, 1967, **23**, 4219.

Nuclear Shielding

the observed shielding will be smaller than that calculated from the equilibrium nuclear geometry by up to several p.p.m., the amount being different for different molecules (see below).

Coupled Hartree–Fock perturbation theory has been employed by Arrighini and co-workers[19] to calculate the shielding constants of the nuclei of H_2O, NH_3, CH_4, and CH_3F. For the proton shieldings their results make clear the gauge dependence of the calculated shielding. Considerable improvement occurs when the origin of the vector potential is at the heavy central nucleus. Thus with $\sigma(CA)$ for the origin at the central nucleus and $\sigma(H)$ for the origin at the proton, the values obtained are:

	H_2O	NH_3	CH_4	CH_3F
$\sigma(CA)$	28·23	32·34	31·70	43·60
$\sigma(H)$	50·24	53·38	33·05	55·84
$\sigma(obs)$	30·30 (±0·33)	30·93 (±0·33)	30·86 (±0·33)	26·86 (±0·33)

The values of $\sigma(obs)$ here differ slightly from those given by Arrighini and have been obtained from more recent literature data. The values of $\sigma(obs)$ given here marginally improve the agreement with $\sigma(CA)$. In addition to the results quoted here, Arrighini *et al.* have obtained values of the components of each of σ^d and σ^p. For the shielding of the central nuclei using a 'best gauge' they obtain the values:

H_2O	NH_3	CH_4	CH_3F
330·18	272·35	193·89	145·39

These results are not in particularly close agreement with those of Ditchfield *et al.*,[16] which are given above. (For H_2O, the best result obtained by Ditchfield *et al.* is 267·3.) Furthermore, the calculated value of $\Delta\sigma$ of CH_3F from CH_4 is −48·5, similar to that of Ditchfield *et al.*, but still far from the experimental value of −77·5. In view of the large number of Slater atomic orbitals employed in these calculations (27 for H_2O, 32 for NH_3, 39 for CH_4, and 47 for CH_3F), the lack of close agreement between calculated and observed shieldings, where known, is a little discouraging.

Coupled Hartree–Fock theory has also been used by Laws *et al.*[20] for the computation of a number of magnetic properties of the molecules AlH and N_2. Calculations were performed with the gauge origin at the electronic centroid. The results for the shielding of AlH are given in Table 2. It is interesting that for this gauge choice σ^p is very small for the proton but quite large for the aluminium nucleus. It is also of interest that the proton shielding in AlH is very similar to that of the hydrogen molecule (26·6). The results of Laws *et al.* for the nitrogen molecule are also given in Table 2. It can be seen that the various parameters are highly sensitive to the bond length. The estimated shielding for the $v = 0$ level is more than 3 p.p.m. less than

---

[19] G. P. Arrighini, M. Maestro, and R. Moccia, *J. Chem. Phys.*, 1970, **52**, 6411; 1971, **54**, 825; *Chem. Phys. Letters*, 1970, **7**, 351.

[20] E. A. Laws, R. M. Stevens, and W. N. Lipscomb, *J. Chem. Phys.*, 1971, **54**, 4269.

that for the equilibrium bond length (2·068 a.u.). The calculated value for σ is negative; that is, the electronic environment produces an 'antishielding' of the nitrogen nuclei. This has been observed experimentally[21] although the calculated σ^p differs by some 80 p.p.m. from that measured. Current-density calculations show the paramagnetic nature of the current densities at the nitrogen nuclei that cause the antishielding.

Table 2 *Calculated[20] nuclear shieldings for the molecules AlH and N_2. The results for N_2 are given for three values of bond length, including the equilibrium bond length of 2·068 a.u., and for the vibrational level v = 0 of the ground electronic state*

	σ_X	σ_X^p	σ_X^d	$\sigma_X^\|$	$\sigma_X^\perp$
AlH					
X = Al	222·71	−572·63	795·35	785·91	−58·59
X = H	26·13	0·32	25·81	34·12	22·13
N_2					
R = 1·868 a.u.	38·45	−321·52(−353·08)[a]	359·97	343·89	−114·27
R = 2·068 a.u.	−19·79	−374·57(−404·06)	354·78	338·42	−198·90
R = 2·268 a.u.	−97·08	−447·90(−475·31)	350·82	334·60	−312·92
v = 0	−23·18	−377·85(−407·27)	354·67	338·32	−203·94

[a] Results in parentheses are the difference between the total shielding (for the gauge origin at the electronic centroid) and σ^d calculated with the gauge origin at the nitrogen nucleus. This should be nearly equal to the experimental value of σ^p, which is −483(±20).

Of particular interest are the contributions made by the various constituent atomic orbitals of a molecule to the nuclear shielding, the subject of a study by Katô.[22] This study concerned the shielding of proton 1 in the acetylene molecule $H_{(1)}$—$C_{(1)}$≡$C_{(2)}$—$H_{(2)}$. Accurate values of the necessary integrals were calculated from SCF wavefunctions obtained from a limited basis of atomic orbitals.[23] The contribution of each orbital to the σ^d and σ^p of the 1-proton was then calculated for several values of the internuclear distances. The origin of the gauge was on proton 1. For σ^d the largest contribution is from the $1s$ orbital on $H_{(1)}$, as would be expected, but the contributions from the $1s$ and $2p_z$ orbitals of $C_{(1)}$ are both substantial. Even the $1s$ orbital of $H_{(2)}$ contributes more than 2% of the total σ^d.

B. Empirical Calculations.—For large molecules *ab initio* calculations are lengthy and involved, so many workers have resorted to empirical methods for the calculation of chemical shifts in such molecules. Here we examine briefly recent work in this area.

CNDO wavefunctions have been used extensively by several authors.

[21] S. I. Chan, M. R. Baker, and N. F. Ramsey, *Phys. Rev.*, 1964, **136**, A1224.
[22] H. Katô, *J. Chem. Phys.*, 1970, **52**, 3723.
[23] W. E. Palke and W. N. Lipscomb, *J. Amer. Chem. Soc.*, 1966, **88**, 2384.

Sadlej[24] has obtained very good results for the diamagnetic shielding constants of some diatomic molecules and has also given a derivation of the general rule of Flygare and Goodisman[25] for estimating σ^d. Detailed CNDO/2 calculations of substituent effects have have been reported by Brownlee and Taft.[26] Much of their work is concerned with electronic distributions at the orbitals of fluorine in conjugated molecules and so lies outside the scope of this review. However, their results permitted comparison with observed ^{19}F chemical shifts. To a good approximation it appears that the ^{19}F substituent chemical shift in *m*- and *p*-substituted fluorobenzenes is the sum of separate effects of σ and π charge densities of the fluorine orbitals. There is no appreciable evidence for an interdependence of σ and π electron effects. For the *meta*-substituent the ^{19}F shift is almost entirely a σ-electron density change, whilst the difference between the ^{19}F shifts for the *p*- and *m*-substituents is largely a π-electron density change at the fluorine.

Davies[27] has employed a self-consistent perturbation method together with the CNDO/2 approximations to calculate shielding constants of the molecules N_2, BF, CO, and F_2. The wavefunctions used comprised a minimal basis set and a slightly extended set with an extra $2p$ orbital. Substantial improvement was found for the latter set, suggesting that the extra $2p$ orbital might give improved results for large molecules for which a large basis is too unwieldy. Sebastian and Grunwell[28] have used the CNDO/2 method to calculate electron densities at the protons in $C_5H_5^-$, C_6H_6, $C_7H_7^+$, and the *para*-protons in $C_5H_5NH^+$ and $C_6H_5CH_2^-$. There is quite a good correlation with the observed proton chemical shifts for these molecules as well as for the *para*-protons of monosubstituted benzenes such as $C_6H_5NO_2$, C_6H_5Cl, and C_6H_5CN. Their correlation is a little surprising in that one might have expected magnetic effects (ring currents, group magnetic anisotropy) and, to a lesser extent, solvent effects to be considerably different among these compounds.

Herring and co-workers[29,30] have applied the INDO method to fluorine and nitrogen shieldings. Their method[29] avoids the average-energy approximation and is concerned only with changes in the paramagnetic term. Quite good agreement was obtained for the ^{19}F shifts of some first-row binary fluorides relative to fluorine[29] and for the nitrogen shifts of some nitrogen–oxygen–halogen compounds. Velenik and Lynden-Bell[31] have applied extended Hückel theory to the calculation of carbon, nitrogen, and oxygen shieldings in a very wide range of compounds. Fair agreement with observation was found but it was limited by the obvious dependence of the experimental

---

[24] A. J. Sadlej, *Org. Magn. Resonance*, 1970, **2**, 63.
[25] W. H. Flygare and J. Goodisman, *J. Chem. Phys.*, 1968, **49**, 3122.
[26] R. F. C. Brownlee and R. W. Taft, *J. Amer. Chem. Soc.*, 1970, **92**, 7007.
[27] D. W. Davies, *Mol. Phys.*, 1971, **20**, 605.
[28] J. F. Sebastian and J. R. Grunwell, *Canad. J. Chem.*, 1971, **49**, 1779.
[29] F. G. Herring, *Canad. J. Chem.*, 1970, **48**, 3498.
[30] F. Aubke, F. G. Herring, and A. M. Qureshi, *Canad. J. Chem.*, 1970, **48**, 3504.
[31] A. Velenik and R. M. Lynden-Bell, *Mol. Phys.*, 1970, **19**, 371.

results on the degree of orbital hybridization on the nucleus of interest and the extra difficulties involved in modifying the integrals to allow properly for this. From a Hückel MO study of pyrocatechol and pyrogallol, Cowherd and Geldard[32] deduced that it is hydrogen-bonding rather than non-bonded interactions that cause the observed anomalous proton chemical shifts. However, this interpretation does rest, to some extent, on some unproved equations concerning solvent effects. The proton chemical shifts in mono-substituted thiophens have been studied by Kamieński and Krygowski[33] using a modification of Hückel theory. Finally, evidence for the above-mentioned conclusions of Sebastian and Grunwell[28] comes from a study[34] of the relation between the proton chemical shifts of aliphatic and aromatic compounds and the electron charge densities on the protons.

5 Transmission of Shielding Effects within Molecules

A. Introduction.—To perform a rationalization of the vast abundance of experimental data on chemical shifts it is necessary to resort to an approximation of Ramsey's equation in which the total shielding is divided up into a number of physically distinct contributions. This approximation, first made by Saika and Slichter,[35] prevents any exact interpretation of observed numerical values of chemical shifts. However, it does facilitate a 'physical' or 'chemical' understanding of observed trends. The total molecular electronic distribution may be divided into a part on the nucleus of interest and a part in the remainder of the molecule. The nuclear shielding is then regarded as being caused by the induced magnetic fields of the local electron distribution and those of the distant electrons. The shielding due to the local electron distribution is made up of a 'local diamagnetic shielding' σ_{loc}^d and a 'local paramagnetic shielding' σ_{loc}^p given, respectively, by[36] equation (30) and equation (31).

$$\sigma_{loc}^d = \frac{\Xi_0 e^2}{3m} \langle 0 | \sum_i r_i^{-1} | 0 \rangle \tag{30}$$

$$\sigma_{loc}^p = - \frac{\Xi_0 \hbar^2}{2m^2} \frac{\langle r^{-3} \rangle_{np}}{\Delta E} \sum_B Q_{AB} \tag{31}$$

In equations (30) and (31), r_i represents the distance of the i'th electron in the local electron distribution from the nucleus, ΔE represents a mean excitation energy, $\sum_B Q_{AB}$ accounts for the amount of imbalance in the populations of the orbitals about the nucleus, and $\langle r^{-3} \rangle_{np}$ denotes the mean inverse cube of the distance of the valence-shell p-electrons from the nucleus. (We do not consider d-electrons here.) Since one is usually interested in shielding changes

---

[32] L. C. Cowherd and J. F. Geldard, *J. Chem. Soc. (A)*, 1971, 486.
[33] B. Kamieński and T. M. Krygowski, *Tetrahedron Letters*, 1971, 103.
[34] P. Lazzeretti and F. Taddei, *Org. Magn. Resonance*, 1971, **3**, 113, 283.
[35] A. Saika and C. P. Slichter, *J. Chem. Phys.*, 1954, **22**, 26.
[36] M. Karplus and J. A. Pople, *J. Chem. Phys.*, 1963, **38**, 2803.

along a series of closely related compounds (*e.g.* substituent chemical shifts, S.C.S.), we see that a substituent may act on the nuclear shielding by directly producing a different magnetic field at the nuclear site from that in the unsubstituted compound and also indirectly by altering either or both of $\sigma_{\text{loc}}^{\text{d}}$ and $\sigma_{\text{loc}}^{\text{p}}$ as compared with the unsubstituted compound. The mechanisms by which these 'direct' and 'indirect' effects take place can be divided into 'through-space' and 'through-bond' effects. A tabulation of this breakdown has been given elsewhere.[37] In the present section we describe these mechanisms, illustrating each one with examples drawn from work published in the review period. Contact and pseudocontact shifts are omitted. However, some account of these topics is included in Chapter 9. We also summarize in the present section recent results concerning shielding anisotropies and isotope shifts.

B. Inductive and Resonance Effects.—These two effects lead to electron density changes at various positions in a molecule and so may be expected to cause shielding changes at the nuclear sites through either $\sigma_{\text{loc}}^{\text{d}}$ or $\sigma_{\text{loc}}^{\text{p}}$ or both. For protons it is usually the case that changes in $\sigma_{\text{loc}}^{\text{p}}$ are negligible on account of very large values of ΔE, whilst for other nuclei it is the changes in $\sigma_{\text{loc}}^{\text{p}}$ which are dominant. We first consider the latter case. An increase in electron density on a nucleus for which $\sigma_{\text{loc}}^{\text{p}}$ is dominant reduces $\langle r^{-3} \rangle_{np}$ since mutual electron repulsion causes the electron distribution to expand. A reduction in $\underset{B}{\Sigma Q_{AB}}$ also ensues with an increased electron density since the additional electrons will tend to fill vacant orbitals, thereby reducing the population imbalance. Whether or not increased electron density changes ΔE depends upon the nature of the molecule, but any change is likely to be a reduction.

We illustrate these ideas for the case of fluorine shielding. The simple ideas above (with ΔE usually held to be invariant for closely related compounds) are so widely accepted that ^{19}F shifts are sometimes used to study bonding properties or group effects. For instance, Wahl and Peterson's work[38] appears very well to substantiate the electron-releasing character of methyl substituents in alkyl fluoroadamantanes (1) by measuring an increase in the fluorine shielding upon steady substitution of methyl groups at the positions shown.

(1)

---

[37] W. T. Raynes, *Mol. Phys.*, 1971, **20**, 321.
[38] G. H. Wahl and M. R. Peterson, *J. Amer. Chem. Soc.*, 1970, **92**, 7238.

Thus for the substituent chemical shifts (*i.e.* relative to $R^1, R^2, R^3 = H$) they find the results shown in Table 3. The simple interpretation of these results is

Table 3

R^1	R^2	R^3	S.C.S.
Me	H	H	+2·8
Me	Me	H	+5·3
Me	Me	Me	+7·8
Et	H	H	+2·3
Bu^t	H	H	+0·15

electron release due to hyperconjugation, leading to an increased electron density at the fluorine, a reduction in σ_{loc}^p and hence a higher chemical shift. However, as pointed out by Wahl and Peterson a more complicated mechanism could be present. The effects of resonance can be detected by considering derivatives of fluorobenzene with substitution at the *meta-* or *para-*position; for a suitable *para*-substituent we expect resonance structures such as:

$$^-F-\langle=\rangle=X^+$$

which are impossible for the *meta*-substituted compound. Consequently, the fluorine nucleus for the *para*-compound should be more highly shielded. Nichols[39] has provided a good example of this for complexes of the type $FC_6H_4AuPR_3$. Relative to fluorobenzene, Nichols finds the following ^{19}F shifts for the *meta-*(δ_m) and *para-*(δ_p) compounds:

R	δ_m	δ_p
Et	2·35	4·04
Bu^n	2·25	3·98
PhMe$_2$	2·31	3·95
Ph	2·27	3·42
OPh	2·10	2·60

The differing values of δ_p for different R reflect differing π-acceptor properties of the PR_3 group which modify the gold–phenyl interaction. A similar study exploiting this idea has been made by Stewart and Treichel[40] involving ^{19}F resonance of FC_6H_4X, where X was a group such as $Mn(CO)_5$, $Mn(CO)_4PPh_3$, *etc.* Here $\delta_p - \delta_m$ was of the order of 8 p.p.m.

Shielding changes due primarily to changes in ΔE are believed to be responsible for the ^{183}W shifts in the series $WF_{6-n}(OMe)_n$ studied by McFarlane, Noble, and Winfield.[41] For the ^{183}W shifts relative to WF_6 they find:

[39] D. I. Nichols, *J. Chem. Soc.* (*A*), 1970, 1216.
[40] R. P. Stewart and P. M. Treichel, *J. Amer. Chem. Soc.*, 1970, **92**, 2710.
[41] W. McFarlane, A. M. Noble, and J. M. Winfield, *J. Chem. Soc.* (*A*), 1971, 948.

	WF$_5$OMe	cis-WN$_4$(OMe)$_2$	trans-WF$_3$(OMe)$_3$	cis-WF$_2$(OMe)$_4$
$\delta(^{183}W)$	-41	-160	-464	-604

Since the ^{19}F shifts increase along this series, implying increased donation of electrons by a methoxy-group as compared with a fluorine atom, the fall in tungsten shielding probably reflects a fall in ΔE values for the σ_{loc}^p of tungsten (but not by any significant amount for σ_{loc}^p of fluorine). This could possibly occur because of overlap by the non-bonded $2p$ electrons for the oxygen of the methoxy-group with vacant $5d$ tungsten orbitals. A similar dominance of ΔE changes is responsible for the ^{14}N shifts of nitrosyl compounds studied by Andersson et al.[42]

For an illustration of inductive and resonance effects on σ_{loc}^d one turns first to proton shielding, where σ_{loc}^p is negligible. However, as is well known, magnetic anisotropy effects are important here so that it is often not obvious to which cause observed trends must be attributed. For instance, Shaw and Allred[43] recently obtained the following proton chemical shifts for the —C(CH$_3$)$_3$ protons of the compounds (CH$_3$)$_3$MC(CH$_3$)$_3$, where M is a Group IVB element:

(CH$_3$)$_3$C—C(CH$_3$)$_3$	(CH$_3$)$_3$Si—C(CH$_3$)$_3$
-0.868	-0.872
(CH$_3$)$_3$Ge—C(CH$_3$)$_3$	(CH$_3$)$_3$Sn—C(CH$_3$)$_3$
-0.926	-1.079

The shifts here are in p.p.m. from tetramethylsilane. A parallel set of results were found by Shaw and Allred[43] for several series obtained by replacing the tertiary carbon atom of the t-butyl group in these compounds successively by Si, Ge, and Sn. The above trend appears to run counter to expectations when bearing in mind the decreasing electronegativity from carbon to tin. However, a weakening of the M—C bond on changing M from carbon to tin in the above series could lead to a greater s-character in the carbon orbitals directed to the methyl group[44] and hence to a greater electronegativity experienced by the protons. An alternative explanation, favoured by Shaw and Allred, invokes magnetic anisotropy effects.

Resonance effects on proton shielding are, of course, well known from studies of benzene derivatives. Here we illustrate their existence with the recent results of Beachley[45] on B-monosubstituted borazine derivatives (2). For substituents such as Me$_2$N, MeO, and F, one expects enhanced shielding at the ortho- and para- (NH) protons due to resonance, as compared with the meta- (BH) protons. The proton shifts observed by Beachley are given in

---

[42] L. O. Andersson, J. Mason, and W. van Bronswijk, J. Chem. Soc. (A), 1970, 296.
[43] C. F. Shaw and A. L. Allred, J. Organometallic Chem., 1971, 28, 53.
[44] R. S. Drago and N. Matwiyoff, J. Organometallic Chem., 1965, 3, 62.
[45] O. T. Beachley, J. Amer. Chem. Soc., 1970, 92, 5372.
[46] H. Spiesecke and W. G. Schneider, J. Chem. Phys., 1961, 35, 731.

Table 4, relative to borazine itself. The proton shifts relative to benzene of the corresponding monosubstituted derivative of benzene are also included.[46]

$$\begin{array}{c} X \\ | \\ B \\ HN \diagup \diagdown NH \\ | \quad \quad | \\ HB \diagdown \diagup BH \\ N \\ | \\ H \end{array}$$
(2)

It is clear that resonance is important in determining the proton shieldings in the *ortho*- and *para*-positions of borazine.

Although often ignored, changes in σ_{loc}^d for nuclei other than protons may be quite significant in some instances. This may be illustrated with the ^{13}C shieldings of some methyl halides relative to methane.[47] The observed shifts are given in Table 5. For substitution by fluorine or chlorine the ^{13}C

Table 4

Protons studied	Me$_2$N	MeO	F	Cl	Br
(*ortho* NH)	+1·17	+0·83	+0·80	+0·12	−0·07
(*ortho* CH)	+0·60	+0·43	+0·31	−0·02	−0·22
(*para* NH)	+0·65	+0·35	+0·24	+0·12	−0·07
(*para* CH)	+0·62	+0·37	+0·22	+0·12	+0·03
(*meta* BH)	+0·06	+0·06	+0·04	+0·09	+0·17
(*meta* CH)	+0·10	+0·04	+0·02	+0·03	+0·08

Table 5 *Carbon chemical shifts of the halogenomethanes with respect to methane (assuming that methane is more shielded by 130·8 p.p.m.[a] than liquid benzene)*

CH$_3$F	CH$_3$Cl			CH$_3$Br	CH$_3$I
−77·5[d]	−27·2[b]			−12·3[b]	+18·4[b]
CH$_2$Cl$_2$		CH$_2$ClBr		CH$_2$Br$_2$	CH$_2$I$_2$
−56·3[b]		−42·1[b]		−23·7[b]	+51·7[b]
CHCl$_3$	CHCl$_2$Br		CHClBr$_2$	CHBr$_3$	CHI$_3$
−79·8[b]	−59·5[b]		−36·7[b]	−14·4[b]	+137·6[b]
CCl$_4$	CCl$_3$Br	CCl$_2$Br$_2$	CClBr$_3$	CBr$_4$	CI$_4$
−98·8[b]	−69·9[b]	−39[a]	−7·2[b]	+26·4[b]	+290·2[c]

[a] P. C. Lauterbur, *Ann. N.Y. Acad. Sci.*, 1958, **70**, 841; [b] W. M. Litchman and D. M. Grant, *J. Amer. Chem. Soc.*, 1968, **90**, 1400; [c] O. W. Howarth and R. J. Lynch, *Mol. Phys.*, 1968, **15**, 431; [d] H. Spiesecke and W. G. Schneider, *J. Chem. Phys.*, 1961, **35**, 722.

[47] J. Mason, *J. Chem. Soc. (A)*, 1971, 1038.

shifts decrease as predicted for inductive effects whereas for bromine an additional effect causes the shielding to increase when more than one or two bromine atoms are present and for iodine the shieldings increase markedly with iodine substitution. Mason suggests that the additional effect here is an enlarged σ_{loc}^d present with heavy atoms. In Table 6 are given some calculated values of σ_{loc}^d and the difference between the observed shift from methane (δ) and σ_{loc}^d, which difference can be attributed to σ_{loc}^p after correcting for the σ_{loc}^d for methane. (The experimental results in Table 6 have been taken from

Table 6

Compound	δ	σ_{loc}^d(calc.)	σ_{loc}^d	Compound	δ	σ_{loc}^d(calc.)	σ_{loc}^p
CH_4	0	295·3	0				
CH_3Cl	−27	376	−108	CH_3I	18	519·5	−206
CH_2Cl_2	−56	458·5	−220	CH_2I_2	52	743·5	−397
$CHCl_3$	−80	542	−327	CHI_3	135	953	−523
CCl_4	−100	624	−429	CI_4	290	1191	−608

Mason's paper and differ slightly in some cases from those in Table 5). It can be seen that upon correction for varying σ_{loc}^d a constant trend in σ_{loc}^p appears. This trend is believed to be caused by a gradually diminishing ΔE as halogenation increases.

C. Magnetic Anisotropy Effects.—We now consider shielding contributions from the magnetic fields of particular chemical bonds or groups. Consider a C—H bond in a molecule. The external magnetic field induces currents in the motions of the bonding electrons which produce a magnetic field at the nucleus of interest elsewhere in the molecule. As the molecule tumbles, this field varies to give a mean field which is non-zero, because the magnetic susceptibility of the bond is anisotropic. If the hydrogen atom of the C—H bond is replaced by another atom or group, the anisotropy will change so that there will be a shielding change for the nucleus of interest. Hence there will be a shielding difference between the substituted and the unsubstituted compound in addition to any difference arising from a change in induction, resonance, or any further effect.

The simplest quantitative formulation is made by using the point-dipole approximation. For a particular group with cylindrical symmetry and possessing a magnetic anisotropy $\Delta\chi$, the shielding contribution σ_m due to this group is given in SI by equation (32), where R is the distance from the nucleus

$$\sigma_m = \frac{\Delta\chi}{12\pi R^3}(1 - 3\cos^2\theta) \qquad (32)$$

of interest to the site of the point-dipole and θ is the angle between the axis of the group and the line joining the site of the point-dipole to the nucleus. For this approximation to be valid it is essential that R be large (in practice, $R > 3$ nm is often considered satisfactory). This requirement assures that the question of where to locate the point-dipole, which is sometimes difficult to answer, is not of critical importance. Since σ_m above does not contain any quantities that involve the electronic environment at the nucleus of interest, this 'direct' effect should not depend on the type of nucleus whose resonance is being obtained. Consequently, it is in the field of proton resonance, where the total range of shifts is small, that magnetic anisotropy effects manifest themselves the most readily.

Provided that the distance R is kept large, it should be possible to use proton resonance to determine values of $\Delta\chi$ for the various functional groups of organic and inorganic chemistry. These could then be used to predict shifts in other compounds. Unfortunately, this ideal has never been realized. There are several reasons for this. Some determinations have used values of R that were too small, so that not only was the point-dipole approximation of doubtful validity but also other effects (induction, conjugation, electric field, *etc.*) were not negligibly small. In addition, solvent effects were unfortunately ignored in some cases. That is, the shielding of the proton being studied was influenced by changes in the surrounding solvent due to the functional group, even though the through-bond and electric-field effects may have been negligible. At an even more fundamental level it is as yet not fully established that the very concept of group anisotropy is justified. For example, in considering the —C≡N group in the usual bonding situation C—C≡N, is it valid to use one $\Delta\chi$ for the C—C≡N group or should one use two $\Delta\chi$'s — one for the C—C bond and one for the C≡N bond? If the former is correct, where does one locate the point-dipole? If the latter is correct, is the C—C bond anisotropy the same as that of, say, ethane? Whatever the answers to these questions, it is certainly true that $\Delta\chi$ values found by proton resonance tend to be larger than those found by other methods. A tabulation of $\Delta\chi$ values determined by proton resonance has been given by the present author.[37]

The most systematic study of magnetic anisotropy effects in recent years is that of ApSimon and co-workers, and in the review period three further papers have appeared.[48-50] By measuring the methyl proton chemical shifts of the keto-androstanes relative to androstane (3), with the carbonyl groups in

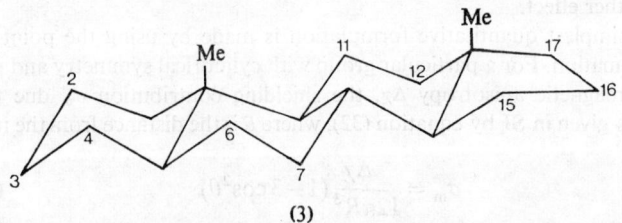

(3)

[48] J. W. ApSimon, P. V. Demarco, D. W. Mathieson, W. G. Craig, A. Karim, L. Saunders, and W. B. Whalley, *Tetrahedron*, 1970, **26**, 119.
[49] J. W. ApSimon and H. Bierbeck, *Canad. J. Chem.*, 1971, **49**, 1328.
[50] J. W. ApSimon, H. Bierbeck, D. K. Todd, P. V. Demarco, and W. G. Craig, *Canad. J. Chem.*, 1971, **49**, 1335.

one of the numbered positions in (3), they were able to obtain values for the two magnetic anisotropies of the carbonyl group. It was reasonably assumed that all through-bond effects on methyl proton chemical shifts due to carbonyl substitution were negligible (except for the 11-position). The methyl proton shifts thus arise from a combination of magnetic anisotropy and electric field effects. Using point-dipole approximations for both mechanisms, calculations were performed with the point-dipoles located at a number of positions along the carbonyl bond. The best result occurs when the dipoles are placed at the bond midpoint and leads to the values:

$\Delta\chi_1 = -352 \times 10^{-36}$ m³ (molecule)$^{-1}$
$(\equiv -28 \cdot 1 \times 10^{-30}$ c.g.s. units per molecule)
$\Delta\chi_2 = -258 \times 10^{-36}$ m³ (molecule)$^{-1}$
$(\equiv -20 \cdot 5 \times 10^{-30}$ c.g.s. units per molecule)

$\Delta\chi_1$ here denotes the difference between the susceptibility perpendicular to the nodal plane of the π-orbitals and along the axis of the carbonyl σ-bond, whereas $\Delta\chi_2$ denotes the difference between the susceptibility along the axis perpendicular to the σ-bond in the nodal plane of the π-orbitals and that along the σ-bond axis. The shielding cone for the carbonyl group, according to ApSimon *et al.*, differs from that previously accepted in that a proton in the nodal plane of the π-orbitals of a carbonyl group is not everywhere deshielded. This remark applies to the C=C bond as well. ApSimon *et al.* give an impressive list of examples of other compounds besides the keto-androstanes to which their $\Delta\chi$ values are applicable with reasonable success.

More recently, ApSimon and Bierbeck[49] have re-examined the work on the keto-androstanes using a more detailed geometrical model. The result of this new calculation indicates that the magnetic anisotropy of the C—H bond is very small and almost negligible — a conclusion at variance with the earlier conclusions. However, this does agree with the work of Zürcher,[51] who earlier studied the proton chemical shifts of substituted steroids. In both

(4)   (5)

these studies the electric field of the polar group is evaluated at the proton of the methyl group rather than at the C—H bond midpoint. Since the latter seems more reasonable [as the parameter A (see Section 5E) is a bond property defined for a uniform electric field], it is perhaps premature to conclude that $\Delta\chi$ for the C—H bond is negligible. A study of the magnetic anisotropy of the ethylene-ketal (4) and ethylene-thioketal (5) groups on substituted androstanes has also been made[50] and values for anisotropies have been deduced.

[51] R. F. Zürcher in 'Progress in N.M.R. Spectroscopy', ed. J. W. Emsley, J. Feeney, and L. H. Sutcliffe, Pergamon Press, Oxford, 1967, vol. 2, p. 205.

However, the authors are dubious of the physical significance of these values in view of the uncertainties in the location of the point-dipoles and the possible uncertainties in the geometrical arrangement of atoms in the steroidal skeleton of the substituted compounds.

The present author[37] has presented a general expression for calculating the contributions of magnetically anisotropic X—Y bonds of a freely rotating —XY_3 group to the shielding of protons of a freely rotating methyl group elsewhere in the same molecule. The expression was applied to the problem of the methyl proton chemical shifts of the methylbenzenes relative to toluene, and to the aryl proton chemical shifts of the methylbenzenes relative to benzene. The magnetic anisotropy values of ApSimon and co-workers were used for the C—H and C—C bonds. The results showed that, apart from the *ortho*-positions where substantial magnetic anisotropy effects are probably present, there is an approximate cancellation so that the shieldings at the *meta*- and *para*-positions are determined entirely by factors other than magnetic anisotropy. However, should it emerge from future work that the C—H bond anisotropy is small or negligible (see above) whilst the C—C bond anisotropy is not negligible, then this statement will have to be modified.

Several other studies have appeared in which magnetic anisotropy is of importance. Matsubayashi *et al.*[52] have examined the proton shielding in 4-*NN*-dimethylaminonitrosobenzene (6) and its 2-methyl and 2,6-dimethyl

(6)

derivatives. They observed shielding differences between the 2,6-, the 3,5-, and the 7,8-protons at low temperature which they attribute entirely to magnetic anisotropy effects of the nitroso-group. Very close agreement is found between the observed shielding differences and the results calculated using equation (32). However, it seems likely that electric fields from the polar nitroso-group would make some contribution to the splittings as well as differential solvent effects at the 2- and 6-positions. Another study in which not only electric fields but also inductive and mesomeric effects were assumed to be small is that of Ficarra *et al.*[53] on the proton chemical shifts of 1,8-diaminoperchloronaphthalene relative to 1,8-naphthalenediamine. In this work the anisotropy of the C—Cl bond was taken to be equal to the diamagnetic susceptibility of the chlorine ion. No evidence was given for this assumption. Zwanenberg *et al.*[54] have studied the proton chemical shifts of aromatic sulphines. In diphenylsulphine (7) a low-field shift for the 1-proton

---

[52] G. Matsubayashi, Y. Takaya, and T. Tanaka, *Spectrochim. Acta*, 1970, **26A**, 1851.
[53] A. Ficarra, S. Huang, and A. A. Reidlinger, *J. Mol. Spectroscopy*, 1970, **33**, 175.
[54] B. Zwanenberg, L. Thijs, and A. Tangerman, *Tetrahedron*, 1971, **27**, 1731.

(7)

clearly indicates a bent arrangement for the CSO group. Magnetic anisotropy effects will be important here, although electric-field effects, solvent effects, and, possibly, steric or van der Waals effects, will also probably be of importance.

A study of the proton chemical shifts of some arylcopper(I) compounds by Baici et al.[55] suggests a very large magnetic anisotropy for the carbon–copper bond. The shielding cone is such as to deshield the *ortho*-positions and to shield the *meta*- and *para*-positions. An interesting suggestion here is that in a compound such as *o*-anisylcopper the electric field of the methoxy-groups may modify the anisotropy of the carbon–copper bond. This phenomenon of intramolecular electric-field-induced changes in bond and group magnetic anisotropies may be more common than is often supposed and be partly responsible for the variable values obtained by proton resonance studies. Other recent studies of magnetic anisotropy effects in proton resonance concern the thione group,[56] the epoxy-group,[57] and certain bicyclic derivatives.[58]

The anomalous proton chemical shifts of some *o*-substituted nitrosobenzenes have been studied by Okazaki and Inamoto[59] and attributed to the magnetic anisotropy and electric-field effects of the nitroso-group. It appears from this work that the shielding cone of the nitroso-group is somewhat different from that previously adopted and more like that of ApSimon et al. for the carbonyl group. However, this conclusion can only be tentatively accepted on account of doubts concerning their electric field calculations (see Section 5E).

D. The Ring Current Effect.—The idea of a 'ring current' induced in the mobile π-electrons of aromatic molecules by an external magnetic field perpendicular to the ring was introduced by Pople[60] to explain the anomalously low shieldings of aryl protons. Pople treated the secondary field created by the ring current as originating from a point magnetic dipole at the centre of the ring.

---

[55] A. Baici, A. Camus, and G. Pellizer, *J. Organometallic Chem.*, 1971, **26**, 431.
[56] R. A. Long and L. B. Townsend, *Chem. Comm.*, 1970, 1087.
[57] A. C. Huitric, V. A. Ruddell, P. H. Blake, and B. J. Nist, *J. Org. Chem.*, 1971, **36**, 809.
[58] I. Tabushi, K. Fujita, and R. Oda, *J. Org. Chem.*, 1970, **35**, 2383.
[59] R. Okazaki and N. Inamoto, *J. Chem. Soc. (B)*, 1970, 1583.
[60] J. A. Pople, *J. Chem. Phys.*, 1956, **24**, 1111.

Waugh and Fessenden[61] and Johnson and Bovey[62] refined this model using the notion of two current loops above and below the ring. Quantum mechanical treatments were then put forward by Pople[63] and McWeeny.[64] There have been further theoretical developments since then, including a criticism of the physical concept of a ring current.[65] However, the idea of a ring current is still used by many chemists as, at least, a working concept. Whatever the description of the effect may be, the origin is identical with that discussed in Section 5C, namely magnetic anisotropy.

During the review period, a particularly important series of papers has appeared on the proton chemical shifts of a large number of unsubstituted, condensed, polycyclic hydrocarbons.[66-70] According to the McWeeny theory the mean secondary field B' at a given proton due to the ring current is given by equation (33) (using c.g.s. units), where β is the resonance integral in

$$B' = \frac{2}{3}\beta \left[\frac{2\pi e}{hc}\right]^2 \frac{S^2 B_0}{a^3} \left\{\sum_i J_i[-K(r_i)]\right\} \qquad (33)$$

Hückel MO theory, S is the area of an aromatic ring, a is the length of a C—C bond, and B_0 is the applied external magnetic field. J_i measures the 'amount' of current in the i'th ring and $K(r_i)$ is a factor dependent entirely on the geometrical position of the proton of interest relative to the ring. Haigh et al. worked in terms of what they called the 'sigma-ratio', i.e. the ratio $B':B'_b$, where B'_b is the value of B' for the benzene molecule. This ratio only involves the variables J_i and $K(r_i)$.

The procedure of Haigh et al.[67] was to fit the calculated sigma-ratios (using simple Hückel theory) to their observed τ values with a linear relationship. They find, for non-overcrowded protons:

$$\tau_{obs} = -1.56(B'/B'_b) + 4.34 \qquad (34)$$

This result implies a ring current effect in benzene of -1.56. This equation fitted very well the sixteen different benzenoid hydrocarbons at low concentration in CCl_4 which they had studied experimentally: the root mean square deviation of calculated from observed τ values was 0·085 p.p.m. However, for certain protons in some of the compounds a considerably larger deviation from equation (34) than this was found. For example, deviations from the predictions of equation (34) for some of the protons of 3,4-benzopyrene (8)

---

[61] J. S. Waugh and R. W. Fessenden, *J. Amer. Chem. Soc.*, 1957, **79**, 846; J. S. Waugh, ibid., 1958, **80**, 6697.
[62] C. E. Johnson and F. A. Bovey, *J. Chem. Phys.*, 1958, **29**, 1012.
[63] J. A. Pople, *Mol. Phys.*, 1958, **1**, 175.
[64] R. McWeeny, *Mol. Phys.*, 1958, **1**, 311.
[65] J. I. Musher, *J. Chem. Phys.*, 1965, **43**, 4081; *Adv. Magn. Resonance*, 1966, **2**, 177; *J. Chem. Phys.*, 1967, **46**, 1219.
[66] C. W. Haigh and R. B. Mallion, *Mol. Phys.*, 1970, **18**, 737.
[67] C. W. Haigh, R. B. Mallion, and E. A. G. Armour, *Mol. Phys.*, 1970, **18**, 751.
[68] C. W. Haigh and R. B. Mallion, *Mol. Phys.*, 1970, **18**, 767.
[69] R. B. Mallion, *J. Mol. Spectroscopy*, 1970, **35**, 491.
[70] R. B. Mallion, *J. Chem. Soc. (B)*, 1971, 681.

are shown alongside the formula. The reason for these discrepancies is not clear although they are thought not to be caused by σ-bond magnetic anisotropy effects.

(8)

For the overcrowded protons very much larger deviations occur between the observed shifts and those predicted from equation (34). Examples of these deviations are shown alongside the formulae for phenanthrene (9), chrysene (10), 1,2-benzpyrene (11), and 1,2;7,8-dibenzanthracene (12). These deviations are very probably caused by the overlap forces between the electronic environments of the adjacent protons.

(9)   (10)   (11)

(12)

Haigh and Mallion[68] have also calculated ring currents relative to that in benzene (taken as unity) for a number of molecules containing six condensed benzene rings. An interesting trend here is for the ring current to become quite small with increasing condensation about the ring in question, although this trend is not obeyed in all compounds. Examples of these ring current intensities are shown alongside the formulae for hexacene (13), 1,2;7,8-dibenztetracene (14), 6,7-benzpentaphene (15) and 2,3-benzperylene (16).

In a more recent paper Mallion[70] has considered the semi-classical theory.[61,62] In this theory an important parameter is the distance $2d$ between the two lobes of the ring current. A value of d is found which enables the calculated ring current effect σ_{sc} from this semi-classical theory for the special

	A	B	C	D	E	F
(13)	1·048	1·291	1·352			
(14)	1·121	0·877	1·299			
(15)	1·104	1·170	0·548	1·044		
(16)	1·077	1·041	0·575	0·236	0·374	0·973

case of benzene to agree with that estimated from experiment. Using this particular value of d and fitting the observed τ values for the non-overcrowded protons of other aromatic molecules, one has the empirical relation of equation (35).

$$\tau_{obs} = 0·597\,\sigma_{sc} + 3·744 \qquad (35)$$

Thus the semi-classical theory over-estimates the ring current effect and must be multiplied by a correction factor of about 0·6. The reason for this is that the original value for the π-electron contribution to the magnetic anisotropy of benzene is over-estimated. Unfortunately, any attempt to adjust d so as to remove the factor 0·597 leads to a physically unreasonable value and produces very large changes in the shieldings of nuclei in the immediate neighbourhood of a given ring. The term 3·744 in equation (35) should equal the τ value of a compound such as cyclohexa-1,3-diene but, in fact, this compound has a τ value of 4·21. All these considerations underline the essentially empirical nature of the semi-classical model. Nevertheless, it does appear for some compounds that this model is more successful than equation (33) for predicting the observed order of τ values.

Variations of the methylene proton chemical shifts in the class of compounds $ArCH_2X$ could possibly be due to changing hyperconjugative effects as Ar is changed or to a changing ring-current effect. From the work of Bentley and Dewar[71] it seems that the latter is in fact the case. In Table 7 are given the methylene proton chemical shifts of $ArCH_2X$ relative to $C_6H_5CH_2X$ for a series of substituents X. For a given Ar there is a remarkable constancy of

Table 7

Ar	X			
	OH	Cl	Br	H
2-naphthyl	0·19	0·18	0·19	0·14
2-phenanthryl	0·25	0·18	0·26	0·23
3-phenanthryl	0·30	—	0·29	0·28
1-naphthyl	0·49	0·48	0·50	0·36
9-phenanthryl	0·54	—	0·54	0·37
1-anthryl	0·63	—	0·67	—
1-pyrenyl	0·70	—	0·75	0·48
9-anthryl	1·00	1·03	1·04	0·65

the shift (excepting X = H) not only for OH, Cl, and Br but for other substituents for which results are not given in the Table. This underlines the probability that ring currents are responsible for these shifts. The anomalous behaviour of the arylmethanes (X = H) is believed to be due to conformational preferences which are absent when X = H but present in all other cases. A similar explanation is applicable to data on the aryl protons of some diarylmethanes reported by Montavdo et al.[72] and to some proton resonance data on phenanthrenes obtained by Bartle et al.[73] Bergman, Chandler, and Moir[74] have presented a technique for the determination of the part of the proton chemical shift arising from ring-current effects on an adjacent aro-

---

[71] M. D. Bentley and M. J. S. Dewar, *J. Org. Chem.*, 1970, **35**, 2707.
[72] G. Montavdo, S. Caccamese, P. Finocchiaro, and F. Battino, *Bull. Chem. Soc. Japan*, 1971, **44**, 1439.
[73] K. D Bartle, P. M. G. Bavin, and D. W. Jones, *Org. Magn. Resonance*, 1970, **2**, 259.
[74] J. J. Bergman, W. D. Chandler, and R. Y. Moir, *Canad. J. Chem.*, 1971, **49**, 225.

matic ring. However, some caution is necessary in making these explanations in view of the results of Frazer and Renaud.[75]

Another class of compounds in which ring currents play an important role are the annulenes. The n.m.r. spectroscopy of these compounds has very recently been reviewed in detail by Haddon et al.[76] Calculations by Dewar and Gleicher[77] indicate that bond alternation and hence the absence of a ring current should occur for all the higher annulenes commencing with [26]annulene, although other calculations of Hultgren[78] disagree with this prediction. During the review period the spectrum of [22]annulene (17) has

(17)

been obtained, thereby providing a step towards the resolution of this problem. McQuilkin et al.[79] find that the proton spectrum at -90 °C consists of two regions of absorption — low-field multiplets at $\tau 0.35$—0.7 and $\tau 0.1$—1.5 and also a high-field multiplet at $\tau 10.4$—11.2. The ratio of the integrated areas of these two regions of absorption was 14:8, clearly indicating the existence of a ring current. Other recent papers in the field include the demonstration of a ring current in a $[4n+3]$annulenone derivative[80] and in the dianion (18).[81] The very large shielding of the inner protons is not only due to

H^1 τ 14·60

H^2 τ 3·02

H^3 τ 3·77

(18)

[75] R. R. Frazer and R. N. Renaud, *Canad. J. Chem.*, 1971, **49**, 800.
[76] R. C. Haddon, V. R. Haddon, and L. M. Jackman, *Topics Current Chem.*, 1971, **16**, 103.

the ring current but probably reflects greater diamagnetic shielding because of additional electron density.

The first synthesis of some [m][n]paracyclophanes (19) has been accomplished by Nakazaki et al.[82] and, as may easily be imagined, the proton resonance data of the compounds with $m = n = 8$ and $m = 8$, $n = 10$ show protons with high shielding ($\tau \sim 10\cdot 5$).

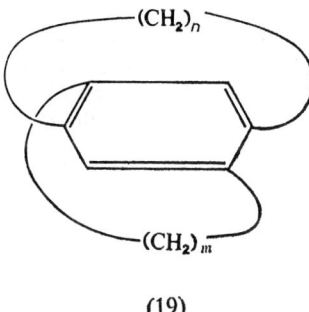

(19)

Although it is on proton chemical shifts that ring currents have the most marked effect, some recent work has indicated that shielding changes on other nuclei may be due to such currents. Thus some evidence for the possible importance of ring-current effects on ^{13}C chemical shifts has been presented by Jones et al.,[83,84] particularly for the compound benzo[ghi]fluoranthene

(20)

(20), where the shielding difference of 6·56 between the quaternary carbon atoms 11 and 12 is much larger than the corresponding difference of 0·28 in acenaphthylene (21). Broad-line ^{13}C studies by Retcofsky and Friedel[85] give some evidence for high aromaticity of anthracite. Onak and Marynick[86]

[77] M. J. S. Dewar and G. J. Gleicher, *J. Amer. Chem. Soc.*, 1965, **87**, 685.
[78] G. O. Hultgren, Ph.D. Thesis, California Inst. Technology, 1966.
[79] R. M. McQuilkin, B. W. Metcalf, and F. Sondheimer, *Chem. Comm.*, 1971, 338.
[80] G. P. Cotterrell, G. H. Mitchell, F. Sondheimer, and G. M. Pilling, *J. Amer. Chem. Soc.*, 1971, **93**, 259.
[81] J. F. M. Oth and G. Schröder, *J. Chem. Soc.* (B), 1971, 904.
[82] M. Nakazaki, K. Yamamoto, and S. Tanaka, *Tetrahedron Letters*, 1971, 341.
[83] A. J. Jones, T. D. Alger, D. M. Grant, and W. M. Litchman, *J. Amer. Chem. Soc.*, 1970, **92**, 2386.
[84] A. J. Jones, P. D. Gardner, D. M. Grant, W. M. Litchman, and V. Boekelheide, *J. Amer. Chem. Soc.*, 1970, **92**, 2395.
[85] H. L. Retcofsky and R. A. Friedel, *Analyt. Chem.*, 1971, **43**, 485.

(21)

have applied a ring-current model to both the proton and ^{11}B spectra of decaborane(14), and by treating the molecule as two fused pentagonal pyramids each sustaining a 'conical ring current' have obtained close agreement between calculated and observed shieldings.

(22)

A considerable amount of controversy has existed concerning the degree of aromatic character to be attributed to the metal chelates of acetylacetone (22), and as one would expect, chemical shift studies have been employed to resolve this problem. Bock et al.[87] synthesized compounds of the types (23) and (24)

(23)

(24)

[86] T. Onak and D. Marynick, *Trans. Faraday Soc.*, 1970, **66**, 1843.
[87] B. Bock, K. Flatau, H. Junge, M. Kuhr, and H. Musso, *Angew. Chem. Internat. Edn.*, 1971, **10**, 225.

and then examined the resonances of the protons which are indicated for comparison with the corresponding truly aromatic compounds in which the chelate ring was replaced by a benzene ring. No evidence for increased shielding existed, thereby indicating that the angular C—C—C part of these chelate rings has no magnetic anisotropy comparable with that of benzene. A similar conclusion was arrived at by Fujii[88] who examined the methyl proton chemical shifts of the complex (25) where X = Cl, Br, or I. Aromatic

(25)

character in the acetylacetone ring ought to be greater than in the imine rings on account of the greater symmetry. Consequently the methyl protons of this ring should be less shielded than those of the imine ring if a ring current is present. Observation showed that the methyl protons of the acetylacetone ring were more shielded than those of the imine ring, thereby indicating a negligible ring current in both types of chelate ring.

An interesting qualitative method for detecting the presence of ring current behaviour has been used by Anet and Schenck.[89] They define a quantity S by equation (36), where $\Delta\sigma_X$ is the chemical shift between acetonitrile and

$$S = \Delta\sigma_X - \Delta\sigma_{\text{cyclohexane}} \qquad (36)$$

cyclohexane in a solvent X and $\Delta\sigma_{\text{cyclohexane}}$ is the same difference in cyclohexane. If we denote the solvent shift of solute Y in a solvent X by σ_X^Y, then S is given by equation (37) (hex = n-hexane)

$$S = [\sigma_X^{\text{MeCN}} - \sigma_X^{\text{hex}}] - [\sigma_{\text{hex}}^{\text{MeCN}} - \sigma_{\text{hex}}^{\text{hex}}] \qquad (37)$$

Acetonitrile is chosen as the solute because of its ability to interact strongly with aromatic compounds, with the methyl group being adjacent to the ring and consequently experiencing a large additional shielding from the ring

[88] Y. Fujii, *Bull. Chem. Soc. Japan*, 1970, **43**, 1722.
[89] F. A. L. Anet and G. E. Schenck, *J. Amer. Chem. Soc.*, 1971, **93**, 556, 3310.

current.[90] Cyclohexane is taken as the reference because of its non-aromatic character.[91] (Cyclohexane has a very small solvent effect on any solute and it has recently been shown by experiment to exhibit very little if any evidence of anisotropic behaviour). By choosing compounds of varying aromatic character, Anet and Schenck were able to obtain a variety of S values. These S values can only be regarded as a qualitative measure since the acetonitrile

(26)

probably interacts to slightly different extents with different aromatic compounds. Their results indicate the presence of a ring current in 1,3,5-cycloheptatriene (26) which has an S value of $+0.33$, about half of that ($S = +1.00$) for benzene. Other compounds showing signs of aromatic behaviour were cyclopentadiene (27) ($S = +0.39$) and norbornadiene (28) ($S = +0.16$). Cyclo-octatetraene (29), however, gave a negative S value of -0.16.

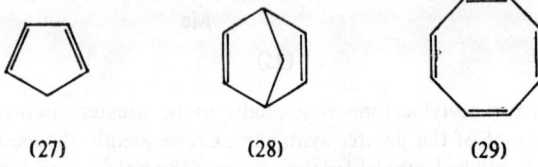

(27)　　　　　(28)　　　　　(29)

A very important result confirming the existence of non-local effects in determining magnetic anisotropy values for cyclopentadiene has been obtained by Benson and Flygare[92] using the method of Zeeman spectroscopy.

E. Electric Field Effects.—In the presence of an electric field there is a distortion of the electron distribution in a molecule which causes changes in the nuclear shielding. If a polar group is substituted into a compound then the electrical asymmetry of the group gives rise to an electric field which alters the shielding of a given nucleus from that of the unsubstituted compound. This phenomenon was first discussed by Buckingham,[93] who showed for an axially symmetric X—H bond that the presence of a uniform electric field causes a change σ_E in the shielding (after averaging over all directions of the magnetic field) given by equation (38), where A and B are bond parameters

$$\sigma_E = -AE_z - BE^2 \qquad (38)$$

[90] W. G. Schneider, *J. Phys. Chem.*, 1962, **66**, 2653.
[91] M. A. Raza and W. T. Raynes, *Mol. Phys.*, 1970, **19**, 199.
[92] R. C. Benson and W. H. Flygare, *J. Amer. Chem. Soc.*, 1970, **92**, 7523.
[93] A. D. Buckingham, *Canad. J. Chem.*, 1960, **38**, 300.

which are independent of the electric field but are dependent on the nature of X. For equation (38), the z-axis is taken along the X—H bond from X to H, so that for an electric field along the bond the shielding will be reduced if A and B are positive — as is believed to be the case for all protons. Equation (38) has also been used a great deal in the interpretation of fluorine chemical shifts, a topic reviewed recently by Emsley and Phillips.[94]

The electric field affecting the shielding may have its source outside the molecule under investigation, so that equation (38), with a proper averaging of the electric field, is applicable to the study of intermolecular effects on chemical shifts. Indeed, most values of A and B presently available were obtained from studies of intermolecular effects on the chemical shifts of gases. Values of A and B for protons in various bonding situations are given in Table 8. Values of these parameters for other nuclei have been given by Emsley and Phillips[94] and Jameson et al.[95]

Electric-field effects are important in areas other than nuclear shielding since such fields can produce changes in other properties that are to some extent 'localized' in a molecule. (We avoid here the contraction to 'field effect' used by many authors since magnetic anisotropy effects and ring current effects are also field effects). The significance of electric-field effects relative to through-bond effects, such as σ-inductive effects, has been the subject of considerable controversy in recent years, and it is not surprising that nuclear shielding studies have been undertaken to resolve this problem. Whilst studies by some authors with certain compounds seem to indicate clearly the existence of significant intramolecular electric fields, studies by other authors with other compounds do not. The reasons for the differences are obscure. However, values of A obtained from studies of intermolecular effects together with well-established values of group dipole moments certainly indicate that in many compounds the resulting shielding changes should be considerably larger than experimental error. Since magnetic anisotropy effects are of importance only in proton shielding, most studies of electric field effects are made using ^{19}F resonance.

Direct experimental evidence for the existence of an electric field effect has been found by Adcock et al.[96] in a study of the ^{19}F shifts of substituted fluorides. In the compounds 1-(p-fluorophenyl)propan-2-one (30) and 1-(m-fluorophenyl)propan-2-one (31), conjugative effects between the ring and carbonyl groups are absent since the carbonyl group is insulated from the ring by the methylene group. Furthermore, the methylene nuclear spin–spin couplings $J(^{13}CH)$ of (30) and (31) differ by only 1 Hz from those of the methyl groups of p- and m-fluorotoluene. This implies that the through-bond inductive effect of the acetyl group has diminished almost to zero before it reaches the ring. However, the ^{19}F shifts of (30) and (31) in benzene solution

---

[94] J. W. Emsley and L. Phillips in 'Progress in N.M.R. Spectroscopy', ed. J. W. Emsley, J. Feeney, and L. Sutcliffe, Pergamon Press, Oxford 1971, vol. 7, p. 1.
[95] A. K. Jameson, C. J. Jameson, and H. S. Gutowsky, *J. Chem. Phys.*, 1970, **53**, 2310.
[96] W. Adcock, P. D. Bettess, and S. Q. A. Rizvi, *Austral. J. Chem.*, 1970, **23**, 1921.

Table 8 *Values of the linear* (A) *and quadratic* (B) *electric field coefficients for proton shielding in the free hydrogen atom and in several bonding situations and the methods by which they were obtained.* A *and* B *are given in the units required for the use of e.s.u. of field in equation* (38). *For conversion to SI:* (a) *multiply* A *given here by* 0.33×10^{-4}, *after which its units are* m V^{-1} (= C m J^{-1}), (b) *multiply* B *given here by* 0.11×10^{-8}, *after which the units are* m^2 V^{-2} (= C^2 m^2 J^{-2}).

Bonding Situation	$A \times 10^{12}$	$B \times 10^{18}$	Method of Determination
H atom	0	0.74	Theory[a]
H—H	2.27	—	Theory[b]
H—H	1.45	—	Theory[c]
H—H	4.5	0.14($B\|$), 1.16($B\perp$)	Theory[d]
H—H	4	—	Theory[e]
C—H	2	—	Theory[f]
C—H	2.8	—	Expt.[g]
C—H	2.9	—	Theory[h]
C—H	2.9, 3.8	—	Theory[i]
C—H (in CHF$_3$)	2.9	0.8	Expt.[j]
C—H (in CH$_3$CN)	3.0, 3.4	—	Expt.[k]
C—H (in styrenes)	3.1	—	Expt.[l]
C—H (in CH$_3$Cl)	8	0.3	Expt.[m]
C—H (in dimethyl ether)	16	1.0	Expt.[m]
N—H (in uracils)	5	—	Expt.[n]
O—H (in CH$_3$OH)	25	—	Expt.[o]
S—H (in H$_2$S)	26	0.65	Expt.[m]
Cl—H	40	0.4	Expt.[p]
Cl—H	33	—	Theory[q]
Br—H	65	1.6	Expt.[m]

[a] T. W. Marshall and J. A. Pople, *Mol. Phys.*, 1958, **1**, 199; [b] J. Gruninger and H. F. Hameka, *J. Chem. Phys.*, 1968, **48**, 4878; [c] J. I. Musher, *Adv. Magn. Resonance*, 1966, **2**, 177; [d] T. Yonemoto, *Canad. J. Chem.*, 1966, **44**, 223; [e] I. V. Aleksandrov, 'The Theory of Nuclear Magnetic Resonance', Academic Press, London and New York, 1966, p. 146; [f] A. D. Buckingham, *Canad. J. Chem.*, 1960, **38**, 300; [g] C. MacLean and E. L. Mackor, *Mol. Phys.*, 1961, **4**, 241 (see ref. *h*); [h] J. I. Musher, *J. Chem. Phys.*, 1962, **37**, 34; [i] R. M. Aminova and R. Z. Gubaidullina, *J. Struct. Chem.*, 1969, **10**, 236; [j] L. Petrakis and H. J. Bernstein, *J. Chem. Phys.*, 1962, **37**, 2731; [k] P. Diehl and R. Freeman, *Mol. Phys.*, 1961, **4**, 39; [l] C. H. Hamer and W. F. Reynolds, *Canad. J. Chem.*, 1968, **46**, 3813; [m] G. Widenlocher and E. Dayan, *Compt. rend.*, 1965, **260**, 6856; [n] P. Laszlo and J. I. Musher, *J. Chem. Phys.*, 1964, **41**, 3906; [o] A. D. H. Clague, G. Govil, and H. J. Bernstein, *Canad. J. Chem.*, 1969, **47**, 625; [p] W. T. Raynes, A. D. Buckingham, and H. J. Bernstein, *J. Chem. Phys.*, 1962, **36**, 3481; [q] W. T. Raynes and B. P. Chadburn, *Mol. Phys.*, 1969, **17**, 543.

relative to *p*-fluorotoluene and *m*-fluorotoluene respectively are 2.18 p.p.m. and 1.78 p.p.m. in the direction of lower shielding. This indicates that the difference is caused by different electric fields at the C—F bonds, as one would predict for the different distances and orientations involved. [For (31) the field is a mean over all conformations whilst for (30) the field is the same for all conformations]. It is necessary to point out here, however, that differential solvent effects could make some contribution to the difference of 0·40 p.p.m.

In another study Della[97] examined the ^{19}F shifts in the *cis*-3-substituted trifluoromethylcyclohexane system, which exists almost exclusively in the conformation (32). For the various substituents X given in Table 9, Della

(32)

obtained the listed ^{19}F shifts relative to the case X = CH_3 in the solvents dimethylformamide (DMF) and methanol.

Table 9

X	$\delta(^{19}F)$, DMF	$\delta(^{19}F)$, methanol
NH_2	−0·33	−0·24
$NH_3{}^+Cl^-$	−0·38	−0·23
$CONH_2$	−0·04	−0·02
CO_2H	−0·21	−0·01
CO_2Me	−0·22	−0·01
O^-Na^+	−0·60	—

The small substituent chemical shifts are interpreted by Della as indicating the absence of any electric-field effect. However, this conclusion is not obvious in view of the large solvent effects, which probably overshadow other effects.

Ceccarelli *et al.*[98] examined the proton chemical shifts of some mono- and di-aryl-substituted ethylene oxides. They found that the oxiran protons H_a and H_b of the *cis*-(33) and *trans*-(34) isomers respectively have considerably different shieldings, the proton in the *trans*-isomer being shielding by about

(33)   (34)

[97] E. W. Della, *Austral. J. Chem.*, 1970, 23, 2421.
[98] G. Ceccarelli, G. Berti, G. Lippi, and B. Macchia, *Org. Magn. Resonance*, 1970, 2, 379.

0·5 p.p.m. more than that of the *cis*-isomer. This is considerably more than can be attributed to the magnetic anisotropy effects of the rings X and Y. The difference was interpreted as originating from the electric field due to the other C—H bond of the oxiran ring, which acts in an opposite direction along the bond CH_a to the one along CH_b. This other C—H bond is believed to have a very high bond moment [possibly as large as 1·84 Debye ($\equiv 6.16 \times 10^{-30}$ C m)].

Okazaki and Inamoto[59] examined the anomalous shifts of the *ortho* aryl protons and the protons in the *ortho*-substituent methyl and t-butyl groups in *o*-substituted nitrosobenzenes. The anomalies were assigned to the magnetic anisotropy of the nitroso-group (see Section 5C) and to the electric field of this polar group. In performing the calculation of the electric-field effect, a value for A of 2×10^{-12} c.g.s. units was employed — a value considerably lower than would seem reasonable (see Table 8). In addition, the nitroso-group was treated not as a point-dipole but as two point charges, of $-0.34e$ on the oxygen atom and $+0.34e$ on the nitrogen atom. These values were obtained from the bond moment and the bond length. This method of dealing with polar groups is rather uncommon.

We conclude here with a brief discussion of a wave-mechanical calculation of the linear electric field dependence of the proton shielding for C—H bonds.[99] The parameter A was determined for C—H bonds with various degrees of carbon hybridization using a variation treatment with a function of the form shown in equation (39), where ψ_0 is the unperturbed, normalized

$$\psi = \psi_0(1 + \mu_\alpha f_\alpha)(1 + B_\alpha g_\alpha)(1 + E_\alpha \xi_\alpha) \qquad (39)$$

ground-state wave-function in the absence of the perturbations μ, B, and E, and f_α, g_α, and ξ_α are initially unknown functions of the co-ordinates. Different results were obtained with different choices of ψ_0 (see reference *i* of Table 8) but it is clear from the work of Aminova and Gubaidullina[99] that the parameter A is sensitive to the degree of hybridization of the carbon bonding orbital. They find for $C(sp^3)$—H, $C(sp^2)$—H, and $C(sp)$—H bonds values of A of 3·9, 2·8, and 2.3×10^{-18}c.g.s. e.s.u. respectively for a particular method of constructing the C—H bond wavefunction.[100] Unfortunately, they do not give values for the various tensor components from which A is obtained by averaging.

F. Intramolecular Dispersion Forces.—Many studies of the effects of solvents on the chemical shifts of solutes have shown that the forces causing the attraction of neighbouring molecules in liquids are also capable of changing (invariably to lower values) the shielding of nuclei in the solute molecule. The most common expression used for calculating the shielding change of a perturbing molecule (2) in a solute molecule (1), due to these forces, is

---

[99] R. M. Aminova and R. Z. Gubaidullina, *J. Struct. Chem.*, 1970, **10**, 236.
[100] D. S. Bartow and J. W. Richardson, *J. Chem. Phys.*, 1965, **42**, 4018.

given[101] by equation (40), where α_2 and I_2 are, respectively, the polarizability

$$\sigma_W = -3B\alpha_2 I_2/r_{12}^6 \qquad (40)$$

and ionization potential of the perturber, r_{12} is the distance between the 'centres' of the two molecules, and B is a parameter dependent only on molecule 1 and is, in fact, usually taken as being identical with the parameter B appearing in equation (38). From the identification of the B's in this way we are regarding the attractive dispersion forces as being caused by the existence of a fluctuating electric field F whose mean value is zero but whose mean square value is non-zero and given by equation (41). It was first suggested by

$$\overline{F^2} = 3\alpha_2 I_2/r_{12}^6 \qquad (41)$$

Schaefer et al.[102] that dispersion forces could produce an intramolecular effect influencing the proton and ^{13}C shieldings of aromatic and aliphatic halogen compounds. Such an effect is often difficult to prove since at the small distances where $\overline{F^2}$ is large steric effects may not always be negligible, and in addition the simple point-dipole models useful for magnetic anisotropy and electric field effects are no longer valid.

During the review period there have been two papers in which the concept of intramolecular dispersion forces has been used. Okazaki and Inamoto,[59] in the paper referred to earlier, assumed that these forces were significant in determining the proton shieldings of some *ortho*-substituted nitrosobenzenes. They employed equation (40), although in their paper the inverse fourth power of the distance is given. This is presumably a misprint. It is not clear what value they adopted for the parameter B.

Homer and Callaghan[103] have investigated the ^{19}F shifts of some bridgehead-substituted fluorobicyclo[2,2,1]heptanes (35). They examined some of

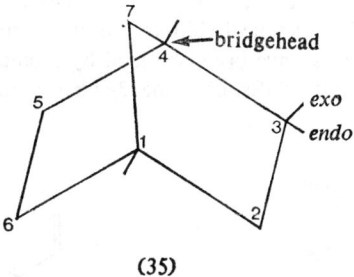

(35)

the compounds containing hydrogen or bromine in either or both of the bridgehead positions and used equation (42) (in their notation) to interpret the shifts (δ) relative to perfluorobicyclo[2,2,1]heptane. X, Y, and Z are

---

[101] W. T. Raynes, A. D. Buckingham, and H. J. Bernstein, *J. Chem. Phys.*, 1962, **36**, 3481.
[102] T. Schaefer, W. F. Reynolds, and T. Yonemoto, *Canad. J. Chem.*, 1963, **41**, 2969.
[103] J. Homer and D. Callaghan, *J. Chem. Soc. (B)*, 1970, 1573.

$$\delta = -X\Delta E_z - Y\Delta(E^2) - Z\Delta\langle E^2\rangle \qquad (42)$$

parameters characteristic of C—F bonds and Δ denotes a difference relative to the bridgehead fluorine shielding in perfluorobicyclo[2,2,1]heptane. This empirical equation is essentially the same as equation (38), with the addition of the final term to represent intramolecular dispersion forces; the quantity $\langle E^2 \rangle$ is identical with $\overline{F^2}$ of equation (41). For the compounds with hydrogen as substituent fairly close agreement was obtained between calculated and observed quantities, whatever the choice of sign for the C—H bond moment. The work of Homer and Callaghan indicates the importance of electric-field effects and dispersion forces on nuclear shielding.

G. Steric Effects.—The introduction of a bulky substituent group into a molecule may often lead to a change in the molecular conformation or the molecular geometry. Such changes will inevitably cause magnetic nuclei elsewhere in the molecule to be differently shielded from those in the unsubstituted compound. Effects such as this are clearly steric in origin but in the present subsection such effects are excluded since the shielding changes actually occur through the various other mechanisms described in the present section. Here we are dealing with the situation when two nuclei in the same molecule, nuclei not bonded to each other, are so close together that their electron clouds overlap. The consequent overlap forces will produce shielding changes for both nuclei. This may take place either indirectly, because of small alterations of molecular geometry, or directly, because of a mutual distortion of the electron clouds, or both. Some experimental proton resonance results for some condensed, benzenoid hydrocarbons that are probably influenced by steric effects have been given earlier in Section 5D. Although, as mentioned there, a deshielding is usually observed, it is not as yet firmly established whether steric effects always cause a deshielding.

Some interesting ^{19}F chemical shift results indicating the very probable presence of steric effects have been obtained by Nomura and Takeuchi.[104] They compared the ^{19}F shifts of some 3-substituted 2,4-dimethylfluoro-

benzenes (36) with those of the corresponding *meta*-substituted fluorobenzenes (37). Some of their results for various substituents X relative to the case X = H are given in Table 10 for solutions in the solvent carbon tetrachloride.

[104] Y. Nomura and Y. Takeuchi, *Chem. Comm.*, 1970, 259.

Table 10

X	$\delta(^{19}F)$, (36)	$\delta(^{19}F)$, (37)
NH$_2$	−2·06	0·4
NMe$_2$	−4·61	−0·1
Cl	−6·42	−2·0
Br	−7·76	−2·30
I	−9·31	−2·4

The implication seems clear. For the compounds (36) the bulky group X forces the methyl group in the 2-position into contact with the fluorine atom, thereby deshielding it.

The only quantitative treatment of steric effects on chemical shifts is that given by Cheney and Grant a few years ago. During the review period there has been no attempt to verify this treatment. Discussion of observed data in terms of steric effects has been given for proton shielding by Bartle et al.,[105] Jones et al.,[106] and Cárdenas,[107] and for fluorine shielding by Sergeev et al.[108]

H. Intramolecular Hydrogen-bonding.—During the review period three papers have appeared which demonstrate the very large deshielding of protons when intramolecular hydrogen-bonding occurs. Yoshida et al.[109] obtained the proton chemical shifts of the enol form of acetylacetone and various derivatives (38) of this tautomer. Their τ values for the enol proton for various substituents X are given in Table 11. The interpretation of these data is that

(38)

because of conjugation electrons are removed from the oxygen atoms of the enol ring so that, relative to acetylacetone (X = H), the enol proton becomes deshielded. The effect of the SCH$_3$ group is surprisingly large and may be a result of the participation of the d-orbitals of sulphur in the conjugation.

Andrews, Rae, and co-workers[110,111] reported investigations of hydrogen-

---

[105] K. D. Bartle, P. M. G. Bavin, D. W. Jones, and R. L'amie, *Tetrahedron*, 1970, **26**, 911.
[106] R. A. Y. Jones, A. R. Katritzky, and P. G. Lehman, *J. Chem. Soc.* (*B*), 1971, 1316.
[107] C. G. Cárdenas, *J. Org. Chem.*, 1971, **36**, 1631.
[108] N. M. Sergeev, O. P. Petrii, and N. N. Shapetko, *J. Struct. Chem.*, 1970, **11**, 767.
[109] Z. Yoshida, H. Ogoshi, and T. Takumitsu, *Tetrahedron*, 1970, **26**, 5691.
[110] J. M. Appleton, B. D. Andrews, I. D. Rae, and B. E. Reichert, *Austral. J. Chem.*, 1970, **23**, 1667.

Table 11

X	τ (OH)	X	τ (OH)
H	−5·84	SCN	−7·10
Cl	−5·55	SCH_3	−7·08
$CH=CHCH_3$	−6·61	$COCH_3$	−7·40
$CH=CHC_2H_5$	−6·60	CO_2CH_3	−7·97
CN	−6·90	$CO_2C_2H_5$	−8·10
NO_2	−6·95	CHO	−8·51

bonding between the NH proton and the substituent X in substituted anilides (39). The degree of joint planarity of the amide group and the ring was monitored by observing the *ortho*-proton signal, which is very sensitive to the orientation of the carbonyl group (which itself is influenced by the nature of the hydrogen-bonding between the NH proton and X).

(39)

I. Isotope Shifts.—The term isotope shift is used to refer to shielding changes occurring upon isotopic substitution. The subject was reviewed in detail fairly recently.[112] The basic physical processes responsible for isotope shifts are simple in principle, although for polyatomic molecules the situation becomes more involved. Let us consider the proton shielding in the molecules H_2 and HD. The individual rotational and vibrational levels of HD lie lower in the potential well than do those of H_2. Consequently the displacement of the nuclei from equilibrium is smaller for HD than H_2. The stretching of a chemical bond usually leads to a reduction in shielding so that (after averaging over all available vibrational and rotational states[113]) HD will be more highly shielded than H_2. This simple argument probably explains the almost universally observed increase in shielding occurring upon substitution by a heavier isotope. It may be that some non-linear polyatomic molecules will exhibit negative proton isotope shifts upon the substitution of a heavier isotope because the shielding effect of smaller nuclear displacements is more

---

[111] B. D. Andrews and I. D. Rae, *Austral. J. Chem.*, 1971, **24**, 413.
[112] H. Batiz-Hernandez and R. A. Bernheim in 'Progress in N.M.R. Spectroscopy', ed. J. W. Emsley, J. Feeney, and L. H. Sutcliffe, Pergamon Press, Oxford, 1967, vol. 3, p. 63.
[113] W. T. Raynes, A. M. Davies, and D. B. Cook, *Mol. Phys.*, 1971, **21**, 123.

Nuclear Shielding

than cancelled by a deshielding effect of magnetic anisotropy changes arising from a slightly different mean molecular geometry. However, so far this has not been observed. Negative isotope shifts found so far are attributed to solvent effects.[114]

With regard to HD, it is of interest to consider the nuclear shielding constants of both proton and deuteron. Within the Born–Oppenheimer approximation they are equal. In reality they are not exactly equal, the proton being slightly less shielded, since the electrons are less able to keep pace with its motion than with that of the deuteron. The question of by how much more the deuteron is shielded is still uninvestigated by both experiment and theory. Conveniently discussed here is some recent work by Williams and co-workers[115,116] and by Breskman and Kanofsky[117] on magnetic resonance studies of the shielding of muons (μ). For the species μH in solution the muon shielding was found to be 0·2 p.p.m. smaller than that of the proton.[115] This must be due to the smaller mass of the muon (about 200 times that of the electron) as compared with that of the proton. The muon shielding in the muonium halides μF, μCl, μBr, and μI was calculated[117] to be some 2 p.p.m. lower than that in the corresponding hydrogen halides (assuming each muonium halide had the same potential well as the corresponding hydrogen halide). This calculation has been disputed by Williams[116] who states that the figure of 2 p.p.m. is too large by a factor of about three.

The important question of whether or not significant differences occur between ^{14}N and ^{15}N shieldings in otherwise identical nitrogen-containing molecules has been resolved with the work of Becker et al.[118] (see also Randall and Gillies[119]). No such shielding differences appear to occur above the level of experimental error. In dealing with isotope shifts, the term 'isotopomers' has been introduced and defined as referring to 'compounds which differ only in isotopic substitution in one position'.[119,120] This usage has not been universally adopted. Some workers (e.g. ref. 114) refer to compounds which differ in isotopic substitution at several positions as isotopomers.

J. Shielding Anisotropies (see also Chapter 10, Section 5).—Apart from some theoretical results given in Section 4A, this Report has hitherto been concerned with mean shielding constants and differences of mean shielding constants. However, for nuclei in most molecules the shielding depends on the direction of the external magnetic field relative to molecule-fixed axes. In conventional experiments in liquids and gases this anisotropy in the shielding does not

---

[114] G. M. Ford, L. G. Robinson, and G. B. Savitsky, *J. Magn. Resonance*, 1970, **4**, 109.
[115] J. F. Hague, J. E. Rothberg, A. Schenck, D. L. Williams, R. W. Williams, K. K. Young, and K. M. Crowe, *Phys. Rev. Letters*, 1970, **25**, 628.
[116] R. W. Williams, *Phys. Letters*, 1971, **34B**, 63.
[117] D. Breskman and A. Kanofsky, *Nuovo Cimento*, 1970, Serie X, **68B**, 147; *Phys. Letters*, 1970, **33B**, 309.
[118] E. D. Becker, R. B. Bradley, and T. Axenrod, *J. Magn. Resonance*, 1971, **4**, 136.
[119] E. W. Randall and D. G. Gillies in 'Progress in N.M.R. Spectroscopy', ed. J. W. Emsley, J. Feeney, and L. H. Sutcliffe, Pergamon Press, Oxford, 1971, vol. 6, p. 119.
[120] E. W. Randall, J. J. Ellner, and J. J. Zuckerman, *J. Amer. Chem. Soc.*, 1966, **88**, 622.

manifest itself in any direct way. Clearly, the shielding anisotropy is derivable from the molecular electronic wavefunctions. A knowledge of the anisotropy (and even more so, the individual components of the shielding tensor) would be of additional value to that supplied by the mean shielding constant when assessing the quality of molecular wavefunctions. In the past decade a number of experimental methods have been devised for measuring shielding anisotropies, and recently an elegant multiple-pulse technique has been developed which permits the accurate determination of the individual components of the shielding tensor[121-123] (see Chapter 3, Section 7).

Most authors have studied shielding anisotropies and shielding tensor components using the fluorine nucleus. For F_2, O'Reilly et al.[124] obtained for $\Delta\sigma (=\sigma_\parallel - \sigma_\perp$, here and below for a cylindrically symmetrical system) a value of $+1050$ (± 50) p.p.m. using five independent methods. They show, using the experimental value of σ^p (-1120 p.p.m.) that the anisotropy in the diamagnetic part of the shielding ($\sigma_\parallel^d - \sigma_\perp^d$) has the small value of -70 (± 50). Yannoni et al.,[125] using liquid-crystal solvents, obtained the fluorine shielding anisotropies of the compounds sym-$C_6F_4Br_3$, CF_3CCl_3, and CF_3I. Their results are relative to the molecular symmetry axes and some difficulty was encountered in relating the results to the Karplus–Das theory of ^{19}F shielding. They attribute this to a lack of cylindrical symmetry of the C—F bond. A similar lack of symmetry for the Xe—F bonds in XeF_4 is implied by the work of Hindermann and Falconer.[126] A preliminary result for the components of the ^{19}F shielding tensor in fluoranil $C_6F_4O_2$ has been given by Mehring et al.[127] and the multiple-pulse techniques have also been applied to MgF_2.[128] Long and Goldstein[129] have measured the ^{19}F shielding anisotropy in tetrafluoro-1,3-dithietan $CF_2S_2CF_2$.

^{13}C shielding anisotropies in HCN[130] and also in CS_2 and $CaCO_3$[131] have been reported recently (see also Chapter 8, Section 3). For CS_2, Pines et al.[131] obtain chemical shifts of $+285$ (± 10) for the shielding parallel to the molecular axis and -140 (± 6) for that perpendicular to this axis, both shifts being referred to the isotropic value of the shielding. A value for the proton shielding anisotropy in benzene has also been obtained by Lindon and Dailey.[132] Chemical shift anisotropies of ^{59}Co and ^{55}Mn have been studies for some cobalt and manganese carbonyls,[133] and discussion concerning the measure-

---

[121] U. Haeberlin and J. S. Waugh, *Phys. Rev.*, 1968, **175**, 453.
[122] J. S. Waugh, C. H. Wang, L. M. Huber, and R. L. Vold, *J. Chem. Phys.*, 1968, **48**, 662.
[123] J. S. Waugh, L. M. Huber, and U. Haeberlin, *Phys. Rev. Letters*, 1968, **20**, 180.
[124] D. E. O'Reilly, E. M. Peterson, Z. M. El Saffar, and C. E. Scheie, *Chem. Phys. Letters*, 1971, **8**, 470.
[125] C. S. Yannoni, B. P. Dailey, and G. P. Ceasar, *J. Chem. Phys.*, 1971, **54**, 4020.
[126] D. K. Hindermann and W. E. Falconer, *J. Chem. Phys.*, 1970, **52**, 6198.
[127] M. Mehring, R. G. Griffin, and J. S. Waugh, *J. Amer. Chem. Soc.*, 1970, **92**, 7222.
[128] L. M. Stacey, R. W. Vaughn and D. D. Elleman, *Phys. Rev. Letters*, 1971, **26**, 1153.
[129] R. C. Long and J. H. Goldstein, *J. Chem. Phys.*, 1971, **54**, 1563.
[130] F. Millett and B. P. Dailey, *J. Chem. Phys.*, 1971, **54**, 5434.
[131] A. Pines, M.-K. Rhim, and J. S. Waugh, *J. Chem. Phys.*, 1971, **54**, 5438.
[132] J. Lindon and B. P. Dailey, *Mol. Phys.*, 1970, **19**, 285.
[133] H. W. Spiess and R. K. Sheline, *J. Chem. Phys.*, 1970, **53**, 3036; 1971, **54**, 1099.

ment of proton shielding anisotropies has made it clear that there are solvent effects present which are difficult to eliminate.[134,135] This latter remark applies to other nuclei as well.

6 Shieldings of Particular Nuclear Species

A. Introduction.—In the present section the reader's attention will be brought to the existence of a large number of papers on chemical shifts additional to those dealt with earlier. In a few of these papers authors have given some explanation of experimental data, have obtained good empirical correlations, or have carried out wave-mechanical calculations. However, for the most part, the aim of the work was the acquisition of new or improved chemical shift data. Unfortunately, due to limitations of space, it will be possible to do little more than merely list the types of compounds studied and refer to the papers concerned.

B. Proton Chemical Shifts.—A number of papers involving discussion of the factors influencing proton shielding in particular classes of compounds have appeared during the review period. In a review of the n.m.r. spectroscopy of amides, Stewart and Siddall[136] have discussed the factors controlling the proton shielding in these compounds. A very extensive empirical correlation of proton shift data for *ortho*-substituted benzenes and naphthalenes has been made by Charton.[137] His conclusion is that, except for the *ortho*-substituted methylbenzenes and a few individual compounds, steric effects are absent and that the observed shifts are governed by inductive and resonance effects. The results of this work are somewhat puzzling in that magnetic anisotropy effects were ignored even though substituents such as COCl, PhCH$_2$, and CH$_2$—CH=CH$_2$ were used. A study of the conformations of *N*-alkyl-*N*-nitroso-anilines has been made in terms of the established picture of the magnetic anisotropy effect of the nitroso-group.[138] Another extensive empirical correlation of the effects upon the shifts of aromatic side-chain protons in terms of substituent constants shows good agreement after correction for magnetic anisotropy effects and ring-current changes.[139] For *ortho*-protons and the *ortho*-methyl protons in substituted benzenes there is a very good correlation of the proton shifts and the parameter Q [$= P(Ir^3)^{-1}$, where P is the C—X bond polarizability for a substituent X, I is the first ionization potential of X, and r is the C—X bond length].[140,141] Turner et al.[142] have examined the ability of three-membered rings to transmit

[134] A. D. Buckingham, E. E. Burnell, and C. A. De Lange, *J. Chem. Phys.*, 1971, **54**, 3242.
[135] K. Hayamizu and O. Yamamoto, *J. Chem. Phys.*, 1971, **54**, 3243.
[136] W. E. Stewart and T. H. Siddall, *Chem. Rev.*, 1970, **70**, 517.
[137] M. Charton, *J. Org. Chem.*, 1971, **36**, 266.
[138] J. T. D'Agostino and H. H. Jaffé, *J. Org. Chem.*, 1971, **36**, 992.
[139] H. Yamada, *Bull. Chem. Soc. Japan*, 1970, **43**, 1459.
[140] G. Socrates, *J. Phys. Chem.*, 1970, **74**, 3141.
[141] G. Socrates and M. W. Adlard, *J. Chem. Soc. (B)*, 1971, 733.
[142] A. B. Turner, R. E. Lutz, N. S. McFarlane, and D. W. Boykin, *J. Org. Chem.*, 1971, **36**, 1107.

substituent effects by studying proton chemical shifts. They find an order of efficiency for such transmissions as follows: cyclopropane ≈ oxiran > aziridine. Other compounds in which observed proton shieldings are believed to be influenced by inductive, magnetic anisotropy, and electric-field effects are the aryl polysulphides[143] and halogenated steroids.[144]

A very large number of studies of substituent effects on proton chemical shifts have appeared. Here we do little more than list the classes of compounds involved: acetophenones,[145] ethoxycarbonylpyrroles,[146] fluorenyl carbanions,[147] aminobornanes,[148] biphenyls,[149] dihalogenobiphenyls,[150] benzothidiazoles,[151] fluorenes,[152] naphthalenes,[153] 2-methyl-benzoxazoles, -benzothiazoles, and -benzoselenazoles,[154] and phenyltrimethylsilanes.[155] New proton chemical shift data have been obtained by studies on the following compounds: cyclopentadiene, cyclohexa-1,3-diene, cyclo-octa-1,3-diene, and 1,2-dihydronaphthalene,[156] naphthalene,[157] 1,4-benzodithiin and 1,4-benzodioxin.[158] New data on organic selenium compounds,[159] a number of beryllium compounds,[160] and barenes and their derivatives[161] have also been obtained.

C. Carbon Chemical Shifts.—With the application of recent experimental developments, carbon is beginning to overtake fluorine as the second most studied nucleus. Before giving a very brief discussion of some of the more recent work on carbon chemical shifts, we list as follows the organic compounds or classes of compounds for which new carbon chemical shift data have become available during the review period: chloro-substituted alkanes,[162] chloro-substituted ethanes and ethylenes,[163] the compounds Ph_3M where M is a Group VB element,[164] diethylphosphonates,[165] triphenyl phospite and

[143] T. Fujisawa and G. Tsuchihashi, *Bull. Chem. Soc. Japan*, 1970, **43**, 3615.
[144] R. E. Lack, J. Nemorin, and A. B. Ridley, *J. Chem. Soc. (B)*, 1971, 629.
[145] J. Bloxbridge, J. R. Jones, and R. E. Marks, *Org. Magn. Resonance*, 1970, **2**, 337.
[146] M. W. Roomi and H. Dugas, *Canad. J. Chem.*, 1970, **48**, 2303.
[147] R. H. Cox, *J. Magn. Resonance*, 1970, **3**, 223.
[148] T. Ahmad, M. N. Anwar, M. Martin-Smith, R. T. Parfitt, and G. A. Smail, *J. Chem. Soc. (C)*, 1971, 847.
[149] P. W. Hickmott and A. T. Hudson, *J. Chem. Soc. (C)*, 1971, 762.
[150] A. R. Tarpley and J. H. Goldstein, *J. Phys. Chem.*, 1971, **74**, 421.
[151] W. H. Poesche, *J. Chem. Soc. (B)*, 1971, 368.
[152] K. D. Bartle, D. W. Jones, and P. M. G. Bavin, *J. Chem. Soc. (B)*, 1971, 388.
[153] J. W. Emsley, S. R. Salman, and R. A. Storey, *J. Chem. Soc. (B)*, 1970, 1513.
[154] A. R. Katritzky and Y. Takeuchi, *Org. Magn. Resonance*, 1970, **2**, 569.
[155] M. E. Freeburger and L. Spialter, *J. Amer. Chem. Soc.*, 1971, **93**, 1894.
[156] M. A. Cooper, D. D. Elleman, C. D. Pearce, and S. L. Manatt, *J. Chem. Phys.*, 1970, **53**, 2343.
[157] R. W. Creceley and J. H. Goldstein, *Org. Magn. Resonance*, 1970, **2**, 613.
[158] K. K. Deb, J. E. Bloor, and T. C. Cole, *Org. Magn. Resonance*, 1970, **2**, 431.
[159] M. Lardon, *J. Amer. Chem. Soc.*, 1970, **92**, 5063.
[160] R. A. Kovar and G. L. Morgan, *J. Amer. Chem. Soc.*, 1970, **92**, 5067.
[161] V. I. Stanko, V. V. Khrapov, A. I. Klimova, and J. N. Shoolery, *J. Struct. Chem.*, 1970, **11**, 497.
[162] C. J. Carman, A. R. Tarpley, and J. H. Goldstein, *J. Amer. Chem. Soc.*, 1971, **93**, 2864.
[163] G. Miyajima and K. Takahashi, *J. Phys. Chem.*, 1971, **75**, 331.
[164] O. A. Gansow and and B. Y. Kimura, *Chem. Comm.*, 1970, 1621.

Nuclear Shielding

phosphate,[166] *para*-substituted triphenylcarbonium ions,[167] 1,3-dioxans,[168] symmetrical *ortho*-dihalogenobenzenes,[169] monosubstituted thiophens,[170] benzimidazole, purine, and their anionic and cationic species,[171] 4-azaindene and related bridgehead nitrogen heterocycles,[172] 1-substituted bicyclo[2,2,2]-octanes,[173] bicyclic hydrocarbons, alcohols, and ketones,[174] borneol,[175] *N*-nitroso-amines and *N*-nitroso-anilines,[176] norbornyl derivatives,[177] perezone and derivatives,[178] carbohydrates,[179,180] methylcyclohexanes,[181] alkaloids,[182] and sulphoxides.[183] Further studies on carbon shieldings have been made for the following classes of organometallic compounds: rhodium–olefin complexes,[184] organoplatinum complexes,[185] organolithium compounds,[186] α-di-imine–metal chelates,[187] and compounds of the type $Me_nSn(SMe)_{4-n}$.[188]

The most extensive experimental investigation of carbon shifts to be published during the review period is that of Lippmaa *et al.*[174] on some fifty bicyclic compounds. The carbon shieldings in these compounds were found to be very sensitive to molecular geometry — reflecting changes in hybridization and consequent changes in electron density affecting σ_{loc}^p. Magnetic anisotropy differences are far too small to have any significant effect on these carbon chemical shifts. Considerable evidence was found by Lippmaa *et al.* for a steric effect on the carbon shieldings. The substitution of a methyl group or the introduction of a double bond leads to a deshielding effect for the carbon at which the substitution occurs and for its neighbours, but to an

---

[165] G. A. Gray, *J. Amer. Chem. Soc.*, 1971, **93**, 2132.
[166] G. C. Levy and J. D. Cargioli, *Chem. Comm.*, 1970, 1663.
[167] G. J. Ray, R. J. Kurland, and A. K. Colter, *Tetrahedron*, 1971, **27**, 735.
[168] G. M. Kellie and F. G. Riddell, *J. Chem. Soc.* (*B*), 1971, 1030.
[169] A. R. Tarpley and J. H. Goldstein, *J. Mol. Spectroscopy*, 1971, **37**, 432.
[170] K. Takahashi, T. Sone, and K. Fujieda, *J. Phys. Chem.*, 1970, **74**, 2765.
[171] R. J. Pugmire and D. M. Grant, *J. Amer. Chem. Soc.*, 1971, **93**, 1880.
[172] R. J. Pugmire, M. J. Robins, D. M. Grant, and R. K. Robins, *J. Amer. Chem. Soc.*, 1971, **93**, 1887.
[173] G. E. Maciel and H. C. Dorn, *J. Amer. Chem. Soc.*, 1971, **93**, 1268.
[174] E. Lippmaa, T. Pehk, J. Paasivirta, N. Belikova, and A. Platé, *Org. Magn. Resonance*, 1970, **2**, 581.
[175] J. Briggs, F. A. Hart, G. P. Moss, and E. W. Randall, *Chem. Comm.*, 1971, 364.
[176] P. S. Pregosin and E. W. Randall, *Chem. Comm.*, 1971, 399.
[177] J. B. Grutzner, M. Jautelat, J. B. Dence, R. A. Smith, and J. D. Roberts, *J. Amer. Chem. Soc.*, 1970, **92**, 7107.
[178] P. Joseph-Nathan, M. P. González, L. F. Johnson, and J. N. Shoolery, *Org. Magn. Resonance*, 1971, **3**, 23.
[179] A. S. Perlin, B. Casu, and H. J. Koch, *Canad. J. Chem.*, 1970, **48**, 2596.
[180] R. Burton, L. D. Hall, and P. R. Steiner, *Canad. J. Chem.*, 1971, **49**, 588.
[181] A. S. Perlin and H. J. Koch, *Canad. J. Chem.*, 1970, **48**, 2639.
[182] W. O. Crain, W. C. Wildman, and J. D. Roberts, *J. Amer. Chem. Soc.*, 1971, **93**, 990.
[183] R. A. Archer, R. D. G. Cooper, P. V. Demarco, and L. F. Johnson, *Chem. Comm.*, 1970, 1291.
[184] G. M. Bodner, B. N. Storhoff, D. Doddrell, and L. J. Todd, *Chem. Comm.*, 1970, 1530.
[185] A. J. Cheney, B. E. Mann, and B. L. Shaw, *Chem. Comm.*, 1971, 431.
[186] L. D. McKeever and R. Waack, *J. Organometallic Chem.*, 1971, **28**, 145.
[187] C. Tänzer, R. Price, E. Breitmaier, G. Jung, and W. Voelter, *Angew. Chem. Internat. Edn.*, 1970, **9**, 963.
[188] E. V. van den Berghe and G. P. van der Kelen, *J. Organometallic Chem.*, 1971, **26**, 207.

increased shielding for carbon atoms three bonds away. This latter effect is interpreted as a steric effect (1,4-non-bonded interaction). The same effect has also been found in the carbon shieldings of the same bicyclic compounds (norbornyl derivatives) by Grutzner et al.[177] and further discussion of the effect for the methylcyclohexanes may be found in the paper of Perlin and Koch.[181]

The carbon chemical shifts of the chlorine-substituted ethanes[163] follow the trend towards lower shielding which is expected from those known (see Table 5) for the corresponding chlorine derivatives of methane. Chlorine substitution lowers the shielding in all cases, but more so for substitution at the carbon atom whose resonance is being obtained than at the second carbon atom. This trend is not fully paralleled for the chlorine-substituted derivatives of ethylene: with increased chlorine substitution at the second carbon atom the shielding of the first atom increases. We have earlier discussed the work of Mason[47] on the carbon chemical shifts of the methyl halides. She has applied her correction for non-zero changes in σ_{loc}^{d} to the shieldings of carbon in a number of other classes of compounds — branched and linear alkanes, methylcyclohexanes, aromatic compounds, alkenes, alkynes, and carbonium ions. With this correction made, the remaining parts of the shieldings can be predicted very well in terms of a small number of additivity parameters. CNDO calculations, not referred to earlier, have been used to rationalize observed carbon chemical shift data by Grant and co-workers[171,172] and by Ray and co-workers.[167]

D. Fluorine Chemical Shifts.—The recently published review of Emsley and Phillips[94] gives a thorough discussion of the various factors which influence fluorine magnetic shielding. They have also included a list of the chemical shifts of fluorine nuclei in a wide variety of organic and inorganic compounds. Although published in 1971, their discussion and tables do not cover any of the work appearing after 1968. Also published recently[189] is a compilation of fluorine chemical shifts of compounds studied up to mid-1967. The subject of fluorine shielding of the ionic fluorides has also been reviewed.[190]

New observed fluorine chemical shift data, in addition to those discussed in earlier sections of the Report, have been obtained for the following compounds or classes of compounds: substituted Dewarbenzenes and their dibromo-adducts,[191] five-co-ordinated phosphorus hydrides,[192] cyclic and bicyclic trifluoroacetates,[193] substituted trifluoroacetanilides,[194] substituted fluoropyridines,[195] fluoroalkyl-platinum(II) and -platinum(IV) compounds,[196]

[189] C. H. Dungan and J. R. van Wazer, 'Compilation of reported ^{19}F N.M.R. Chemical Shifts', Wiley–Interscience, New York, 1970.
[190] Y. V. Gagarinskii and S. P. Gabuda, *J. Struct. Chem.*, 1970, **11**, 897.
[191] L. Cavalli, *J. Chem. Soc. (B)*, 1970, 1616.
[192] V. V. Sheluchenko, G. I. Drozd, M. A. Landau, S. S. Dubov, and S. Z. Ivin, *J. Struct. Chem.*, 1970, **11**, 580.
[193] H. J. Schneider, G. Jung, E. Breitmaier, and W. Voelter, *Tetrahedron*, 1970, **26**, 5369.
[194] H. W. Johnson and Y. Iwata, *J. Org. Chem.*, 1970, **25**, 2822.
[195] S. Giam and J. L. Lyle, *J. Chem. Soc. (B)*, 1970, 1516.

tin tetrafluoride diadducts of various aromatic amine oxides,[197] fluoroaromatic compounds,[198] compounds of the type $FArMR_3$ where M is a Group IVB element,[199] and carboranes, carborane anions, and metallocarboranes.[200]

E. **Nitrogen Chemical Shifts.**—This topic has been reviewed recently by Randall and Gillies.[119] New, observed, chemical shift data have been published for the following compounds or classes of compounds: substituted anilines,[201] acetonitrile and methylamines,[202] Group IV isothiocyanates,[203] isocyanates and cyanates,[204] and cobalt(III) complexes with nitrogen-containing ligands.[205] Becker[3] has proposed a scale for nitrogen chemical shifts with the standard reference nucleus being the nitrogen nucleus of $Me_4N^+\ I^-$ used as an internal reference. Three papers by Mason and co-workers presenting and discussing nitrogen chemical shift data have been published.[206,207]

F. **Chemical Shifts of Nuclei Other Than H, C, F, and N.**—New publications on the chemical shifts of nuclei other than H, C, F, and N and not discussed in earlier sections of this Report are listed below for the nuclei indicated, with the compounds or classes of compounds in which the resonances were obtained also given. ^{31}P: tervalent phosphorus derivatives of metal carbonyls,[208] substituted phenylphosphonic acids,[209] 1-phospha-2,6,7-trioxabicyclo[2,2,1]heptane,[210] rhodium–phosphine complexes;[211] ^{11}B: 2-carba-*nido*-hexaborane(9),[212] boron hydrides,[213] aminoboranes and borates,[214] barenes and derivatives;[215] ^{183}W: phosphine–tungsten complexes;[216]

[196] H. C. Clark and J. D. Ruddick, *Inorg. Chem.*, 1970, **9**, 2556.
[197] C. E. Michelson and R. O. Ragsdale, *Inorg. Chem.*, 1970, **9**, 2718.
[198] M. A. Cooper, H. E. Weeber, and S. L. Manatt, *J. Amer. Chem. Soc.*, 1971, **93**, 2369.
[199] A. J. Smith, W. Adcock, and W. Kitching, *J. Amer. Chem. Soc.*, 1970, **92**, 6140.
[200] R. G. Adler and M. F. Hawthorne, *J. Amer. Chem. Soc.*, 1970, **92**, 6174.
[201] P. Hampson, A. Mathias, and R. Westhead, *J. Chem. Soc.* (B), 1971, 397.
[202] M. Alei, A. E. Florin, W. M. Litchman, and J. F. O'Brien, *J. Phys. Chem.*, 1971, **75**, 932.
[203] K. M. Mackay and S. R. Stobart, *Spectrochim. Acta*, 1971, **27A**, 923.
[204] K. F. Chew, W. Derbyshire, N. Logan, A. H. Norbury, and A. I. P. Sinha, *Chem. Comm.*, 1970, 1708.
[205] B. M. Fung and S. C. Wei, *J. Magn. Resonance*, 1970, **3**, 1.
[206] J. Mason and W. van Bronswijk, *J. Chem. Soc.* (A), 1970, 1763; 1971, 791.
[207] R. Grinter and J. Mason, *J. Chem. Soc.* (A), 1970, 2196.
[208] R. Mathieu, M. Lenzi, and R. Poilblanc, *Inorg. Chem.*, 1970, **9**, 2030.
[209] C. C. Mitsch, L. D. Freedman, and C. G. Moreland, *J. Magn. Resonance*, 1970, **3**, 446.
[210] R. D. Bertrand, J. G. Verkade, D. W. White, D. Gagnaire, J. B. Robert, and J. Verrier, *J. Magn. Resonance*, 1970, **3**, 494.
[211] B. E. Mann, C. Masters, and B. L. Shaw, *J. Chem. Soc.* (A), 1971, 1104.
[212] T. Onak and J. Spielman, *J. Magn. Resonance*, 1970, **3**, 122.
[213] J. B. Leach, T. Onak, J. Spielman, R. R. Rietz, R. Schaeffer, and L. G. Seddon, *Inorg. Chem.*, 1970, **9**, 2170.
[214] F. A. Davis, I. J. Turchi, and D. N. Greeley, *J. Org. Chem.*, 1971, **36**, 1300.
[215] V. I. Stanko, V. V. Khrapov, A. I. Klimova, and J. N. Shoolery, *J. Struct. Chem.*, 1970, **11**, 585.
[216] P. J. Green and T. H. Brown, *Inorg. Chem.*, 1971, **10**, 206.

^{55}Mn: organotin–manganese compounds;[217]
^{119}Sn: methyl(methylthio)stannanes;[188]
^{77}Se: organic selenium compounds;[159]
^{9}Be: inorganic and organic beryllium compounds;[160]
and finally, ^{3}H: tritium-labelled organic compounds.[218]

A large part of this Report was written during the tenure of a Visiting Associateship at the California Institute of Technology, Pasadena, California. The author would like to thank Professor Sunney I. Chan for inviting him to spend the summer of 1971 at the Institute.

[217] S. Onaka, Y. Sasaki, and H. Sono, *Bull. Chem. Soc. Japan.*, 1971, **44**, 726.
[218] J. Bloxbridge, J. A. Elvidge, J. R. Jones, and E. A. Evans, *Org. Magn. Resonance*, 1971, **3**, 127.

2
Nuclear Spin–Spin Coupling

BY R. GRINTER

1 Introduction

This chapter is concerned only with the indirect or electron-coupled nuclear spin–spin interactions. The direct, or dipole–dipole, interactions such as are measured in orientated samples, are only discussed where they impinge directly upon some aspect of indirect coupling, *e.g.* in the question of direct coupling anisotropies. The chapter has been split up in such a way as to make the material phenomenon- rather than compound-orientated. It is appreciated that this mode of division, just like any other, is in many respects artificial and inevitably results in the same work appearing in several places, divorces theory from experiment, *etc*. Nevertheless, it is hoped that such an organization of the material has its redeeming features and, at the expense of repetition, the attempt has been made to make each section substantially self-contained and dependent only upon the basic theory and the literature prior to the period under review. Thus, an observed correlation between $^1J(^{13}\text{CH})$ and $^2J(\text{HCH})$ may appear in both Section 4 and Section 5, but not necessarily so. In particular, many couplings in π-electron systems are discussed only in Section 7.

2 Basic Theory

A. Ramsey's Equations.—Until very recently, the interpretation and calculation of indirect nuclear spin–spin coupling was based almost entirely on the fundamental work of Ramsey,[1] according to which the Hamiltonian operator $\mathcal{H}'$, corresponding to the energy of the electron-coupled spin–spin interaction of two nuclei A and B, is given by:

$$\mathcal{H}' = \mathcal{H}_1^{(a)} + \mathcal{H}_1^{(b)} + \mathcal{H}_2 + \mathcal{H}_3 \tag{1}$$

The individual terms in the above Hamiltonian have the following names and forms.
The orbital terms:

$$\mathcal{H}_1^{(a)} = (e\hbar\beta/c)\underset{A\,B\,k}{\Sigma\Sigma\Sigma}\gamma_A\gamma_B r_{kA}^{-3} r_{kB}^{-3} \times \{(\mathbf{I}_A \cdot \mathbf{I}_B)(\mathbf{r}_{kA} \cdot \mathbf{r}_{kB}) - (\mathbf{I}_A \cdot \mathbf{r}_{kB})(\mathbf{I}_B \cdot \mathbf{r}_{kA})\} \tag{2}$$

---

[1] N. F. Ramsey, *Phys. Rev.*, 1953, **91**, 303.

$$\mathcal{H}_1^{(b)} = (2\beta\hbar/i)\underset{A\ k}{\Sigma\Sigma}\gamma_A r_{kA}^{-3}\mathbf{I}_A \cdot (\mathbf{r}_{kA} \times \nabla_k) \tag{3}$$

The spin-dipolar term:

$$\mathcal{H}_2 = 2\beta\hbar\underset{A\ k}{\Sigma\Sigma}\gamma_A \times \{3(\mathbf{S}_k \cdot \mathbf{r}_{kA})(\mathbf{I}_A \cdot \mathbf{r}_{kA})r_{kA}^{-5} - (\mathbf{S}_k \cdot \mathbf{I}_A)r_{kA}^{-3}\} \tag{4}$$

The Fermi contact term:

$$\mathcal{H}_3 = (16\pi\beta\hbar/3)\underset{A\ k}{\Sigma\Sigma}\gamma_A \delta(\mathbf{r}_{kA})\mathbf{S}_k \cdot \mathbf{I}_A \tag{5}$$

In these equations β (the Bohr magneton) $= e\hbar/2mc$, $\mathbf{I}$ and $\mathbf{S}$ are the nuclear and electron spin angular momenta respectively (in units of $\hbar$), γ is the magnetogyric ratio, and r is the electron–nuclear distance. $\delta(\mathbf{r}_{kA})$ is a Dirac delta function which picks out the value at $\mathbf{r}_{kA} = 0$ in any integration over the co-ordinates of electron k. The other symbols conform to standard usage.

The energy corresponding to $\mathcal{H}'$ is much smaller than the sum of the electronic potential and kinetic energies, which are normally the only ones considered in calculating quantum mechanical wavefunctions, so that the energy of nuclear spin–spin coupling, E_{AB}, was expressed by Ramsey[1] in terms of perturbation theory. It is necessary to go to second order:

$$E_{AB} = \Sigma_n \langle 0|\mathcal{H}'|n\rangle\langle n|\mathcal{H}'|0\rangle/(E_0 - E_n) \tag{6}$$

$|0\rangle$ represents the wavefunction for the molecular electronic ground state and $|n\rangle$ that of an excited state, their respective energies being E_0 and E_n. The summation should be taken over all excited states, including those of the continuum.

When equation (1) is substituted into equation (6), terms of second order in $\mathcal{H}_1^{(a)}$, $\mathcal{H}_1^{(b)}$, $\mathcal{H}_2$, and $\mathcal{H}_3$ result, together with their various cross-terms. However, as Ramsey showed, states $|n\rangle$ which give non-zero values for $\mathcal{H}_1^{(a)}$ and $\mathcal{H}_1^{(b)}$ give zero values for $\mathcal{H}_2$ and $\mathcal{H}_3$, and *vice versa*, so that the only cross-term arising is that between $\mathcal{H}_2$ and $\mathcal{H}_3$. Ramsey showed that this term too would average to zero under conditions of frequent intermolecular collisions so that for most work it may be neglected. It becomes important, however, in the nuclear spin–spin coupling in orientated molecules, especially with regard to the anisotropy of such coupling.

Thus, for coupling measured in the isotropic fluid phase we may write, in an obvious notation,

$$E_{AB} = E_{AB}^{1a} + E_{AB}^{1b} + E_{AB}^2 + E_{AB}^3 \tag{7}$$

and, following Ramsey's demonstration that for hydrogen E_{AB}^3 constitutes at least 90% of E_{AB}, attention has been focused very heavily on this term. Recently it has become clear that significant contributions to E_{AB} may also arise from other terms, particularly in couplings involving fluorine.

The connection of E_{AB} to the measured coupling constant in Hz, $J(AB)$, may now be made through the equation

$$E_{AB} = hJ(AB)\mathbf{I}_A \cdot \mathbf{I}_B \qquad (8)$$

It is important to note that only under conditions of rapid molecular motion will $J(AB)$ be isotropic; in orientated molecules anisotropic contributions to the electron-coupled nuclear-spin interactions occur.[2]

Thus the calculation of $J(AB)$ following Ramsey's approach reduces to the evaluation of the terms on the right-hand side of equation (7) through infinite summations such as are represented by equation (6). Valence bond (VB) and molecular orbital (MO) formulations of $|0\rangle$ and $|n\rangle$ have both been extensively used. Since this work has been the subject of two detailed reviews,[3,4] it seems appropriate here to mention only a few points which are rather important in the discussion which follows.

B. The Average Energy Approximation.—Since the evaluation of the infinite sum in equation (6) involves wavefunctions $|n\rangle$ which are unknown, the equation has often been simplified by removing the term $(E_0 - E_n)^{-1}$ from the summation, after which the closure theorem can be used giving:

$$E_{AB} \simeq (E_0 - E_n)^{-1} \sum_n \langle 0|\mathscr{H}'|n\rangle\langle n|\mathscr{H}'|0\rangle$$

$$= (\Delta E)^{-1} \langle 0|\mathscr{H}' \cdot \mathscr{H}'|0\rangle \qquad (9)$$

Some average energy must then be assigned to ΔE, but the choice of this energy is very difficult. A significant advance was made by Pople and Santry[5] when they showed how the average-energy approximation could be avoided, or made at a much later stage, in a molecular orbital formulation in which the excited states were formed by promotion of electrons from the occupied to the virtual MO, and the summation in equation (6) was carried out explicitly. More recently, a similar improvement in the VB theory of nuclear spin–spin coupling has been made by Barfield.[6]

C. Useful General Expressions.—Although the average energy approximation has many deficiencies, it has made possible the formulation of some simple relations for the Fermi contact contribution to nuclear spin–spin coupling, which have been much used in the interpretation of experimental results. Thus, in the linear combination of atomic orbitals (LCAO) MO formalism, Pople and Santry[5] were able to show that:

$$J^3(AB) = \frac{16h\beta^2}{9} \gamma_A \gamma_B S_A^2(0) S_B^2(0) \pi_{S_A S_B} \qquad (10)$$

---

[2] A. D. Buckingham and I. Love, *J. Magn. Resonance*, 1970, **2**, 338.
[3] M. Barfield and D. M. Grant, *Adv. Magn. Resonance*, 1965, **1**, 149.
[4] J. N. Murrell, *Progr. N. M. R. Spectroscopy*, 1970, 6, 1.
[5] J. A. Pople and D. P. Santry, *Mol. Phys.*, 1964, **8**, 1.
[6] M. Barfield, *J. Chem. Phys.*, 1967, **46**, 811; 1968, **48**, 4458.

where $\pi_{s_A s_B}$ is the mutual polarizability of the valence s orbitals of atoms A and B and $S_A^2(0)$, $S_B^2(0)$ are the electron densities of these same orbitals at the nuclei A and B. Making the average-energy approximation, equation (10) reduces to:

$$J^3(AB) = (16h\beta^2/9\Delta E)\gamma_A\gamma_B S_A^2(0)S_B^2(0)P_{S_A S_B}^2 \qquad (11)$$

where $P_{S_A S_B}$ is the MO bond order between the orbitals S_A and S_B. Finally, when working with the VB theory, it is convenient to use hybrid orbitals such as:

$$\phi_X = \alpha_X s_X + (1-\alpha_X^2)^{\frac{1}{2}} p_X \qquad (12)$$

where α_X^2 is the s-character of the hybrid orbital ϕ_X. $J^3(AB)$ is then given by the equation:[7]

$$J^3(AB) = (64h\beta^2/9\Delta E)\gamma_A\gamma_B \eta^2 \alpha_A^2 \alpha_B^2 S_A^2(0) S_B^2(0) \qquad (13)$$

where η is a normalizing factor for the VB function describing the AB bond using the hybrid orbitals ϕ_A and ϕ_B and allowing, if necessary, for the polarity of the bond. Equation (13) shows clearly the dependence of $J^3(AB)$ upon s-character, a point often made in the discussion of experimental coupling constants.

On the whole, equations (10), (11), and (13) have provided the most frequently used bases for the non-empirical discussion of coupling constants, especially those between directly-bonded nuclei. For other couplings, more specific relationships have been derived. Particularly important examples of these are the MO treatment of $^2J(HCH)$ by Pople and Bothner-By[8] and the VB analysis of $^3J(HCCH)$ by Karplus.[9,10]

D. The Finite Perturbation Theory (FPT).—A theoretical advance which has not yet been the subject of review, and which is important in that it provides a completely different approach to the calculation of spin–spin coupling, has been described by Pople and his co-workers.[11] They calculate the Fermi contact term only, and essentially the method entails the evaluation of a set of self-consistent-field molecular orbitals (SCFMO) which are unrestricted, *i.e.*, electrons of different spin occupy two different sets of MO, they are not paired in the same MO as is usually the case. These unrestricted molecular orbitals are evaluated in the presence of a finite perturbation d_A at one nucleus A which represents the effect of the nuclear spin and is such that, at that atom, the energy of an electron of α spin is higher than that of an electron having β spin. The balance of electron spin at that atom is thus disturbed and a spin density ρ_A results at atom A and also at all the other atoms. In this way, spin information is passed through the molecule. Since we are considering only

---

[7] C. Juan and H. S. Gutowsky, *J. Chem. Phys.*, 1962, **37**, 2198.
[8] J. A. Pople and A. A. Bothner-By, *J. Chem. Phys.*, 1965, **42**, 1339.
[9] M. Karplus, *J. Chem. Phys.*, 1959, **30**, 11.
[10] M. Karplus, *J. Amer. Chem. Soc.*, 1963, **85**, 2870.
[11] J. A. Pople, J. W. McIver, and N. S. Ostlund, *J. Chem. Phys.*, 1968, **49**, 2965.

the Fermi contact term, only s-orbital spin density is important and the equation for $J(AB)$ is:[11]

$$J(AB) = (16h\beta^2/9)\gamma_A\gamma_B S_A^2(0)S_B^2(0)\left\{\frac{\partial}{\partial d_A}\rho_{S_B}^2(d_A)\right\}_{d_A = 0} \quad (14)$$

$\rho_{S_B}^2$ is the diagonal spin density matrix element calculated for the orbital S_B when the perturbation d_A is applied at atom A. Thus, we see that the required coupling is simply proportional to the derivative of $\rho_{S_B}^2$, and this is evaluated numerically by taking finite differences.

Although the FPT work of Pople and his collaborators has been confined to the Fermi contact term, applications to the spin–dipolar and orbital terms have recently been reported.[12,13]

E. Coupling by π-Electrons.—Special adaptations of the general formulae to the problems of π-electron coupling have been described. Up until very recently, all these methods have relied upon introducing σ–π electron exchange by means of perturbation theory[14,15] and, following McConnell,[16] on the use of hyperfine splittings from the e.s.r. spectra of related free radicals. During the period under review, some FPT approaches to this problem have been described for the first time.

F. The Variation Method.—Although the perturbation theories outlined above have been used almost exclusively in the interpretation and calculation of coupling constants, the variation method does offer a potential solution to the problem of avoiding the infinite sum in equation (6) and some of the problems of the FPT.[17] However, the difficulties which beset a variation calculation, even for the hydrogen molecule, are very severe and little progress has been made in this direction. Nevertheless, two interesting papers on this subject have appeared in the recent months and they are discussed in the theoretical section.

G. SI Units.—The coupling constant $J(AB)$ as defined by equation (8) has the same value and is measured in the same units, Hz, in both the c.g.s. system and SI. For the discussion of nuclear spin–spin coupling in relation to molecular structure, it is useful to define[5,18] a reduced coupling constant $K(AB)$ which is independent of the nuclear magnetic moments of A and B and depends only on their electronic environment, *i.e.*

---

[12] A. C. Blizzard and D. P. Santry, *Chem. Comm.*, 1970, 1085.
[13] H. Nakatsuji, K. Hirao, H. Kato, and T. Yonezawa, *Chem. Phys. Letters*, 1970, **6**, 541.
[14] M. Karplus, *J. Chem. Phys.*, 1960, **33**, 1842.
[15] R. Ditchfield and J. N. Murrell, *Mol. Phys.*, 1968, **15**, 533.
[16] H. M. McConnell, *J. Mol. Spectroscopy*, 1957, **1**, 11; *J. Chem. Phys.*, 1959, **30**, 126.
[17] R. Ditchfield, N. S. Ostlund, J. N. Murrell, and M. A. Turpin, *Mol. Phys.*, 1970, **18**, 433.
[18] R. M. Lynden-Bell and N. Sheppard, *Proc. Roy. Soc.*, 1962, **A269**, 385.

$$K(AB) = (2\pi/\hbar\gamma_A\gamma_B)J(AB) \tag{15}$$

In the c.g.s. system, the units of $K(AB)$ are cm^{-3} and in SI they are N A^{-2} m^{-3}. Numerically,

$$K(AB)(\text{c.g.s.}) = K(AB)(\text{SI}) \times 10 \tag{16}$$

In general, however, numerical values of K have not been quoted in this report.

The equations used in the theory of n.m.r. are quoted above in their c.g.s. form. Conversion to the correct form for use in SI is, however, straightforward: equations (2)—(5), (10), (11), (13), and (14) need to be multiplied by $\mu_0/4\pi$, where μ_0 is the permeability of a vacuum. This factor is 10^{-7} kg m s^{-2} A^{-2} exactly.

3 Recent Theoretical Work

In this section, only those calculations based on the fundamental theory of nuclear spin–spin coupling will be described. Correlations with other parameters, charge densities, bond orders, *etc.* will be discussed under the headings corresponding to the coupling constants involved.

A. Calculations on H_2 or HD.—The exact calculation of the spin–spin coupling of the simplest molecule, hydrogen deuteride, remains an objective of considerable importance in the theory of magnetic resonance, particularly with regard to the effects on this calculation of inadequate wavefunctions or the truncation of the infinite summations, since these provide guides to the reliability of the theory in more complex situations. Variational methods have been applied and De-Jeu[19] has pointed out some of the problems associated with this type of approach, particularly the fact that there is no minimum principle, only a stationary principle, for the energy of nuclear spin–spin coupling. As a result one can only be confident about the outcome of a variational calculation if, on expanding the basis of the calculation, clear evidence of convergence is obtained. Agreement with experiment can arise fortuitously and is no criterion of the accuracy of the wavefunction used. Unfortunately, De-Jeu observed no evidence of convergence in an attempt to calculate the Fermi contact contribution to $^1J(\text{HD})$ using simple MO and VB and James-and-Coolidge-type wavefunctions.

Although De-Jeu's result was unsatisfactory, Paviot and Hoarau[20] have made what appears to be a rather more successful calculation of $^1J(\text{HD})$ using a variation method. Following an earlier calculation by Das and Bersohn,[21] they have attempted to minimize the nuclear self-coupling energy as opposed to the internuclear coupling energy. It has been argued that this can

---

[19] W. H. De-Jeu, *Mol. Phys.*, 1971, **20**, 573.
[20] J. Paviot and J. Hoarau, *Compt. rend.*, 1971, **272**, C, 1718.
[21] T. P. Das and R. Bersohn, *Phys. Rev.*, 1959, **115**, 897.

be a valid technical procedure even though the self-coupling energy does not converge.[21] Using a trial function of the form:

$$|\psi\rangle = |\psi_0\rangle + \frac{4e}{3c}\sum_A\sum_j \gamma_A f_{jA} \mathbf{S}_j \cdot \mathbf{I}_A |\psi_0\rangle \tag{17}$$

where

$$f_{jA} = a_1 U_{jA} + a_2 \log r_{jA} + a_3 r_{jA} + a_4 z_{jA} + a_5 r_{jA}^2$$

$$U_{jA} = \int_{r_{jA}}^{\infty} (1 - e^{-S/r_0})/S^2 dS$$

$z_{jA} = z$ component of r_{jA}. $r_0 = 1.4089 \times 10^{-13}$ cm, a constant of the order of nuclear dimensions.

Paviot and Hoarau find the results quoted in Table 1 using Weinbaum's function[22] for $|\psi_0\rangle$.

Table 1 *The results of Paviot and Hoarau[20] for a variational calculation of coupling and self-coupling in* HD

Variation function	J(HD)/Hz	J(HH) × 10⁻⁷/Hz
1 term	+11·00	−0·290801358
2 terms	+39·44	−0·290828786
3 terms	+50·76	−0·290830380
4 terms	+61·64	−0·290836758
5 terms	+63·92	−0·290836944
Experiment	+42·7[a]	−0·199819 $\alpha = 1\cdot 0^b$
		−0·229508 $\alpha = 1\cdot 193^c$

[a] T. F. Wimett, *Phys. Rev.*, 1953, **91**, 476
[b] Calculated for atomic hydrogen with orbital exponent, $\alpha = 1\cdot 0$
[c] Calculated for atomic hydrogen with $\alpha = 1\cdot 193$, Weinbaum's optimized value

This calculation shows strong evidence of convergence in both the internuclear coupling and the self coupling. That the latter should converge seems to be at variance with the work of Das and Bersohn.[21] In any event, the calculation would be worth repeating with a more accurate wavefunction for $|\psi_0\rangle$ since the authors suggest that the difference between the calculated and experimental ^1J(HD) may be due to the inadequacy of Weinbaum's wavefunction.

Emanuel[23] has carried out a detailed investigation of the Fermi contact contribution to the spin–spin coupling in H_2 starting with Ramsey's equation for this term.[1] He has calculated the value of either the first or the first two terms in the summation using a wide variety of ground- and excited-state

[22] S. Weinbaum, *J. Chem. Phys.*, 1933, **1**, 593.
[23] R. V. Emanuel, *Mol. Phys.*, 1970, **19**, 399.

wavefunctions, including the most accurate available and some commonly used approximate functions. The results are not encouraging. For example, the first term is calculated to be 199 Hz with James-and-Coolidge-type wavefunctions. The experimental value is 278 ± 5 Hz,[24] and since the calculated result is quite stable with respect to an increase of the number of terms in the wavefunction, and with respect to changing orbital exponents, Emanuel believes that inadequacies in the wavefunction used are unlikely to be responsible for this discrepancy. He suggests that truncation of the summation, the possible inapplicability of second-order perturbation theory to the problem, the neglect of terms other than the Fermi contact, or the approximate form of the Hamiltonian used in the Ramsey expression, may be the cause of the difficulties.

With regard to these points, the work of Dutta, Dutta, and Das[25] is both encouraging and informative. They have calculated the Fermi contact contribution to ^1J(HD) using the perturbation theory in a form which enables them to overcome the major difficulty; that of a knowledge of all the excited states in the molecule, including those of the continuum. Their method is essentially a multiple-perturbation approach applied by means of linked-cluster many-body perturbation theory.[26] They are thus enabled to use as their starting point the wavefunctions of the hydrogen-molecule ion (with an internuclear distance corresponding to that of H_2), a complete set of which may be obtained. Their perturbing Hamiltonian $\mathcal{H}'$ is then given by

$$\mathcal{H}' = 1/r_{12} + \mathcal{H}'_H + \mathcal{H}'_D \qquad (18)$$

where r_{12} is the interelectronic distance and $\mathcal{H}'_H$ and $\mathcal{H}'_D$ are the Fermi contact operators at the proton and deuteron respectively. The calculation is taken to first order in $\mathcal{H}'_H$ and $\mathcal{H}'_D$, and to as many orders in $1/r_{12}$ as are required for satisfactory convergence. Dutta, Dutta, and Das find, taking the calculation to fourth order in $1/r_{12}$, a calculated value of ^1J(HD) of 42·57 Hz, and estimate that fifth-order terms will contribute less than 1 Hz. This result is in excellent agreement with the experimental value[24] of $+42·7 \pm 0·7$ Hz and the authors suggest that the neglected orbital and spin-dipolar terms do not contribute more to the observed coupling than the quoted experimental error. The importance of this result to the theory of spin–spin coupling is considerable, but it appears unlikely that this type of calculation can be extended to molecules containing several electrons. Semi-empirical theories with truncated summations, in spite of their manifest problems, will remain important tools for chemists.

A treatment of spin–spin coupling using second quantization and a Green's function method has been described by Aono.[27] Unfortunately, this paper is in Japanese and at the time of writing efforts to obtain a translation

[24] T. F. Wimett, *Phys. Rev.*, 1953, **91**, 476.
[25] C. M. Dutta, N. C. Dutta, and T. P. Das, *Phys. Rev. Letters*, 1970, **25**, 1695.
[26] J. Goldstone, *Proc. Roy. Soc.*, 1957, **A239**, 267.
[27] S. Aono, *J. Chem. Soc. Japan*, 1971, **92**, 115.

Nuclear Spin–Spin Coupling

have failed; it appears, however, since no tables of figures are given, that no comparison with experiment has yet been made.

B. Ab Initio Calculations of Coupling in Small Molecules.—The value of ^1J(HF) in hydrogen fluoride has been thoroughly investigated by two groups[28,29] following the earlier work of Kato and Saika.[30] Both have concentrated upon the infinite summation (for the Fermi contact term only) and its convergence as the basis set of atomic orbitals is increased[29] or when configuration interaction is included.[28] In both cases, no improvement in the convergence was found over the very unsatisfactory situation first reported by Kato and Saika.[30] Murrell et al.[28] also investigated the change in ^1J(HF) with internuclear distance, and concluded that there should be an isotope effect in the reduced coupling constant which is largely determined by the anharmonic terms of the potential.

Hinchliffe and Cook[31] have also made *ab initio* MO calculations of the Fermi contact contribution to coupling in some small molecules. Agreement with experiment is poor and very erratic. As they point out, it is very unfortunate that the more elaborate theories give such poor results, since so much important material regarding the nature of chemical bonding and structure will become available when the factors controlling the magnitudes and signs of coupling constants are qualitatively and quantitatively understood.

C. Semi-empirical Calculations.—Despite, or perhaps because of, the discouraging results of attempts to improve the rigour of coupling constant calculations, many calculations using semi-empirical MO and truncated summations or average-energy approximations continue to appear.[32–35] It should be clear from the discussion above that the results of such calculations, though often useful, should always be treated with some degree of suspicion. They are obviously at their best when used to correlate trends in a related series of molecules, substituent effects *etc*. Gil and Geraldes,[36] for example, have considered the effects of substituents on ^2J(HH) in substituted methanes using the Pople–Santry theory.[5] They have presented their theoretical results paying particular attention to the effect of substituents with lone pairs. They have, as yet, reported no numerical results. On the subject of trends in the coupling constants of directly bonded nuclei, Emanuel[37] has commented upon the earlier paper on this subject by Jameson and Gutowsky.[38]

---

[28] J. N. Murrell, M. A. Turpin, and R. Ditchfield, *Mol. Phys.* 1970, **18**, 271.
[29] W. Adam, A. Grimison, and P. A. Sprangle, *Theor. Chim. Acta*, 1970, **18**, 385.
[30] Y. Kato and A. Saika, *J. Chem. Phys.*, 1967, **46**, 1975.
[31] A. Hinchliffe and D. B. Cook, *Theor. Chim. Acta*, 1970, **17**, 91.
[32] C. S. Cheung, M. A. Cooper, and S. L. Manatt, *Tetrahedron*, 1971, **27**, 701.
[33] J. A. Varga and S. S. Zumdhal, *Theor. Chim. Acta*, 1971, **21**, 211.
[34] K. G. R. Pachler, *Tetrahedron*, 1971, **27**, 187.
[35] R. R. Fraser and R. N. Renaud, *Canad. J. Chem.*, 1971, **49**, 755.
[36] V. M. S. Gil and C. F. G. C. Geraldes, *Rev. Port. Quim.*, 1970, **12**, 32.
[37] R. V. Emanuel, *J. Chem. Phys.*, 1970, **53**, 856.
[38] C. J. Jameson and H. S. Gutowsky, *J. Chem. Phys.*, 1969, **51**, 2790.

He points out that their analysis necessarily predicts a positive one-bond coupling constant between a pair of directly-bonded equivalent nuclei, whereas negative values of $^1J(^{31}P^{31}P)$ are known.

The finite perturbation theory has been systematically applied to a wide range of molecules and coupling constants (mostly H–H but also ^{13}C–H and ^{13}C–F) by Pople and his associates. Maciel, McIver, Ostlund, and Pople have used FPT with INDO (intermediate neglect of differential overlap) MO to investigate $^2J(HH)$ in saturated and unsaturated CH_2 groups,[39] values of $^3J(HH)$ in substituted ethanes and ethylenes,[40] and all proton–proton couplings in substituted benzenes.[41] This work is discussed in later sections dealing with the specific coupling constants involved but the general conclusions may be usefully summarized here. They are:

(a) General trends are well reproduced, i.e. the dependence of $^2J(HH)$ on substituent and structural effects and that of $^3J(HCCH)$ on dihedral angle.

(b) Less success is achieved in calculating actual values and reproducing experimental substituent effects in certain cases.

(c) An adjustment of the value of $S^2(0)$, according to Slater's rules, to account for the variation of electron density at the nucleus with atomic charge, does not give a significant improvement in the results.

(d) Uncertainties in conformations may be responsible for some of the differences between theory and experiment but cannot explain all of these.

Ellis and Maciel have also used FPT with INDO wavefunctions to calculate $^1J(^{13}CH)$ in small-ring compounds[42] and $^2J(HH)$ and $^3J(HH)$ in monosubstituted cyclopropanes.[43] Again, the general conclusions are that trends are reproduced, and gross geometrical effects are better accounted for than substituent effects. Particular difficulties arise with substituents of the $-I$-type, and it is suggested[43] that 'through-space' effects may be important in such cases. Ellis and Maciel also conclude[42] that direct computation of $^1J(^{13}CH)$ is more reliable than attempting to correlate the coupling with the C–H bond order [equation (11)] since that type of approach is found to be particularly unsuccessful in accounting for substituent effects.

The effect of substituents on $^1J(^{13}CH)$ and $^2J(HH)$ in methane has been investigated by Maciel and Summerhays.[44] Using FPT and INDO MO they calculate the couplings in CH_3X where X, which they call the 'pseudo atom' is represented by a varying series of INDO parameters. They find that the values and trends of the calculated coupling constants conform to the established experimental patterns.

---

[39] G. E. Maciel, J. W. McIver, N. S. Ostlund, and J. A. Pople, *J. Amer. Chem. Soc.* 1970, **92**, 4151.
[40] G. E. Maciel, J. W. McIver, N. S. Ostlund, and J. A. Pople, *J. Amer. Chem. Soc.*, 1970, **92**, 4497.
[41] G. E. Maciel, J. W. McIver, N. S. Ostlund, and J. A. Pople, *J. Amer. Chem. Soc.*, 1970, **92**, 4506.
[42] P. D. Ellis and G. E. Maciel, *J. Amer. Chem. Soc.*, 1970, **92**, 5829.
[43] P. D. Ellis and G. E. Maciel, *Mol. Phys.*, 1971, **20**, 433.
[44] G. E. Maciel and K. D. Summerhays, *J. Amer. Chem. Soc.*, 1971, **93**, 520.

The FPT calculations described above have been confined to the Fermi contact term. For atoms such as fluorine this is not valid, as Blizzard and Santry[12] have shown using their variant of the FPT which permits ready calculation of orbital and spin-dipolar terms in the INDO approximation. They find that, even with optimum scaling of parameters, the contact term alone is quite unable to predict the correct order of $^1J(^{13}C^{19}F)$ in the four fluorine-substituted methanes. However, inclusion of the neglected spin–dipolar and orbital terms gives results in good agreement with experiment, though parameter scaling is still necessary. The amount of parameter scaling required is a cause for some concern. Approximate calculations of terms other than the Fermi contact have been made for $^1J(^{13}C^{13}C)$ by Barbier, Faucher, and Berthier[45] using Pople and Santry's[5] formulation of Ramsey's equation and wavefunctions from the literature. They report calculations for ethane (staggered and eclipsed), ethylene, and acetylene. Some of their results are given in Table 2. The results indicate that, for $^1J(^{13}C^{13}C)$ at least, spin-dipolar

Table 2 Contributions to $^1J(^{13}C^{13}C)$, in Hz[45]

		J^{1a}	J^{1b}	J^2	J^3	J^{total}	J^{exp}
Ethane	Staggered	1·10	−1·34	1·38	36·2	37·34	34·6
	Eclipsed	1·10	−0·75	1·39	38·1	39·84	
Ethylene		1·61	−6·44	3·67	69·2	68·04	67·6
Acetylene		1·89	0·0	0·35	190·7	192·94	171·5

and orbital terms play a much less important role than they do for $^1J(^{13}C^{19}F)$. It would be interesting to know if fluorine substitution has the marked effect on the orbital contribution to $^1J(^{13}C^{13}C)$ that it does[12] on the orbital part of $^1J(^{13}C^{19}F)$.

D. Theoretical Work on π-Electron Coupling.—The coupling of protons in conjugated systems *via* the π-electrons remains a topic of considerable interest and a number of calculations of such couplings have been reported. Cunliffe, Grinter, and Harris[46] have described calculations of π-coupling for a variety of non-aromatic conjugated systems, using three different methods; a modification of the Karplus method[14] which does not suffer from the drawbacks described by Ditchfield and Murrell,[15] a FPT approach based on the work of Pople, McIver, and Ostlund,[11] and a variation of McConnell's original method.[16] In all calculations, use was made of the proton hyperfine splitting constants, a_H, of related π-electron radicals to evaluate certain parameters or to relate π-electron spin density to spin density at the protons. The results obtained agree well with estimated values of π-electron coupling, and in particular they substantiate the prevailing view concerning the magnitudes of such couplings which was questioned by Ditchfield and Murrell.[15]

Bacon and Maciel[47] have also used FPT to calculate proton–proton

---

[45] C. Barbier, H. Faucher, and G. Berthier, *Theor. Chim. Acta*, 1971, **21**, 105.
[46] A. V. Cunliffe, R. Grinter, and R. K. Harris, *J. Magn. Resonance*, 1970, **3**, 299.
[47] M. Bacon and G. E. Maciel, *Mol. Phys.*, 1971, **21**, 257.

coupling in substituted butadienes and to investigate the effect upon them of changes of molecular geometry. They have used the INDO approximation and have tried to separate σ- and π-electron effects by neglecting specific exchange integrals centred on two carbon atoms, the idea being that spin information cannot pass from a proton to the π-system, or *vice-versa*, if the exchange integrals of the carbon atom attached to that proton are set to zero, (or at least that this will block the most important path for the transmission of such information). The results for ^4J(HH) and ^5J(HH) are in good agreement with experimental estimates and with the calculations of Cunliffe *et al.*[46] The values of ^3J(HH) across the formal single and double bonds, however, are puzzling. They are illustrated in the Figure. It seems strange, if σ-effects have really been removed and it was the exchange integrals at C_a and C_b which were set to zero, that the two ^3J(HH) π-electron couplings across the formal double bond should be so different in magnitude, and that one of them should be

Figure *Some of the π-electron contributions to* J(HH) (*in* Hz) *calculated for butadiene*

only about one-half of the corresponding coupling across the formal single bond. In discussing the value of the latter coupling, Bacon and Maciel suggest that a direct transmission of spin density from the hydrogen bound to C_c into the C_a–C_b π-system might be possible. The calculations indicate that in-plane deformation, out-of-plane twist, and alkyl substituent effects are important in determining the values of the coupling constants.

It would be inappropriate, in a discussion of the applications of the FPT, to neglect to mention the work of Ditchfield *et al.*[17] who have explained the large differences which are sometimes observed in the results of the finite perturbation and the conventional sum-over-states perturbation methods. They show that these differences are essentially due to the fact that in FPT one cannot introduce singly-excited states into the perturbed wavefunction without at the same time introducing doubly-excited states in a restricted way. The two methods are expected to give very different results when configuration interaction is important for the unperturbed wavefunctions.

Van der Hart has developed some general formulae within the framework of the group function method and has applied these to the particular problem of proton coupling by π-electrons.[48] Although very general in principle, it appears that the application of these formulae to particular molecules will still

---

[48] W. J. Van der Hart, *Mol. Phys.*, 1971, **20**, 385, 399.

involve the same types of assumption and use of semi-empirical data as has been the case hitherto. Nevertheless, it is interesting to note that the values calculated for π-coupling in ethylene and butadiene reflect the larger magnitudes earlier assumed for these quantities. It would be useful if Van der Hart would apply his method to a wider range of molecules.

E. The Anisotropy of Electron-coupled Nuclear Spin Interactions.—(See also Chapter 10, Section 4.)

Although the Fermi contact contribution to nuclear spin–spin coupling is isotropic, the other terms and the Fermi contact/spin-dipolar cross-term are not, and recent interest in the n.m.r. spectra of oriented molecules has stimulated the theoretical investigation of the anisotropy of the coupling tensor. Krugh and Bernheim[49,50] have measured the n.m.r. spectrum of methyl fluoride in a liquid-crystal solvent and, by subtracting the direct coupling anisotropy (calculated from the known gas-phase geometry) from the experimentally observed anisotropy, they have obtained estimates of the anisotropy of the indirect couplings. Attempts to interpret these results theoretically have been made by a number of workers.

Buckingham and Love[2] have discussed indirect coupling anisotropy in considerable detail, basing their work upon the MO formalism.[5] They have given general formulae and have applied these to the calculation of directly-bonded coupling constants and their anisotropies. Calculated anisotropies are found to be far too small to explain the discrepancies between molecular structures deduced from microwave spectroscopy and from n.m.r. studies of solutes in nematic solvents. However, Buckingham and Love point out that, because of the non-rigidity of molecules, large uncertainties result from the determination of indirect coupling anisotropies from n.m.r. splittings in oriented molecules. They also give a very useful discussion of the symmetry properties of the coupling tensor.

Nakatsuji and his collaborators have calculated the anisotropy of $^1J(^{13}C^{19}F)$ and $^1J(^{13}CH)$ in CH_3F using[51] CNDO/2 (complete neglect of differential overlap/two) and INDO[13] MO. Although the two calculations show considerable differences, all results are of similar magnitude and very much smaller than those deduced[49,50] from experiment. It might appear from these results and those of Buckingham and Love[2] that molecular motion may be responsible for the discrepancies. Bulthius and MacLean[52] have investigated this possibility but they find that, far from improving the position, the inclusion of vibrational effects leads to a poorer agreement between the calculated direct couplings and the experimental total anisotropic coupling. That is, the anisotropic indirect couplings become larger than they appear to be when the

---

[49] T. R. Krugh and R. A. Bernheim, *J. Amer. Chem. Soc.*, 1969, **91**, 2385.
[50] T. R. Krugh and R. A. Bernheim, *J. Chem. Phys.*, 1970, **52**, 4942.
[51] H. Nakatsuji, H. Kato, I. Morishima, and T. Yonezawa, *Chem. Phys. Letters*, 1970, **4**, 607.
[52] J. Bulthius and C. MacLean, *Chem. Phys. Letters*, 1970, **7**, 242; *J. Magn. Resonance*, 1971, **4**, 148.

vibrational corrections are not made. The authors suggest that the difficulty may lie in the change of the magnitudes of the indirect couplings on going from the isotropic to the nematic phase. They point out that solvent effects on J of three or four Hz would be sufficient to explain the observed differences between the experimental total anisotropic couplings and the calculated direct couplings. This idea and the theoretical work receive support from the observation by Long and Goldstein[53] that the anisotropic contribution to $^1J(^{13}C^{19}F)$ in tetrafluoro-1,3-dithietan (1) is negligible.

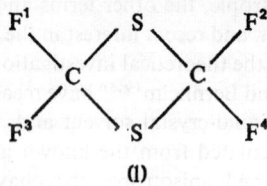

(1)

Finally, on the subject of indirect coupling and anisotropy, Love[54] has re-examined the relationship between nuclear screening tensors and the orbital contribution to the nuclear spin-coupling tensor. He has shown that, within a dipole approximation, the orbital contribution to nuclear spin–spin coupling J^1, and its anisotropy ΔJ^1, may be written in terms of the nuclear screening σ, and screening anisotropy $\Delta \sigma$, for the coupled nuclei A and B as follows:

$$J^1 = (2/3)(\hbar \gamma_A \gamma_B / 2\pi)(\Delta \sigma_A + \Delta \sigma_B)/R^3 \quad (19)$$

$$\Delta J^1 = (\hbar \gamma_A \gamma_B / 2\pi)\{3(\sigma_A + \sigma_B) + (\Delta \sigma_A + \Delta \sigma_B)\}/R^3 \quad (20)$$

for an axially symmetric molecule. This corrects an earlier expression due to Pople.[55] The agreement between values of J^1 and ΔJ^1 calculated from the above equations and from the second-order perturbation formula using LCAO MO wavefunctions is rather poor; in particular, the purely theoretical values of ΔJ^1 are much larger than the semi-empirical ones. Love attributes this to the failure of the dipole approximation used in deriving the equations (19) and (20); it indicates that short-range effects are important.

F. Solvent Effects on Coupling.—A theoretical study of the effect of solvent on nuclear spin–spin coupling has been made by Johnson and Barfield.[56] They use the Onsager reaction-field model and represent both the effect of the solvent and the coupling itself as perturbations on a set of INDO MO. When both perturbations are expressed by means of FPT the results, though poor in absolute magnitude, show trends in agreement with experimental

---

[53] R. C. Long and J. H. Goldstein, *J. Chem. Phys.*, 1971, **54**, 1563.
[54] I. Love, *Mol. Phys.*, 1970, **19**, 733.
[55] J. A. Pople, *Mol. Phys.*, 1958, **1**, 216.
[56] M. D. Johnson and M. Barfield, *J. Chem. Phys.*, 1971, **54**, 3083.

observations. When a sum-over-states method was used to evaluate the perturbation of the spin–spin coupling, no satisfactory results could be obtained.

4 Coupling of Directly Bonded Nuclei

This topic has recently been reviewed by McFarlane.[57] As earlier sections have indicated, the basic theory of nuclear spin–spin coupling is, in principle, well developed, but lack of accurate wavefunctions and energies, particularly for excited states, means that the theory is no more than a vital adjunct to our understanding at present. Though much progress is being made, correlation of experimental coupling constants over a wide range of the Periodic Table relies heavily on theoretical concepts rather than on direct calculation, and within this field coupling constants involving one hydrogen atom, *i.e.* $^1J(XH)$, are of particular interest since they focus attention on the properties of X. Among these properties, special attention has been paid to the state of hybridization and the effective nuclear charge.[58] Essentially, it is suggested that the magnitude of $^1J(XH)$ will increase with increasing *s*-character of ϕ_X [equations (12) and (13)] and with increasing effective nuclear charge of X, Z_X. An important adjunct to the assessment of the former are the ideas of Bent,[59] in particular the concept that substitution of X with substituents more electronegative than H increases the participation of the valence *s*-orbital of X in the X–H bond. Nevertheless, it has been pointed out[60] that many theoretical discussions, in concentrating too heavily upon the properties of X, have tended to ignore the important consequences which the change of the hydrogen 1*s* orbital exponent from molecule to molecule can have.

Pople and Santry[5] have shown that in a simple MO theory of coupling constants the only significant contributions to $^1J(XH)$ come from the Fermi contact term. This conclusion is supported by recent work for $^1J(^{13}CH)$[13] and $^1J(^{19}FH)$,[2] and there seems no reason to doubt that even more elaborate calculations would indicate only very small contributions to the total coupling from other sources. In what follows on $^1J(XH)$, we shall consider only the Fermi contact term since the calculations[2,13] noted above have been described in the theoretical section under anisotropy.

A. Coupling between ^{13}C and ^1H.—The most numerous $^1J(XH)$ values reported are those for which X = ^{13}C and, since these exemplify many of the important points, they will be considered first. Correlations of $^1J(^{13}CH)$ and $^1J(^{13}C^{13}C)$ with the *s*-character of the carbon hybrid-orbitals calculated by the maximum-overlap method have been given by Maksić *et al.*[61] They find:

$$^1J(^{13}CH) = 1079\alpha_{CH}^2/(1+S_{CH}^2) - 54\cdot9 \text{ (Hz)}$$

[57] W. McFarlane, *Quart. Rev.*, 1969, **23**, 187.
[58] D. M. Grant and W. M. Lichtman, *J. Amer. Chem. Soc.*, 1965, **87**, 3994.
[59] H. A. Bent, *Chem. Rev.*, 1961, **61**, 275.
[60] J. C. Hammel and J. A. S. Smith, *J. Chem. Soc. (A)*, 1970, 1852.
[61] Z. B. Maksić, M. Eckert-Maksić, and M. Randić, *Theor. Chim. Acta*, 1971, **22**, 70.

where S_{CH} is the carbon hybrid–hydrogen overlap integral. The standard deviation of this relation is 0·9 Hz and the authors suggest that the large constant term is due to the neglect of ionic structures. Servis and his co-workers[62] have noted a non-additive decrease in $^1J(^{13}CH)$ with increasing trimethylsilyl substitution of methane. They suggest that the deviation from additivity results from sterically-induced rehybridization.

In a direct approach to the calculation of $^1J(^{13}CH)$ in small-ring compounds, Ellis and Maciel[42] used FPT, their object being to assess the capability of this method to predict the observed structural dependence of the coupling. They also investigated the performance in this respect of equations which relate the coupling to the square of the carbon-2s–hydrogen-1s bond order, *i.e.* equation (11). They conclude, tentatively, that the FPT is somewhat more successful in accounting for gross geometrical effects than substituent effects, and that for the latter case the direct computation of $^1J(^{13}CH)$ is much more reliable than the calculation through $P^2_{C_{2s}H_{1s}}$. Also using FPT, Maciel and Summerhays[44] have investigated the effect of changing X on the coupling constants in CH_3X. (References 42 and 44 are also discussed in Section 3.)

Pursuing the study of electronegativity effects, Lunazzi, Macciantelli, and Taddei[63] have extended earlier work by measuring $^1J(^{13}CH)$ for the four halogeno-acetylenes. The thirteen measurements now available show a good correlation of J with substituent electronegativity, E_X, under conditions in which changes due to other effects should play a very subordinate role. They find

$$^1J(^{13}CH)_X = 208\cdot4\,(\pm4\cdot7) + 17\cdot4\,(\pm1\cdot8)E_X \text{ (Hz)}$$

Other correlations of $^1J(^{13}CH)$ with electronegativities have been reported,[64] and Lacey *et al.*[65] have described a correlation of $^1J(^{13}CH)$ with $^2J(HH)$ in a range of methyl derivatives which will be discussed further in the following section.

Lazzeretti and Taddei[66] have reported correlations of $^1J(^{13}CH)$ and $^3J(HH)$ in substituted aliphatic hydrocarbons with charge densities calculated by Del Re's method.[67] In particular, $^1J(^{13}CH)$ shows a linear relationship with the product of the charge density on the carbon atom and that on the hydrogen atom, different lines being obtained for substituents from different rows of the Periodic Table. Multiply substituted molecules lie on the same line as the corresponding singly substituted. The authors suggest that differing hybridization of carbon is the source of the different slope obtained for different rows of the Periodic Table. [Similar results are found for $^3J(HH)$—see Section 6].

[62] K. L. Servis, W. P. Weber, and A. K. Willard, *J. Phys. Chem.*, 1970, **74**, 3960.
[63] L. Lunazzi, D. Macciantelli, and F. Taddei, *Mol. Phys.*, 1970, **19**, 137.
[64] C. H. Yoder, D. R. Griffith, and C. D. Schaeffer, *J. Inorg. Nuclear Chem.*, 1970, **32**, 3689.
[65] M. J. Lacey, C. G. Macdonald, A. Pross, J. S. Shannon, and S. Sternhell, *Austral. J. Chem.*, 1970, **23**, 1421.
[66] P. Lazzeretti and F. Taddei, *Org. Magn. Resonance*, 1971, **3**, 113.
[67] G. Del Re, *J. Chem. Soc.*, 1958, 4031.

These results are stimulating but they may contain an element of deceptive simplicity since, if we think in terms of Slater's rules, the orbital exponent is linearly dependent on charge whereas electron density at the nucleus depends on the cube of the exponent.[58] A linear dependence of J upon charge is not therefore to be expected on simple grounds.[68] Also, the work of Pople et al.[39,69] has suggested that changes in electron density at the nucleus due to changing atomic charge are probably significant but not dominant.

Ramaswamy and Devarajan[70] have investigated the application of the two types of additivity rule introduced by Malinowsky[71] to substituent effects on $^1J(^{13}CH)$ in vinyl compounds. The agreement between observed and calculated J values is poor and for this the π-electrons may be partially responsible. In neutral and protonated azines, however, Gil and Pinto[72] have shown that H–H and ^{13}C–H couplings may, in the main, be quite accurately calculated by assuming an additivity of the effects of N and NH$^+$ on the corresponding hydrocarbons. This changes the problem to one of interpreting the relative coupling constants of pyridine, its ion, and benzene. Gil and Pinto conclude that the greater part of the differences in both H–H and ^{13}C–H couplings can be attributed to the delocalization of the nitrogen lone pair.

Correlations of $^1J(^{13}CH)$ in phenyltrimethylsilanes[73] and N-methylpyridinium salts[74] with Hammett σ and σ^+ values have been reported. The value of $^1J(^{13}CH)$ for the series PhMMe$_n$ (M = Si, C, N, O; n = 3, 3, 2, 1) increases (119·5, 125·5, 134·4, 143·0 Hz) and these results have been interpreted in terms of $p\pi \rightarrow d\pi$ back-bonding.[73]

Comparisons of $^1J(^{13}CH)$ and the CH stretching force constant have been made[75] for a number of dimethyl metals and dimethyl metal ions. $^1J(^{13}CH)$ has been measured in phenylcyclobutadienetricarbonyliron[76] and in a variety of monosubstituted thiophens.[77] Anomalous $^1J(^{13}CH)$ values have been reported for α-halogeno-aldehydes.[78]

B. Coupling between ^{15}N and ^{1}H.—Double-resonance experiments[79] have shown that all observable couplings for the molecules CH$_3$^{15}NH$_2$ and ^{13}CH$_3$^{15}NH$_2$ have the same sign when expressed in the reduced form. Thus $^1K(^{15}NH)$ is positive if $^1K(^{13}CH)$ is positive, in agreement with the prediction of Pople and Santry[5] and the generally accepted view.[57] An attempt

[68] G. A. Gray, *J. Amer. Chem. Soc.*, 1971, **93**, 2132.
[69] G. E. Maciel, J. W. McIver, N. S. Ostlund, and J. A. Pople, *J. Amer. Chem. Soc.*, 1970, **92**, 1, 11.
[70] K. Ramaswamy and V. Devarajan, *Indian J. Pure Appl. Phys.*, 1970, **8**, 173.
[71] E. R. Malinowsky, *J. Amer. Chem. Soc.*, 1961, **83**, 4479; E. R. Malinowsky and T. Vladimiroff, *J. Amer. Chem. Soc.*, 1964, **86**, 3575.
[72] V. M. S. Gil and A. J. L. Pinto, *Mol. Phys.*, 1970, **19**, 573.
[73] M. E. Freeburger and L. Spialter, *J. Amer. Chem. Soc.*, 1971, **93**, 1894.
[74] F. W. Wehrli, W. Giger, and W. Simon, *Helv. Chim. Acta*, 1971, **54**, 229.
[75] C. W. Hobbs and R. S. Tobias, *Inorg. Chem.*, 1970, **9**, 1998.
[76] H. A. Brune, H. Hanebeck, and H. Hüther, *Tetrahedron*, 1970, **26**, 3099.
[77] K. Takahashi, T. Sone, and K. Fujieda, *J. Phys. Chem.*, 1970, **74**, 2765.
[78] D. F. Ewing, *Org. Magn. Resonance*, 1971, **3**, 279.
[79] L. Paolillo and E. D. Becker, *J. Magn. Resonance*, 1970, **3**, 200.

has been made[80] in the case of $^1J(^{15}NH)$ to distinguish between the effects of hybridization and effective nuclear charge on this coupling. Nitro-group substitution in 2-methylindole and the quinolinium and anilinium ions has been found to have virtually no effect on the magnitude of $^1J(^{15}NH)$ in either the N sp^2 or the N sp^3 systems.[80] This, in view of the marked effect of substituents on other properties of these molecules, suggests that the effect of substituents on $^1J(^{15}NH)$ is largely due to changes in the hybridization of N. On this basis, percentage N 2s-character in the N–H bonds of some P–N, As–N, and S–N compounds has been estimated from $^1J(^{15}NH)$ values by Cowley and Schweiger.[81]

The value of $^1J(^{15}NH)$ in the ammonium ion and its methyl-substituted derivatives has been measured.[82] It is found that the increase in $^1J(^{15}NH)$ per methyl group is about one-third as large in the ions as it is in the free amines.

The spectrum of [^{15}N]pyrrole has been carefully re-analysed.[83] The magnitude of $^1J(^{15}NH)$ (96·54 Hz), for which quite diverse values have been quoted in the literature,[80] is in good agreement with the result given by Axenrod et al.[80] Furthermore, the hybridization of the nitrogen determined from two J–s-character relations[84] is in good accord with an ab initio calculation by Clementi et al.[85]

A study of the effect of solvent on $^1J(^{15}NH)$ in [^{15}N]aniline shows a marked increase in the magnitude of the coupling with increase in solvent hydrogen-bonding power,[86] (78·0 Hz in C_6D_{12}, 82·3 Hz in [$^2H_6$]DMSO). The authors suggest that this could be due either to the effect of the solvent on the hybridization of N, or on the lone pair, or both, but they do not suggest a mechanism for such effects. Presumably this will depend upon the way in which the amine protons and the nitrogen lone pair take part in the hydrogen-bonds which are formed.

C. Coupling between ^{31}P and ^{1}H.—The ^{1}H and ^{19}F n.m.r. spectra of the ions HPF_5^-, $CF_3PF_4H^-$, and $(CF_3)_2PF_3H^-$ have been discussed.[87,88] The spectra are in accord with octahedral co-ordination about phosphorus and the large $^1J(^{31}PH)$ values are attributed to the high phosphorus s-character of the P–H bond and the high effective nuclear charge of the phosphorus, both these results being a consequence of the highly electronegative fluorine ligands.

$^1J(^{31}PH)$ values which show the influence of changes in P-hybridization in

[80] T. Axenrod, M. J. Wieder, G. Berti, and P. L. Barili, *J. Amer. Chem. Soc.*, 1970, **92**, 6066.
[81] A. H. Cowley and J. R. Schweiger, *Chem. Comm.*, 1970, 1492.
[82] M. Alei, A. E. Florin, and W. M. Lichtman, *J. Phys. Chem.*, 1971, **75**, 1758.
[83] E. Rahkamma, *Ann. Acad. Sci. Fennicae*, 1970, **A354**, 3; *Mol. Phys.*, 1970, **19**, 727.
[84] (a) G. Binsch, J. D. Lambert, B. W. Roberts, and J. D. Roberts, *J. Amer. Chem. Soc.*, 1964, **86**, 5564; (b) A. J. R. Bourn and E. W. Randall, *Mol. Phys.*, 1964, **8**, 567.
[85] E. Clementi, H. Clementi, and D. R. Davies, *J. Chem. Phys.*, 1967, **46**, 4725.
[86] L. Paolillo and E. D. Becker, *J. Magn. Resonance*, 1970, **2**, 168.
[87] W. McFarlane, J. F. Nixon, and J. R. Swain, *Mol. Phys.*, 1970, **19**, 141.
[88] J. F. Nixon and J. R. Swain, *J. Chem. Soc. (A)*, 1970, 2075.

the effective absence of other types of substituent effect have been reported.[89] For a variety of groups Y, $^1J(^{31}PH)$ of (2) is always some 40 Hz less than that of (3), indicating increasing phosphorus *s*-character in the P–H bond on going from (2) to (3). This is in agreement with the change to be expected in the NPN angle when the ring is closed. This conclusion is confirmed by a comparison of the chemical shifts of the phosphorus atom and the hydrogen bonded to it.

```
        Me                              H   Me
        |                               |   |
   Me—N\        N—Y               H—C—N\        N—Y
         \    //                      |    \    //
          P                           |     P
         /   \                        |    /   \
   Me—N/      H                  H—C—N/      H
        |                            |   |
        Me                           H   Me

        (2)                             (3)
```

Values of $^1J(^{31}PH)$ and $^1J(^{15}NH)$ in the two series PhMHX, X = CMe$_3$, SiMe$_3$, GeMe$_3$, or SnMe$_3$ and in (Me$_3$Si)$_2$MH, M = ^{31}P or ^{15}N, decrease in a manner which is consistent with changes in σ-electron distribution and is opposite to that expected if $p\pi \to d\pi$ bonding to phosphorus is important.[90] $^1J(^{31}PH)$ values for some protonated phosphorus oxyacids have been given by Olah and McFarland[91] and for many di-t-butylphosphine–transition-metal complexes by Shaw and his associates.[92]

D. Coupling between ^{1}H and Other Nuclei.—Some directly bonded couplings between ^{11}B and ^{1}H have been reported.[93] Schumann and Dreeskamp[94] have themselves measured or have collected together a large number of ^1J(XH) values for substituted XH$_4$ molecules where X = ^{13}C, ^{29}Si, ^{73}Ge, ^{119}Sn, and ^{207}Pb. There are good grounds for believing that every ^1K(XH) in this series is positive.[94] An isotope effect was detected[94] in the coupling $^1J(^{29}SiH) = -239.5 \pm 0.1$ Hz for SiH$_3$I and -239.2 ± 0.1 Hz for SiH$_2$DI. However, the ratio of $^1J(^{29}SiH)$ to $^1J(^{29}SiD)$ does not differ significantly from the ratio of γ_H to γ_D. $^1J(^{29}SiH)$ has also been measured[95] in P(SiH$_3$)$_3$ and Sb(SiH$_3$)$_3$, and in a variety of trihalogenosilanes, where the expected dependence of the coupling on substituent electronegativity is found.[96] The magnitudes of

---

[89] A. Schmidpeter, H. Rossknecht, and K. Schumann, *Z. Naturforsch.*, 1970, **25b**, 1182.
[90] P. G. Harrison, S. E. Ulrich, and J. J. Zuckerman, *J. Amer. Chem. Soc.*, 1971, **93**, 2307.
[91] G. A. Olah and W. McFarland, *J. Org. Chem.*, 1971, **36**, 1374.
[92] A. Bright, B. E. Mann, C. Masters, B. L. Shaw, R. M. Slade and R. E. Stainbank, *J. Chem. Soc. (A),* 1971, 1826.
[93] G. Jugie, J.-P. Laussac, and J.-P. Laurent, *Bull. Soc. chim. France*, 1970, 2542, 4238.
[94] C. Schumann and H. Dreeskamp, *J. Magn. Resonance*, 1970, **3**, 204.
[95] K. D. Crosbie and G. M. Sheldrick, *Mol. Phys.*, 1971, **20**, 317.
[96] E. Hengge and F. Hoefler, *Z. Naturforsch.*, 1971, **26a**, 768.

directly bonded couplings of ^{1}H to ^{187}Os[97] and ^{195}Pt[98] have been determined in osmium and platinum hydride complexes.

E. $^1J(^{13}C^{13}C)$ and $^1J(^{13}C^{14/15}N)$.—The calculations of the spin-dipolar and orbital contributions to $^1J(^{13}C^{13}C)$ have been described in Section 3. As for $^1J(^{13}C^{14}N)$, calculation shows[99] that the dominant term in carbon–nitrogen coupling across the triple bond in cyanides and isocyanides is the Fermi contact, and the calculations predict positive values of $^1J(^{13}C^{14}N)$ for cyanides but negative values for isocyanides. Experimental measurements of $^1J(^{13}C^{14}N)$ have been made using proton-decoupled ^{13}C n.m.r.[99] and substantial differences between the values for cyanide and isocyanide bonds were found. In fact, the magnitude of the coupling for a bond involving an sp^3-hybridized carbon atom is comparable to that of the isocyanide bond involving an sp carbon, but is only about one-half of the value for the C≡N bond in cyanides. Thus the proportionality found [84a] for cyanides between $^1J(^{13}C^{15}N)$ and the s-character of the carbon and nitrogen atom orbitals does not appear to hold for isocyanides. Calculations by Gray[68] also cast doubt on the connection between $^1J(^{13}C^{15}N)$ and nitrogen and carbon s-characters, since the couplings show no simple relationship with $P^2_{s_C s_N}$. Values of $^1J(^{13}C^{13}C)$ showed[68] more correlation with $P^2_{s_C s_C}$ but the direct computation of the coupling using FPT was much more reliable, especially with regard to changing carbon hybridization.

Carhart and Roberts[100] have analysed the ^{1}H and ^{13}C spectra of Br^{13}CH$_2$-^{13}CH$_2$Br and shown that $^1J(^{13}C^{13}C)$ and both the vicinal ^3J(HH) couplings have the same sign. $^1J(^{13}C^{13}C) = +38 \cdot 9 \pm 0 \cdot 2$ Hz. Using the maximum-overlap method of determining the s-character of carbon hybrid-orbitals, Maksić et al.[61] find the following relationship between $^1J(^{13}C\text{-}1^{13}C\text{-}2)$ and $\alpha^2_{C\text{-}1}$ and $\alpha^2_{C\text{-}2}$ to be applicable.

$$^1J(^{13}C\text{-}1^{13}C\text{-}2) = 1020 \cdot 5\, \alpha^2_{C\text{-}1}\alpha^2_{C\text{-}2}/(1+S^2_{C\text{-}1C\text{-}2}) - 8 \cdot 2 \text{ (Hz)}$$

where $S_{C\text{-}1\,C\text{-}2}$ is the overlap integral for the two carbon hybrid-orbitals forming the C(1)–C(2) bond.

F. **Coupling between ^{13}C and ^{31}P.**—As Dreeskamp, Schumann, and Schmutzler have pointed out,[101] this coupling is very sensitive to the co-ordination number of the phosphorus, going from small positive or negative values in three-co-ordinate compounds to large positive values in four-co-ordinate molecules. The changes of sign found in the former systems have led to incorrect assumptions concerning sign in such compounds, and Albrand and Gagnaire have shown[102] that, in methyldichlorophosphine, $^1J(^{13}C^{31}P)$ is

---

[97] B. E. Mann, C. Masters, and B. L. Shaw, *Chem. Comm.*, 1970, 1041.
[98] M. J. Church and M. J. Mays, *J. Chem. Soc.* (A), 1970, 1938.
[99] I. Morishima, A. Mizuno, T. Yonezawa, and K. Goto, *Chem. Comm.*, 1970, 1321.
[100] R. E. Carhart and J. D. Roberts, *Org. Magn. Resonance*, 1971, 3, 139.
[101] H. Dreeskamp, C. Schumann, and R. Schmutzler, *Chem. Comm.*, 1970, 671.
[102] J. P. Albrand and D. Gagnaire, *Chem. Comm.*, 1970, 874.

Nuclear Spin–Spin Coupling

negative (-45 ± 1 Hz), relative to a positive $^1J(^{13}CH)$, contrary to earlier assumptions.

Dreeskamp et al.[101] report values of $^1J(^{13}C^{31}P)$ for Me_3PF_2 ($+128 \cdot 1 \pm 0 \cdot 5$ Hz) and $[MePF_5]^-$ ($+262 \pm 3$ Hz) [signs relative to $^1J(^{13}CH) > 0$] and conclude that the increase of the coupling from $+56 \cdot 5$ Hz in $[Me_4P]^+$ to 128 Hz in Me_3PF_2 is consistent with the assumption of a dominating contact mechanism for the coupling, sp^3-character for the phosphorus in $[Me_4P]^+$, and nearly sp^2-character for the three equatorial phosphorus orbitals in Me_3PF_2. These conclusions are also supported by the behaviour of $^1J(^{13}CH)$ in the same molecules. Again, the very large value of $^1J(^{13}C^{31}P)$ in $[MePF_5]^-$ seems explicable in terms of Bent's rules.[59] A careful review of the $^1J(^{13}C^{31}P)$ data to date has been given by Gray,[68] who has investigated the spectra of some diethylphosphonates in great detail. He finds that FPT calculations are successful in predicting magnitudes, signs, and relative ordering of $^1J(^{13}C^{31}P)$ values in a related series of phosphonates. The experimental couplings also correlate with $P^2_{S_C S_P}$, though the latter varies over about half of the percentage range of the former.

Jakobsen and Manscher[103] have measured the natural-abundance ^{13}C spectra of some heteroaromatic phosphine derivatives and report a number of ^{13}C–^{31}P coupling constants. The values of $^1J(^{13}C^{31}P)$ found are smaller than the $^2J(^{13}CC^{31}P)$ couplings in all cases, and are often smaller than the coupling over three bonds.

G. Coupling between ^{13}C and ^{19}F.—Work on carbon–fluorine coupling has been theoretically orientated with the emphasis on non-Fermi contact contributions to the coupling and anisotropy. This work has been described in Section 3.

Long and Goldstein[53] have measured all the coupling constants for tetrafluoro-1,3-dithietan (1), and have also determined their relative signs. The measurements were made in both the isotropic and the nematic phase. The effect of p-substituents on $^1J(^{13}C^{19}F)$ in benzotrifluoride has been described,[104] and the ^{19}F n.m.r. spectra of some fluoroaromatic compounds have been measured, using noise decoupling of the protons[105] to give a number of $^1J(^{13}C^{19}F)$ values. Weigert and Roberts[106] have determined the proton-decoupled ^{13}C n.m.r. spectra of a series of substituted fluorobenzenes and have thus obtained carbon–fluorine coupling constants over one to four bonds. The one-bond couplings in the p-derivatives are proportional to the fluorine chemical shifts, but a similar correlation is poor for the m-derivatives and does not hold at all for ortho. In cyclic geminal difluorides, $^1J(^{13}C^{19}F)$ varies with ring size.[106] Attempts to calculate the couplings using extended Hückel wavefunctions and the Pople–Santry[5] formalism were made for

---

[103] H. J. Jakobsen and O. Manscher, *Acta Chem. Scand.*, 1971, **25**, 680.
[104] C. G. Moreland and C. L. Bumgardner, *J. Magn. Resonance*, 1971, **4**, 20.
[105] M. A. Cooper, H. E. Weber, and S. L. Manatt, *J. Amer. Chem. Soc.*, 1971, **93**, 2369.
[106] F. J. Weigert and J. D. Roberts, *J. Amer. Chem. Soc.*, 1971, **93**, 2361.

fluorobenzene. The result was quite good for $^1J(^{13}C^{19}F)$ but poor otherwise, and the authors conclude that, at present, it seems to be more practical to attempt to find correlations between the C–F couplings and other couplings or molecular properties.

H. $^1J(^{13}CM)$, where M is a Metal.—Hildenbrand and Dreeskamp[107] have reported the ^{13}C-metal one-bond couplings in $CdMe_2$, $TlMe_3$, and $TlMe_2Br$. The couplings to ^{111}Cd and ^{113}Cd are negative relative to $^1J(^{13}CH) > 0$, whereas those to ^{203}Tl and ^{205}Tl are positive so that all the reduced coupling constants are positive. The couplings to thallium are also very temperature-sensitive. Values of $^1J(^{195}Pt^{13}C)$ have been given.[108]

I. Couplings to ^{19}F, except $^1J(^{19}F^{31}P)$.—The theoretical work on $^1J(^{19}FH)$ has been described in the previous section, as has that of Blizzard and Santry[12] on $^1J(^{19}F^{13}C)$. The latter shows the importance of spin-dipolar and orbital terms in fluorine coupling and this makes the interpretation of such coupling very uncertain. On the whole, rather few correlations of $^1J(^{19}FX)$ have been made,[57] and those that have show a dependence of the coupling upon $S_X(0)^2$ and upon the effective nuclear charge of X which indicates that the Fermi contact term is playing an important part. Until recently, in all cases where signs had been determined, $^1K(^{19}FX)$ was reported to be negative.[57] However, $^1K(^{19}F^{11}B)$ can be zero[109] in BF_4^-. $^1K(^{19}FH)$ is now known to be positive[110] and McFarlane, Noble, and Winfield have shown[111] that $^1J(^{19}F-^{183}W)$ can change sign even without a change of valence. In fact, the two $^{183}W-^{19}F$ coupling constants have the same sign in $MeOWF_3$, (and in many other tungsten hexafluoride and tungsten oxytetrafluoride derivatives), but opposite signs in the WOF_5^- and the $(F_4OWFWOF_4)^-$ anions. Unfortunately, the absolute signs remain unknown. The explanation of the different signs is not easy, particularly in view of the fact that a significant orbital contribution to the coupling cannot be ruled out. The authors offer an interpretation based upon the dominance of the Fermi contact term, and an analysis like that of Pople and Santry.[5]

$^1J(^{19}F^{51}V)$ has been observed[112] in VOF_4^- but no sign determination was made. An interesting series of $^1J(^{19}F^9Be)$ measurements has been made by Reeves and his co-workers.[113] They report values for BeF_4^{2-}, BeF_3^-, BeF_2, and BeF^+ in aqueous solution and, in spite of the various equilibria involved, the authors feel that there can be no doubt about the assignment of the measured couplings. No rationalization of the values by any theoretical method was attempted.

[107] K. Hildenbrand and H. Dreeskamp, Z. phys. Chem. (Frankfurt), 1970, 69, 171.
[108] A. J. Cheney, B. E. Mann, and B. L. Shaw, Chem. Comm., 1971, 431.
[109] R. J. Gillespie and J. S. Hartmann, J. Chem. Phys., 1966, 45, 2712.
[110] W. Klemperer, private communication quoted in ref. 28.
[111] W. McFarlane, A. Noble, and J. Winfield, Chem. Phys. Letters, 1970, 6, 547; J. Chem. Soc. (A), 1971, 948.
[112] J. A. S. Howell and K. C. Moss, J. Chem. Soc. (A), 1971, 270.
[113] M. G. Hogben, K. Radley, and L. W. Reeves, Canad. J. Chem., 1970 48, 2960.

$^1J(^{19}F^{29}Si)$ has been determined[114] in a number of fluoro-complexes of Si. The same coupling has been shown[115] to be pressure-dependent in the gas phase. No dependence on pressure was found for one-bond couplings in CH_4, SiH_4, and PF_3 and the case cited is the only example so far found. The result indicates that all solvent effects do not have to be due to reaction field effects, and valence or van der Waals interactions may be equally important. $^1J(^{19}F^{119}Sn)$ values for fluorostannate complexes[116] and for pyridine 1-oxide diadducts of SnF_4[117] have been reported.

J. Coupling between ^{31}P and ^{19}F.—Phosphorus–fluorine directly bonded coupling constants are generally believed to be negative, though their values cover a wide range, depending on the state of co-ordination of the phosphorus.[101] Although a good deal of data on $^1J(^{31}P^{19}F)$ are available for three- and four-co-ordinate phosphorus, less is known about five- and six-co-ordinate compounds. However, this information is now becoming available

(4; R, R' = alkyl, phenyl, or benzyl)   (5)

(6)   (7; R, R' = alkyl, phenyl, or benzyl)

and a very comprehensive paper containing over eighty $^1J(^{31}P^{19}F)$ values has recently been published by Reddy and Schmutzler.[118] They report values of the coupling for all types of phosphorus co-ordination and it is unfortunate that, even with such a mass of data, the authors were unable to establish any meaningful substitution rules for the $^1J(^{31}P^{19}F)$ values. In addition to this work, Schmutzler and his collaborators have also given phosphorus–fluorine couplings for some heterocyclic fluorophosphoranes of general formula (4),[119] for (5) and (6), and for compounds of general formula (7).[120] Some other

[114] P. A. W. Dean and D. F. Evans, *J. Chem. Soc. (A)*, 1970, 2569.
[115] A. K. Jameson and J. P. Reger, *J. Phys. Chem.*, 1971, **75**, 437.
[116] Yu. A. Buslaev and S. P. Petrosyants, *Zhur. strukt. Khim*, 1970, **11**, 443.
[117] C. E. Michelson and R. O. Ragsdale, *Inorg. Chem.*, 1970, **9**, 2718.
[118] G. S. Reddy and R. Schmutzler, *Z. Naturforsch.*, 1970, **25b**, 1199.
[119] R. E. Dunmur and R. Schmutzler, *J. Chem. Soc. (A)*, 1971, 1289.
[120] G. O. Doak and R. Schmutzler, *J. Chem. Soc. (A)*, 1971, 1295.

recent contributions to phosphorus–fluorine one-bond couplings are summarized in Table 3. The ion cis-$[CF_3PF_4H]^-$ seems to present an exception to the rule that coupling of phosphorus to an equatorial fluorine is always larger than to an axial fluorine.

Harris, Woplin, and Schmutzler[121] have given the $^1J(^{31}P^{19}F)$ values for a series of fac-trisubstituted fluorophosphine- and cis-fluorophosphine-molybdenum complexes.

K. Other One-bond Couplings involving ^{31}P.—The important couplings, $^1J(^{31}P^{13}C)$ and $^1J(^{31}P^{19}F)$, have been discussed above. Among those remaining, the most important is probably $^1J(^{31}P^{31}P)$, for which both positive and negative values are known.[122] It has been suggested[123] that $^1J(P^{III}P^{III})$ may be, in general, negative whereas $^1J(P^VP^V)$ may be, in general, positive, and some support for this has recently been obtained by the determination of the signs of $^1J(^{31}P^{31}P)$ in (8)[124] and (9).[122]

Table 3 Directly bonded $^1J(^{31}P^{19}F)$ values in Hz. The signs are given only if they have been experimentally established

Co-ordination at phosphorus	Compound formula	Coupling constant	Reference
3	O,O-P—F (cyclic)	1226	a
	Me-N, Me-N-P—F (cyclic)	1055	a
	S,S-P—F (cyclic)	1120	a
	Me_3CPF_2	1219	b
	$(Me_3C)_2PF$	848	b
4	$Me_3CP(:O)F_2$	1195	b
	$(Me_3C)_2P(:O)F$	1090	b
	$Me_3CP(:S)F_2$	1209	b
	$(Me_3C)_2P(:S)F$	1093	b

[121] R. K. Harris, J. R. Woplin, and R. Schmutzler, *Ber. Bunsengesellschaft Phys. Chem.*, 1971, **75**, 134.
[122] C. W. Schultz and R. W. Rudolph, *J. Amer. Chem. Soc.*, 1971, **93**, 1898.
[123] R. W. Rudolph and R. A. Newmark, *J. Amer. Chem. Soc.*, 1970, **92**, 1195.
[124] R. K. Harris, J. R. Woplin, and W. Stec, *Chem. Comm.*, 1970, 1391.

Table 3 (contd.)

Co-ordination at phosphorus	Compound formula	Coupling constant	Reference
5	Me$_3$CPF$_4$	1060	b
	(Me$_3$C)$_2$PF$_3$	910 (ax), 980 (eq)	b
	(Me$_3$C)$_3$PF$_2$	808	b
	Me$_2$PF$_2$NMe$_2$	654	c
	Me$_2$PF$_2$NEt$_2$	663	c
	MePhPF$_2$NEt$_2$	693	c
	Ph$_2$PF$_2$NMe$_2$	709	c
	Ph$_2$PF$_2$NEt$_2$	730	c
	MePF$_2$HCl	785 (ax)	d
	MePF$_2$HNEt$_2$	832 (ax)	d
	MePF$_2$HPr^i	725 (ax)	d
	MePF$_3$H	800 (ax), 1140 (eq)	d
	MePF$_4$	$-967 \cdot 7^e$	f
	Me$_2$PF$_3$	$-781 \cdot 6^e$ (ax), $-976 \cdot 2^e$ (eq)	f
	Me$_3$PF$_2$	$-552 \cdot 2^e$	f
6	Cs$^+$[PhPF$_5$]$^-$	687 (ax), 820 (eq)	c
	Cs$^+$[Me$_2$NPF$_5$]$^-$	664 (ax), 767 (eq)	c
	Cs$^+$[Ph$_2$PF$_4$]$^-$	947 (ax)	c
	K$^+$[HPF$_5$]$^-$	-729^g(ax), -817^g(eq)	h
	trans-Cs$^+$[(CF$_3$)$_2$PF$_4$]$^-$	-903^g(eq)	h
	trans-K$^+$[CF$_3$PF$_4$H]$^-$	858 (eq)	i
	cis-K$^+$[CF$_3$PF$_4$H]$^-$	784 (ax)	i
	trans?-K$^+$[(CF$_3$)$_2$PF$_3$H]$^-$	767 and 759 (eq) 725 and 834 (eq)	i
	j [MePF$_5$]$^-$	$-675 \cdot 7^e$ (ax), $-832 \cdot 4^e$ (eq)	f
	Me$_3$PPF$_5$	784 (ax), 900 (eq)^{k,l}	m
	Me$_2$HPPF$_5$	783 (ax), 873 (eq)^{k,l}	m
	MeH$_2$PPF$_5$	800 (ax), 867 (eq)^{k,l}	m

[a] J.-P. Albrand, A. Cogne, D. Gagnaire, J. Martin, J.-B. Robert, and J. Verrier, *Org. Magn. Resonance*, 1971, **3**, 75.
[b] M. Fild and R. Schmutzler, *J. Chem. Soc. (A)*, 1970, 2359.
[c] S. C. Peake, M. J. C. Hewson, and R. Schmutzler, *J. Chem. Soc. (A)*, 1970, 2364.
[d] V. V. Sheluchenko, G. I. Drozd, M. A. Landau, S. S. Dubov, and S. Z. Ivin, *Zhur. strukt. Khim.*, 1970, **11**, 623.
[e] Signs relative to ^1J(CH) > 0.
[f] H. Dreeskamp, C. Schumann, and R. Schmutzler, *Chem. Comm.*, 1970, 671.
[g] Signs relative to ^1J(PH) > 0.
[h] W. McFarlane, J. F. Nixon, and J. R. Swain, *Mol. Phys.*, 1970, **19**, 141.
[i] J. F. Nixon and J. R. Swain, *J. Chem. Soc. (A)*, 1970, 2075.
[j] Gegenion not recorded.
[k] Signs determined but relative to ^1J(PF).
[l] ^{19}F n.m.r. values; ^{31}P gave slightly different results.
[m] C. W. Schultz and R. W. Rudolph, *J. Amer. Chem. Soc.*, 1971, **93**, 1898.

The coupling constants, relative to $^3J(^{31}$PH) in (8) and $^1J(^{31}$P^{19}F) in (9), were found to be positive, having values of 475 ± 40, 715 ± 20, and 723 ± 20 Hz for (8), (9a), and (9b), respectively. However, a very small $^1J(^{31}$P^{31}P) value, 22·5 Hz, has been measured in (10), and no reason could be found for the surprising size of this coupling.[125] A determination of its sign would be of considerable interest. It appears[122] that two P^V atoms bound by a single bond and doubly bonded to either oxygen or sulphur have large $^1J(^{31}P^{31}P)$ values if the remaining bonds are to oxygen, but small values if those bonds are to carbon.

Schultz and Rudolph[122] have also collected together a large number of $^1J(^{31}P^{31}P)$ values, from which they conclude that (a) when the P–P bond contains at least one P atom having a lone pair of electrons, then the coupling constant is negative and greater in magnitude than 100 Hz, and (b) for singly bonded P atoms all the known J values are positive.

(8).

(9) a; R=Me
b; R=H

(10)

In view of these facts and the large positive value for $^1J(^{31}P^{31}P)$ in Me$_3$P–PF$_5$, they suggest that the neglected orbital and spin-dipolar terms may be much more important in determining the coupling than has been believed hitherto. If this is so, these terms contribute a substantial negative term to the coupling.

$^1J(^{31}P^{15}N)$ has been shown to be positive if $^1J(^{15}NH)$ is negative.[126] Thus, $^1K(^{15}N^{31}P)$ has the same sign as $^1K(^{31}P^{31}P)$ in P^{III}–P^{III} compounds, suggesting perhaps that ^{15}N resembles ^{31}P in the pattern of its coupling constants. It has also been suggested[124] that $^1J(^{31}P^{31}P)$ and $^1J(^{13}C^{31}P)$ may show some similarity in their behaviour (but see ref. 127). The study of coupling between the four elements C, N, Si, and P might prove helpful in understanding the connections between coupling constant and molecular structure in this part of the Periodic Table.

$^1J(^{29}Si^{31}P)$ has recently been found[95] to be opposite in sign to $^1J(^{29}SiH)$,

[125] J. Koketsu, M. Okamura, Y. Ishii, K. Goto, and S. Shimuzu, *Inorg. Nuclear Chem. Letters*, 1971, 7, 15.
[126] A. H. Cowley, J. R. Schweiger, and S. L. Manatt, *Chem. Comm.*, 1970, 1491.
[127] C. Schumann, H. Dreeskamp, and O. Steltzer, *Chem. Comm.*, 1970, 619.

Nuclear Spin–Spin Coupling

which is virtually certain to be negative[94] ($\gamma^{29}Si<0$); $^1J(^{29}Si^{31}P)$ would therefore appear to be positive and the corresponding reduced coupling negative. Values of $^1J(^{117}Sn^{31}P)$ and $^1J(^{119}Sn^{31}P)$ have been reported by Norman[128] and of $^1J(^{111}Cd^{31}P)$ and $^1J(^{113}Cd^{31}P)$ by Mann.[129]

Couplings of phosphorus to directly bonded transition metals have been reported by several groups, (Rh, Pt, and Hg),[130,131] (Rh),[132] and (Pt).[108,134] Pidcock and his associates[130] show how the $^1J(^{195}Pt^{31}P)$ values for the phosphonate and tributylphosphine groups in $trans$-[PtX{(PhO)$_3$PO}(Bu$_3$P)$_2$] correlate linearly with the $^2J(^{195}PtH)$ and $^1J(^{195}Pt^{31}P)$ values in $trans$-[PtMeX(Et$_3$P)$_2$], respectively. However, the coupling constants cis to X vary with X in a rather different way from the couplings $trans$ to X, and this implies that changes in $S_{Pt}(0)^2$ and ΔE, which would affect both types of coupling approximately equally, cannot be solely responsible for the variations in these coupling constants. Comparisons of metal-^{31}P couplings in corresponding ^{199}Hg and ^{195}Pt compounds show that the former are between 2·16 and 2·40 times as large as the latter.[131] The authors show that this is the figure to be expected if this ratio depends upon $S_{Hg}(0)^2/S_{Pt}(0)^2$ and conclude therefore that the coupling is dominated by the Fermi contact term. Values of $^1J(^{103}Rh^{31}P)$ for seventeen complexes are all found[132] to lie between 115·9 and 126·4 Hz, with the exception of one containing a P(OMe)$_3$ unit which is found to have a coupling of 193·9 Hz, showing again the important influence of OR groups on phosphorus coupling constants. It is noted[132] that a relation such as that found for tungsten complexes between $^1J(^{31}P^{183}W)$ and i.r. frequencies is not found in similar rhodium complexes.

Shaw and his co-workers[135] have also reported measurements of $^1J(^{187}Os$-$^{31}P)$ in OsH$_4$(PEt$_2$Ph)$_3$ and cis-OsCl$_2$(CO)$_2$(PBu^{t}Pr$_2^n$)$_2$. Reduction of the couplings shows that, as far as magnitude is concerned, ^{187}Os lies between ^{183}W and ^{195}Pt.

5 Coupling between Atoms Separated by Two Chemical Bonds, ^2J

These couplings, though less amenable in general to semi-empirical calculation than the one-bond couplings, have received much attention on account of the information which they can yield concerning molecular structure. The coupling most thoroughly investigated and for which theoretical treatments are available is that between two protons separated by a carbon atom, ^2J(HCH) or ^2J(HH).

---

[128] A. D. Norman, *J. Organometallic Chem.*, 1971, **28**, 81.
[129] B. E. Mann, *Inorg. Nuclear Chem. Letters*, 1971, **7**, 595.
[130] F. H. Allen, A. Pidcock, and C. R. Waterhouse, *J. Chem. Soc. (A)*, 1970, 2087.
[131] J. Bennett, A. Pidcock, C. R. Waterhouse, P. Coggon, and A. T. McPhail, *J. Chem. Soc. (A)*, 1970, 2094.
[132] B. E. Mann, C. Masters, and B. L. Shaw, *J. Chem. Soc. (A)*, 1971, 1104.
[133] R. K. Harris, N. C. Pyper, R. E. Richards, and G. W. Schulz, *Mol. Phys.*, 1970, **19**, 145.
[134] F. H. Allen and S. N. Sze, *J. Chem. Soc. (A)*, 1971, 2054.
[135] D. F. Gill, B. E. Mann, C. Masters, and B. L. Shaw, *Chem. Comm.*, 1970, 1269.

A. **Geminal Proton–Proton Coupling across Carbon.**—This subject has been reviewed recently by Sternhell[136] and other important reviews and compilations of data are cited there. The coupling can be either positive or negative and the most important influences on it are the hybridization of the central carbon atom, the electronegativity of any substituents, and their orientation with respect to the HCH plane. Groups with π-bonds and lone pairs have particularly marked effects. To date, the most useful theoretical rationalization of this coupling has been provided by Pople and Bothner-By.[8]

In the period under review, theoretical treatments have been largely based upon the FPT.[11] The most comprehensive of these papers[39] reports calculations of geminal proton coupling constants in 19 compounds with saturated CH_2 groups and 22 compounds having unsaturated CH_2 groups. The calculations were based on the Fermi contact term only and used INDO molecular orbitals; they are quite successful in reproducing the known experimental trends. The experimental negative sign of $^2J(HH)$ in saturated hydrocarbons and the positive sign in $H_2C=$ fragments is reproduced, as well as the intermediate values found for cyclopropyl CH_2 groups. Useful information is also obtained about one effect in the absence of others, a situation impossible to achieve experimentally. Thus $^2J(HH)$ is shown to increase algebraically with increasing HCH angle in both H_3C- and $H_2C=$ fragments. However, the calculations are less successful in reproducing the actual values observed in particular molecules, though part of this problem may be due to uncertainties concerning the molecular conformations. The qualitative results of Pople and Bothner-By's earlier work[8] are substantiated.

Maciel and his associates[43,44] have also used FPT and INDO wavefunctions to investigate the effect of substituents on $^2J(HH)$ in methane[44] and cyclopropane.[43] The calculations on methane have been described in Section 3. The cyclopropane work[43] considered the effect of a wide range of substituent groups and gave results qualitatively consistent with experimental observations. However, problems were encountered for substituents for which through-space interactions may be important, particularly if the substituent is of the $-I-$ type.

Gil and Geraldes[36] have investigated the effect of substituents on $^2J(HH)$ in methane, starting with equation (10). They show that the change in the mutual polarizability, $\Delta\pi_{AB}$, resulting from a substituent with a lone-pair orbital n is given by:

$$\Delta\pi_{AB} = \pm\beta_{nt_A}(\beta_{AB}+\beta_{t_At_B}\mp\beta_{At_B})\beta_{At_A}^{-4}+\ldots \text{higher terms} \quad (21)$$

t_A and t_B are the carbon hybrid orbitals directed towards hydrogens A and B respectively and the β_{ij} are resonance integrals between orbitals i and j. The upper signs are to be taken if the lone-pair orbital is doubly-occupied and the lower signs if it is vacant. Gil and Geraldes show that this simple expression gives results in qualitative agreement with experiment, but unfortunately no numerical applications of it are made.

[136] S. Sternhell, *Quart. Rev.*, 1969, **23**, 236.

On the experimental and correlative side, Sternhell and his collaborators[65] have measured ^2J(HD) and from this calculated ^2J(HH) for the series (Me)$_n$X where X = C, Ge, Sn, Pb, O, S, I, N, and Li. The relation between the magnitudes of ^2J(HH), ^1J(^{13}CH), and the electronegativity of X is explored using these results and other data from the literature. The plot of ^2J(HH) against Muller electronegativity shows a considerable scatter, which is not surprising since theoretical treatments do not predict a simple relationship between those properties. However, recalling the work of Douglas,[137] it is tentatively suggested[65] that four straight lines may be drawn through these points, each line joining compounds which have X in the same group of the Periodic Table. The lithium compound is anomalous. It is also shown that linear relationships exist between ^2J(HH) and ^1J(^{13}CH) and between ^2J(HH) and the product of the Muller electronegativity and the C–X bond distance. Finally, the authors demonstrate that plots of ^2J(HH) against the period of X show pairs of intersecting straight lines having their largest negative values at period three, where they join. The lines for each group of the Periodic Table are parallel.

Chivers and Crabb[138] have considered in detail the influence of nitrogen lone-pair electrons on the geminal coupling constant of an adjacent methylene group. Considering coupling constants in both five- and six-membered rings, they tentatively suggest a graphical relationship between ^2J(HH) and the angle of the nitrogen lone pair to one C–H bond. They point out that, as with other correlations of this nature, great care must be exercised in its use. The influence of lone-pair orientation on ^2J(HH) has also been studied by Katritzky and his associates.[139] They find that the variation of the geminal coupling shows no systematic correlation with the conformational equilibrium of adjacent N-alkyl groups in a variety of saturated heterocycles.

Swaelens, Anteunis, and Travernier[140] have investigated the influence of β-substitution on ^2J(HH) [the geminal coupling J(OCH$_2$O)] in a large number of 1,3-dioxans. They find that the influence of the substituent depends upon its orientation. The effects seem to be more pronounced with axial than with equatorial substituents and some possible models for the action of these effects are presented. Anteunis, Swaelens, and Gelan[141] have related ^2J(HH) in XCH$_2$Y to the Pauling electronegativities of X and Y, the C–X and C–Y bond lengths, and the mutual orientation of lone-pair orbitals and the C–H bonds.

The conformational dependence of the effect of substituents on ^2J(HH) has been investigated by Fraser and Renaud[142] for the systems (11) and (12), in which the benzylic methylene group is maintained in a definite orientation with respect to the adjacent benzene ring. They find that the slope, ρ, of the

---

[137] A. W. Douglas, *J. Chem. Phys.*, 1966, **45**, 3465.
[138] P. J. Chivers and T. A. Crabb, *Tetrahedron*, 1970, **26**, 3389.
[139] P. J. Halls, R. A. Y. Jones, A. R. Katritzky, M. Snarey, and D. L. Trepanier, *J. Chem. Soc. (B)*, 1971, 1320.
[140] G. Swaelens, M. Anteunis, and T. Travernier, *Bull. Soc. chim. belges*, 1970, **79**, 441.
[141] M. Anteunis, G. Swaelens, and J. Gelan, *Tetrahedron*, 1971, **27**, 1917.
[142] R. R. Fraser and R. N. Renaud, *Canad. J. Chem.*, 1971, **49**, 755.

(11) a; Y=O
b; Y=NMe
c; Y=⁺NHMe

(12)

linear relationship between ^2J(HH) and the σ(para) value of the substituent X for each series depends primarily on the angle, ϕ, between the plane of the benzene ring and the plane which contains the Ar–CH$_2$ bond and bisects the HCH angle. The dependence upon electronegativity was found to be extremely small. The results are summarized in Table 4. These results were interpreted as follows. Since the ρ values for series (11a—c) are quite similar, and in these molecules ϕ must be very near zero, the conclusion must be that the differing electronegativities of O, NMe, and ⁺NHMe, have a very small effect on the coupling constant. The value of ϕ for (12) can be estimated from Dreiding models to be about 72°, so that the small positive value of ρ obtained for this

Table 4 *Slopes of correlations between σ(para) and ^2J(HH)*

Series	ρ	Correlation coefficient	Standard deviation
(11a)^x	−1·9	0·976	0·25
(11b)^x	−1·7	0·979	0·23
(11c)^y	−1·6	0·982	0·17
(12)^x	+0·19	0·619	0·10
(12)^z	+0·17	0·698	0·07

^{x} CD$_3$Cl solvent; ^{y} CF$_3$CO$_2$H solvent;
^{z} MeCOMe solvent

series is in qualitative agreement with a dependence of ^2J(HH) on a cos$^2\phi$ relationship. The authors compare their results with theoretical predictions and conclude that they are in agreement with either the VB results of Barfield and Grant[143] or a cosϕcosϕ' relation. However, since the angles ϕ and ϕ' (neither of which is the angle ϕ used earlier in the paper) are not defined, and the method of calculation is not described or given a reference, this conclusion cannot be properly assessed. Fraser and Renaud[142] point out that Chow's results[144] for *N*-benzyl-2-methylpyridines, in which ρ = 1·4 for CCl$_4$ solutions and ρ = 0·0 in 1N-DCl solutions, for which correlations of ^2J(HH) against σ were observed, seem to be best interpreted as a change of ϕ from 90° to 0° on going from the former to the latter solvent.

[143] M. Barfield and D. M. Grant, *J. Amer. Chem. Soc.*, 1963, **85**, 1899.
[144] Y. L. Chow, S. Black, J. E. Blier, and M. M. Tracey, *Canad. J. Chem.*, 1970, **48**, 2134.

Some anomalous geminal couplings in ethylene episulphoxides have been noted.[145] Large negative values of $^2J(\text{HH})$ ($-6\cdot4$ and $-6\cdot0$ Hz) are found for the episulphoxides while the corresponding episulphides have much smaller values ($-0\cdot7$ and $-0\cdot9$ Hz). (The signs were determined by tickling experiments.) These large negative values are surprising in view of the fact that the electronegativity of the S→O group is greater than that of S. The authors suggest that the transfer of electrons from the antisymmetric CH_2 orbital into the π-system of the S→O group may be the reason for the anomaly.

Careful analyses of the spectra of series of compounds having an unsaturated methylene group have been made in three laboratories.[146–148] For systems of general formula (13), $^2J(\text{HH})$ showed a correlation with the chemical shifts of the coupled protons, increasing as the shifts increased.[146] The same coupling in molecules of general formula (14) showed no relationship with the electronegativity of Y, though such a correlation was found for the $^4J(\text{CH}_3\text{–H})$ values.[147]

```
   H       H                H       H
    \     /                   \     /
     C = C                     C = C
    /     \                   /     \
   H       S—R               H       Y

  (13)  R = alkyl            (14)
```

The magnitude of $^2J(\text{HH})$ has been measured for methylamine ($12\cdot1 \pm 0\cdot1$ Hz) and methylamine hydrochloride ($12\cdot3 \pm 0\cdot1$ Hz).[149]

B. Geminal Proton–Proton Coupling across Other Elements.—Data on $^2J(\text{HXH})$, where X is a Group IVB element, have been reported by Schumann and Dreeskamp.[94] The signs of many of these coupling constants have been determined and the authors find that $^2J(\text{HXH})$ is negative for X = C but positive for the other four elements. The data clearly show a regular algebraic increase in $^2J(\text{HH})$ in XH_4 on descending the group, and the earlier interpretation[150] of this observation in terms of the MO theory of Pople and Bothner-By[8] is confirmed. Descending the group, the energy separation between the occupied orbitals ψ_1 and ψ_2 and the unoccupied orbitals ψ_3 and ψ_4 decreases, giving an increased positive contribution to $^2J(\text{HH})$ from the excitation $\psi_2 \rightarrow \psi_3$, in agreement with experiment. A study of the effects of substituents on the coupling confirms this interpretation. The theory[8] is also substantiated with regard to the effect on the coupling of inductive and hyperconjugative electron withdrawal by the substituents.

For the symmetrically substituted dialkylstannanes a very good linear

---

[145] M. Ohtsuru, K. Tori, and M. Fukuyama, *Tetrahedron Letters*, 1970, 2877.
[146] G. Ceccarelli and E. Chiellini, *Org. Magn. Resonance*, 1970, **2**, 409.
[147] D. G. de Kowalewski, *J. Magn. Resonance*, 1971, **4**, 249.
[148] D. F. Ewing and K. A. W. Parry, *J. Chem. Soc. (B)*, 1970, 970.
[149] J. L. Sudmeier and D. R. Murayama, *Org. Magn. Resonance*, 1970, **2**, 625.
[150] H. Dreeskamp and C. Schumann, *Chem. Phys. Letters*, 1968, **1**, 555.

correlation of $^2J(\text{HSnH})$ with $^1J(^{119}\text{SnH})$ is observed[94] and Ph_2SnH_2 also falls on this line. In general terms, this work demonstrates just how useful the Pople–Bothner-By analysis can be in interpreting the salient features of geminal coupling between protons, and Schumann and Dreeskamp even go as far as to suggest[94] that the theory may also be applicable to $^2J(^{13}\text{CH})$, $^2J(^{13}\text{C}^{13}\text{C})$, $^2J(^{13}\text{C}^{119}\text{Sn})$, and $^2J(^{13}\text{C}^{207}\text{Pb})$ if suitable semi-empirical estimates of the electron densities at the nuclei are used.

The magnitude of $^2J(\text{H}^{15}\text{NH})$ in some amino-acids has been found to increase with pH.[151]

C. $^2J(^{13}\text{CCH})$.

—Couplings of this type have been extensively investigated by Goldstein and his colleagues using high-resolution ^{13}C n.m.r. They have recently reported values for cyclopropanes,[152] substituted ethylenes,[153] and symmetrical *ortho*-dihalogenobenzenes;[154] for the last, a determination of the relative signs of the coupling constants was possible. Correlations of $^2J(^{13}\text{CCH})$ with $^2J(\text{HCH})$ and $^3J(\text{HCCH})$ are found for the cyclopropanes[152] and with

(15)   (16)   (17)

$^1J(^{13}\text{CH})$ and $^3J(\text{HCCH})$ for the vinyl compounds.[153] The effects of substituents are discussed. For the dihalogenobenzenes, correlations of $^2J(^{13}\text{CCH})$ with substituent electronegativity are found.[154] In all cases $^2J(^{13}\text{CCH})$ is smaller than $^3J(^{13}\text{CCCH})$.

Correlations of $^2J(^{13}\text{CCH})$ with $^2J(\text{HCH})$ and *cis*-$^2J(\text{HCCH})$ in (15)—(17) have also been noted.[155] The theoretical basis of these relationships is far from clear at present, but their application is believed to be promising[152] and their practical use in the analysis of complex spectra has been indicated.[155] Anomalous values of $^2J(^{13}\text{CCH})$ in α-halogeno-aldehydes have been reported[78] and values of $^2J(^{13}\text{CCH})$ for monosubstituted thiophens[77] and in halogenoacetylenes[63] have been discussed.

D. Coupling of Other Elements to Hydrogen through Carbon.

—This subject

[151] R. L. Lichter and J. D. Roberts, *Spectrochim. Acta*, 1970, **26A**, 1813.
[152] K. M. Crecely, R. W. Crecely, and J. H. Goldstein, *J. Phys. Chem.*, 1970, **74**, 2680.
[153] K. M. Crecely, R. W. Crecely, and J. H. Goldstein, *J. Mol. Spectroscopy*, 1971, **37**, 252.
[154] A. R. Tarpley and J. H. Goldstein, *J. Mol. Spectroscopy*, 1971, **37**, 432.
[155] D. M. McKinnon and T. Schaefer, *Canad. J. Chem.*, 1971, **49**, 89.

has been discussed by Johannesen, Ferretti, and Harris[156] for compounds of the type R_mX, where R is methyl, ethyl, or isopropyl, X is the heteroatom, and m is its valence state. With the exception of R_3P, increased alkyl substitution causes ^2J(HX) to become more positive; R_3P is also of interest in that ^2J(H^{31}P) changes sign on going from trimethylphosphine to tri-isopropylphosphine. Particular interest attaches to ^2J(^{31}PCH) in terms of its sign and magnitude, the former having in the past been incorrectly assumed to be negative in dichloromethylphosphine. This coupling has now been shown to be positive relative to ^1J(^{13}CH)>0.[102] In contrast, negative values of ^2J(^{31}PCH) have been determined in some P^V compounds by Dreeskamp and his co-workers.[101] Some recent values of ^2J(^{31}PCH) are summarized in Table 5.

Table 5 *Some values, in Hz, of ^2J(PCH). Signs are given only where they have been experimentally determined for the molecule in question or a very similar structure*

P^{III}			P^V		
Compound	J(PCH)	Ref.	Compound	J(PCH)	Ref.
MePCl$_2$	+17.6	a	MePF$_4$	−20.3	c
Me$_3$P→PF$_5$	13.5	b	Me$_2$PF$_3$	−17.0	c
Me$_2$HP→PF$_5$	15.7	b	Me$_3$PF$_2$	−17.5	c
MeH$_2$P→PF$_5$	17	b	[MePF$_5$]$^−$	−20.5	c
Et$_3$P→BH$_{3−n}$X$_n$^{d}	9.5–12.5	e	Me$_3$P=CH$_2$	HC=P H$_3$CP	
				7 12.5	g
			Me$_3$P=CH—CH=CH$_2$	22 12.8	g
			Me$_3$P=CH—Ph	20 12.7	g
			Me$_2$PF$_2$NMe$_2$	17.5	i
	PCH −2.0		Me$_2$PF$_2$NEt$_2$	19.0	i
	PCH$_3$ 3.9	f	MePhPF$_2$NEt$_2$	17.0	i

	PCH	+22.7	
	PCH$_3$	3.4	f

| [Me$_2$CH]$_3$P | −2.27 | h |

^{a} J. P. Albrand and D. Gagnaire, *Chem. Comm.*, 1970, 874.
^{b} C. W. Schultz and R. W. Rudolph, *J. Amer. Chem. Soc.*, 1971, **93**, 1898.
^{c} H. Dreeskamp, C. Schumann, and R. Schmutzler, *Chem. Comm.*, 1970, 671.
^{d} X = halogen, not necessarily all the same.
^{e} G. Jugie, J.-P. Laussac, and J.-P. Laurent, *Bull. Soc. chim. France*, 1970, 2542, 4238.
^{f} J. P. Albrand, D. Gagnaire, M. Picard, and J. B. Robert, *Tetrahedron Letters*, 1970, 4593.
^{g} W. Malisch, D. Rankin, and H. Schmidbaur, *Chem. Ber.*, 1971, **104**, 145.
^{h} R. B. Johannesen, J. A. Ferretti, and R. K. Harris, *J. Magn. Resonance*, 1970, **3**, 84.
^{i} S. C. Peake, M. J. C. Hewson, and R. Schmutzler, *J. Chem. Soc. (A)*, 1970, 2364.

[156] R. B. Johannesen, J. A. Ferretti, and R. K. Harris, *J. Magn. Resonance*, 1970, **3**, 84.

The couplings in large numbers of ylides have been reported.[157,158] The structures of the two isomeric cyclic phosphines (Table 5) were assigned by Albrand et al.[159] on the basis of the values of $^2J(^{31}PCH)$ and the connection of this with the orientation of the phosphorus lone pair.

It has been found[160] that in compounds such as XCH_2PXNMe_2 there are two distinct values of $^2J(^{31}PCH)$ which have the same sign if X = F but different signs if X = Cl.

The value of $^2J(^{15}NCH)$ depends strongly upon the orientation of the nitrogen lone pair,[161] absolute values of the coupling being much larger for an *anti*-proton (10—16 Hz) than for a *syn*-proton (2—4 Hz). Similar results have now been reported[162] for 2-(α-naphthyl)[^{15}N]aziridine (18). For this molecule, the relative signs of all $^2J(^{15}NH)$ values were found to be the same, and for a variety of solvents $^2J(^{15}NH_C)$ was always larger than $^2J(^{15}NH_B)$. The authors suggest that this indicates that the sign of the coupling is negative if the influence of electronegative substituents on $^2J(^{15}NH)$ is the same as on $^2J(HH)$, since the α-naphthyl group is more electronegative than hydrogen. The results are in agreement with earlier calculations.[163]

(18)

The sign of $^2J(^{31}PNH)$ relative to $^1J(^{15}NH) < 0$ has been found to be negative[126] in $(CF_3)_2PNH_2$; its value is −14·21 Hz.

Two-bond couplings of metals to hydrogen through carbon have been widely reported.[94,107,128,164-174] The reduced couplings to methyl groups are

---

[157] W. Malisch, D. Rankin, and H. Schmidbaur, *Chem. Ber.*, 1971, **104**, 145.
[158] H. Schmidbaur and W. Malisch, *Chem. Ber.*, 1971, **104**, 150.
[159] J. P. Albrand, D. Gagnaire, M. Picard, and J. B. Robert, *Tetrahedron Letters*, 1970, 4593.
[160] J. E. Bissey, H. Goldwhite, and D. G. Rowsell, *Org. Magn. Resonance*, 1970, **2**, 81.
[161] F. G. Riddell and J. M. Lehn, *J. Chem. Soc. (B)*, 1968, 1224.
[162] M. Ohtsuru and K. Tori, *Tetrahedron Letters*, 1970, 4043.
[163] T. Yonezawa, I. Morishima, K. Fukuta, and Y. Ohmori, *J. Mol. Spectroscopy*, 1969, **31**, 341.
[164] C. F. Shaw and A. L. Allred, *J. Organometallic Chem.*, 1971, **28**, 53.
[165] A. Bugge, B. Gestblom, and O. Hartmann, *Acta Chem. Scand.*, 1970, **24**, 1953.
[166] N. W. G. Debye, D. E. Fenton, S. E. Ulrich, and J. J. Zuckermann, *J. Organometallic Chem.*, 1971, **28**, 339.
[167] M. Gielen, J. Nasielski, and G. Vandendunghen, *Bull. Soc. chim. belges*, 1971, **80**, 175.

uniformly negative, as far as signs are known, and they increase in magnitude on descending Group IVB.[94] As Hildenbrand and Dreeskamp note,[107] the ratio $^2J(XCH):{}^1J(XC)$ changes in a systematic manner with the position of X in the Periodic Table. Such observations are in accord with the supposition that these couplings are dominated by the Fermi contact interaction. The vinyl compounds all show positive $^2K(XCH)$ values, in agreement with the VB theory of the coupling.[175]

Onaka, Sasaki, and Sano[171] have discussed the relationship between the ^{119}Sn Mössbauer isomer shift and the $^2J(^{119}SnCH)$ coupling in compounds of general formula $R_{3-n}X_nSnMn(CO)_5$ (X = Cl or Br, R = Me or Ph) and Visser et al.[170] have noted some effects in the $^3J(^{199}HgCH)$ couplings of mercury alkyls which do not correspond with the predictions of the usual electronegativity arguments. Simonnin and Lequan[169] have observed the $^2J(XCH)$ values in allenic derivatives of Pb, Sn, and Hg, and relative signs of two-bond and four-bond couplings are deduced from satellite spectra in the proton n.m.r. They conclude that $^2K(XCH)$ is probably positive for ^{207}Pb, ^{119}Sn, ^{117}Sn, and ^{199}Hg. $^2J(^{195}PtCH)$ values have been reported for a large number of platinum complexes by Clark and Ruddick.[176]

E. $^2J(^{19}FP^{19}F)$ and $^2J(^{19}FPH)$.
—Values of these couplings[87,88,177,178] have been reported from several laboratories. $^2J(^{19}FPH)$ shows an interesting variation with FPH angle,[88] being zero or very small for $F\hat{P}H = 180°$ and maximal (values of the order of 100 Hz) for $F\hat{P}H = 90°$. The data from reference 178 also fit this scheme of $F\hat{P}H$ versus $^2J(FPH)$. Though fewer data are available it would appear that $^2J(^{19}FP^{19}F)$ also decreases in magnitude as the FPF angle increases and a through-space mechanism could be of importance. The low values of 19, 28, and 25 Hz in CH_3PF_3H,[178] cis-$[CF_3PF_4H]^-$,[88] and $[(CF_3)_2PF_3H]^{-88}$ respectively, require some further explanation but it is worth noting that they all involve a fluorine atom which has a small $^2J(^{19}FPH)$ value. Both the signs and magnitudes of $^2J(^{19}F_a{}^{19}F_e)$ and $^2J(^{19}F_eH)$ have been determined for $HPF_5{}^-$.[87]

F. $^2J(^{19}FM^{19}F)$, where M is a Metal.
—A number of groups have reported

---

[168] M. Gielen, M. R. Barthels, M. De-Clercq, and J. Nasielski, *Bull. Soc. chim. belges*, 1971, **80**, 189.
[169] M.-P. Simonnin and M. Lequan, *Org. Magn. Resonance*, 1970, **2**, 369.
[170] H. D. Visser, L. P. Stodulski, and J. P. Oliver, *J. Organometallic Chem.*, 1970, **24**, 563.
[171] S. Onaka, Y. Sasaki, and H. Sano, *Bull. Chem. Soc. Japan*, 1971, **44**, 726.
[172] T. Kamitani, H. Yamamoto, and T. Tanaka, *J. Inorg. Nuclear Chem.*, 1970, **32**, 2621.
[173] J. M. Homan, J. M. Kawamoto, and G. L. Morgan, *Inorg. Chem.*, 1970, **9**, 2533.
[174] P. L. Goggin, R. J. Goodfellow, and F. J. S. Reed, *J. Chem. Soc. (A)*, 1971, 2031.
[175] E. Sackmann and H. Dreeskamp, *Spectrochim. Acta*, 1965, **21**, 2005.
[176] H. C. Clark and J. D. Ruddick, *Inorg. Chem.*, 1970, **9**, 2556.
[177] S. C. Peake, M. J. C. Hewson, and R. Schmutzler, *J. Chem. Soc. (A)*, 1970, 2364.
[178] V. V. Sheluchenko, G. I. Drodz, M. A. Landau, S. S. Dubov, and S. Z. Ivin, *Zhur. strukt. Khim.*, 1970, **11**, 623.

values of $^2J(^{19}FM^{19}F)$ where M = Sb, Ge, Te, W, Mo, and Sn. The couplings are broadly characterized by the fact that they increase in magnitude as the FMF angle decreases. The oxoperoxofluoromolybdates, for example, have the pentagonal-bipyramidal structure (19) so that the F_eMoF_e angle is approximately 70° and the F_aMoF_e angle 90°. This is reflected in the corresponding couplings which lie between 128 and 136 Hz for $^2J(F_eF_e)$ and between 42 and 68 Hz for $^2J(F_eF_a)$.[179] Similarly, in $WF_{6-n}(OR)_n$ [R = Me, n = 1—5; R = Ph, n = 1, 2], WOF_4L [L = Me_2O or $(MeO)_2P(O)Me$], and $(MeO)_3PMe^+$-WOF_5^-, couplings between F atoms *cis* and *trans* to the oxygen atom have values between 53 and 67 Hz. They are all of the same sign, and of the same sign as all the $^1J(^{19}F^{183}W)$ couplings except that in WOF_5^- and in $(F_4OWFWOF_4)^-$.[111,180] Values of $^2J(^{19}FTi^{19}F)$ in TiF_4 diadducts are also of approximately the same magnitude, lying between 35 and 38 Hz.[181]

O
O—‖—F_e
⫽ Mo
O   F_e
  \\   /
   F_e
   |
   F_a
(19)

In octahedrally co-ordinated tin,[116] and germanium and titanium derivatives[114] too, $^2J(^{19}FM^{19}F)$ values were found to lie in similar ranges, 32—60 Hz for Sn (7 compounds), 35—45 Hz for Ge (12 compounds), and 36—46 Hz for Ti (13 compounds). In fact, the range for tin is only 32—48 Hz if the ion *cis*-$[SnF_3Cl_3]^{2-}$ is omitted, and the similarity of these couplings, covering as they do such a wide range of molecules, suggests that the central atom is of little importance in $^2J(^{19}F^{19}F)$ coupling. Nevertheless, a remarkable deviation from this general rule obtains with silicon, for hexaco-ordinate complexes of which only one value of $^2J(^{19}FSi^{19}F)$ (= 2·4 Hz) could be observed.[114] For three other complexes no fine structure corresponding to this coupling could be detected. The authors point out that such striking differences between silicon and germanium complexes have been observed before. Very much larger values of $^2J(^{19}F^{19}F)$ are found in octahedral tellurium derivatives,[182] the range of values for five compounds being 176—183 Hz.

For octahedrally co-ordinated antimony a wide range of geminal fluorine–fluorine couplings are found. Gillespie and his associates[183] have measured

---

[179] Yu. A. Busalaev, S. P. Petrosyants, and U. P. Tarasov, *Zhur. strukt. Khim.*, 1970, **11**, 616.

[180] A. M. Noble and J. M. Winfield, *J. Chem. Soc. (A)*, 1970, 2574.

[181] C. E. Michelson, D. S. Dyer, and R. O. Ragsdale, *J. Chem. Soc. (A)*, 1970, 2296.

[182] P. Bladon, D. H. Brown, K. D. Crosbie, and D. W. A. Sharp, *Spectrochim. Acta*, 1970, **26A**, 2221.

[183] J. Bacon, P. A. W. Dean, and R. J. Gillespie, *Canad. J. Chem.*, 1970, **48**, 3413.

the spectra of antimony pentafluoride in sulphuryl chlorofluoride and have identified in these solutions the ions $Sb_2F_{11}^-$ and $Sb_3F_{16}^-$, and the complexes $(SbF_5)_2SO_2ClF$, $(SbF_5)_2SOF_2$, and $(SbF_5)_2SO_2$. The n.m.r. evidence indicates that these compounds exist as *cis*-fluorine-bridged structures. The necessary beginning to this work involved the re-interpretation of the n.m.r. spectrum of liquid SbF_5, which had previously been incorrectly assigned. The coupling constants were obtained by first calculating line positions (LAOCN3) followed by the building up of a spectrum upon these lines as a sum of individual Lorentzian curves of appropriate height and width. The process was repeated with varying parameters until a very good correspondence of theoretical and experimental spectra was obtained.

G. $^2J(^{19}FC^{19}F)$.—Two careful analyses of spectra in which this coupling occurs have been reported. The ^{19}F n.m.r. spectrum of octafluorostyrene has been analysed[184] and with the help of tickling experiments the magnitudes obtained by direct spectral analysis have all been assigned signs. $^2J(^{19}FC^{19}F)$ is found to be +62·5 Hz. Long and Goldstein[53] have examined the spectrum of tetrafluoro-1,3-dithietan (1). They find $^2J(F_1F_3)$ to be +137·06 ± 0·03 Hz relative to a negative $^1J(^{13}C^{19}F)$.

H. $^2J(^{13}CO^{31}P)$ and $^2J(^{13}CC^{31}P)$.—Levy and Cargioli[185] have shown that the carbon–phosphorus couplings in triphenyl phosphate and triphenyl phosphite show significant differences. They find $^2J(^{13}CO^{31}P) = 3\cdot2$ Hz in the phosphite and 7·2 Hz in the phosphate. The latter value agrees well with the range of −5·9 to −7·3 Hz found by Gray[68] for some fifteen organophosphonates in which the phosphorus atom is quinquevalent. Jakobsen and Manscher[103] have found that the two-bond $^{13}C-C-^{31}P$ coupling in heteroaromatic phosphine derivatives is larger than the one-bond $^{13}C-^{31}P$ coupling.

I. $^2J(^{31}PO^{31}P)$ and $^2J(^{31}PS^{31}P)$.—Values of these couplings have been reported by Kang, Servis, and Burg.[186] They find the following values: $(CD_3CF_3P)_2S$, 243·4; $[(CF_3)_2P]_2S$, 234·2; $[(CF_3)_2P]_2O$, 108·3 Hz.

J. $^2J(^{31}PM^{31}P)$, where M is a Metal.—The factors influencing this coupling have recently been discussed by Harris, Woplin, and Schmutzler[121] with particular reference to the case where M = Mo in fluorophosphine complexes of molybdenum. They find that the coupling becomes more negative as the electronegativity of the substituents bonded to the phosphorus nuclei increases, and suggest that this effect may be due to changing mutual polarizability of the two ^{31}P valence *s*-orbitals. $^{31}P-M-^{31}P$ couplings have also been measured[130] in phosphite and phosphate complexes of platinum, rhodium, and palladium; all were determined directly from the phosphorus spectra. These

---

[184] E. Lustig and E. A. Hansen, *Chem. Comm.*, 1970, 661.
[185] G. C. Levy and J. D. Cargioli, *Chem. Comm.*, 1970, 1663.
[186] Dai-Ki Kang, K. L. Servis, and A. B. Burg, *Org. Magn. Resonance*, 1971, **3**, 101.

complexes show very large differences in $^2J(^{31}PM^{31}P)$ depending on whether the phosphorus atoms are *cis* or *trans* to each other. These results support the suggestion[187] that $|^2J(^{31}PM^{31}P)|$ for both *cis* and *trans* P–P complexes increases with increasing electronegativity of the groups attached to phosphorus. However, exceptions to this rule have been reported[188] for *trans* Cr complexes. This point is taken up by Harris *et al.*[121] The order of the effect of the metal on the magnitude of the *trans* coupling seems likely to be Pd > Pt,[130] but the differences appear to be much smaller than those in Group VIA. $^2J(^{31}PM^{31}P)$ values have also been given by Allen and Sze[134] and by Bright *et al.*[92] for Pt, Ru, Ni, Rh, Ir, and Pd. The latter suggest that one can account for these coupling constants by considering the changes of ΔE. More determinations of sign are very much needed in this field.

K. Other Two-bond Couplings.—$^2J(^{31}PPtH)$ values have been reported by Church and Mays.[98] Measurement of ^{13}C n.m.r. with proton decoupling has allowed the determination of $^2J(^{13}CC^{19}F)$ for a variety of *ortho*-substituted fluorobenzenes.[106] Norman[128] has given some values of $^2J(^{119}SnPH)$ and $^2J(^{117}SnPH)$. Values of $^2J(^{15}NC^{19}F)$ have been obtained for three fluoropyridines by observation of ^{15}N satellites in their ^{19}F spectra.[133]

6 Coupling between Atoms Separated by Three Chemical Bonds, ^3J

^3J or vicinal coupling constants have attracted much theoretical and experimental attention in the past with the emphasis lying very much on $^3J(HCCH)$, as is shown by two important reviews.[3,136] More recently, a brief discussion and an extensive list of references concerning $^3J(HCCH)$ have been given.[40]

However, with the improving experimental accessibility of other nuclei, an increasing number of papers describing three-bond couplings involving nuclei other than hydrogen have appeared. It seems as if certain of the theoretical concepts and empirical relationships which have been applied to $^3J(HCCH)$ may also be relevant to other three-bond couplings, but there remains considerable uncertainty in this area, not least because the theoretical interpretation of $^3J(HCCH)$ is still at a very unsophisticated level. In this Report it therefore seems appropriate to discuss $^3J(HCCH)$ first and to follow this with an account of vicinal couplings involving other nuclei.

A. Coupling of Hydrogen Atoms through Two Carbon Atoms.—On the theoretical side, two important papers have described the application of the FPT with INDO wavefunctions to coupling in substituted ethanes, ethylenes, and related molecules,[40] and to coupling in substituted benzenes.[41] As with the similar calculations described in Section 5, the experimental trends are well reproduced. However, difficulties are experienced in accounting for individual

---

[187] A. Pidcock, *Chem. Comm.*, 1968, 92.
[188] R. D. Bertrand, F. B. Ogilvie, and J. G. Verkade, *J. Amer. Chem. Soc.*, 1970, **92**, 1908; F. B. Ogilvie, J. M. Jenkins, and J. G. Verkade, *ibid.*, p. 1916.

substituent effects and it is worth noting that a variation of $S_H(0)^2$ (the electron density at the proton) according to the atomic charge and Slater's rules does not improve the results. The well-established ideas concerning the relationship of the coupling to molecular structure and conformation are substantiated by these detailed calculations.[40] For example, ^3J(HCCH) is generally positive and protons in *trans* relationships have larger coupling constants than those in *cis* or *gauche* configurations. The dependence of ^3J(HCCH) on the hybridization or co-ordination of the intervening carbon atoms is also well reproduced; the couplings $J_{average}$(ethane), J_{cis} and J_{trans}(ethylene), and $J_{average}$(propene) are given in the correct order. A reduction of the coupling by electronegative substituents is also found, in agreement with experiment.

The dependence of the coupling on the HCCH dihedral angle[40] closely resembles the results obtained previously, particularly by Karplus.[9,10] The curves are symmetrical about maxima at 180° but the minima are somewhat displaced above and below 270° and 90° respectively, depending on the molecule in question. This displacement is about 25° for acetaldehyde and of the order of 5° for ethane, which emphasises the degree of uncertainty in this type of correlation. The square of the INDO bond orders between the coupled hydrogens was also found to follow an angular dependence similar to that of the coupling constant, with similar displacement of the minima from 90° and 270°, substantiating McConnell's early work.[189] In fact, several of the FPT–INDO calculations recently reported show good correlations between coupling and bond order, though it is usually found that the best correlation with the experimental coupling constants is obtained by a direct calculation of that property, rather than *via* the bond order.

In the work on substituted benzenes[41] good agreement was found for di-substituted molecules between coupling constants calculated directly and those determined from the mono-substituted systems by means of the additivity relationship of Hayamizu and Yamamoto.[190] For several cases the deviations from additivity in the experimental values of the coupling constants were much greater than the deviations in the corresponding calculated values. The authors suggest that this indicates that the calculations do not give proper weight to experimentally important conjugative effects. Evidence for the importance of π-electron effects is also found in the influence of substituents on $^3J_{cis}$ and $^3J_{trans}$ in the substituted ethylenes.[40] These effects are found to be particularly large where the substituent places a conformationally variable π-bonded atom, or one with a lone pair, directly adjacent to one of the unsaturated carbon atoms. Such observations imply an appreciable involvement of π-electrons in determining substituent effects on ^3J(HCCH). Finally, the calculations[40] show a qualitative agreement with the experimental relationships between ^2J(HCH) and ^3J(HCCH) in ethylenes and, although the computed values of ^3J(HCCH) in the substituted molecules are not always given in the correct order, their *sensitivities* are.

[189] H. M. McConnell, *J. Chem. Phys.*, 1956, **24**, 460.
[190] K. Hayamizu and O. Yamamoto, *J. Mol. Spectroscopy*, 1968, **25**, 422.

Several papers reporting correlations of vicinal proton couplings with parameters calculated by some form of molecular orbital theory have appeared. Cox[191] has found a correlation of ^3J(HCCH) values in fluorenyl cations with the bond orders between the intervening carbon atoms calculated by the ω technique (the Hückel theory proved unsatisfactory), and Taddei et al.[192] have found a similar result for trithienylmethyl carbonium ions. Skála and Kuthan[193] and Haigh and Mallion[194] have both found satisfactory correlations of ^3J(HCCH) with Hückel theory bond orders, provided that the hydrogen atoms involved are not sterically hindered.[194] The basis of this relationship is now believed[195] to lie in the connection between bond order and bond length, and here the results of Skála and Kuthan[193] are of interest since their good correlation includes 14 points out of a total of 26 which do not correspond to directly bound carbons. As they found for $^1J(^{13}$CH), Lazzeretti and Taddei[66] note surprisingly good correlations of ^3J(HCCH) in aliphatic compounds with the product of the electron densities on the coupled protons. Parallel lines are found for substituents which belong to different rows of the Periodic Table. The slopes of these relationships are negative, as opposed to the corresponding ones for $^1J(^{13}$CH) which are positive. The authors can find no interpretation of these results at present.

CNDO/2 calculations of ^3J(HCCH) in strained benzocycloalkanes[32] and extended Hückel theory calculations of the effect of substituents on the vicinal coupling in ethanes[34] have been described. The influence of electronegativity on vicinal coupling in ethylene has been investigated by Yalymova and Samitov,[198] and various empirical correlations have been recorded.[199,200]

Cooper and Manatt, and their collaborators,[201–203] have carried out very careful investigations of the factors controlling ^3J(HCCH) in a variety of molecular systems. They have been particularly interested in the effect of the CCH angle and have examined this for cis-alkenes[201] and cyclic dienes.[202] In the case of the cis-alkenes they find that, whereas cis ^3J(HH) decreases rapidly with increasing C=C—H angle, the opposite seems to be the case for trans ^3J(HH). A complete analysis of the proton spectrum of cyclopentadiene has been given[202] together with analyses of the vinyl proton spectra of 1,3-cyclohexadiene and 1,3-cyclo-octadiene, which were obtained under decoupling of the methylene protons. The spectrum of the non-aromatic ring of 1,2-

[191] R. H. Cox, J. Magn. Resonance, 1970, 3, 223.
[192] F. Taddei, P. Spagnolo, and M. Tiecco, Org. Magn. Resonance, 1970, 2, 159.
[193] V. Skála and J. Kuthan, Coll. Czech. Chem. Comm., 1970, 35, 2378.
[194] C. W. Haigh and R. B. Mallion, Mol. Phys., 1970, 18, 737.
[195] M. A. Cooper and S. L. Manatt, J. Amer. Chem. Soc., 1969, 91, 6325.
[196] D. J. Sardella and G. Vogel, J. Phys. Chem., 1970, 74, 4532.
[197] M. Barfield, J. Chem. Phys., 1968, 42, 4458, 4463.
[198] S. V. Yalymova and U. U. Samitov, Zhur. org. Khim., 1970, 6, 1945.
[199] G. K. Hamer, W. F. Reynolds, and D. J. Wood, Canad. J. Chem., 1971, 49, 1755.
[200] A. R. Katritzky and Y. Takeuchi, Org. Magn. Resonance, 1970, 2, 569.
[201] M. A. Cooper and S. L. Manatt, Org. Magn. Resonance, 1970, 2, 511.
[202] M. A. Cooper, D. D. Elleman, C. D. Pearce, and S. L. Manatt, J. Chem. Phys., 1970, 53, 2343.
[203] M. A. Cooper and S. L. Manatt, J. Amer. Chem. Soc., 1970, 92, 4646.

dihydronaphthalene has also been analysed.[202] For these and some other cyclic molecules certain geminal couplings behave as expected, whereas others have surprisingly similar values which do not seem to be in accord with the known molecular geometries. Further evidence on the details of the molecular structures is required before these couplings are completely understood. It is of interest to compare this work with that of Rummens and de Haan[204] who have noted the effects of steric hindrance on ^3J(HH) and ^4J(HH) in disubstituted olefins. Distortions to relieve strain have also been shown[203] to account for the unexpected ^3J(HH) values which occur in overcrowded molecules, cf. Haigh and Mallion.[194]

Bramwell and Randall[205] have used double- and triple-resonance techniques to determine many coupling constants, among them ^3J(HCCH) in some substituted pyridines and some relative sign information has been obtained. Ewing and Parry[144] have given ^3J(HCCH) values for a variety of chloro- and di-chloro-olefins.

Vicinal coupling in conformationally labile systems depends upon the solvent, and Abraham and his associates[206-208] have continued their investigations of this subject, as have also Reynolds and Wood.[209] The latter, investigating eleven 1-phenyl-1,2,2-trihalogenoethanes in various solvents, find that the changes in ^3J(HH) with these solvents do not obey the electrostatic model for solvent effects developed by Abraham.[210] They conclude that this failure is due to solute–solvent hydrogen-bonding specifically favouring the *trans*-rotamer of the molecules studied, and discuss in general terms the conditions under which such theories would be expected to break down.

Pachler and Wessels[211] have discussed the relationship of ^3J(HH) to rotational isomerism, particularly with respect to 2-fluoroethanol and ethylene glycol. Carhart and Roberts[100] have analysed the ^{1}H and ^{13}C spectra of 1,2-dibromoethane completely, and find that the ^3J(HH) couplings have the same sign, positive, as the $^1J(^{13}C^{13}C)$. Abraham and Kemp[208] have given a detailed analysis of the ^{19}F and ^{1}H spectra of 1,2-difluoro- and 1,1,2-tri-fluoro-ethane.

Seeking further information concerning the factors which control ^3J(HCCH) values, McKinnon and Schaefer[155] and Abraham, Parry, and Thomas[212] have analysed the spectra of some five-membered ring systems. Abraham *et al.* considered some five-membered heterocyclic olefins and related compounds. They were able to observe some weak transitions which had been hitherto ignored, thus removing the deceptive simplicity of the spectra and leading to well-determined and non-equal values of $^3J_{cis}$ and $^3J_{trans}$. The

[204] F. H. A. Rummens and J. W. de Haan, *Org. Magn. Resonance*, 1970, **2**, 351.
[205] M. R. Bramwell and E. W. Randall, *Spectrochim. Acta*, 1970, **26A**, 1877.
[206] R. J. Abraham, *J. Mol. Struct.*, 1970, **6**, 49.
[207] L. Cavalli and R. J. Abraham, *Mol. Phys.*, 1970, **19**, 265.
[208] R. J. Abraham and R. H. Kemp, *J. Chem. Soc. (B)*, 1971, 1240.
[209] W. F. Reynolds and D. J. Wood, *Canad. J. Chem.*, 1971, **49**, 1209.
[210] R. J. Abraham, *J. Phys. Chem.*, 1969, **73**, 1192.
[211] K. G. R. Pachler and P. L. Wessels, *J. Mol. Struct.*, 1970, **6**, 471.
[212] R. J. Abraham, K. Parry, and W. A. Thomas, *J. Chem. Soc. (B)*, 1971, 446.

mechanisms contributing to vicinal proton–proton coupling were considered very carefully and the following conclusions were drawn:

(a) Ring buckling and the known dihedral angle dependence satisfactorily account for the values in carbocyclic rings.

(b) The above contributions, plus angle deformations and the orientation of electronegative substituents, account for the values in saturated five-membered heterocyclics.

(c) No reasonable combination of these mechanisms provides a quantitative explanation of the large couplings observed in the five-membered heterocyclic olefins investigated.

The authors suggest that lone pairs on the heteroatom may be partially responsible for these large couplings.

The heterocycles examined by McKinnon and Schaefer[155] were vinyl carbonate (15), 1,3-dithiole-2-thione (16), and 1,3-dithiol-2-one (17), for which precise analyses of the proton spectra gave all proton–proton and

(20)   (21)

carbon–proton coupling constants. For these compounds, benzene, thiophen, furan, and other ethylene derivatives, an interesting linear correlation between $^3J(HCCH)$ and $^2J(^{13}CCH)$ was found. The correlation consists in fact of two straight and almost parallel lines, one for benzene, thiophen, and furan, and the other for the remaining molecules, including (15), (16), and (17). Similar correlations have been noted by Crecely, Crecely, and Goldstein[152] and their application as an aid to spectral analysis has been pointed out.[155]

The ^{1}H n.m.r. spectra of Δ^3- and Δ^4-pyrrolin-2-one, (20) and (21) respectively, have been measured and analysed, and the signs and magnitudes of all proton–proton couplings have been reported.[213]

Brune, Hanebeck, and Hüther[76] have analysed the spectrum of phenyl-cyclobutadienetricarbonyliron to obtain all observable H–H and ^{13}C–H couplings. They find, as they have previously observed in similar systems, that the vicinal proton–proton coupling in the cyclobutadiene ring is less than 0·2 Hz in magnitude. This is confirmed by observation of ^{13}C satellite spectra.

[213] R. Mondelli, V. Bocchi, G. P. Gardini, and L. Chierici, *Org. Magn. Resonance*, 1971, 3, 7.

The authors suggest that this observation can be interpreted in terms of Dixon's theory of spin–spin coupling[214] by which a zero vicinal coupling is predicted for protons attached to carbons by bonds which make an angle of 90° with each other.

Various couplings, including ^3J(HCCH) in three-membered rings, have been given by Pierre and his collaborators.[215]

Proton–proton coupling in aromatic hydrocarbons in general[194] and in monosubstituted naphthalenes,[216] has been the subject of very detailed investigation. The theoretical aspects of the work of Haigh and Mallion[194] have been outlined above; on the experimental side their work is noteworthy since it provides details on fifteen different unsubstituted condensed benzenoid hydrocarbons, the ^{1}H n.m.r. spectra of which have all been measured in a common solvent (CCl_4) and carefully analysed by iterative second-order analysis. In the work on the naphthalenes,[216] 100 MHz spectra, 220 MHz spectra, double resonance, triple resonance, and computation were used in order to extract as much information as possible with the best possible accuracy. Even so, in order to use computational methods effectively it is essential to identify observed lines with calculated transitions, and in some cases this proved to be impossible. Nevertheless, a very comprehensive list of coupling constants and an estimation of their probable errors has been given.

Tarpley and Goldstein[217] have analysed the spectra of the twelve symmetrically substituted dihalogenobiphenyls. They describe a number of correlations of coupling constants, among them ^3J(HCCH), with electronegativity parameters. Investigations of the effect of substituents on the ^3J(HCCH) values in the phenyl ring(s) of aryl phosphines[218] and triphenylsilanes[219] have been published. The ^{1}H n.m.r. spectra of 1,4-benzodithiin and 1,4-benzodioxin have been analysed.[220]

Among much work on the connection between vicinal coupling and the geometries of six-membered saturated rings, the work of Buys[221] is of particular interest. Buys has analysed the methylene group spectra of the molecules (22a—e) using the iterative programme LAME. He then uses the vicinal couplings thus determined, plus those of *trans*-2,3-dichloro-1,4-dioxan and *trans*-2,3-dichloro-1,4-oxathian, in order to test the relationship:

$$R = \frac{J_{trans}}{J_{cis}} = \frac{\frac{1}{2}(J_{aa}+J_{ee})}{\frac{1}{2}(J_{ae}+J_{ea})} = \frac{3-2\cos^2\psi}{4\cos^2\psi} \qquad (22)$$

---

[214] W. T. Dixon, *Theor. Chim. Acta*, 1966, **6**, 359; *J. Chem. Soc. (A)*, 1967, 1879.
[215] (*a*) J. L. Pierre, M. Vidal, and P. Arnaud, *Bull. Soc. chim. France*, 1970, 1544; (*b*) J. L. Pierre, R. Perraud, and P. Arnaud, *ibid*, 1970, 1539, 4459; (*c*) J. L. Pierre, M. Vincens, and M. Vidal, *ibid*, 1971, 1775.
[216] J. W. Emsley, S. R. Salman, and R. A. Storey, *J. Chem. Soc. (B)*, 1970, 1513.
[217] A. R. Tarpley and J. H. Goldstein, *J. Phys. Chem.*, 1971, **75**, 421.
[218] H. Goetz, H. Hadamik, and H. Juds, *Annalen*, 1970, **737**, 132.
[219] R. H. Cox and W. K. Austin, *J. Organometallic Chem.*, 1971, **26**, 331.
[220] K. K. Deb, J. E. Bloor, and T. C. Cole, *Org. Magn. Resonance*, 1970, **2**, 431.
[221] H. R. Buys, *Rec. Trav. chim.*, 1970, **89**, 1244, 1253.

on six-membered rings which do not exist as an equilibrium mixture of two equivalent conformers. ψ is the ring torsion angle for the $-CH_2-CH_2-$ moiety. The agreement between the values of ψ calculated from equation (22) and determined by X-ray diffraction is very good. The method could have considerable use for the rapid determination of the geometry of part of a ring system.

(22a) (22b) (22c) (22d) (22e)

R = p-chlorophenyl
R¹ = phenyl

A large amount of data on $^3J(HCCH)$ in (−)-quinic acid and its derivatives has been given by Haslam and Turner.[222]

B. $^3J(HNCH)$ and $^3J(HOCH)$.—Henold[223] has measured $^3J(HNCH)$ in secondary amines after drying them with sodium–potassium alloy to inhibit intermolecular hydrogen exchange with water. He finds: PhNHMe, 5·21; HNMe₂, 6·11; HNEt₂, ~6·8 Hz. The couplings are presumably positive since $^3J(HNCH)$ was found[79] to be positive relative to $^1J(^{13}CH) > 0$ in MeNH₂, its magnitude being 7·1 ± 0·1 Hz. Pierre, Vincens, and Vidal[215c] have studied $^3J(HOCH)$ in (23).

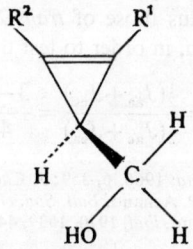

(23) R¹ and R² are alkyl groups or hydrogen.

[222] E. Haslam and M. J. Turner, *J. Chem. Soc. (C)*, 1971, 1496.
[223] K. L. Henold, *Chem. Comm.*, 1970, 1340.

C. $^3J(^{13}CCCH)$.—Couplings of this type have been measured by Goldstein and his colleagues in substituted ethylenes[153] and o-dihalogenobenzenes,[154] and by Takahashi, Sone, and Fujieda[77] in substituted thiophens. In almost all cases, the three-bond ^{13}C–H couplings are larger in magnitude than the two-bond couplings. Both geminal and vicinal ^{13}C–H couplings to the carbonyl carbon in acrolein and acrylic acid were found[153] to be positive and the *trans* couplings are larger than the *cis*, which gives further support to the idea that the behaviour of these couplings may parallel that of $^3J(HCCH)$. For the o-dihalogenobenzenes[154] the $^3J(^{13}CCCH)$ values are related to substituent electronegativity; in all cases, halogen substitution increases the absolute value of $^3J(^{13}CCCH)$ over the corresponding value in benzene.

D. **Coupling to Fluorine through Two Carbon Atoms.** $^3J(HCC^{19}F)$, $^3J(^{13}CCC^{19}F)$, and $^3J(^{19}FCC^{19}F)$.—A number of interesting accounts of couplings of the above type have recently appeared. McFarlane[224] has used heteronuclear $^1H-\{^{19}F\}$ double-resonance experiments to obtain the magnitudes and some of the signs of the aromatic H–^{19}F and ^{19}F–^{19}F coupling constants in 2,3,5,6-tetrafluoroaniline. The spectrum is complex, but the precision obtained is shown to be equivalent to that of single resonance with direct observation of the ^{19}F spectrum. Ambiguities in the signs of certain couplings are resolved by use of triple-resonance tickling techniques. The ^{1}H n.m.r. of some fluorinated pyridine derivatives has been investigated by Thomas and Griffin;[225] they find that the magnitudes and signs of $J(H^{19}F)$ follow those of $J(HH)$ but with an exaggerated range. Large changes in the n.m.r. parameters are found when the nitrogen atom is protonated, $^3J(HCC^{19}F)$ changing sign in some α-fluoropyridines. In fluorinated benzenes, $^3J(^{13}CCC^{19}F)$ seems to be dependent on substituent electronegativity; it is increased in magnitude by substituents more electronegative than hydrogen.[106] A complete analysis giving signs and magnitudes of all ^{19}F–^{19}F couplings in octafluorostyrene has been published[182] and $^3J(^{19}FCC^{19}F)$ has been observed in a number of fluoroaromatic molecules,[105] where it was found to show no correlation with solvent dielectric constant.

Cavalli[226] has measured the spectra of many derivatives of hexafluorobicyclo[2,2,0]hexa-1,5-diene and its dibromo-adduct, (24) and (25) respectively. The results show low $^3J(F_2F_3)$ values (1—4 Hz) which are fairly constant and not greatly affected by the presence of substituents. Unfortunately, signs could not be determined and there is some ambiguity concerning the assignment of F(5) and F(6) (see footnote on page 1 of Cavalli's paper).

Hall and his co-workers[227] have measured vicinal fluorine couplings in six-membered rings. They have examined the molecules (26)—(28) and have

[224] W. McFarlane, *Mol. Phys.*, 1970, **18**, 817.
[225] W. A. Thomas and G. E. Griffin, *Org. Magn. Resonance*, 1970, **2**, 503.
[226] L. Cavalli, *J. Chem. Soc. (B)*, 1970, 1616.
[227] L. D. Hall, R. N. Johnson, J. Adamson, and A. B. Foster, *Canad. J. Chem.*, 1971, **49**, 118; *Chem. Comm.*, 1970, 463.

obtained the results shown in Table 6. All relative signs were determined by means of heteronuclear double resonance and absolute signs are relative to $^1J(H^{19}F) > 0$. In view of these results, the authors suggest that vicinal $^{19}F-^{19}F$ couplings may follow a $\cos^2 3\phi/2$ or $\cos 3\phi/2$ rule rather than the $\cos^2\phi$ relationship which has been suggested for H–H and H–^{19}F three-bond couplings. It is of interest that in (26) and (27) all the $^3J(^{19}F^{19}F)$ values are negative in sign, in contrast to the positive values of $^3J(H^{19}F)$ and $^3J(HH)$. Also noteworthy are the facts that $J(FF)_{trans}$ is approximately equal to $J(FF)_{gauche}$ and that the former seems to be more susceptible to substituent electronegativity than the latter.

Nuclear Spin–Spin Coupling

The use of $^3J(\mathrm{H^{19}F})$ values in the conformational analysis of cyclohexanones has been suggested by Cantacuzène and Jantzen.[228] Doddrell, Charrier, and Roberts[229] have examined the proton-decoupled ^{13}C n.m.r. spectrum of 1,1-difluorocyclohexane. They find that the triplet observed for C(3) at room temperature $[^3J(^{13}\mathrm{C^{19}F}) = 4\cdot7$ Hz$]$ collapses to a doublet $[^3J(^{13}\mathrm{C^{19}F}) = 9\cdot5$ Hz$]$ at -90 °C. They propose that only the equatorial fluorine is coupled to C(3) because of back-lobe interaction.

The ^{1}H and ^{19}F spectra of 1,1,2,2-tetrafluoroethane, in the neat liquid and in ten different solvents, have been completely analysed by Cavalli and Abraham.[207] The values of all coupling constants were obtained by iterative analysis and absolute signs were assigned relative to $^3J(\mathrm{HH}) > 0$ and $^2J(\mathrm{H^{19}F}) > 0$. The only remaining uncertainty was the assignment of the two different $^3J(^{19}\mathrm{F^{19}F})$ values. Consideration of the effects of solvent upon these couplings permits two alternative assignments for the *gauche*- and *trans*-type couplings in the *gauche*- and *trans*-rotamers, but although the evidence favours one of these, no unambiguous assignment could be made.

Table 6 *The coupling constants* $^3J(^{19}\mathrm{FCC^{19}F})$ *measured for molecules* (26)—(28)

Compound	(26a)	(26b)	(27a)	(27b)	(28a)		(28b)	
					1,3	2,3	1,3	2,3
$^3J(\mathrm{FF})$/Hz	$-20\cdot0$	$-13\cdot5$	$-18\cdot8$	$-15\cdot8$	$-2\cdot2$	$+11\cdot4$	$-1\cdot2$	$+9\cdot4$
Nominal dihedral angle (ϕ)	180	60	60	60	120	0	120	0

The signs and magnitudes of ^{19}F–^{19}F couplings in a number of dichlorotetrafluorocyclopropanes have been examined by Cavalli.[230] No useful correlations of $^3J(^{19}\mathrm{FCC^{19}F})$ with solvent dielectric constant could be detected, but variation of the coupling with temperature was useful for assigning the vicinal couplings and for defining their signs. The three-bond ^{19}F–^{19}F couplings increase algebraically with increasing electronegativity of substituents in the cyclopropane ring.

E. Three-bond Couplings involving ^{31}P and ^{1}H.—^3J (^{31}PCCH). Couplings of phosphorus to hydrogen through two carbon atoms have been measured by several groups.[221,231–233] Jakobsen,[233] in an investigation of ^{1}H–^{31}P coupling in tri-3-furylphosphine derivatives, finds that all the $^3J(\mathrm{HCC^{31}P})$ values are positive. The coupling constants are affected by the electronegativity of the

---

[228] J. Cantacuzène and R. Jantzen, *Tetrahedron*, 1970, **26**, 2429.
[229] D. Doddrell, C. Charrier, and J. D. Roberts, *Proc. Nat. Acad. Sci. U.S.A.*, 1970, **67**, 1649.
[230] L. Cavalli, *Org. Magn. Resonance*, 1970, **2**, 233.
[231] V. V. Negrebetskii, N. P. Ignatova, A. V. Kessenikh, N. N. Mel'nikov, and N. I. Shvetsov-Shilovskii, *Zhur. strukt. Khim.*, 1970, **11**, 633.
[232] K. C. Hansen, C. H. Wright, A. M. Aguiar, C. J. Morrow, P. M. Turkel, and N. S. Bhacca, *J. Org. Chem.*, 1970, **35**, 2820.
[233] H. J. Jakobsen, *J. Mol. Spectroscopy*, 1971, **38**, 243.

substituent on the phosphorus atom, the $^3J(\text{HCC}^{31}\text{P})$ couplings increasing as the substituent electronegativity decreases.

$^3J(^{31}\text{POCH})$, $^3J(^{31}\text{PNCH})$, $^3J(^{31}\text{PSCH})$, and $^3J(^{31}\text{PNPH})$. Considerable interest has been shown in these couplings in the past year, particularly in five-membered rings containing various combinations of P, N, O, and S atoms. Albrand and his collaborators[234-236] have shown that the signs of $^3J(^{31}\text{POCH})$ and $^3J(^{31}\text{PNCH})$ are the same and positive[235] relative to $^1J(^{31}\text{P}^{19}\text{F}) < 0$. $^3J(^{31}\text{PSCH})$ is smaller in magnitude and can have either sign. Verkade and his associates[237-239] have reported the spectra of many molecules, which include $^3J(^{31}\text{POCH})$ values. They note[238] values of this coupling ranging from 0·29—24 Hz in twenty compounds and correlate these with the POCH dihedral angle. Separate correlations are found for P^V and P^III derivatives and the correlations are broad bands rather than lines. Reasons for this are suggested.

Dependence of $^3J(^{31}\text{POCH})$ on dihedral angle for P^III compounds has also been suggested by Kainosho and Nakamura.[240] More recently, Kainosho[241] has shown that P^V compounds do not fall on the same curve as the P^III, but he suggests that the same type of relationship may hold. He further proposes that the coupling is also dependent upon the s-electron density of atoms in the coupling path and includes a contribution by π-electrons. A very similar suggestion regarding s-electron density in the coupling path has been made with regard to $^3J(^{31}\text{PNPH})$.[89]

The temperature dependence of $^3J(^{31}\text{POCH})$ values in trineopentyl phosphate and of $^3J(\text{HH})$ and $^3J(^{31}\text{POCH})$ in tris(β-chloroethyl) phosphate has been examined by Bothner-By and Trautwein.[242] Both $^3J(^{31}\text{POCH})$ values show monotonic increases in magnitude with increasing temperature and, though there are difficulties of interpretation, the authors conclude that the dependence of $^3J(^{31}\text{POCH})$ on dihedral angle and on substitution is similar to that well established for $^3J(\text{HH})$. Hägele[243] has made measurements of $^3J(^{31}\text{POCH})$ in a spiro-compound and finds these to be in agreement with a dependence of the coupling on the dihedral angle. Solvent effects on $^3J(^{31}\text{POCH})$ have been studied.[244]

Walker and his co-workers[245,246] have measured a variety of $^3J(^{31}\text{PNCH})$

[234] J. Deviller, J. Navech, and J. P. Albrand, *Org. Magn. Resonance*, 1971, **3**, 177.
[235] J. P. Albrand, A. Cogne, D. Gagnaire, J. Martin, J. B. Robert, and J. Vernier, *Org. Magn. Resonance*, 1971, **3**, 75.
[236] J. P. Majoral, R. Pujol, and J. Navech, *Compt. rend.*, 1971, **272**, C, 1913.
[237] D. W. White, G. K. McEwen, R. D. Bertrand, and J. G. Verkade, *J. Magn. Resonance*, 1971, **4**, 123.
[238] D. W. White and J. G. Verkade, *J. Magn. Resonance*, 1970, **3**, 111.
[239] R. D. Bertrand, J. G. Verkade, D. W. White, D. Gagnaire, J. B. Robert, and J. Vernier, *J. Magn. Resonance*, 1970, **3**, 494.
[240] M. Kainosho and A. Nakamura, *Tetrahedron*, 1969, **25**, 4071.
[241] M. Kainosho, *J. Phys. Chem.*, 1970, **74**, 2853.
[242] A. A. Bothner-By and W.-P. Trautwein, *J. Amer. Chem. Soc.*, 1971, **93**, 2189.
[243] G. Hägele, *Z. Naturforsch.*, 1971, **26b**, 1.
[244] L. I. Vinogradov, U. U. Samitov, E. G. Yarkova, and A. A. Muratova, *Optics and Spectroscopy*, 1970, **29**, 493.
[245] J. Nelson, R. Spratt, and B. J. Walker, *Chem. Comm.*, 1970, 1509.
[246] D. C. H. Bigg, R. Spratt, and B. J. Walker, *Tetrahedron Letters*, 1970, 107.

values, and conclude that this coupling also shows a dihedral angle dependence. It is also very sensitive not only to the orientation of the phosphorus lone-pair electrons but also to the nature of the nitrogen substituents in aminophosphines.[245] In substituted aromatic aminophosphines they find[245] an approximate correlation of the value of $^3J(^{31}PNCH)$ with Hammett substituent constants and suggest that the π-bonding ability of the nitrogen substituents is important. $^3J(^{31}PNCH)$ values have also been reported by Schmutzler and his associates.[119,177]

$^3J(^{31}PMCH)$, *where M is a Metal*. Such couplings have been measured for Pt,[133,247] for Au,[248] and for Sn.[128]

F. Miscellaneous Three-bond Couplings.—$^3J(^{15}NCCH)$ values in a number of ^{15}N-enriched amino-acids have been reported by Lichter and Roberts.[249] They use the $^3J(HCCH)$ values which they also measured to estimate the populations of the various rotational isomers and are thus able to draw conclusions concerning the dependence of $^3J(^{15}NCCH)$ on geometrical factors. They conclude that, provided all the couplings have the same sign, then a somewhat skewed and fairly shallow type of Karplus correlation between ^3J and dihedral angle obtains. The minimum of the curve is uncertain but seems to lie between 80 and 120°.

Shaw and Allred[164] report values of $^3J(^{119}SnMCH)$ in Me_3MSnMe_3 where M = C, Si, Ge, and Sn. $^3J(^{119}SnSnCH)$ has been measured.[167] Measurements of $^3J(^{77}SeCCH)$ have been reported by Dahl and Nielsen,[250] by Lardon,[251] and by Bugge, Gestblom, and Hartmann.[165] Other measurements include $^3J(^{199}HgCCH)$,[252] $^3J(^{195}PtSCH)$,[174] $^3J(^{31}PMPH)$[92] where M = Ru, Rh, Ir, Ni, Pd, or Pt, $^3J(HCN^{19}F)$,[177] $^3J(^{13}CCO^{31}P)$,[68,185] $^3J(HCP^{19}F)$,[101] and $^3J(^{31}PMoP^{19}F)$.[121]

7 Coupling between Nuclei Separated by Four or More Chemical Bonds

Many complications attend the measurement and interpretation of coupling constants between nuclei separated by four or more chemical bonds. The couplings are frequently small and difficult to measure and the spectral interpretation is made complex by the fact that molecules in which four-bond couplings may be observed usually contain many absorbing nuclei. In addition, particularly in cyclic systems, the coupling path is not unique and significant contributions to the coupling may arise from paths which formally traverse the same or even more chemical bonds. In fact, the concept of *n*-bond

---

[247] J. H. Nelson, J. J. R. Read, and H. B. Jonassen, *J. Organometallic Chem.*, 1971, **29**, 163.
[248] A. Shiotoni, H. F. Klein, and H. Schmidbaur, *J. Amer. Chem. Soc.*, 1971, **93**, 1555.
[249] R. L. Lichter and J. D. Roberts, *J. Org. Chem.*, 1970, **35**, 2806.
[250] B. M. Dahl and P. H. Nielson, *Acta Chem. Scand.*, 1970, **24**, 1468.
[251] M. Lardon, *J. Amer. Chem. Soc.*, 1970, **92**, 5063.
[252] L. Lunazzi, M. Tiecco, C. A. Boicelli, and F. Taddei, *J. Mol. Spectroscopy*, 1970, **35**, 190.

coupling becomes less and less useful as n exceeds three, or even two. As in earlier sections, it seems best to discuss couplings involving only protons first and then to proceed to other nuclei.

A. **Long-range Coupling between Protons.**—This subject has been discussed in several of the well-known reviews on nuclear spin–spin coupling[3,4,136] and recently has received exclusive and detailed analysis by Barfield and Chakrabarti.[253] We attempt to separate the field largely but not exclusively into saturated and unsaturated systems.

Long-range H–H Coupling in Saturated Systems. Within this field, coupling between protons in six-membered rings continues to be the object of discussion and disagreement.[254,255] What seem to be needed in this difficult area are more detailed analyses of representative molecules, such as have been given by Albriktsen[256] for trimethylenesulphite and by Buys,[221] who has analysed the spectra of some 1,3-dioxans, 1,3-oxathians, and 1,3-dithians. Allingham, Crabb, and Newton[257] have also discussed the latter systems and note in comparing (29) and (30) that there is a marked long-range coupling (2·1 Hz) between H_{2e} and H_{4e} in (29) but not between H_{2e} and H_{6e} in (30).

(29) (30)

The observation of coupling through sulphur but not through oxygen has been made before.[258] Also of interest is the contrast between $^4J(H_{4e}H_{6e})$ in (22a) and (22c), which Buys[221] finds to be 2·6 Hz and 0 Hz respectively. The problems associated with the interpretation of ^4J coupling in these systems are summed up by Allingham, Crabb, and Newton[257] as follows. 'If, as seems certain, ^4J depends upon the electronegativity of substituents, then the rather similar values of ^4J (2·0—2·5 Hz) in O–CH–C–CH–O and O–CH–C–CH–S systems are puzzling. So are the only slightly smaller values of $^4J(H_{4e}H_{6e})$ (1·6 Hz) in the 1,3-dithians as compared to the $^4J(H_{4e}H_{6e})$ values (1·6—2·0 Hz) in tetrahydro-1,3-oxazines. There would seem to be no simple relationship between $^4J(HH)$ and inductive and lone-pair effects of an adjacent heteroatom such as that existing for geminal coupling constants.'

Multiple-resonance experiments at 100 MHz have been used to demonstrate coupling through epoxides by protons separated by four and by six σ-bonds,[259]

[253] M. Barfield and B. Chakrabarti, *Chem. Rev.*, 1969, **69**, 757.
[254] C. W. Jefford and B. Waegell, *Bull. Soc. chim. belges*, 1970, **79**, 427.
[255] M. Anteunis and N. Schamp, *Bull. Soc. chim. belges*, 1970, **79**, 437.
[256] P. Albriktsen, *Acta Chem. Scand.*, 1971, **25**, 478.
[257] Y. Allingham, T. A. Crabb, and R. F. Newton, *Org. Magn. Resonance*, 1971, **3**, 37.
[258] Y. Allingham, R. C. Cookson, T. A. Crabb, and S. Vary, *Tetrahedron*, 1968, **24**, 4625.
[259] P. Joseph-Nathan and E. Diaz, *Org. Magn. Resonance*, 1971, **3**, 193.

and a ^6J(HH) coupling of 1 Hz has been observed in the Diels–Alder adduct of bicyclo[2,1,0]pent-2-ene and cyclopentadiene.[260] Derivatives also show the same long-range coupling between protons separated by six σ-bonds and the results have been confirmed by tickling experiments and detailed analysis. The coupled protons appear to be joined by a zigzag path, indicating again the very special nature of this geometrical arrangement in nuclear spin–spin coupling.

Crosbie and Sheldrick[95] have measured the magnitude and relative signs of all coupling constants involving ^{1}H, ^{29}Si, and ^{31}P in trisilylphosphine and trisilylstibine. They find ^4J(HSiPSiH) = -0.70 Hz and ^4J(HSiSbSiH) = -0.57 Hz relative to negative ^1J(^{29}SiH).

Long-range H–H Coupling in Unsaturated Systems. Papers in which the emphasis lies on the theoretical interpretation of coupling are first considered, and this is followed by an account of work concerning four-bond couplings. Finally, publications describing a variety of coupling paths and the general interpretation of these couplings are discussed.

The theoretical aspects of long-range H–H couplings, particularly in non-aromatic conjugated systems where π-electron contributions to the coupling can be large, continue to arouse interest. A very useful concept in the theory of coupling in π-electron systems is the idea of dividing the coupling into two parts, one transmitted by the σ- and the other by the π-electrons. For a meaningful comparison with experiment therefore, it is necessary to attempt to identify the separate σ- and π-electron contributions to the experimental values. Albriktsen, Cunliffe, and Harris[261] have discussed this problem in connection with the coupling in penta- and hexa-dienes. Bacon and Maciel[47] have calculated σ- and π-electron contributions to proton–proton coupling in cis,cis-1,4-dialkyl-1,3-butadienes using FPT and INDO MO. In this type of semi-empirical MO theory, exchange integrals between atomic orbitals centred on the same atom are retained and not neglected as they are, for example, in the CNDO/2 theory. π-Electron contributions to spin–spin coupling are therefore included in this theory, in contrast to many others, and by explicitly setting certain exchange integrals to zero Bacon and Maciel attempted to determine the calculated π- and σ-electron contributions to J(HH) separately. They also considered the effects of changes in molecular geometry on the coupling. In general, good agreement with experiment and with other calculations by different methods[46] is obtained and all the major trends in butadiene n.m.r. spectra are accounted for. Nevertheless, there are some strange features concerning the π-electron contributions to ^3J(HH); these have been described in Section 3. With regard to the geometrical factors, the authors conclude[47] that in-plane deformation of the diene skeleton is at least partly responsible for the increased magnitude of proton–proton coupling in these substituted molecules, as Albriktsen, Cunliffe, and Harris[261]

[260] J. E. Baldwin and R. K. Pinschmidt, *J. Amer. Chem. Soc.*, 1970, **92**, 5247.
[261] P. Albriktsen, A. V. Cunliffe, and R. K. Harris, *J. Magn. Resonance*, 1970, **2**, 150.

had earlier suggested. They also find, however, that an alkyl substituent and an out-of-plane twist may have effects of comparable importance.

Cunliffe, Grinter, and Harris[46] have calculated the π-electron contributions to J(HH) in a variety of unsaturated non-aromatic hydrocarbons using three different MO methods (see Section 3). They find that for calculating π-electron couplings there is little to choose between the modified Karplus method and the FPT approach, though the latter is the more straightforward in its application. The modified McConnell treatment shows qualitative agreement with experiment except where π-electron configuration interaction is included, in which case the signs of some small coupling constants are incorrectly predicted. Some comparisons of π-electron coupling through three, four, and five chemical bonds are made. The FPT using INDO MO has been used to discuss proton–proton couplings in substituted benzenes[41] with the general conclusions outlined in Section 3. A spin polarization method for the calculation of π-electron contributions to spin–spin coupling has been described.[262]

Values of ^4J(HH) have been measured in many laboratories. A careful analysis and detailed discussion of the n.m.r. spectra of cyclopentadiene, 1,3-cyclohexadiene, and 1,3-cyclo-octadiene has been given.[202] The long-range couplings between vinyl protons in the first two compounds, which are planar and nearly planar respectively, are significantly different from those in the non-planar cyclo-octadiene, and the authors are able to rationalize these differences in terms of the differing molecular structures. Some five-bond couplings are also discussed.

Some typical values, *i.e.* approximately 6·5 Hz, have been found for the ^4J(HH) coupling of the allenic protons in substituted ethylallenes.[263] Coupling constants for a large number of chloro- and dichloro-olefins have been given by Ewing and Parry,[148] and ^4J(HH) in four *cis–trans* pairs of 1,2-disubstituted olefins has been measured and discussed by Rummens and de Haan.[204] They find that in three of the four isomer pairs the *trans*-allylic coupling is more negative than the *cis*-allylic coupling. For crotonic acid only the difference is reversed and is therefore in accord with Barfield's prediction.[264] These couplings and also some ^3J(HH) values are discussed in terms of sterically induced rehybridization at the sp^2 carbon atoms. (The data in Table 1 of reference 204 appear to be at variance with the discussion in the text; the assumption here is that the text is correct and that the words *cis* and *trans* in the table have been reversed.)

The spectra of seven 2-substituted propenes have been analysed by de Kowalewski,[147] and double-irradiation experiments were used to obtain the relative signs of the coupling constants. A correlation was found between the electronegativity of the substituent and the two ^4J(CH$_3$–H) couplings. The author

[262] P. V. Schastnev, N. D. Chuvylkin, and G. M. Zhidomirov, *Teor. i eksp. Khim.*, 1971, **7**, 86.
[263] R. S. Macomber, *J. Org. Chem.*, 1971, **36**, 999.
[264] M. Barfield, *J. Chem. Phys.*, 1964, **41**, 3825.

suggests that this indicates that the substituent is operating *via* a σ-electron mechanism if Barfield and Chakrabarti's[253] analysis is correct.

The spectra of Δ^3- and Δ^4-pyrrolin-2-one, (20) and (21), have been analysed[213] and the signs of all coupling constants have been determined by tickling and triple-resonance experiments. The four-bond couplings are discussed in some detail. $^4J(H_XH_Z)$ in (21) is $+1.0$ to $+1.5$ Hz, in agreement with the expected value for a σ-mechanism[253] and a 'W' arrangement. The similar value of $^4J(H_YH_Z) = +1.5$ Hz found in (20) for the same path but without the double bond supports the conclusion that the π-electron contribution to $^4J(H_XH_Z)$ in (21) is zero, in agreement with theory.[253] The other allylic transoid couplings are negative, as expected.

Shinokawa, Fukui, and Sohma[265] have determined all the magnitudes and relative signs of H–H coupling constants in pyrrole-2-carboxylic acid and pyrrole-2-aldehyde. Nitrogen decoupling revealed the spectrum of the *N*-H proton. The relative signs of all the couplings in both molecules were found to be the same, except for $^4J(CHO-NH)$ in the aldehyde. They were assumed positive on the basis of the five-bond long-range coupling of the aldehyde proton being π-dominated and therefore positive.

(31) a; R=CN
b; R=H

General accounts of coupling in π-electron systems have been numerous. Lequan and Simonnin[266] have studied solvent effects on proton–proton coupling in a series of vinyl ethers. Their results demonstrate the stereospecificity of $^4J(HCOCH)$ and of $^5J(HCCOCH)$ and the importance of the 'W' arrangement for the former. Sardella and Vogel[196] have determined the signs and magnitudes of a number of four-, five-, and six-bond couplings in two anhydrides, (31a) and (31b). The signs were found by a combination of spin-decoupling and tickling experiments and depend upon the assumptions: $^4J_{13}$ (31a) $= -1.39$ Hz, $^3J_{24}$(31b) $= +7.60$ Hz. In Table 7 the results are compared with the calculations of Barfield using his truncated matrix sum method.[253,197] As the Table shows, all signs are correctly predicted and agreement is also very good with regard to magnitudes. The allylic coupling ($^4J_{13}$)

[265] S. Shinokawa, H. Fukui, and J. Sohma, *Mol. Phys.*, 1970, **19**, 695.
[266] R.-M. Lequan and M.-P. Simonnin, *Bull. Soc. chim. France*, 1970, 4419.

discrepancy is clearly due to a positive σ-contribution which the authors estimate to be about +0·3 Hz, thus bringing calculated and experimental results into excellent accord. No simple explanation can be advanced for the $^5J_{14}$ discrepancy in (31b). The effects of substituents on long-range coupling constants are briefly discussed.

Long-range coupling in substituted methyl pyridines has been investigated by Bramwell and Randall[205] and by Rowbotham, Wasylishen, and Schaefer.[267] The former use multiple-resonance experiments to obtain relative sign information and find results consistent with data for other molecules. Rowbotham et al.[267] interpret their results for the four methyl derivatives of 2-fluoropyridine in terms of σ- and π-contributions. They find that the signs and magnitudes of the long-range coupling constants between the methyl protons and the ring protons, and between the methyl protons and the fluorine, are consistent with a model in which the nitrogen atom polarizes the σ-electron system but leaves the π-electron contribution to the coupling relatively unchanged.

Table 7 Comparison of the measured and calculated proton–proton couplings (in Hz) for (31a) and (31b)

	(31a)			(31b)	
Coupling	Exp.[a]	Calc.[b]	Coupling	Exp.[a]	Calc.[b]
$^6J_{12}$	−0·32	−0·4	$^6J_{12}$	−0·20	−0·4
$^6J_{23}$	−0·77	−0·7	$^6J_{23}$	−0·66	−0·7
			$^5J_{14}$	<0·1	+0·4
			$^5J_{34}$	+0·92	+1·0
$^4J_{13}$	−1·39	−1·7	$^4J_{13}$	−1·30	−1·7
			$^3J_{24}$	+7·60	—

[a] Ref. 196; [b] Refs 197 and 253.

Pivcová and Kahovec[268] have examined long-range coupling in o-alkyl phenols. They find that the coupling between protons on the α-carbon of the alkyl group and the protons in the ring does not depend upon the nature of the alkyl group. Müllen[269] too has investigated phenols and also SH and Te substituted benzenes. He discusses the coupling constants in terms of additive substituent contributions.

Janzen and Schaefer[270] have examined the effect of α-substituents, X, on benzylic spin–spin coupling constants in 2,6-dichlorotoluene. They conclude that the six-bond coupling is dominated by a π-electron mechanism and depends upon the conformation of the $-CH_2X$ group with respect to the benzene ring, and on the electronegativity of X. The five-bond coupling is remarkably independent of X, and the authors suggest that the effect of the

[267] J. B. Rowbotham, R. Wasylishen, and T. Schaefer, Canad. J. Chem., 1971, 49, 1799.
[268] H. Pivcová and J. Kahovec, Coll. Czech. Chem. Comm., 1971, 36, 1388.
[269] K. Müllen, Org. Magn. Resonance, 1971, 3, 331.
[270] A. F. Janzen and T. Schaefer, Canad. J. Chem., 1971, 49, 1818.

substituent on this coupling can be explained in terms of a decreasing π-contribution being balanced by an increasing σ-contribution. Wasylishen and Schaefer[271] have suggested that in benzene ^5J(HH) is purely a π-electron coupling, the σ-contribution vanishing. They propose further, basing their argument upon a model in which the σ-core of the molecule is polarized by the presence of substituents, that the σ-contribution to ^5J(HH) is negative in benzene derivatives and positive in substituted pyridines. Roomi and Dugas[272] have discussed the effect of substituent electronegativity on the coupling between ring protons in ethoxycarbonylpyrroles.

Coupling between ring protons in aromatic molecules has received considerable attention (see also Section 6). A complete analysis of the spectrum of naphthalene has been published,[273] and Haigh and Mallion[194] have described a very careful and complete investigation of fifteen benzenoid hydrocarbons. The latter authors emphasize the effect of steric overcrowding on proton coupling constants, and this point is also discussed in detail by Cooper and Manatt[203] who present analyses of the spectra of 1,4-dimethylnaphthalene, 1,4-di-t-butylnaphthalene, and benzo[c]phenanthrene. The proton n.m.r. spectra of substituted naphthalenes have also been investigated by Crecely et al.[274] and by Emsley and his associates[216] (see Section 6).

Tarpley and Goldstein[217] have analysed the spectra of the twelve symmetrically substituted dihalogenobiphenyls. The effects of substituents on the couplings are shown to correlate quite well with substituent electronegativity. Sciacovelli and von Philipsborn[275] have measured the 100 MHz spectra of acridine and its five monomethyl derivatives and have analysed them with the help of multiple-resonance and computational techniques. They discuss their results in terms of the geometries of the coupling paths, and some π-contributions, large for condensed π-systems, are indicated. Hilbers and Maclean[276] have investigated the proton resonance spectra of nitrobenzene and 2,4-dimethylnitrobenzene while applying an external electric field. They are able to determine all the magnitudes and absolute signs of the proton–proton couplings in nitrobenzene; the latter turn out to be uniformly positive. Correlations of J(HH) with π-electron MO theory bond orders have been reported.[193]

Long-range couplings in thiophen-2-aldehyde and furan-2-aldehyde have been measured[277] and values of ^6J(HH) have been given for substituted 2-vinylfurans and 2-vinylthiophens.[278] Brinkmann et al.[279] have described

---

[271] R. Wasylishen and T. Schaefer, *Canad. J. Chem.*, 1971, **49**, 94.
[272] R. W. Roomi and H. Dugas, *Canad. J. Chem.*, 1970, **48**, 2303.
[273] R. W. Crecely and J. H. Goldstein, *Org. Magn. Resonance*, 1970, **2**, 613.
[274] R. W. Crecely, S. L. Baughcum, K. R. Long, and J. H. Goldstein, *J. Magn. Resonance*, 1970, **3**, 103.
[275] O. Sciacovelli and W. von Philipsborn, *Org. Magn. Resonance*, 1971, **3**, 339.
[276] C. W. Hilbers and C. Maclean, *Chem. Phys. Letters*, 1970, **7**, 587.
[277] B. Roques, S. Combrisson, C. Riche, and C. Pascard-Billy, *Tetrahedron*, 1970, **26**, 3555.
[278] T. N. Huckerby, *Tetrahedron Letters*, 1971, 353.
[279] A. W. Brinkmann, M. Gordon, R. G. Harvey, P. W. Rabideau, J. B. Stothers, and A. L. Ternay, *J. Amer. Chem. Soc.*, 1970, **92**, 5912.

homoallylic couplings between protons at the 9- and 10-positions in 9-alkyl-9,10-dihydroanthracenes, and some long-range couplings in selenothiophens have been determined.[165]

Visser and Oliver[280] have observed long-range couplings over six bonds and a mercury atom in divinylmercury: $^6J_{tt} = 0.6$, $^6J_{cc} = 0.4$, $^6J_{ct} = 0.2$ Hz. The ^5J and ^4J couplings were less than 0.2 Hz in magnitude. The authors suggest that these couplings are π-electron dominated and that the coupling is transmitted through mercury orbitals of suitable energy and symmetry.

B. **Long-range Couplings involving at least One Nucleus other than H.**—Couplings involving fluorine will be included here if they do not appear to be of particular interest with regard to the question of 'through-space' coupling. Couplings relevant to that problem will be discussed in subsection 7C.

Saturated Systems — Coupling between ^1H and ^{19}F. Some very large values of $^4J(H^{19}F)$ and $^5J(H^{19}F)$ have been reported[281] in the methyl and ethyl esters of fluorosulphinic acid, CH_3OSF and C_2H_5OSF. The former has $^4J(H^{19}F) = 46.5$ Hz while the latter has $^4J(H^{19}F)$ and $^5J(H^{19}F) = 46.5$ and 25.8 Hz respectively. Shapiro and Sardella[282] have investigated the rotational isomerism of some fluorinated esters. They find that $^4J(H^{19}F)$ in 1,1-difluoroacetone and $^5J(H^{19}F)$ in methyl difluoroacetate have the same sign, and this is probably positive since $^4J(H^{19}F)$ in 1,3-difluoroacetone and 1,1,3,3-tetrafluoroacetone is positive. Albrand and his collaborators[235] have measured the magnitudes and signs of $^4J(H^{19}F)$ in phosphorus-containing five-membered rings. The paths involved are FPOCH, FPNCH, and FPSCH.

Saturated Systems — Coupling between other Nuclei. Long and Goldstein[53] have determined values of $^4J[^{19}F(1)^{19}F(2)] = +5.19$ Hz and $^4J[^{19}F(1)^{19}F(4)] = +31.91$ Hz for tetrafluoro-1,3-dithietan, (1). The signs are relative to a negative $^1J(^{13}C^{19}F)$ and a positive $^2J[^{19}F(1)^{19}F(3)]$. $^4J(^{31}PH)$ and $^4J(^{31}P^{31}P)$ values have been reported for complexes of platinum and palladium[174] and $^4J(^{31}P^{19}F)$ values, through phosphorus and oxygen or phosphorus and sulphur, in oxy- and thio-biphosphines have been measured.[186]

Unsaturated Systems. Values of $^4J(^{13}C^{19}F)$ for a variety of substituted fluorobenzenes have been obtained from their proton-decoupled ^{13}C n.m.r. spectra by Weigert and Roberts.[106] The analysis of the spectra allowed accurate (± 0.2 Hz) values of the $^{19}F-^{19}F$ couplings in *o*-, *m*-, and *p*-difluorobenzene to be determined. No consistent interpretation of the effect of substituents on the four-bond $^{13}C-^{19}F$ coupling or on the corresponding three- and two-bond couplings could be found. Cooper, Weber, and Manatt[105] have observed the ^{19}F n.m.r. spectra of a number of fluoroaromatic molecules under conditions

---

[280] H. D. Visser and J. P. Oliver, *J. Magn. Resonance*, 1970, **3**, 117.
[281] F. Seel, R. Budenz, and W. Gambler, *Z. Naturforsch.*, 1970, **25b**, 885.
[282] B. L. Shapiro and D. J. Sardella, *J. Magn. Resonance*, 1970, **3**, 336.

of complete noise decoupling of protons. Using ^{13}C satellite spectra where necessary, they have been able to obtain values of three-, four-, and five-bond $^{19}F-^{19}F$ couplings in these systems. They have also investigated the effect of solvent upon $^{19}F-^{19}F$ couplings in 1,2,4-trifluorobenzene. In this respect it is interesting that whereas the value of $^3J(^{19}F^{19}F)$ shows no correlation with solvent dielectric constant, the ^4J and ^5J couplings do, suggesting an importance of the reaction field effect in the last two but not in the first. A seven-bond fluorine–fluorine coupling of 0·8 Hz is found[105] in 1-methyl-4,7-difluorophenanthrene. A complete analysis of the spectrum of octafluorostyrene has been made and the signs of all couplings have been determined.[184]

Values of $^4J(H^{19}F)$ and $^5J(H^{19}F)$ in conjugated molecules have been reported.[283,284] Wasylishen and Schaefer[271] have fully analysed the ^1H n.m.r. spectrum of p-fluorotoluene and use the value of $^6J(CH_3^{19}F)$ thus obtained as a measure of the π-electron contribution to $^5J(H^{19}F)$ in fluorobenzene. That is, since $^6J(CH_3^{19}F)$ in p-fluorotoluene = 1·15 Hz, $^5J^\pi(H^{19}F)$ in fluorobenzene = −1·15 Hz. A good correlation ($r = 0.990$) is found between the $^5J(H^{19}F)$ values of monosubstituted fluorobenzenes and the electronegativity of the substituent. This is interpreted as an inductive effect on the σ-system with negligible effect on the π-electron contribution to the coupling. The authors conclude that $H-^{19}F$ couplings in aromatic molecules show the same qualitative dependence on substituent electronegativity as do H–H couplings. However, the range of values is greater for the former than for the latter. An inter-ring $^7J(H^{19}F)$ value of 0·16—0·21 Hz (depending on solvent) has been reported[217] in 4,4′-difluorobiphenyl. No such coupling was observed for the corresponding 3,3′- and 2,2′-compounds. Values of ^4J ($H^{19}F$), $^4J(^{19}F^{19}F)$, and $^5J(^{19}F^{19}F)$ have been obtained by McFarlane[224] from a detailed investigation of 2,3,5,6-tetrafluoroaniline, and relative signs have also been established (see also Section 6). Cavalli[226] has measured many $^{19}F-^{19}F$ couplings in substituted Dewar benzenes. High values of $^5J[F(2)F(5)]$ and $^5J[F(3)F(6)]$ of the order of 13—14 Hz are observed [for numbering of fluorine atoms see structure (24)]. Harris, Pyper, Richards, and Schulz[133] have determined a number of $^{19}F-^{19}F$ couplings in three fluoropyridines, demonstrating the usefulness of bandshape analysis in the assignment of coupling constants. Double-resonance experiments show that the *meta* $^{19}F-^{19}F$ coupling constants in 3-chloro-2,4,5,6-tetrafluoropyridine differ in sign, $^4J(^{19}FCCC^{19}F)$ being positive and $^4J(^{19}FCNC^{19}F)$ negative.

Long-range $^{13}C-H$,[154] $^{13}C-^{31}P$,[185] and $^{31}P-H^{233}$ couplings over four bonds have been observed. Coupling of hydrogen to metals over four or more bonds has been investigated in several laboratories. Three- and four-bond mercury–hydrogen couplings have been measured[252] for the two isomers of difuryl- and dithienyl-mercury. An excellent correlation of these couplings with the corresponding $J(HH)$ values in furan and thiophen indicates that the Fermi contact mechanism is also the dominant one in long-range mercury–

---

[283] H. Günther and J. B. Pawliczek, *Org. Magn. Resonance*, 1971, **3**, 267.
[284] F. Terrier, J.-C. Hallé, and M.-P. Simonnin, *Org. Magn. Resonance*, 1971, **3**, 361.

hydrogen coupling. The relative signs of ^2J(X–C–H) and ^4J(X–C=C=C–H) have been deduced by Simonnin and Lequan[169] for allenic derivatives of Pb, Sn, and Hg (X = metal). ^4K(X–C=C=C–H) is probably negative for X = ^{207}Pb, ^{119}Sn, and ^{117}Sn, and positive for ^{199}Hg. Petrosyan and Reutov[285] have measured H–^{199}Hg couplings through five, six, and seven bonds in o-, m-, and p-methyl- and 2,6- and 3,5-dimethyl-benzylmercuric chlorides. The seven-bond coupling (38·0 Hz) is larger than the six-bond (4·0 and 15·0 Hz), and the five-bond is intermediate (26·0 and 28·0 Hz). The authors suggest that the sign of the coupling changes with each additional bond.

C. 'Through-space' Coupling.—Couplings involving at least one fluorine atom are often observed to be particularly large even when the atoms in question are separated by five or more chemical bonds. In such cases it is invariably found that the atoms are, or can possibly be, very close to each other in space and it has been suggested[286] that in such cases there is an important direct or through-space contribution to the coupling. Further examples of this type of phenomenon in ^{19}F–H coupling have been reported in fluorine-substituted styrenes,[287] stilbenes,[287] and indoles,[288] and Vasileff and Koster[289] have noted the $^5J(^{19}F^{19}F)$ couplings in maleoyl fluoride (4·78 Hz) and fumaryl fluoride (0·22 Hz).

Stock and Wasielewski[290] have measured $^5J(H^{19}F)$ in the molecules (32)—(35). The coupling has a value of 0·88 ± 0·05 Hz in (32) but is undetectable in the other compounds. The authors conclude therefore that the through-bond contribution to $^5J(H^{19}F)$ is very small and thus the large values observed for this coupling in other molecules must be dominated by a through-space contribution.

Ng[291] has examined the temperature dependence of the ^{1}H n.m.r. spectrum of NN-dimethyltrifluoroacetamide. At ambient probe temperature the two methyl resonances show couplings of 1·5—1·68 Hz and 0·60—0·80 Hz with the –CF$_3$ group. In chloroethane solution at 130 °C the coupling averages to 1·20 Hz, which is exactly the rotational average of the couplings found at lower temperature, indicating that the two couplings have the same sign. Ng concludes that all the evidence is compatible with a dominant through-space contribution to the coupling and that therefore the methyl resonance showing the largest $^5J(H^{19}F)$ coupling belongs to the –CH$_3$ group nearest to the –CF$_3$.

In attempts to obtain less ambiguous evidence on the subject of through-space coupling than hitherto, two groups have investigated ^6J(CH$_3^{19}$F) and

[285] V. S. Petrosyan and O. A. Reutov, *Izvest. Akad. Nauk, S.S.S.R., Ser. khim.*, 1970, 1403.
[286] L. Petrakis and C. H. Sederholm, *J. Chem. Phys.*, 1961, **35**, 1243.
[287] N. M. Sergeev, O. P. Petrii, and N. N. Shapet'ko, *Zhur. strukt. Khim.*, 1970, **11**, 828.
[288] E. I. Berus, V. A. Barkhash, and Yu. N. Molin, *Zhur. strukt. Khim.*, 1970, **11**, 819.
[289] T. P. Vasileff and D. F. Koster, *J. Org. Chem.*, 1970, **35**, 2461.
[290] L. M. Stock and M. R. Wasielewski, *J. Org. Chem.*, 1970, **35**, 4240.
[291] S. Ng, *J. Chem. Soc. (A)*, 1971, 1586.

$^5J(^{19}F^{19}F)$ in bridged biphenyl derivatives.[292,293] Their results are summarized in Table 8. The absence of coupling in 1-fluoro-8-methylbiphenylenes (36), as opposed to the very significant values for fluorene and its derivatives speaks very strongly in favour of a through-space coupling in these systems. Gribble and Douglas[292] also make the point that, if through-bond coupling were making a significant contribution, then one would expect a larger range of $^6J(CH_3{}^{19}F)$ values than 1·0 Hz in the fluorene derivatives in view of their considerable electronic differences. Servis and Jerome[293] directed their work primarily to showing whether long-range H–^{19}F and ^{19}F–^{19}F coupling

(32) (33) (34)

(35)

constants have comparable stereochemical dependence in related compounds. The results in Table 8 show that

$$^5J(^{19}F^{19}F)(36c)/^5J(^{19}F^{19}F)(36b) = 0·57 \text{ and}$$
$$^6J(CH_3{}^{19}F)(36e)/^6J(CH_3{}^{19}F)(36d) = 0·64.$$

The close similarity of these figures implies, in the authors' opinion, a similarity of coupling mechanism in the two cases, pointing to a through-space mechanism.

Very strong evidence for the existence of through-space coupling between mercury and fluorine has been obtained by McFarlane[294] from an investigation of o-, m-, and p-trifluoromethylphenylmercury derivatives. He compares the couplings $J(^{199}Hg^{19}F)$ and $^3J(^{199}HgH_o)$ (H_o is the proton ortho to the

[292] G. W. Gribble and J. R. Douglas, *J. Amer. Chem. Soc.*, 1970, **92**, 5764.
[293] K. L. Servis and F. R. Jerome, *J. Amer. Chem. Soc.*, 1971, **93**, 1535.
[294] W. McFarlane, *Chem. Comm.*, 1971, 609.

Table 8 Coupling constants (Hz) in bridged biphenyls

(36)

Compound	R	R^1	X	$^5J(^{19}F^{19}F)$	$^6J(CH_3{}^{19}F)$	Ref.
(36a)	Direct bond	H	Me	—	≈0	292
	CH_2	H	Me	—	8·3	292
	$MeCO_2CH$	H	Me	—	7·8	292
	CHCl	H	Me	—	8·1	292
	CHOH	H	Me	—	7·8	292
	>C=O	H	Me	—	7·3	292
(36b)	CH=CH	Me	F	170	—	293
(36c)	$(CHOH)_2$	Me	F	98	—	293
(36d)	CH=CH	Me	Me	—	11·9	293
(36e)	$(CHOH)_2$	Me	Me	—	7·7	293
	$(C=O)_2$	Me	Me	—	8·2	293
	O=COC=O	Me	Me	—	3·7	293

mercury atom) in each pair of compounds diarylmercury and monoaryl mercuric bromide. With one exception, the replacement of the aryl group (trifluoromethylphenyl) with bromine results in an increase of the above couplings by a factor very close to two. This is a well-established result interpreted as due to the diversion of mercury s-orbital character into the Hg–C bond by the more electronegative bromine. The notable exception to this behaviour in the case of the *ortho*-compound, where the $^4J(^{199}Hg^{19}F)$ coupling increases only from 26·5 to 28·8 Hz on replacement of the aryl group by bromine, strongly suggests that there is a substantial through-space contribution to this coupling.

A large value of $^6J(CH_3H) \simeq 0.5$ Hz has been observed[295] in compounds such as (37). The authors suggest tentatively that there may be some through-

(37) R = acyl or methyl

[295] W. T. L. Sidwell, H. Fritz, and C. Tamm, *Helv. Chim. Acta*, 1971, **54**, 207.

Nuclear Spin–Spin Coupling

space contribution to this coupling and they propose to investigate other similar systems.

8 Experimental Work of Significance for the Measurement or Interpretation of Spin–Spin Coupling Constants

In this section some recent experimental advances are described briefly (see also Chapter 4). The criterion for their selection is that they should help, or show promise of helping, the measurement and/or interpretation of nuclear spin–spin coupling. The summary is not expected to be comprehensive.

A. Multiple-resonance Experiments. (see also Chapter 5)—Ziessow[296] has described the use of transient nutations in $^{19}F-\{^{13}C\}$ double resonance for the measurement of ^{13}C line positions in fluorinated organic compounds. The reproducibility of the measurements is claimed to be ± 0.3 Hz. Sciacovelli *et al.*[297] have shown that the INDOR method when applied to protons is capable of a precise determination of the frequencies of hidden lines and hence of chemical shifts, and they use the method to establish a complex structure. The nuclear Overhauser enhancement of ^{15}N resonances has been investigated by Lichter and Roberts[298] who discuss the theoretical and practical implications of the results for natural-abundance ^{15}N n.m.r. spectroscopy. Fung and Olympia[299] have examined ^{19}F double resonance in the presence of chemical exchange. They find that strong irradiation of one component of a strongly-coupled spin multiplet in the presence of intramolecular exchange causes a direct saturation of the transitions connecting energy levels that have the same m_1 values as the one being irradiated, whereas the intensities of the other transitions are only slightly affected. They suggest that this may provide an important complement to the spin-tickling technique for determining energy level diagrams and the relative signs of coupling constants. McFarlane[224] has described heteronuclear $^{1}H-\{^{19}F\}$ double-resonance experiments and shows that, even in the case of a complex spectrum, the precision obtained is equal to that of single resonance with direct observation of ^{19}F. He uses some novel triple-resonance experiments to resolve ambiguities with regard to the signs of some couplings.

B. Bandshape Analyses. (see also Chapter 6)—Yamamoto and Kamezawa[300] have investigated the effect of the exchange of the OH proton upon the lineshape of the β-proton absorption in alcohols. They show that the lineshape can be used to determine the relative signs of $^3J(H_AH_A)$ and $^4J(H_BH_X)$ in (38).

Harris and his associates[133,301] have also demonstrated how bandshape

---

[296] D. Ziessow, *Chem. Comm.*, 1971, 463.
[297] O. Sciacovelli, W. von Philipsborn, C. Amith, and D. Ginsburg, *Tetrahedron*, 1970, **26**, 4589.
[298] R. L. Lichter and J. D. Roberts, *J. Amer. Chem. Soc.*, 1971, **93**, 3200.
[299] B. M. Fung and P. L. Olympia, *Mol. Phys.*, 1970, **19**, 685.
[300] O. Yamamoto and N. Kamezawa, *J. Magn. Resonance*, 1970, **3**, 269.
[301] R. K. Harris and N. C. Pyper, *Mol. Phys.*, 1971, **20**, 467.

E

analysis can provide coupling constants and their relative signs in the case of n.m.r. spectra of molecules containing quadrupolar nuclei, and Anderson and Lee[302] have shown that the spectrum of a two-spin system undergoing mutual exchange is sensitive to the relative sign of the direct (dipole–dipole) and indirect (electron-coupled) coupling over a wide range of exchange rates.

C. Effects of Paramagnetic Materials.—Correlations between contact shifts and nuclear spin–spin coupling constants have been observed and discussed by Morishima.[303,304] In studies of the ^{1}H and ^{13}C contact shifts induced in chloroform, methylene chloride, and phenylacetylene by the di-t-butyl-nitroxyl radical, they find[303] upfield and downfield contact shifts for ^{1}H and ^{13}C respectively. The relative contact shifts $\Delta\delta^{13}C/\Delta\delta^1H$ show a good linear correlation with $^1J(^{13}CH)$, which the authors interpret in terms of the FPT of coupling. Similar results are found for a variety of other molecules, where it is shown[304] that a good correlation exists between the relative value of the proton contact shifts induced by nickel acetylacetonate co-ordination and those of the proton spin–spin coupling along the corresponding σ-bonds.

$$\begin{array}{c} \text{Cl} \quad \text{H} \\ | \quad\quad | \\ \text{Cl}-\text{C}-\text{C}-\text{O}-\text{H}_A \\ | \quad\quad | \\ \text{H}_X \quad \text{H}_B \end{array}$$

(38)

Gillies and Baird[305] have noted that the addition of cobalt acetylacetonate to solutions of pyridine and α-, β-, and γ-picoline causes decoupling of the β- from the α-protons. Similar observations of the decoupling of ^{31}P and ^{1}H in phosphorus esters by transition metals have been made by Engel and Jung.[306]

D. Other Experimental Techniques.—The useful experiments with oriented molecules are discussed in Chapter 10. Gray and Ozier[307] have described a technique for the measurement of n.m.r. splittings with an accuracy of 0·001 Hz, and Allerhand and Doddrell[308] have discussed the use of partially relaxed Fourier transform spectra as an aid to assignment in ^{13}C spectra.

Elvidge and his collaborators[309] have given an account of some preliminary studies in triton (^{3}H) spectroscopy. They show that routine measurements may be undertaken, and useful results derived, even with relatively simple

[302] J. M. Anderson and A. C.-F. Lee, *J. Magn. Resonance*, 1970, **3**, 427.
[303] I. Morishima, K. Endo, and T. Yonezawa, *Chem. Phys., Letters*, 1971, **9**, 203.
[304] I. Morishima and T. Yonezawa, *J. Chem. Phys.*, 1971, **54**, 3238.
[305] E. Gillies and M. C. Baird, *Chem. Phys. Letters*, 1970, **7**, 451.
[306] R. Engel and A. Jung, *J. Chem. Soc. (C)*, 1971, 1761.
[307] K. W. Gray and I. Ozier, *Phys. Rev. Letters*, 1971, **26**, 161.
[308] A. Allerhand and D. Doddrell, *J. Amer. Chem. Soc.*, 1971, **93**, 2777.
[309] J. Bloxidge, J. A. Elvidge, J. R. Jones, and E. A. Evans, *Org. Magn. Resonance*, 1971, **3**, 127.

n.m.r. equipment. They state that couplings to tritons can differ appreciably from the corresponding couplings to protons and that there is no constant simple relationship between the two. This surprising result was not fully substantiated and deserves further investigation, particularly in view of the general lack of evidence on this subject and of the results of Murrell and his co-workers.[28,310]

[310] C. N. Banwell, J. N. Murrell, and M. A. Turpin, *Chem. Comm.*, 1968, 1466.

3
Nuclear Spin Relaxation

BY N. BODEN

1 Introduction

The study of nuclear spin relaxation is important for two reasons. Firstly, it provides information about the dynamics of the spin-bearing molecules and also the spin-dependent interactions which couple the nuclear spins to the molecular degrees of freedom (lattice). Secondly, it provides a knowledge of spin relaxation times which is necessary for the operation of spectrometers and in the interpretation of many spectroscopic observations. In this chapter we report the results of relaxation studies by principally pulse methods; all references to relaxation for fluids as studied by high-resolution lineshape analysis and double-resonance techniques are excluded as they are reviewed, respectively, in Chapters 6 and 7. Relaxation studied in gases, liquids, and solids is comprehensively covered; all references to metals and macromolecule systems are excluded, the latter being reported in Chapter 8. We also review in this chapter the development of the multiple-pulse experiment as a line-narrowing technique in solids and the spin-locked double-resonance experiment for the ultrasensitive detection of weakly resonant spin systems: both these techniques are of current interest and look particularly promising for the study of material in the solid state.

The theory of nuclear spin relaxation is well developed and is described in the texts by Abragam,[1] Slichter,[2] and, with special reference to relaxation in solids, in the new text by Goldman.[3] The theory expresses the spin–lattice, the spin–spin, and the rotating-frame spin–lattice relaxation times (T_1, T_2, and $T_{1\rho}$) in terms of Fourier transforms of the correlation functions for the spin-dependent interactions which couple the nuclear spins to the lattice. Before information about molecular parameters and dynamics can be obtained from measured relaxation times, these correlation functions must be evaluated; this is by far the most difficult problem in relaxation theory.[4] It is frequently assumed that the time-dependence of the correlation function is

---

[1] A. Abragam, 'The Principles of Nuclear Magnetism', Oxford University Press, London, 1961.
[2] C. P. Slichter, 'Principles of Magnetic Resonance', Harper, New York, 1963.
[3] M. Goldman, 'Spin Temperature and Nuclear Magnetic Resonance in Solids', Oxford University Press, London, 1970.
[4] The calculation of time correlation functions in n.m.r. is discussed in: J. M. Deutch and I. Oppenheim, *Adv. Magn. Resonance*, 1968, 3, 43; R. G. Gordon, *ibid.*, p. 1.

exponential; in this case the decay constant or the correlation time, τ_c, can be determined from the measured relaxation time. Of course, there still remains the equally difficult problem of obtaining a satisfactory interpretation of τ_c in terms of molecular properties. There are a number of physically plausible models which lead to exponential correlation functions, e.g. isotropic small-step rotational diffusion in liquids and reorientation of a molecule between geometrically equivalent orientations in a solid. In these cases we observe the predicted ω_0^2 dependence of T_1 in the long-correlation-time limit and the frequency independence in the short-correlation-time limit. On the other hand, there are many systems encountered where relaxation measurements as a function of frequency show quite clearly that the lattice correlation functions are non-exponential. Examples of current interest are the long-range collective orientational order fluctuations in nematic liquid crystals, the critical damping of certain lattice modes in the neighbourhood of second-order phase transitions, and translational diffusion. On the theoretical side, current interest is primarily in the development of techniques for the calculation of appropriate correlation functions for the spin-dependent interactions.

2 Nuclear Spin Relaxation in Gases

A. Introduction.—The work reported this year is noted for the absence of any reference to the H_2 gas system which, because of its simplicity, has previously been extensively studied.[5] It is apparent that attention is now being directed to more complex molecules. The main motivation behind spin relaxation studies in gases is that T_1 and T_2 can be interpreted to provide quantitative information about the anisotropic part of the intermolecular potential which is not readily available by measurement of conventional transport properties. The interactions predominantly responsible for spin relaxation in dilute molecular gases are the intramolecular dipole–dipole interaction, the spin–rotational interaction, and, for nuclei having spin $I > \frac{1}{2}$, the intramolecular quadrupole interaction.[6] Collisions between the molecules cause transitions among the molecular rotational states, making these intramolecular spin-dependent interactions time-dependent. The correlation functions for these interactions must be evaluated before information about the molecular properties of the gas can be obtained from T_1 data. It is usually assumed,[6] and in some cases experimentally verified,[7] that the correlation function can be approximated as the product of the free molecule correlation function and an exponential function of time. T_1 is dependent on the Larmor frequency ω_0, the correlation time τ_c (in a dilute gas τ_c will be associated with the time between molecular collisions and therefore related to the number of molecules per unit volume ρ, i.e. $\tau_c \propto \rho^{-1}$), and the strength of the intermolecular spin-dependent interactions. A plot of T_1 vs. ρ goes through a characteristic mini-

---
[5] J. M. Deutch and I. Oppenheim, *Adv. Magn. Resonance*, 1966, **2**, 225; M. Bloom and I. Oppenheim, *Adv. Chem. Phys.*, 1967, **12**, 549.
[6] M. Bloom, F. Bridges, and W. N. Hardy, *Canad. J. Phys.*, 1967, **45**, 3533.
[7] R. Y. Dong and M. Bloom, *Canad. J. Phys.*, 1970, **48**, 793.

Nuclear Spin Relaxation

mum at $\omega_0 \tau_c \approx 1$; for $\omega_0 \tau_c \gg 1$, $T_1 \propto \rho^{-1}$ while for $\omega_0 \tau_c \ll 1$, $T_1 \propto \rho$. T_2 is a monotonic function of ρ even for $\omega_0 \tau_c \gg 1$. The temperature-dependence of T_1 is by the kinetic theory, according to $T_1/\rho \propto T^{-\frac{1}{2}}$ for the dipolar and quadrupolar interactions, and $T_1/\rho \propto T^{-3/2}$ for the spin–rotational interaction. Experimentally, the dependence of T_1 or T_2 on temperature and density is usually investigated.

B. Theoretical Developments.—Nuclear spin relaxation by spin–rotational interaction in tetrahedral and octahedral molecules in the gas phase has been treated by McCourt and Hess,[8] using the Waldman–Snyder[9] kinetic equation for polyatomic gases. By this treatment, the relaxation times are expressed not in terms of correlation functions but in terms of collision integrals similar to those obtained in describing transport processes in polyatomic gases, as follows:

$$\frac{1}{T_1} = \frac{2}{3} C_a^2 \langle J^2 \rangle \left\{ \frac{1}{\omega_{\text{coll}}} \frac{1}{(1+\varphi^2)} + \frac{4}{45} \left(\frac{\Delta C}{C_a}\right)^2 \frac{q}{4\tilde{\omega}_{\text{coll}}} \frac{1}{(1+\tilde{\varphi}^2)} \right\}$$

$$\frac{1}{T_2} = \frac{1}{3} C_a^2 \langle J^2 \rangle \left\{ \frac{1}{\omega_{\text{coll}}} \left[1 + \frac{1}{(1+\varphi^2)} \right] + \frac{4}{45} \left(\frac{\Delta C}{C_a}\right)^2 \frac{q}{4\tilde{\omega}_{\text{coll}}} \left[1 + \frac{1}{(1+\tilde{\varphi}^2)} \right] \right\}$$

where $\varphi = \omega/\omega_{\text{coll}}$, $\tilde{\varphi} = \omega/\tilde{\omega}_{\text{coll}}$, $\omega = \omega_0 - \omega_J$ and, for $J > 2$, q is 3 for tetrahedral molecules and 2 for octahedral molecules. The quantities $C_a = \frac{1}{3}(C_\parallel + 2C_\perp)$ and $\Delta C = C_\parallel - C_\perp$ describe the spin–rotation tensor. Explicit expressions are given for the two collision integrals ω_{coll} and $\tilde{\omega}_{\text{coll}}$. It is important to note that owing to the explicit consideration of molecular symmetry, the 'effective' spin–rotation coupling coefficient found in tetrahedral and octahedral molecules is not the same. It can be seen that measurement of T_1 and T_2 under the conditions φ and $\tilde{\varphi} \gg 1$, where $T_1 \neq T_2$, should allow ω_{coll} and $\tilde{\omega}_{\text{coll}}$ to be determined provided values for C_a and $|\Delta C|$ are available. Alternatively, if ω_{coll} and $\tilde{\omega}_{\text{coll}}$ could be evaluated and a value of C_a were available, a value for $|\Delta C|$ could be calculated from T_1 measured in the limit φ, $\tilde{\varphi} \ll 1$, where $T_1 = T_2$. This is important because accurate values of C_a may be obtained with relative ease from an analysis of molecular beam spectra, but $|\Delta C|$ is more difficult to measure. At present, there are, unfortunately, no collision calculations available for tetrahedral or octahedral molecules. Nevertheless, it is still possible to estimate $|\Delta C|$ using the method developed by Dong and Bloom.[10] They use physical reasoning to estimate the ratio of the two collision integrals; then $|\Delta C|$ is obtained by fitting the T_1 vs. ρ curve in the vicinity of the T_1 minimum, using a molecular-beam value for C_a. In methane, from molecular-beam measurements $C_a = 10.4$ kHz, but values of 18·2 kHz[11] and 21·0 kHz[12] have been claimed for $|\Delta C|$. Dong and Bloom[10]

---

[8] F. R. McCourt and S. Hess, *Z. Naturforsch.*, 1970, **25a**, 1169.
[9] L. Waldman, *Z. Naturforsch.*, 1957, **12a**, 660; 1958, **13a**, 609; R. F. Snyder, *J. Chem. Phys.*, 1960, **32**, 1051.
[10] R. Y. Dong and M. Bloom, *Canad. J. Phys.*, 1970, **48**, 793.
[11] P. Yi, I. Ozier, A. Khosla, and N. F. Ramsey, *Bull. Amer. Phys. Soc.*, 1967, **12**, 509.
[12] I. Ozier, L. M. Crapo, and S. S. Lee, *Phys. Rev.*, 1968, **172**, 63.

report that the former value gives the better fit to their T_1 data, but McCourt and Hess show that, using their expression for T_1, the best-fit value is not 18·2 kHz but rather 21·0 kHz, and they therefore claim that the latter value is the definitive one.

As it is now feasible[10] to measure T_1 in gases under conditions where $\omega_0 \tau_c \gg 1$, we mention here the results of Hubbard's[13] calculation of T_1 and T_2 due to intramolecular quadrupole interactions in this limit. He shows that the relaxations of the longitudinal and transverse components of a spin I ($I \geqslant 1$) are different, but each is the sum of I decaying exponential terms if I is an integer, or the sum of $(I+\frac{1}{2})$ decaying exponential terms if I is half integral. When $\omega_0 \tau_c \ll 1$, both the longitudinal and transverse relaxations are simple exponential decays, as shown previously (ref. 1, pp. 313–315).

C. Diatomic Gases.—In dilute gases and gas mixtures, where the translational and rotational motions may be described classically, the spin–rotational relaxation time for a nucleus in a linear molecule undergoing binary collisions in the extreme narrowing limit is, as obtained by Gordon,

$$T_1 = (3\hbar^2/4C^2 IkT)\rho \, \bar{V} \sigma_{\text{tot}}$$

where C is the spin–rotation interaction constant, $\bar{V}$ is the relative velocity of a colliding pair of molecules, and σ_{tot} is the total cross-section for changes in $\langle J^2 \rangle$. Cattolica et al.[14] have determined, by T_1^H measurements in HCl–Ar mixtures, values of T_1^H appropriate to pure HCl and to HCl infinitely dilute in Ar in the density regions where only binary collisions are important. As relaxation is dominated by the spin–rotational interaction, and rotational states up to about $J = 9$ are occupied at the temperature of the investigation (304—423 K), the above equation was used to calculate values of σ_{tot} for HCl–HCl and HCl–Ar collisions: $\sigma_{\text{tot}}^{\text{HCl–HCl}}$ varies from 0·81 to 0·51 nm² between 304 K and 423 K, and $\sigma_{\text{tot}}^{\text{HCl–Ar}}$ varies between 0·105 and 0·077 nm². The value calculated from kinetic theory is about 0·314 nm² for both HCl–Ar and HCl–HCl interactions. The differences in these values reflect the nature of the anisotropic potential: the anisotropic part of the HCl–HCl potential is caused by the dipole–dipole interaction and varies as r^{-3}, whereas the anisotropic part of the HCl–Ar potential arises from dipole–induced-dipole interaction and follows an r^{-6} law. The observed temperature dependence, $\sigma_{\text{tot}} \propto T^{-1}$ in both cases, is the familiar one occurring in molecules like CH_4, C_2H_2, NH_3, BF_3, SiF_4, CF_4, etc., and suggests that in the temperature region of these measurements the HCl molecules behave classically with respect to the transfer of angular momentum. It is also clear that the T_1 data are not sensitive to the radial dependence of the anisotropic part of the intermolecular potentials (if they were, the difference in the radial dependence for the two interactions would be reflected by a marked difference in the temperature dependence of $\sigma_{\text{tot}}^{\text{HCl–HCl}}$ and $\sigma_{\text{tot}}^{\text{HCl–Ar}}$). Nevertheless, the rotational

[13] P. S. Hubbard, J. Chem. Phys., 1970, **53**, 985.
[14] A. M. L. Cattolica, K. O. Prins, and J. S. Waugh, J. Chem. Phys., 1971, **54**, 769.

transfer cross-sections do contain important information about the strength of the anisotropic interactions averaged over a collision.

D. Polyatomic Gases.—*Pure Gases.* Values of T^H in CH_4 and its deuterio-derivatives and in CH_4–CD_4 mixtures have been measured as a function of temperature and density by Trappeniers and co-workers;[15] the various contributions to the relaxation are separated. The spin–rotational interaction is the main contributor in the gas phase, the only other sizeable contribution coming from the intramolecular dipolar interaction. For example, in pure CH_4 at 108 amagat,* the contribution from the latter interaction varies from 4% of the total relaxation rate at 300 K to 8% at 140 K. The behaviour of the rotational correlation times is, even up to liquid densities, in agreement with the transient approximation of the kinetic model.[6] Mohanty and Bernstein[16] report T_1^F, determined by linewidth measurement, in pure CF_4 and its mixtures with SF_6, SiF_4, Xe, Kr, and Ar. In the pure gas $T_1^F/\rho \propto T^{-1.5}$, indicating that the relaxation is *via* the spin–rotational interaction. In the gas mixtures distinct 'medium' effects were observed; a qualitative explanation in terms of collision rate and viscosity is given. Unfortunately, no attempt was made to obtain T_1^F values appropriate to CF_4 infinitely dilute in solvent gas and in density regions where only binary collisions are important. If this were done, cross-sections for the transfer of rotational angular momentum could be calculated and this would be more meaningful. Tison and Hunt[17] have found that the temperature dependence of T_1^F/ρ ($\propto T^{-1.5 \pm 0.1}$) in the vapour phase of SF_6 below T_c does not differ radically from that in the gas phase ($\propto T^{-1.7 \pm 0.1}$) above T_c; thus, relaxation here is predominantly due to the spin–rotational interaction. In BF_3 gas between 270 and 350 K and at densities up to 80 amagat, the ^{19}F relaxation is according to $T_1/\rho \propto T^{-1.5 \pm 0.2}$, as shown by Hinshaw and Hubbard.[18] The value of T_1 of ^{11}B ($I = 3/2$) in the same gas was found to increase linearly with density up to about 40 amagat, then less rapidly than linear, and eventually to reach a maximum and then decrease at about 100 amagat. At densities up to 22 amagat, $T_1^B/\rho \propto T^{-0.5 \pm 0.2}$ in accordance with relaxation by the quadrupole interaction. A similar effect in the T_1^N vs. ρ plot has previously been observed[19] in N_2 gas at 145 and 200 K, where the relaxation is again due to quadrupole interaction. The cause of this curious behaviour remains to be explained.

Stepped T_1/ρ *vs.* ρ *Plot.* In the previous section we saw how the relaxation of spin-½ nuclei in the gas phase is usually *via* the spin–rotational interaction,

---

* The density in amagat is the density (in $g l^{-1}$) at the experiment temperature and pressure divided by the density (in $g l^{-1}$) at 273 K and 1 atm.

[15] C. J. Gerritsma and N. J. Trappeniers, *Physica*, 1971, **51**, 365; C. J. Gerritsma, P. H. Oosting, and N. J. Trappeniers, *ibid.*, p. 381; P. M. Oosting and N. J. Trappeniers, *ibid.*, pp. 395, 418.
[16] S. Mohanty and H. J. Bernstein, *J. Chem. Phys.*, 1970, **53**, 461; 1971, **54**, 3730.
[17] J. K. Tison and E. R. Hunt, *J. Chem. Phys.*, 1971, **54**, 1526.
[18] W. S. Hinshaw and P. S. Hubbard, *J. Chem. Phys.*, 1971, **54**, 428.
[19] P. A. Speight and R. L. Armstrong, *Canad. J. Phys.*, 1969, **47**, 1475.

giving a density dependence of the form $T_1 = A\rho$, where A is a function of temperature. Dong and Bloom[20] found that in gaseous CHF_3 and CH_3F the plot of T_1^F/ρ vs. ρ was horizontal as expected, but in the corresponding plot for the protons T_1/ρ decreased with increasing ρ in a step-like manner. This new effect is thought to be due to an appreciable contribution from the intramolecular dipole–dipole interaction to the relaxation of the protons. Theoretically, the effect must be due to the behaviour of the correlation functions of the dipole–dipole interaction. Dong and Bloom suggest that the steps arise when groups of rotational levels of the molecules are collision-broadened by an amount of the order of the characteristic frequencies of the oscillating terms of the dipolar interactions. To determine whether or not there is a similar step-like dependence of T_1/ρ on ρ for a nucleus relaxed by intramolecular quadrupole interactions, for which the correlation functions are essentially the same as those for the dipole–dipole interaction, Hinshaw and Hubbard[18] measured T_1 for the ^{11}B nuclei in BF_3 gas. The value of T_1 was found to increase linearly with density at low densities, up to about 40 amagat, and less rapidly than linear at high densities, but no evidence of a step-like dependence on density was found. Of course, the step-like behaviour might not occur in all molecules, since the occurrence of the effect presumably depends on the spacing and the collisional broadening of the molecular rotational levels.

CH_4–O_2 and CH_4–NO *Mixtures.* Siegel and Lipsicas[21] have measured T_1^H in these mixtures between 195 and 350 K and x_{O_2}, x_{NO} up to 0·03. Deviations from the linear dependence of T_1 on ρ occur at densities above 60 amagat owing to short-circuiting of the intramolecular spin–rotational interaction by the transient intermolecular magnetic dipolar interaction between the proton magnetic moment and the O_2 or NO molecular magnetic moment during each collision. In a previous study of the CH_4–O_2 system, Johnson and Waugh[22] treated the transient dipolar relaxation using a hard-sphere intermolecular potential for the collision and the expression for T_1 given by Bloembergen (ref. 1, p. 322). These new experimental data for the CH_4–O_2 and CH_4–NO mixtures are at variance with this simple treatment. This discrepancy arises from the use of the static magnetic susceptibility value for the magnetic moment of O_2 and NO rather than the time autocorrelation of the paramagnetic moment over the period of a collision, which is what is really effective in the transient dipolar interaction. The details of this correlation are quite different for O_2 and NO and result in a striking difference in the T_1 values in the two cases, even though the static paramagnetic susceptibilities of the two species only differ by a factor of approximately 3/2. The experimental data also suggest that chemical effects may play a minor role in the relaxation at very short range. Clearly, the study of transient dipolar relaxation in adulterated hydrocarbons should provide interesting information about collision dynamics.

[20] R. Y. Dong and M. Bloom, *Phys. Rev. Letters*, 1968, **20**, 981.
[21] M. M. Siegel and M. Lipsicas, *Chem. Phys. Letters*, 1970, **6**, 259.
[22] C. S. Johnson and J. S. Waugh, *J. Chem. Phys.*, 1961, **35**, 2020.

E. Self-diffusion Measurements.

The self-diffusion coefficient, D, in moderately dense gases may be expressed in terms of the density[23] as:

$$D(\rho, T) = D_0(T) + D_1(T)\rho + D_2(T)\rho^2 \ln\rho + D_3(T)\rho^2 + \ldots$$

The zero-order density coefficient can be calculated by the Chapman–Ensog theory, but the higher coefficients have not yet been calculated. Oosting and Trappeniers[15] have used the spin-echo technique to measure D in methane over a wide range of temperature ($90 < T < 300$ K) and density ($10 < \rho < 500$ amagat). As the temperature dependence of D was very similar over the total density range investigated, the data for D are expressed as:

$$T^{-0.9}D\rho = 10^{-2}(0.1181 + 0.785 \times 10^{-4}\rho - 0.520 \times 10^{-7}\rho^2 \ln\rho)$$

We note that up to 200 amagat, D is given by the zero-order density coefficient. By measuring D in ^{3}He gas doped with 4% O_2 to shorten T_1, Karra and Kemmerer[24] have obtained, at 296 K, $D_0 = 0.56 \pm 1.4$ and $D_1 = 0.056 \pm 0.05$. Note that the density coefficients depend on composition in a gas mixture.

3 Spin Relaxation in Liquids

A. Pure Liquids.—*Molecular Rotation.* In recent years there has been considerable interest, both theoretical and experimental, in the time correlation functions for the rotational variables of molecules in the liquid state. The correlation functions are described by the correlation times τ_1, τ_2, and τ_J defined by

$$\tau_1 = \int_0^\infty \langle Y_{lm}^*[\Omega(0)] Y_{lm}[\Omega(t)] \rangle \, dt$$

and

$$\tau_J = \int_0^\infty \frac{\langle J(0) \cdot J(t) \rangle}{\langle J(J+1) \rangle} \, dt$$

The Y_{lm} are the normalized spherical harmonics and Ω is defined by the polar and azimuthal angles θ and ϕ relative to a laboratory co-ordinate system. For dielectric relaxation $l = 1$, while for nuclear spin relaxation by intramolecular dipolar or quadrupolar interactions $l = 2$. For isotropic small-step rotational diffusion $\tau_2 = \tau_1/3$.[25] On the other hand, for large-step angular rotation $\tau_1 = \tau_2$. The random-walk theory of rotational diffusion outlined by Ivanov[26] predicts $\tau_1 = g\tau_2$, where g varies from 3 to 1 if the jump angle varies from 0° to about 120°. A comparison of τ_1 and τ_2 therefore enables one to characterize the kind of molecular motion. As an example, for [^{2}H$_2$]aniline in [^{2}H$_7$]aniline solution at 293 K, $\tau_2 = 1.9 \times 10^{-11}$ s as determined by Bock *et*

---
[23] R. Zwanig, *Phys. Rev.*, 1963, **129**, 486; K. Kawasaki and I. Oppenheim, *ibid.*, 1965, **139**, A1763.
[24] J. S. Karra and G. E. Kemmerer, *Phys. Letters*, 1970, **33A**, 105.
[25] P. Debye, 'Polar Molecules', Dover Publications, New York, 1945.
[26] E. N. Ivanov, *Zhur. eksp. teor. Fiz.*, 1963, **45**, 1509 (*Sov. Phys. JETP*, 1964, **18**, 1041).

al.[27] from the concentration dependence of the proton T_1, whereas in pure liquid aniline $\tau_1 = 2\cdot2 \times 10^{-11}$ s. Thus, it would appear that in pure liquid aniline molecular reorientation is by large-angle jumps. A similar result is obtained[27] for the reorientation of the aniline molecule in dilute benzene solution: thus, τ_2 determined from the proton T_1 of [^{2}H$_2$]aniline at infinite dilution in C_6D_6 at 293 K is $6\cdot3 \times 10^{-12}$ s and $\tau_1 \approx 8 \times 10^{-12}$ s. τ_2 has an Arrhenius temperature-dependence with $E_a = 18\cdot8$ kJ mol^{-1}.

Values for τ_2 and τ_J can be calculated from T_1 data if the strength of the relevant magnetic interaction is known. In the case of nuclei with spin $I > \frac{1}{2}$, T_1 is usually dominated by the intramolecular quadrupole interaction, and if the nuclear quadrupole coupling constant is known τ_2 can be determined (ref. 1, pp. 313–315). For spin-$\frac{1}{2}$ nuclei, τ_2 can be obtained from the intramolecular dipole–dipole interaction (ref. 1, pp. 289–305) or the anisotropic chemical shielding interaction (ref. 1, pp. 315–316) and τ_J from the spin–rotational interaction.[28] To extract these data from the measured relaxation times, the contributions from the various mechanisms must first be separated, and this is frequently a difficult experimental problem. The relaxation from the intermolecular dipolar interaction must also be separated; for ^{1}H relaxation, dilution studies in deuteriated analogues are often used for this purpose.[29] It is important to realize, too, that when intramolecular dipole–dipole, quadrupole, and anisotropic chemical shielding interactions are important in a nuclear relaxation process, interference effects between them, or any two of them, can cause a non-exponential time-dependence of the longitudinal and transverse magnetizations. This arises because these interactions all depend upon correlation functions of the spherical harmonics $Y_{2m}[\Omega(t)]$. On the other hand, the spin–rotational interaction depends on the correlation function of the rotational angular momentum operator $J(t)$ of the molecule and interference effects do not arise. Blicharski[30] presents results of numerical calculations, based on his earlier theoretical treatments,[30a] which demonstrate the conditions under which the interference of different intramolecular interactions in one- and two-spin systems gives rise to non-exponential relaxation. In the case of a system containing two identical spin-$\frac{1}{2}$ nuclei undergoing isotropic reorientation, interference between the dipolar and anisotropic chemical shielding interactions gives a two-exponen-decay process for both longitudinal and transverse relaxation. For a single spin $I = 1$, quadrupole–chemical shielding interference causes two-exponential relaxation. In a two-spin system ($I = \frac{1}{2}$, $S = 1$), for the quadrupole–dipolar interference one can expect two-exponential and three-exponential processes for longitudinal and transverse relaxation, respectively. Theoretical

---

[27] E. Bock, J. Czubryt, and E. Tomchuk, *Canad. J. Chem.*, 1970, **48**, 2814.
[28] R. J. C. Brown, H. S. Gutowsky, and K. Shimomura, *J. Chem. Phys.*, 1963, **38**, 76; P. S. Hubbard, *Phys. Rev.*, 1963, **131**, 1155.
[29] D. W. G. Smith and J. G. Powles, *Mol. Phys.*, 1966, **10**, 451; J. C. Duplan, A. Brignet, and J. Delmau, *J. Chem. Phys.*, 1971, **54**, 3648.
[30] J. S. Blicharski, *Acta Phys. Polon.*, 1970, **A38**, 19.
[30a] J. S. Blicharski, *Phys. Letters*, 1967, **24A**, 608; *Acta Phys. Polon.*, 1969, A36, 211.

Nuclear Spin Relaxation

treatments of interference between the intramolecular dipole–dipole and anisotropic chemical shielding interactions are also given for molecules containing three magnetically equivalent spin-½ nuclei[31] and for molecules containing four spin-½ nuclei located at the corners of a tetrahedron, a square, and a rectangle.[32] The absence of experimental data demonstrating these effects is a clear indication of the rarity of systems in which interference effects are important, and in general it does not appear to be important to account for them when calculating correlation times.

A large number of relaxation studies have been concerned with the experimental determination of either, or both, τ_2 and τ_J over a wide temperature range in order to check models for rotational motion in liquids. The relationship between τ_2 and τ_J has, in particular, attracted considerable interest. For spherical-top molecules τ_2 and τ_J are related in the small-step diffusion limit $[\bar{\omega} \tau_J \ll 1$, with $\bar{\omega} = (kT/I)^{\frac{1}{2}}$ where I is the moment of inertia of the molecule] where $\tau_2 \gg \tau_J$, by Hubbard's relation:[28]

$$\tau_2 \tau_J = I/6kT$$

and in the dilute-gas or perturbed-free-rotor limit ($\bar{\omega} \tau_J \gg 1$) by:[6]

$$\tau_2 = \tau_J/5$$

No simple analytical relationship exists for the intermediate region between these two limits; this is a particularly important region since many of the experimental data for τ_J are obtained in liquid gases, to which Hubbard's relation is inapplicable. In a recent paper, McClung[33] has discussed the interrelation of τ_2 and τ_J for spherical-top molecules in terms of the J- and M-diffusion cases of the extended diffusion model (angular diffusive steps of arbitrary size are allowed) introduced by Gordon[34] to describe rotational diffusion of linear-top molecules. The extended J-diffusion model reduces to the small-step diffusion model when $\bar{\omega} \tau_J \ll 1$ and the J- and M-diffusion models closely resemble the perturbed-free-rotor model when $\bar{\omega}\tau_J \gg 1$; numerical calculations for the intermediate range show how τ_2 goes through a minimum value when τ_2, τ_J, and $\bar{\omega}^{-1}$ are approximately equal. Mountain[35] has drawn attention to the exact equivalence of the J- and M-diffusion models to two special cases of Fixman and Rider's[36] two-parameter relaxation model: the J-diffusion model corresponds to the case ($\beta_1 = 0$, $\beta_0 = \beta$) and M-diffusion to ($\beta_1 = \beta$, $\beta_0 = 0$). Maryott et al.[37] have shown how values of τ_2 and τ_J calculated from the ^{35}Cl and ^{19}F relaxation times measured in liquid ClO$_3$F (130—368 K) are correlated using the J-diffusion model: ClO$_3$F is treated as a quasi-spherical molecule because of the similarity in its two

---

[31] J. S. Blicharski and W. Nosel, *Acta Phys. Polon.*, 1970, **A38**, 25.
[32] J. S. Blicharski, W. Nosel, and H. Schneider, *Ann. Physik*, 1971, **27**, 17.
[33] R. E. D. McClung, *J. Chem. Phys.*, 1969, **51**, 3842; 1971, **54**, 3248.
[34] R. G. Gordon, *J. Chem. Phys.*, 1966, **44**, 1830.
[35] R. D. Mountain, *J. Chem. Phys.*, 1971, **54**, 3243.
[36] M. Fixman and K. Rider, *J. Chem. Phys.*, 1969, **51**, 2425.
[37] A. A. Maryott, T. C. Farrar, and M. S. Malmberg, *J. Chem. Phys.*, 1971, **54**, 64.

moments of inertia and its small dipole moment. Below 200 K, where $\tau_J/\tau_2 <$ 0·1, the results agree with Hubbard's relation, but at the highest temperatures $\tau_2 \approx \tau_J$ and the small-step diffusion model is no longer applicable. On the other hand, the results agree over the entire temperature range with the J-diffusion model. τ_2 follows an Arrhenius temperature-dependence: $\tau_2/s = 1·06 \times 10^{-13}$ exp(3·75 kJ mol$^{-1}/RT$). The τ_2 values are compared with those calculated from the Debye–Stokes–Einstein relation as modified by Gierer and Wirtz,[38] $\tau_2 = 4\pi a^3 \eta f / 3kT$, where f is the microviscosity factor ($f = 0·163$ for pure liquids); as is often found, τ_2 is not a linear function of η/T; the 'apparent' microviscosity factor varies from 1/4 at the highest temperature to 1/120 at the lowest. τ_J does not exhibit an Arrhenius temperature-dependence, but approximately $\tau_J \propto T^2$. Thus, although the extended diffusion model gives an oversimplified picture of the rotational motion of molecules in liquids, it clearly represents a considerable improvement over the small-step diffusion model proposed by Debye.[25] Most studies of τ_J have been made either in liquids at high temperatures or in liquid gases, where the Debye model is inapplicable. Strange and Morgan[39] have collated values of τ_2 and τ_J measured in the liquids HCl, HBr, PCl$_3$, and PBr$_3$; at their melting points, $\tau_2 \gg \tau_J$ and $\tau_2 \tau_J$ approximates to the Hubbard relation, but, at their critical points $\tau_2 \approx \tau_J$. Clearly, comparison of these and other data with extended diffusion models should be rewarding. Theoretical studies of the relation of τ_2 and τ_J to the molecular properties of the liquid in this region are also needed.

O'Reilly et al.[40] have measured T_1^F in liquid and solid F$_2$. In the liquid between 53·5 K and 105 K, T_1 is determined by the spin–rotational interaction, from which values for τ_J were calculated using the known value of C_F^2 [$C^2 = \frac{1}{3}(C_\parallel^2 + 2C_\perp^2)$]. At 56 K, $\tau_J = 9·5 \times 10^{-14}$ s; τ_J has an Arrhenius temperature-dependence with apparent activation energy 1·25 kJ mol^{-1}. A value for τ_2 was not obtained experimentally, but calculated from the quasi-lattice random-flight model[41] (at 56 K, $\tau_2 = 6·8 \times 10^{-12}$ s); the product $\tau_2 \tau_J$ is a factor of 20 longer than predicted by the Hubbard relation. It was therefore assumed that reorientation is by large jumps. Using the reorientational jump model, as proposed by Brown et al.,[28] for the spin–rotation interaction, they obtain for a diatomic molecule:

$$\tau_2 \tau_J = I\langle\theta^2\rangle/2kT$$

The experimental data are consistent with a root-mean-square angle of jump $\langle\theta\rangle^{\frac{1}{2}} \approx 2·6$ rad. Of course, this result rests on the applicability of the quasi-lattice random-flight model to τ_2; in this case $\langle\theta^2\rangle^{\frac{1}{2}}$ corresponds to the angular jump of the molecule in an 'interstitial' lattice site. Upon solidification T_1 is essentially unchanged while T_2 and $T_{1\rho}$ decrease discontinuously. This type of behaviour is frequently observed in plastic crystals and results

---

[38] A. Gierer and K. Wirtz, Z. Naturforsch., 1953, **8a**, 532.
[39] J. H. Strange and R. E. Morgan, J. Phys. (C), 1970, **3**, 1999.
[40] D. E. O'Reilly, E. M. Peterson, D. L. Hogenboom, and C. E. Scheie, J. Chem. Phys., 1971, **54**, 4194.
[41] D. E. O'Reilly, J. Chem. Phys., 1968, **49**, 5416.

from a large decrease in D with essentially no change in τ_2 and τ_J. This result is consistent with the quasi-lattice random-flight model if it is assumed that excitations of molecules to interstitial positions, and hence molecular rotation, continue to take place at essentially the same rate as in the liquid. If this is the case, the entire increase ($\approx$ 4 kJ mol^{-1}) in the activation energy for D must result from the increase in work to form a vacancy in the solid. Blicharski[42] shows that in liquid CF_2Cl_2, $T_1^F T^2/\eta$ is a constant at temperatures above 293 K; this behaviour is consistent with relaxation by spin–rotational interaction and the reorientational jump model of Brown et al.[28]

Other studies of relaxation by the spin–rotational interaction have been concerned with the calculation of spin–rotational interaction constants making use of Hubbard's relation to estimate τ_J from a value of τ_2 calculated from a measured intramolecular dipolar relaxation rate. Seymour and Jonas[43] have estimated values of C_P in the pure liquids dichlorophenylphosphine (7·5 kHz) and chlorophenylphosphine (3·5 kHz): although there is the possibility of internal rotation about the C—P bonds in these compounds, the spin-over-all-rotation coupling is thought to provide the spin–rotational relaxation mechanism for the ^{31}P nuclei; a value for τ_2 calculated from T_1^H data was therefore considered reliable in the calculation of τ_J. The values obtained for C_P are a factor of 2–3 times larger than those calculated from ^{31}P chemical shifts; similar discrepancies have been observed by Deverell.[44] It is not obvious that the source of this discrepancy lies entirely in the use of the Hubbard relation, but could well be in the chemical-shift scale suggested for ^{31}P.[44] Chien and Wade[45] have reported T_1^H and T_1^F values for the pure liquids 1,3,5- and 1,2,4-trifluorobenzene. The relaxation was exponential, so it was assumed that the dipolar interaction between unlike spins can be treated as an interaction between like spins. A value of $C_F \approx 2\cdot3$ kHz was estimated from the derived ^{19}F spin–rotational relaxation rate. Despite the numerous assumptions made, C_F is in reasonable agreement with the molecular-beam value (3·5 kHz) for C_6H_5F.

Studies of Anisotropic Rotation. Although the rotational motion of a non-spherical molecule in a liquid is in general anisotropic, the results of nuclear magnetic relaxation experiments have, however, often been analysed assuming isotropic rotation. A number of investigators[46] have recently described how, in the case of anisotropic rotation, τ_2 can, in the small-step diffusion limit, be expressed in terms of the three principal rotational diffusion constants for the molecule, and how these can be determined by measurement of the

[42] J. S. Blicharski and B. Blicharski, *Acta Phys. Polon.*, 1970, **A38**, 289.
[43] S. J. Seymour and J. Jonas, *J. Chem. Phys.*, 1971, **54**, 487.
[44] C. Deverell, *Mol. Phys.*, 1970, **18**, 319.
[45] M. Chien and C. G. Wade, *J. Chem. Phys.*, 1971, **54**, 3562.
[46] H. Shimizu, *J. Chem. Phys.*, 1964, **40**, 754; D. E. Woessner, B. S. Snowden, and E. T. Strom, *Mol. Phys.*, 1968, **14**, 265; T. T. Bopp, *J. Chem. Phys.*, 1967, **47**, 3621; J. Jonas and T. M. DiGennaro, *J. Chem. Phys.*, 1969, **50**, 2392; W. T. Huntress, *J. Phys. Chem.*, 1969, **73**, 103; A. Allerhand, *J. Chem. Phys.*, 1970, **52**, 3596; W. T. Huntress, *Adv. Magn. Resonance*, 1970, **4**, 1.

relaxation times of three quadrupolar nuclei in geometrically non-equivalent positions in the molecule. Clearly, it is of great importance to determine the full anisotropy of the motion in order to investigate the effects of intermolecular interactions. Gillen and Noggle[47] have determined values for both $D_\parallel$ and $D_\perp$ from the quadrupole relaxation times of ^{51}V and ^{35}Cl in VOCl$_3$, ^{11}B and ^{35}Cl in BCl$_3$, and ^{14}N and ^{35}Cl in CCl$_3$CN, using quadrupole coupling constants known for the solid. Within the temperature ranges investigated, there is an Arrhenius behaviour of the diffusion constants and $0.5 \leqslant (D_\parallel/D_\perp) \leqslant 2$. Good agreement is obtained between D calculated from the modified Debye–Stokes–Einstein relation, $D = kT/8\pi a^3 \eta f$, and $D_\perp$ for VOCl$_3$, CCl$_3$CN, CD$_3$CN, CDCl$_3$, CD$_3$C⋮CD, and ND$_3$ whereas D seems to be totally unrelated to $D_\parallel$; $D_\parallel > D$, indicating that the C_3 rotation is probably dominated by inertial effects. In the non-polar compound BCl$_3$, there is no correlation between D and either $D_\parallel$ or $D_\perp$; both motions seem to have strong inertial effects. On the basis of these observations, it is concluded that molecular motions of small molecules in pure liquids which reorientate an electric dipole are likely to be in the small-step rotational diffusion limit, and the reorientation rates can be calculated with reasonable accuracy using the modified Debye–Stokes–Einstein relation. Rotations which do not reorientate a dipole will be dominated by inertial effects, and their reorientation rates will be faster than predicted. It should be noted that it is possible to study the anisotropic rotation of molecules by measurement of intramolecular spin–rotational and dipolar relaxation rates, but, because of the experimental difficulties in determining these interactions, the study of quadrupolar relaxation is usually more convenient in practice.

Relaxation by internal rotation of functional groups. The relaxation of nuclei in a functional group may occur as a result of both overall molecular motions and internal motions.[48] Bull *et al.*[49] have used the effect of pressure on the ^{19}F spin–rotational relaxation rate in liquid benzotrifluoride to differentiate between relaxation due to spin–internal-rotation and spin–overall-rotation interactions. Increasing the pressure should quench the spin–overall-rotation interaction by viscous damping, while the contribution from the spin–internal-rotation interaction should be independent of pressure as the rotation of the CF$_3$ groups is dynamically coherent. Experimentally, the spin–rotational relaxation rate was found to decrease smoothly with pressure and to be in good qualitative agreement with Burke and Chan's[50] model for ^{19}F relaxation in liquid benzotrifluoride, which assumes that relaxation is due to uncorrelated spin–internal-rotation and spin–overall-rotation coupling. The results are in contradiction to the previous proposal by Bull *et al.*[49a] that relaxation was due solely to spin–internal-rotation coupling.

Another interesting study of relaxation by spin–internal-rotation inter-

---

[47] K. T. Gillen and J. H. Noggle, *J. Chem. Phys.*, 1970, **53**, 801.
[48] D. E. Woessner, B. S. Snowden, and G. H. Meyer, *J. Chem. Phys.*, 1969, **50**, 719.
[49] T. E. Bull, J. S. Barthel, and J. Jonas, *J. Chem. Phys.*, 1971, **54**, 3663.
[49a] T. E. Bull and J. Jonas, *J. Chem. Phys.*, 1970, **52**, 1978.
[50] T. E. Burke and S. I. Chan, *J. Magn. Resonance*, 1970, **2**, 120.

action has been made by Gillen et al.;[51] their objective was to assess the importance of this relaxation mechanism for the ^{1}H and ^{13}C nuclei in methyl groups. T_1 was measured between 210 K and 350 K for the ^{1}H, ^{2}D, and ^{13}C nuclei in liquid methyl iodide, [^{2}H$_3$]methyl iodide, and [^{13}C]methyl iodide. The intermolecular proton–proton relaxation ($E_a = 10\cdot0$ kJ mol^{-1}) was determined by a dilution study in [^{2}H$_3$]methyl iodide. The intramolecular relaxation of the protons and ^{13}C — assumed to be exponential — was found to be due to both dipole–dipole and spin–rotational interactions. The spin-rotational relaxation times were separated from the intramolecular dipolar contribution by calculating the latter theoretically. To do this, values were required for the various dipolar correlation times which, since the motion of the molecule is anisotropic, required evaluating values for $D_\parallel$ and $D_\perp$. A value for $D_\perp$ was determined from the dielectric relaxation time and its activation energy ($E_a = 7\cdot9$ kJ mol^{-1}) was calculated from the viscosity and density of the liquid. $D_\parallel$ was then determined from the deuteron relaxation times ($E_a = 3\cdot4$ kJ mol^{-1}). The derived spin–rotational relaxation times were found to be significantly smaller than the values calculated assuming rotational diffusion of the whole molecule. The apparent activation energies found for the spin–rotational interaction are $2\cdot9$ and $3\cdot4$ kJ mol^{-1} for ^{13}C and ^{1}H respectively. It was therefore concluded that the spin–rotational relaxation of the ^{1}H and ^{13}C nuclei is due to spin–internal-rotation of the methyl group. The anisotropic reorientational motion of the methyl iodide and acetonitrile[47] molecules is found to be very similar, as might be expected since the molecules have similar shapes and moments of inertia and the compounds have similar melting points and boiling points.

Assink and Jonas[52] have employed T_1^D measurements to study the nature of the internal rotation of the [^{2}H$_3$]methyl groups in [^{2}H$_6$]isobutene. The valve of τ_2 for the overall, assumed isotropic, reorientation of isobutene was calculated from T_1^D measured in [^{2}H$_2$]isobutene, assuming the deuteron quadrupole coupling constant to be the same as in [^{2}H$_6$]benzene, namely 194 kHz. Using $\tau_2 = 1/6D$, at 273 K, $D = 0\cdot28 \times 10^{12}$ s^{-1} and $\bar{\omega}\tau_2 \approx 1$, showing that the molecular reorientation is relatively free, probably due to the molecular shape and small dipole moment. The activation energy for T_1^D is $5\cdot8$ kJ mol^{-1} in both [^{2}H$_2$]- and [^{2}H$_6$]-isobutene. Assuming spherical-top behaviour and a quadrupole coupling constant of 165 kHz, a value of $D \ll 0\cdot2 \times 10^{12}$ s^{-1} was obtained for the rotation of the [^{2}H$_3$]methyl group. It is not possible from the data to differentiate between a diffusional or a random-jump model for the internal motion.

Relaxation in water. A number of mechanisms have been proposed for the processes which lead to reorientation and relaxation of water molecules in liquid water.[53] In principle, information about both the nature of the relaxing

---

[51] K. T. Gillen, M. Schwartz, and J. H. Noggle, *Mol. Phys.*, 1971, **20**, 899.
[52] R. A. Assink and J. Jonas, *J. Chem. Phys.*, 1970, **53**, 1710.
[53] D. Eisenberg and W. Kauzmann, 'The Structure and Properties of Water', Oxford University Press, London and New York, 1969.

species and the details of the molecular motion can be obtained from spin-lattice relaxation measurements on water containing the isotopic species ^{1}H, ^{2}H, and ^{17}O. Previous measurements (ref. 54, pp. 190–193) show that the relaxation processes in water generally exhibit a non-Arrhenius temperature-dependence. To investigate its cause, Hindman et al.[55] have carried out very detailed T_1 measurements of the ^{2}H and ^{17}O nuclei in ^{2}H$_2$O and H$_2$^{17}O, respectively, between 255 and 451 K. The temperature dependence of T_1 can be expressed as the sum of two Arrhenius terms: the apparent activation energies determined from the deuteron T_1 data are $42 \cdot 8 \pm 1 \cdot 2$ kJ mol^{-1} and $13 \cdot 8 \pm 0 \cdot 2$ kJ mol^{-1} at low and high temperatures, respectively. The relative magnitudes of these activation energies suggest that the high-temperature process is that involving isotropic rotational diffusion of unbonded water molecules, whereas the low-temperature process is associated with the breaking of hydrogen-bonds. The data have therefore been interpreted in terms of an equilibrium between a hydrogen-bonded 'lattice' and 'free' or 'defect' molecules which relax by an isotropic rotational diffusion process. The enthalpies and entropies derived from the D$_2$O data are, for the equilibrium, $\Delta H = 28 \cdot 5 \pm 0 \cdot 8$ kJ mol^{-1} and $\Delta S = 110 \cdot 0 \pm 3 \cdot 8$ J K^{-1} mol^{-1}, and for the rotational rate process of a 'free' water molecule $\Delta H^* = 10 \cdot 5 \pm 0 \cdot 2$ kJ mol^{-1} and $\Delta S^* = 15 \cdot 1 \pm 0 \cdot 4$ J K^{-1} mol^{-1}. Similar results are obtained from the H$_2$^{17}O data. Quadrupole coupling constants are calculated from the T_1 data using a τ_2 calculated from the dielectric relaxation time. The values obtained, 258 kHz for ^{2}H and 7·85 MHz for ^{17}O, are intermediate between those previously measured for molecules in the solid and gas phase.

Determination of nuclear quadrupole coupling constants. Values for nuclear quadrupole coupling constants in liquids are conveniently calculated from T_1 data provided appropriate values for τ_2 are available. One approach is to estimate τ_2 from the Debye–Stokes–Einstein expression, but this is notoriously unsatisfactory. Jenks[56] suggests that satisfactory values for τ_2 can, however, be obtained in this way if an 'apparent molecular volume', which he defines as the volume traced out by the molecular rotation causing the relaxation, is used; he quotes values of ^{14}N quadrupole coupling constants in a variety of liquids obtained in this way. A more reliable approach is to calculate τ_2 from an intramolecular dipolar relaxation rate. In the case of non-spherical molecules, whose motion might be anisotropic, the internuclear dipole–dipole vector should be parallel to the direction of the electric field gradient in order that τ_2 for the two interactions be the same. Assink and Jonas[57] have done just this: they have calculated the deuteron quadrupole coupling constant to be 181 ± 10 kHz in chloro[4-^{2}H]benzene and 230 ± 14 kHz in 1,4-di[^{2}H]ethynylbenzene; in both cases τ_2 was obtained from the intra-

---

[54] H. G. Hertz, *Progr. N.M.R. Spectroscopy*, 1967, **3**, 159.
[55] J. C. Hindmann, A. J. Zielen, A. Svirmickas, and M. Wood, *J. Chem. Phys.*, 1971, **54**, 621.
[56] G. J. Jenks, *J. Chem. Phys.*, 1971, **54**, 658.
[57] A. Assink and J. Jonas, *J. Magn. Resonance*, 1971, **4**, 347.

molecular dipolar interaction between the two protons *ortho* to each other on each side of the ring.

Studies of Scalar Relaxation. Modulation of the scalar coupling $\hbar A \mathbf{I} \cdot \mathbf{S}$ by rapid relaxation of the spin S provides a mechanism for coupling the spin I to the lattice (ref. 1, pp. 306—312). Provided $(\omega_I - \omega_S)T_{2S} \geqslant 1$, the scalar relaxation contribution to T_{1I} may be determined by its field dependence, enabling values for both A and T_{2S} to be determined. Studies of scalar relaxation, where S is not an electron, are relatively rare. This year two detailed studies are reported. Blicharski *et al.*[42] have measured T^F in liquid CF_2Cl_2, at 293 K, from 1×10^{-4} T to 2·4 T. Values of T_1 are independent of field above 1·0 T, but decrease rapidly with the field below this value. Spin–rotational interaction is the dominant relaxation mechanism at high fields; the field-dependent contribution observed at low fields is due to the modulation of the F–Cl scalar coupling by the rapid relaxation of the chlorine nucleus: $A/2\pi = (12 \pm 3)$ Hz. A more elegant way to investigate scalar relaxation is by measurement of $T_{1\rho}$; this is particularly convenient as in most cases the modulation rates are less than 10^6 s^{-1}. Strange and Morgan[39] have obtained an expression for $T_{1\rho}$ which is similar to that quoted by Solomon[58] for $T_2(B_1)$,

$$\frac{1}{T_{1\rho}} = \frac{A^2 S(S+1)}{3}\left\{\frac{T_{1S}}{(1+\omega_1^2 T_{1S}^2)} + \frac{T_{2S}}{1+(\omega_I - \omega_S)^2 T_{2S}^2}\right\}$$

Clearly, when $\omega_1 T_{1S} \to 0$, this expression reduces to the expression for T_2 due to scalar relaxation (ref. 1, p. 311). By measuring $T_{1\rho}$ as a function of ω_1 under conditions where $\omega_1 T_{1S} \gg 1$ and $(\omega_I - \omega_S)T_{2S} \gg 1$, values for A and T_{1S} are readily obtained. Strange and Morgan have used this procedure to determine values for A in $H^{35}Cl$ (41 Hz), $H^{79}Br$ (57 Hz), and $P^{35}Cl_3$ (120 Hz). The application of the $T_{1\rho}$ method to the indirect determination of scalar coupling constants and fast relaxation times is discussed in some detail.

Relaxation in glasses and viscous liquids. Molecular motions in glasses and viscous liquids are complex and their experimental study is difficult. Measurement of T_1 as a function of temperature and, in particular, frequency provides a useful way to investigate this problem, but the method has not yet been widely used. Measurements[59] of the temperature- and frequency-(4—32 MHz) dependence of the proton relaxation times in raw polyisoprene and polybutadiene in the amorphous state, above the glass transition temperature, show that in the limit $\omega_0 \tau_c \gg 1$, $1/T_1 \propto \omega_0^{-1.3} \tau_c^{-0.35}$ whereas in the limit $\omega_0 \tau_c \ll 1$, $1/T_1 \propto \omega_0^{-0.5} \tau_c^{0.5}$. The experimental data are not explicable by intramolecular dipolar relaxation and a Cole–Cole distribution of correlation times. Lenk[60] points out that the frequency dependence observed on the low-temperature side of the T_1 minimum in these systems suggests that relaxation is by modulation of the intermolecular dipolar interaction by a stochastic isotropic translation of the molecules, for which $1/T_1 \propto \omega_0^{-3/2} \tau_D^{-1/2}$. An alternative, and

[58] I. Solomon, *Compt. rend.*, 1959, **249**, 1631; 1959, **248**, 92.
[59] R. Lenk and J. P. C. Addad, *Solid State Comm.*, 1970, **8**, 1869.
[60] R. Lenk, *J. Phys. (C)*, 1971, **4**, L21.

equally plausible, interpretation is given by Glarum's 'defect'-diffusion model,[61] which corresponds to a non-exponential correlation function for molecular reorientation: this model predicts $1/T_1 \propto \omega_0^{-3/2} \tau_c^{-1/2}$ in the limit $\omega_0 \tau_c \gg 1$ and $1/T_1 \propto \omega_0^{-1/2} \tau_c^{1/2}$ when $\omega_0 \tau_c \ll 1$. The striking agreement between the experimental data and this model suggests that relaxation is by configurational fluctuations of the molecules at 'defect' sites.

It is not always realized that the complex time-dependence of the intermolecular dipolar interaction in liquids, due to the relative diffusion of the molecules, gives rise to a frequency-dependent relaxation rate. This arises because, whereas for nearest neighbours $\omega_0 \tau_D \ll 1$, the interactions between spins further removed will, because of their longer correlation times, be relatively more efficient and produce an appreciable frequency dependence in T_1. In the low-frequency limit, the intermolecular relaxation rate is approximated by:[62]

$$1/T_1 = A - B\omega_0^{1/2}/D^{3/2}$$

where A and B are known constants and $D = \langle r^2 \rangle / 6\tau_D$. Such a frequency dependence has indeed been observed in low-temperature liquid ethane.[62] Harmon[63] now demonstrates that the frequency dependence of T_1 observed in glycerol at frequencies below 6 MHz and at 296 K is due to this intermolecular interaction. The plot of $1/T_1$ vs. $\omega_0^{1/2}$ is linear, and from the slope he obtains $D = (1.9 \pm 0.1) \times 10^{-12}$ m² s⁻¹, a value which compares favourably with that of $(2.0 \pm 0.2) \times 10^{-12}$ m² s⁻¹ obtained directly by the pulsed-gradient spin-echo method.[64]

To study the molecular reorientation in liquid glycerol, Drake and Meister[65] have measured T_1^D in [²H₈]glycerol at 5 MHz between 248 and 343 K. The deuteron relaxation is controlled by the intramolecular quadrupole interaction, and the general behaviour of the relaxation times as a function of temperature agrees with the theory for nuclear quadrupole interaction in liquids (ref. 1, pp. 313-315). The experimental data could not be accounted for by a single correlation time, but were equally well represented by a Cole–Davidson distribution of correlation times with a width parameter $\beta = 0.3$ and a Gaussian distribution with width $b = 0.7$. The value of the deuteron quadrupole coupling constant determined directly from the T_1 minimum (equal to 5.1×10^{-4} s at 259 K) using the Gaussian distribution is 139 kHz, and using the Cole–Davidson distribution, 158 kHz. These values and the measured relaxation times were employed to determine the average correlation times, $\bar{\tau}_2$, for the reorientational motions responsible for the relaxation. Comparison of these values with the ones determined by Noack and Preissing[66] from the proton relaxation times in glycerol shows that both their

[61] B. I. Hunt and J. G. Powles, *Proc. Phys. Soc.*, 1966, **88**, 513.
[62] J. F. Harmon and B. H. Muller, *Phys. Rev.*, 1969, **182**, 400.
[63] J. F. Harmon, *Chem. Phys. Letters*, 1970, **7**, 207.
[64] E. O. Stejskal and J. E. Tanner, *J. Chem. Phys.*, 1965, **42**, 288.
[65] P. W. Drake and R. Meister, *J. Chem. Phys.*, 1971, **54**, 3046.
[66] F. Noack and G. Preissing, *Z. Naturforsch*, 1969, **24a**, 143.

magnitudes and temperature-dependence are considerably different (they exhibit, respectively, non-Arrhenius and Arrhenius temperature-dependences. The contribution to the proton relaxation from the intramolecular dipolar interaction is shown to be negligible, indicating that the proton relaxation is primarily due to diffusional modulation of the intermolecular interactions. A comparison of $\bar{\tau}_2$ with the average dielectric correlation time $\bar{\tau}_1$ yields a ratio $\bar{\tau}_1/\bar{\tau}_2$ between 1·2 and 1·8; it is therefore concluded that molecular reorientation in glycerol is by large-angle jumps. T_2 is also measured: for temperatures above 278 K the ratio of T_1 to T_2 is a constant equal to 2·0, an anomaly which is not explained.

B. Molecular Motion and Association in Mixtures of Liquids.

—Values of τ_2 for solute molecules, usually referred to infinite dilution conditions, are in general more reliably interpreted by the Hill[67] mutual viscosity model than by the Debye–Stokes–Einstein or Gierer–Wirtz hydrodynamic models or Steele's inertial model [$\tau_2 = (\pi I/12kT)^{\frac{1}{2}}$].[68] All these models assume that the molecules in the liquid are only weakly interacting with one another and that any specific interaction leading to structural arrangements or aggregates is absent. We might, however, expect the structural properties of a liquid to be reflected in τ_2. The inter-relation of τ_2, and also D, to the structural properties of both pure liquids and liquid mixtures has been discussed in some detail by Hertz.[69]

Pajak et al.[70] have measured the relaxation times of the ring and methyl protons in toluene as a function of concentration in CCl_4 at 293 K. From the extrapolated values at infinite dilution they calculate $(\tau_2^{ring}/\tau_2^{methyl}) = 2\cdot5$, which is significantly larger than the corresponding value of 1·3 obtained by Pritchard and Richards[71] in CS_2 at 298 K. Using this ratio, the intramolecular rotational and intermolecular translational contributions to the relaxation were separated, enabling values for τ_2 to be derived for the entire composition range. Assuming the molecule to be rigid, they used the Hill[67] mutual viscosity model to calculate correlation times for reorientation about each of the three principal inertial axes of the molecule. Good agreement is obtained, for the entire range of concentrations, between the experimental values of τ_2^{methyl} and values calculated for reorientation of the rigid molecule about the molecular axis coinciding with the C_{3v} axis of the methyl group, and also between values of τ_2^{ring} and the values calculated for reorientation about the two axes perpendicular to the C_{3v} axis. It is therefore concluded that, at least at the temperature of these experiments, the relaxation of the methyl protons is not affected by internal rotation of the methyl group, contrary to what has previously been suggested (ref. 54, p. 200).

[67] N. E. Hill, *Proc. Phys. Soc.*, 1954, **B67**, 149; 1955, **B68**, 209.
[68] W. A. Steele, *J. Chem. Phys.*, 1963, **38**, 2404, 2411; W. Moniz, W. Steele, and J. Dixon, *ibid.*, p. 2418.
[69] H. G. Hertz, *Ber. Bunsengesellschaft Phys. Chem.*, 1970, **74**, 666; E. V. Goldammer and H. G. Hertz, *J. Phys. Chem.*, 1970, **74**, 3734.
[70] Z. Pajak, E. Szczesiak, J. Angerer, and K. Jurga, *Acta Phys. Polon.*, 1970, **A38**, 767.
[71] A. M. Pritchard and R. E. Richards, *Trans. Faraday Soc.*, 1966, **62**, 1388.

Measurements[72] of the concentration dependence of the proton relaxation rate of pyridine in CS_2 and CCl_4 have been interpreted, on the basis of the Debye–Stokes–Einstein model, in terms of molecular association between pyridine and solvent. More realistically, the results probably represent the inapplicability of this model in these liquids.

The proton relaxation times have been determined[73] for chloroethanes (CH_3CH_2Cl, CH_3CHCl_2, CH_2ClCH_2Cl, CH_3CCl_3, $CH_2ClCHCl_2$, CH_2ClCCl_3, $CHCl_2CHCl_2$, and $CHCl_2CCl_3$) at infinite dilution in CS_2 at room temperature. Assuming relaxation is due entirely to intramolecular dipolar interactions, reorientational correlation times were evaluated and were found to increase with the number of chlorine atoms in the molecule. The experimental values of τ_2 are compared with values predicted by the Debye–Stokes–Einstein model and Steele's inertial model. Both models predict the observed trends equally well, but the Debye model overestimated τ_2 by a factor of 10, while the Steele model underestimated it by a factor of 2. Thus, although neither model satisfactorily explains the experimental data, the reorientation of the chloroethanes in non-viscous CS_2 is apparently controlled more by the moments of inertia of the molecules than by the frictional forces of the solvent. Burke and Chan[74] have made a detailed study of the effects of solvent (CS_2, CCl_4, a CS_2–CCl_4 mixture, diethyl ether, CH_3CN, and $CHCl_3$) on T_1^F values in a series of substituted 1,1,1-trifluoroethanes (CF_3CH_2Cl, CF_3CH_2Br, CF_3CH_2I, and CF_3CCl_3) measured at 12·3 MHz and 298 K. The intramolecular relaxation rates obtained at infinite dilution are interpreted in terms of intramolecular dipole–dipole and spin–rotational interactions and assuming the applicability of Hubbard's relation, so that $1/T_1 = C\,\tau_2 + C'\,\tau_2^{-1}$. The experimental data are in all cases found to depend on the nature of the solute–solvent system; Steele's inertial model is, therefore, obviously not applicable. With the sole exception of CF_3CCl_3, the Debye–Stokes–Einstein model does not satisfactorily explain the experimental results; on the other hand, the Hill model provides a good explanation of all the results and is obviously the more relevant model for molecular rotation in liquid mixtures.

A number of investigators have studied molecular motion and structure in aqueous solutions of non-electrolytes. The work of Goldammer and Hertz[69] is by far the most extensive. First, the inter-relation between the experimental quantities τ_2 and D and the structural properties of a liquid are discussed. A set of rules for these quantities are given which must hold if certain structural aggregates are to exist in the liquid. These rules are then applied to experimental data (T_1 and D as a function of composition and temperature) obtained for aqueous mixtures with acetonitrile, pyridine, methanol, ethanol, t-butanol, acetone, and tetrahydrofuran. Three important results are obtained: (i) long-lived rigid hydration cages do not exist for the organic solutes studied. (Here, rigid implies that the structural configuration exists

[72] V. G. Kher and S. G. Modak, *Indian J. Pure Appl. Phys.*, 1970, **8**, 409.
[73] C. R. Miller and S. L. Gordon, *J. Chem. Phys.*, 1970, **53**, 3531.
[74] T. E. Burke and S. I. Chan, *J. Magn. Resonance*, 1970, **3**, 55.

Nuclear Spin Relaxation

longer than the reorientational correlation time of the aggregate, *i.e.* typically $>10^{-11}$—10^{-10} s); (ii) at low concentration of the organic component, there is an increase in the degree of structure in the solution as compared with pure water; (iii) in some mixtures, there is evidence that the organic components, as well as the water, associate preferentially among themselves in their respective low-concentration ranges, leading to a certain degree of microheterogeneity. Reeves and Yue[75] have also studied T_1 in water–acetone mixtures and they have obtained similar results. The dimethyl sulphoxide–water system has been investigated in detail by Packer and Tomlinson[76] *via* T_1, T_2, and D measurements. Strikingly, the rotational correlation times of both components are very similar, suggesting that their motions are not entirely independent. All measurements indicate a minimum in molecular mobility at a mole fraction of water of 0·65. The experimental data are seen to be consistent with the results obtained by Goldammer and Hertz.

Aqueous biopolymer solutions are an important class of aqueous non-electrolyte solutions (see also Chapter 8). However, because of the very restricted temperature interval experimentally accessible, it is necessary to make T_1 measurements over an extended frequency range in order to investigate the complex molecular motions and structure in these solutions. The utility of this approach is nicely demonstrated by the T_1 measurements in solutions of serum albumin and poly-γ-benzyl-L-glutamate by Kimmich and Noack.[77] The frequency dependence of T_1^H in highly concentrated aqueous serum albumin solution shows a 'high-frequency' and a 'low-frequency' dispersion, indicating two relaxation processes with a short and long correlation time, respectively. Since the free induction decay signal shows both 'solid' and 'liquid' type components for the protein protons and the T_1 values in D_2O solution show the same relaxation behaviour as the related H_2O solution, it is concluded that all effective relaxation mechanisms take place inside the protein molecule. The protons in the rather rigid and therefore slowly moving regions of the molecule are considered to act as the relaxation centre for all protons in the low-frequency region, while the relatively mobile side-chains act as the relaxation centre in the high-frequency region. As there is no diminution in T_1 in D_2O solution as compared with H_2O solution, the relaxation rates are clearly controlled by the spin diffusion process. In poly-γ-benzyl-L-glutamate solutions remarkable differences in relaxation behaviour are obtained for the helix and random-coil conformations; these are explicable in terms of spin diffusion between the polypeptide and water protons. In the flexible random-coil conformation a non-exponential T_1 decay is observed, showing that the spin-diffusion rate is slower than the relaxation rates. In contrast, simple exponential T_1 relaxation is observed in the relatively rigid helix conformation consistent with the expected faster spin diffusion.

[75] L. W. Reeves and C. P. Yue, *Canad. J. Chem.*, 1970, **48**, 3307.
[76] K. J. Packer and D. J. Tomlinson, *Trans. Faraday Soc.*, 1971, **67**, 1302.
[77] R. Kimmich and F. Noack, *Z. Naturforsch.*, 1970, **25a**, 1680; *Ber. Bunsengesellschaft Phys. Chem.*, 1971, **75**, 269.

T_1 measurements have also been used[78] to study molecular association through inter-amide hydrogen-bonding in N-methylacetamide (NMA) and NN-dimethylacetamide (DMA) in D_2O and CCl_4 solutions. In D_2O solution at 298 K, $\tau_2 = 2 \cdot 5 \times 10^{-12}$ s for the C-methyl and the N-methyl groups in both NMA and DMA and is characteristic of 'free molecule' rotation. On the other hand, in CCl_4 solution, the molecular association of NMA is strong enough to appreciably affect T_1; $\tau_2 = 1 \cdot 4 \times 10^{-11}$ s and so the lifetime of a molecule in a hydrogen-bonded aggregate must be longer than 10^{-11} s. For DMA, association effects are much weaker and only detectable at high concentrations.

Wang and White[79] have described how T_1 measurements may be used for the accurate ($\approx 0 \cdot 2\%$) analysis of the ortho–para content in liquid D_2 samples. Their method depends upon the fact that the ortho- and para-deuterium relaxation times differ by something like three orders of magnitude. The longer T_1 is for the ortho ($I = 2, J = 0$) molecules and is a measure of the cross-relaxation time between the two spin systems which are nearly decoupled in the liquid; the shorter T_1 occurs for the para ($I = 1, J = 1$) molecules since they are in rotationally excited states.

C. **Electrolyte Solutions.**—*Diamagnetic Solutions.* The interpretation of spin relaxation data for nuclei in polar solvent molecules in electrolyte solutions requires a model for molecular reorientation in the electric field of the ion. O'Reilly[80] has obtained a formal solution of the Debye rotational diffusion equation for a polar molecule in an electric field[25] and applied it to relaxation by intramolecular dipolar or quadrupolar interactions. It is found that, in general, the rotational autocorrelation function decays more rapidly owing to the field, the dipolar or quadrupolar interaction is not averaged completely by the rotational diffusion, and translational diffusion or rotation of the molecule about the ion is required to average the nuclear interaction to zero. O'Reilly,[81] together with Peterson, has also investigated the relationship between τ_2, D, and viscosity in solutions of electrolytes. Values for D and τ_2 were measured for the solvent in $NaNO_3$ solution in liquid ammonia by spin-echo and T_1^D measurements, respectively. τ_2 is only weakly dependent on concentration, and at low concentrations (< 1 mol l^{-1}), D obeys the Stokes–Einstein relation ($D = kT/6\pi a \eta$). Corresponding measurements in aqueous solutions of CsI show that the Stokes–Einstein relationship is reasonably valid, but, in contrast to the behaviour observed in ammonia, τ_2 varies markedly with concentration. The linear concentration dependence of viscosity (B coefficient) in this and seven other aqueous electrolyte solutions studied is shown to correlate well with the concentration dependence of τ_2 as given by both the quasi-lattice random-flight model and the Debye–Stokes–Einstein relation. The difference in the behaviour of τ_2 in aqueous and ammonia solu-

---

[78] K. Sato and A. Nishioka, *Bull. Chem. Soc. Japan*, 1971, **44**, 52.
[79] R. Wang and D. White, *Rev. Sci. Instr.*, 1971, **45**, 887.
[80] D. E. O'Reilly, *J. Phys. Chem.*, 1970, **74**, 3277.
[81] D. E. O'Reilly and E. M. Peterson, *J. Phys. Chem.*, 1970, **74**, 3280.

Nuclear Spin Relaxation

tions is explained in terms of the unusual structural properties of liquid water.

Relaxation studies of quadrupolar nuclei provide a convenient way of obtaining rather detailed information about the solvation of ions in solution. The principles behind the method have been outlined in a recent review article by Deverell.[82] Nakamura and Kawamura[83] report an investigation of the hydration co-ordination numbers of the La^{3+} ion in aqueous solutions of halides, nitrate, sulphate, and perchlorate by measurement of T_1 values of ^{139}La. The relaxation rate at infinite dilution is shown to be determined by rapid exchange between symmetrical $La(H_2O)_8^{3+}$ and unsymmetrical $La(H_2O)_9^{3+}$ species. Lindblom et al.[84] have measured T_1 values of ^{81}Br in the isotropic solution phase extending from the hexanol corner in the three-component system cetyltrimethylammonium bromide–n-hexanol–water as a function of temperature, frequency, and composition. It is shown that, except for high hexanol concentrations, the bromide ions are located in the water cores of reversed micelles. At the highest hexanol concentrations, the relaxation rate is very fast owing to ion pair formation.

Paramagnetic Solutions. The nuclear spin relaxation rates of protons in the hydration shell of paramagnetic ions in aqueous solution are usually controlled by the dipole–dipole and scalar interactions with the unpaired electrons of the paramagnetic ion (ref. 1, Chap. 8). The correlation time for the scalar interaction can often be identified with the electronic relaxation times T_{1e} and T_{2e}. The usual expressions[86] for the nuclear T_1 and T_2 due to this mechanism assume a single T_{1e} and T_{2e}. However, in a paper by Rubenstein, Baram, and Luz,[85] it is pointed out that electronic relaxation of $S = 3/2$ and $S = 5/2$ ions (e.g. Cr^{3+}, Mn^{2+}, Fe^{3+}) in aqueous solution is controlled by modulation of the quadratic zero-field splitting interaction by collisions of the hydrated complex with bulk solvent molecules and for relaxation by this mechanism there is more than one relaxation time involved when $\omega_e \tau_e \gg 1$, but when $\omega_e \tau_e \ll 1$ only one T_{1e} and T_{2e} is involved and $T_{1e} = T_{2e}$. Theoretical expressions are given for T_{1e} and T_{2e} and the usual equations for T_1 and T_2 are modified to take into account the existence of several values of T_{1e} and of T_{2e}. These equations were used to calculate the scalar relaxation rates for the protons in the hydration shells of Cr^{3+}, Fe^{3+}, and Mn^{2+} ions using data obtained from e.s.r. measurements. Comparison of the results with experimental n.m.r. relaxation data shows that good agreement is obtained for Cr^{3+} and Fe^{3+}, but only moderate agreement for Mn^{2+}. This theoretical development is a considerable improvement over the earlier Bloembergen and Morgan[86] treatment and should be important for the interpretation of nuclear T_1 and T_2 data in solutions of paramagnetic ions when relaxation is controlled by the electron-spin relaxation times.

[82] C. Deverell, *Progr. N.M.R. Spectroscopy*, 1969, **4**, 235.
[83] K. Nakamura and K. Kawamura, *Bull. Chem. Soc. Japan*, 1971, **44**, 330.
[84] G. Lindblom, B. Lindman, and L. Mandell, *J. Colloid Interface Sci.*, 1970, **34**, 262.
[85] M. Rubenstein, A. Baram, and Z. Luz, *Mol. Phys.*, 1971, **20**, 67.
[86] N. Bloembergen and L. O. Morgan, *J. Chem. Phys.*, 1961, **34**, 842.

Spin-relaxation methods provide a convenient tool for investigating the structure of solvated paramagnetic ions in solution and the exchange between free and co-ordinated solvent molecules. The study by Richards and co-workers[87] of dilute solutions of Ni^{2+} in CH_3CN nicely demonstrates the use of these methods and the wealth of information obtainable. By measurement of T^H and T^H (using the large pulse spacing limit in a Carr–Purcell spin-echo experiment), values were obtained for the exchange rate, the electron-spin relaxation time, the rotational correlation time of the $Ni-CH_3CN$ complex and their respective activation energies; values for the proton–electron hyperfine coupling constant and the distance of closest approach of the protons to the nickel are also obtained. A feature of particular interest in their work is the first reported application of the Carr–Purcell experiment to the study of chemical exchange between free and co-ordinated solvent molecules; this is possible whenever the exchange rate is comparable with the chemical shift difference between the two sites since the echo decay rate is then a function of the pulse-repetition frequency (see Section 4c). In addition to the exchange rate, the number of exchanging co-ordinated molecules is obtained. Using this technique, it was found that there are only four exchanging CH_3CN molecules per nickel ion. This result appears contradictory to the electronic spectra of these solutions, which are strongly indicative of octahedral co-ordination to Ni^{2+}. The results of ^{14}N linewidth measurements[88] are also in accordance with there being four equivalent exchanging molecules and also indicate that the two other co-ordinated molecules are exchanging faster still. The evidence is then that Ni^{II} forms a distorted octahedron with methyl cyanide. As a check, the Carr–Purcell experiment was used to determine the number of exchanging co-ordinated solvent molecules in a dilute solution of Co^{2+} in CH_3OD. It was found to be six, in agreement with previous results obtained by high-resolution n.m.r. under conditions where both bound and unbound resonances could be observed. The expression for the echo decay rate[89] as a function of exchange rate, chemical shift, and pulse-repetition frequency has been investigated numerically.

The wealth of detailed information obtainable by relaxation measurements of spins in the solvation shell of paramagnetic ions in solution has led to the use of paramagnetic ions as probes for the investigation of active sites in metalloenzymes and is potentially capable of yielding unique information regarding configuration dynamics at this site. Two excellent review articles have been published[90] which describe the principles and applications of this technique. The interpretation of experimental data in terms of the various relaxation mechanisms is not always a simple problem; here frequency

---

[87] I. D. Campbell, J. P. Carver, R. A. Dwek, A. J. Nummelin, and R. E. Richards, *Mol. Phys.*, 1971, **20**, 913; I. D. Campbell, D. E. Nixon, and R. E. Richards, *ibid.*, p. 923.

[88] I. D. Campbell, R. A. Dwek, R. E. Richards, and M. N. Wiseman, *Mol. Phys.*, 1971, **20**, 933.

[89] J. P. Carver, *J. Amer. Chem. Soc.* (to be published).

[90] A. S. Mildvan and M. Cohen, *Adv. Enzymol.*, 1970, **31**, 1; M. Cohen and J. Reuben, *Accounts Current Res.*, 1971, **4**, 214.

dependence as well as temperature dependence of T_1 and T_2 is useful. Several groups of workers[91-93] have demonstrated how measurements of T_1 as a function of frequency can be used to correctly interpret the correlation time for the nuclear–electron dipolar interaction; this is possible since the electron relaxation time is the only frequency-dependent one. Using this procedure, it has been found that the number of exchanging water molecules co-ordinated to Mn^{II} bound to carboxypeptidase A is one,[91] to transfer-RNA two,[92] and contradictory values of two[91] and three[93] have been reported for pyruvate kinase. A computer procedure, more reliable than the previously used graphical methods, has been described[94] for interpreting the relaxation data for ternary metal–substrate–enzyme complexes: this is a complicated problem when the bound ion is readily dissociable since there are then four metal-ion-containing species in solution. It is important to note that, thus far, all experimental data obtained in these systems have been interpreted using existing theories for relaxation in solutions of small paramagnetic ions, without experimental or theoretical justification for their validity.

D. Studies of Critical Phenomena in Liquids.—So far, very few n.m.r. investigations of critical phenomena in binary liquid mixtures have been reported.[95] In an attempt to study the local order which is known to exist in these systems in the vicinity of the critical solution temperature, Anderson and Gerritz[96] measured proton chemical shifts and T_1^H(inter) values as a function of composition in the cyclohexane–aniline system at a temperature about 4 degrees above $T_c = 302.6$ K. Their results are at best qualitatively consistent with the existence of local order involving aniline–aniline and cyclohexane–cyclohexane interactions. Self-diffusion coefficients were also measured using the spin-echo method and indicate that the basic translational process involves the motion of individual molecules; furthermore, there is no indication that molecular self-diffusion is affected to any observable extent by closeness to the critical temperature; similar results were obtained by Neronov and Drabkin[95] for the triethylamine–water system. The absence of any discernible change in D at the critical temperature in these studies might be due to inadequate temperature control; this point is demonstrated by the work of Arata and Fukumi.[97] These workers show that the plot of linewidth of the water resonance ($= 1/\pi T_1$) versus temperature for a critical mixture of 2,6-lutidine and water contaminated with $\approx 0.2\%$ of tris(acetylacetonato)chromium(III), exhibits a sharp anomaly in the vicinity of the critical temperature — the width of this anomaly is about 4 mK. This behaviour probably reflects the anomaly in the mutual diffusion coefficient. The chemical shift between the

[91] S. Navon, *Chem. Phys. Letters*, 1970, **7**, 390.
[92] A. Danchin and M. Gueron, *J. Chem. Phys.*, 1970, **53**, 3599.
[93] J. Reubin and M. Cohen, *J. Biol. Chem.*, 1970, **245**, 6539.
[94] G. H. Reed, M. Cohen, and W. J. O'Sullivan, *J. Biol. Chem.*, 1970, **245**, 6547.
[95] Yu. I. Neranov and G. M. Drabkin, *Zhur. fiz. Khim.*, 1965, **39**, 2691; J. G. Powles and Z. Pajak, *Mol. Phys.*, 1963–64, **7**, 579.
[96] J. E. Anderson and W. H. Gerritz, *J. Chem. Phys.*, 1970, **53**, 2584.
[97] Y. Arata and T. Fukumi, *Mol. Phys.*, 1970, **19**, 135.

water and methyl resonances shows similar effects. Thus, it appears that T_1 shows a critical behaviour in the immediate vicinity of T_c in binary liquid mixtures.

Kawasak[98] has suggested that, although the self-diffusion coefficient of a fluid remains finite at the critical temperature T_c, $D\rho$ could have a positive infinite slope as T approaches T_c along the liquid–vapour coexistence curve. Noble and Bloom[99] have obtained results consistent with this suggestion in ethane, but they unfortunately had to add oxygen to reduce the T_1. Attempts to observe this anomalous behaviour in methane[100] and xenon[101] have been unsuccessful. Tison and Hunt[102] have made a very careful search for this critical phenomenon in SF_6. They have measured D and T_1^F in a very pure sample along the liquid–vapour coexistence curve in the vicinity of the critical point. The density ρ was also measured in the same samples, using the amplitude of the free induction decay signal as a measure of density. Samples were contained in critically filled double-bulbed tubes so that measurements could be taken separately on each fluid phase along the coexistence curve; temperature stability was better than 1 mK. No anomalous behaviour was found in the critical region. The ^{19}F spin–rotational interaction is, as expected, the dominant relaxation mechanism.

E. ^{13}C **Relaxation Studies.**—In recent years numerous studies of ^{13}C n.m.r. spectra have been published and extensive data on coupling constants and chemical shifts accumulated, but, in contrast, studies of T_1^C and T_2^C have been relatively rare. A knowledge of T_1^C and T_2^C and their mechanisms is important both as a tool for probing molecular dynamics in liquids and also for spectroscopic purposes. In view of the current interest in ^{13}C n.m.r. spectroscopy we have collected together the results of ^{13}C relaxation studies.

Hitherto, the interactions controlling relaxation of ^{13}C have not been characterized; here we report a number of studies which have had this objective. Jaeckle et al.[103] report results of T_1^C measurements in a variety of liquids, including benzene, derivatives of benzene, saturated rings, and CS_2, obtained at 62 MHz using a superconducting solenoid and a digital signal averager; the apparatus is described. Intramolecular dipole–dipole interactions are found to provide by far the most important spin–lattice relaxation mechanism for protonated carbon atoms, even when the samples contain dissolved oxygen; also, intermolecular dipolar interactions are seen to be relatively unimportant. Clearly, T_1^C data should be easier to interpret than T_1^H data. The anisotropic chemical shielding interaction does not appear to be important, even in CS_2. Spin–rotational relaxation is only important for $^{13}CH_3$ and arises from the spin–internal-rotation interaction. Gillen et al.[51] have also

[98] K. Kawasaki, *Phys. Rev.*, 1966, **150**, 285.
[99] J. D. Noble and M. Bloom, *Phys. Rev. Letters*, 1965, **14**, 250.
[100] N. J. Trappeniers and P. Oosting, *Phys. Letters*, 1966, **23**, 445.
[101] R. S. Ehrlich and H. Y. Carr, *Phys. Rev. Letters*, 1970, **25**, 314.
[102] J. K. Tison and E. R. Hunt, *J. Chem. Phys.*, 1971, **54**, 1526.
[103] H. Jaeckle, U. Haeberlen, and D. Schweitzer, *J. Magn. Resonance*, 1971, **4**, 198.

demonstrated that the methyl-group internal-rotation is the principal source of the ^{13}C spin–rotational interaction in ^{13}CH$_3$I. This result has important implications for ^{13}C nuclear Overhauser effects (n.o.e.); it implies that there will be a reduction in the enhancements of ^{13}C nuclei in internally rotating methyl groups. Indeed, recent n.o.e. measurements by Lyerla et al.[104] in ^{13}CH$_3$I, ^{13}CH$_3$CN, and ^{13}CH$_3$OH confirm this conclusion. These authors also demonstrate how, since the n.o.e. is solely dependent upon the dipolar relaxation mechanism, the consideration of ^{13}C–{^{1}H} n.o.e. in conjunction with T_1^C data enables the dipolar contribution to ^{13}C relaxation to be separated from the other interactions. An alternative procedure for the determination of the ^{13}C–^{1}H dipolar relaxation rate has been described by Briguet et al.[105] for ^{13}CHCl$_3$ in natural abundance in chloroform. It involves studying the relaxation, by the saturation-recovery method, of the proton lines of ^{13}CHCl$_3$ (considered as an AX system) and T_1^H for ^{12}CHCl$_3$ in the same sample.

Since the relaxation of protonated carbons in liquids is usually dominated by dipolar interaction with the attached protons, the ^{13}C relaxation rates will be simply related to the number of directly bonded hydrogen atoms and the reorientational correlation time for this interaction. In simple molecules this will correspond to that for the molecule as a whole, but in complex molecules, where internal motions of functional groups and side-chains are possible, it may be shorter and will then be a sensitive function of the location of the carbon atom in the molecule. Clearly, measurements of T_1^C could be used diagnostically in the assignment of resonances in proton-decoupled ^{13}C spectra. T_1^C values of individual resonances are conveniently measured by the 'partially relaxed Fourier transform' (PRFT) technique,[106] which was first described by Vold et al.[107] Allerhand and his co-workers[106] have demonstrated the utility of this procedure not only for the assignment of resonances in complex molecules of biological interest, but also as a tool for studying the conformation and segmental motion of biopolymer molecules in solution. Freeman and Hill[108] have also discussed the utility of this technique.

Shoup and VanderHart[109] have shown how T_2^C, measured by the Carr–Purcell spin-echo method, is substantially shorter than T_1^C in ^{13}CH$_3$I (3·9 and 13·4 s, at 298 K) and ^{13}CH$_3$CO$_2$CD$_3$ (6·1 and 19·2 s, at 298 K), while in ^{13}CS$_2$, $T_2^C \approx T_1^C$. $T_2^C \ll T_1^C$ will occur whenever relaxing spin-$\frac{1}{2}$ nuclei, e.g. protons, are coupled to ^{13}C such that $A \equiv 2\pi J \gg 1/T_1^H$. This scalar contribution to $1/T_2^C$ arises from the relaxation of the proton spins, which causes an irreversible dephasing of the precessing ^{13}C spins and a consequent reduction in T_2^C. Clearly, $1/T_2^C$ will be related to $1/T_1^H$ but must be greater than the latter. It follows therefore that the resolution in ^{13}C spectra cannot be any better

[104] J. R. Lyerla, D. M. Grant, and R. K. Harris, *J. Phys. Chem.*, 1971, **75**, 585.
[105] A. Briguet, J. C. Duplan, and A. Erbeia, *J. Phys. (Paris)*, 1971, **32**, 23.
[106] A. Allerhand, D. Doddrell, V. Glushko, D. W. Cochran, E. Wenkert, P. J. Lawson, and F. Gurd, *J. Amer. Chem. Soc.*, 1971, **93**, 544; A. Allerhand and D. Doddrell, *ibid.*, pp. 2777, 2779.
[107] R. L. Vold, J. S. Waugh, M. P. Klein, and D. E. Phelps, *J. Chem. Phys.*, 1968, **48**, 3831.
[108] R. Freeman and H. D. W. Hill, *J. Chem. Phys.*, 1970, **53**, 4103.
[109] R. R. Shoup and D. L. VanderHart, *J. Amer. Chem. Soc.*, 1971, **93**, 2053.

than in the corresponding proton spectra. An important consequence of this behaviour is that the advantage predicted by the use of refocussing schemes such as driven-equilibrium and spin-echo Fourier transform (DEFT and SEFT) techniques[110] will be greatly reduced, since the promised sensitivity enhancement was based on the assumption that $T_2^c \approx T_1^c$. However, for proton-decoupled spectra at decoupling r.f. power levels such that $A \ll$ (the reciprocal of the lifetime of the proton spin state), $T_2 \to T_1^c$ and the refocusing methods might still be useful.

F. Methods for Selective Measurement of T_1 and T_2.

Although the importance of selective measurement of T_1 and T_2 for the individual resonances in a high-resolution n.m.r. spectrum has long been recognised, extensive studies have not been undertaken to date, no doubt owing to lack of suitable experimental techniques. In this section we review developments in techniques for this purpose.

Improvements of the adiabatic fast-passage experiment for the selective measurement of T_1 and T_2 of chemically shifted resonances have been described.[111] This method is only suitable for well-separated resonances, i.e. $\geqslant 1$ p.p.m.; the PRFT method referred to in Section 3E is less restricted. In this method, the free precession signal following a 180°—τ—90° pulse sequence is digitally recorded and then Fourier-transformed to yield a partially relaxed frequency-domain spectrum in which signal amplitudes are given by $A_i(\tau) = A_i(\infty) [1 - 2 \exp(-\tau/T_1^i)]$. Values of T_1 are determined by repeating the experiment as a function of τ. Freeman and Hill[112] have described an alternative Fourier transform method based on a classical pulse experiment for measuring T_1 in solid samples. A sequence of equally spaced 90° pulses is applied to the sample and the 'steady state' free precession signal, established after only a few pulses, is digitally recorded and then Fourier-transformed to give a frequency spectrum in which $A_i(\tau) = A_i(\infty)[1 - \exp(-\tau/T_1^i)]$. The condition $T_2 < \tau < T_1$ must be fulfilled. Since in liquids $T_2 \approx T_1$, an artificial method must be employed to destroy the phase coherence of the precessing spins; possible ways for accomplishing this are discussed. It is demonstrated that, for proton-noise-decoupled ^{13}C resonances, this is conveniently effected by residual proton–carbon splittings which act as a line-broadening mechanism. This difficulty is not encountered when a 180°—τ—90° pulse sequence is employed. On the other hand, one advantage the Freeman–Hill[112] method has over the latter is that it is not necessary to wait $5 \times T_1(\text{max})$ between successive applications of the pulse sequence, eliminating the need to estimate $T_1(\text{max})$ prior to setting up the experiment; but, note, its resolving power is limited to $1/\tau$ Hz. This 'progressive-saturation' Fourier transform (PSFT) method is particularly convenient when extensive signal averaging is required,

---

[110] E. D. Becker, J. A. Ferretti, and T. C. Farrar, *J. Amer. Chem. Soc.*, 1969, **91**, 7784; A. Allerhand, *ibid.*, 1970, **92**, 4482; J. S. Waugh, *J. Mol. Spectroscopy*, 1970, **35**, 298.
[111] Z. Pajak, K. Jurga, and J. Jurga, *Acta Phys. Polon.*, 1970, **A38**, 263; H. Lütje, *Z. Naturforsch.*, 1970, **25a**, 1764.
[112] R. Freeman and H. D. W. Hill, *J. Chem. Phys.*, 1971, **54**, 3367.

but the sensitivity of the two methods is approximately the same. We suggest that the first method should be referred to as the 'inversion-recovery' Fourier transform (IRFT) method and the two indiscriminately as PRFT methods.

A number of papers have been published which describe applications of the well-known Carr–Purcell spin-echo experiment[113] to the selective measurement of T_2 in multi-line spectra. Muir and Turner[114] have discussed the use of Carr–Purcell experiments employing low-level r.f. pulses ($\tau_{90°}$ = 30 ms) for the selective measurement of T_2 for chemically shifted proton resonances. The inherent limitation of this technique is that relaxation effects, including spin diffusion in r.f. field gradients, may occur during the long pulses and, of course, resolution is limited. A more promising approach has been described by several groups of workers[115-117] and involves taking the Fourier transform of the spin-echo envelope obtained using conventional strong r.f. pulses. A coupled homonuclear spin system yields a spin-echo envelope which is, in general, modulated in a manner determined by the spectral parameters and the pulse-repetition frequency.[134] The Fourier transform of this envelope yields a multi-line spectrum in the frequency domain, the 'spin-echo spectrum'. This technique has been discussed theoretically by Vold and Chan[115] and demonstrated experimentally by Freeman and Hill[116] and by Stempfle and Hoffman.[117] Linewidths are independent of static field inhomogeneity and other instrumental instabilities, being determined by the relevant values of T_2. Line frequencies are determined by the modulation components of the spin-echo decay. In the case of 'first-order' spin coupling and in the small pulse-repetition-frequency limit, these frequencies correspond to one-half the sums and differences of the spin coupling constants, *i.e.* it is the equivalent of the continuous wave (c.w.) spectrum with the chemical shifts dropped out. Freeman and Hill[116] have called this special case of a spin-echo spectrum a '*J*-spectrum' and have shown how a full assignment may be made by use of 'partial *J*-spectra' which correspond to the Fourier transforms of the spin-echo responses appropriate to single groups of chemically shifted nuclei and observed selectively by tuning the detector to each chemical shift in turn and using a narrow-band filter. This procedure enables otherwise unresolvable small coupling constants to be measured. Of course, when the spin-echo envelope is not modulated, the spin-echo spectrum is a single line, and selectivity is lost. Vold and Chan[115] have suggested that if the pulse train is stopped part way down the echo decay, Fourier transformation of the subsequent free precession signal yields a multi-line spectrum, and measurement of the intensities of these lines as a function of pulse-train duration yields the desired T_2 values. In this case, it might be simpler to extract each resonance in turn from the spin-echo decay by narrow-band filtration, as suggested by Freeman and Hill. Vold and Chan also discuss how selective r.f. pulse experi-

[113] H. Y. Carr and E. M. Purcell, *Phys. Rev.*, 1954, **94**, 630.
[114] A. R. Muir and D. W. Turner, *Chem. Comm.*, 1971, 286.
[115] R. L. Vold and S. O. Chan, *J. Magn. Resonance*, 1971, **4**, 208.
[116] R. Freeman and H. D. W. Hill, *J. Chem. Phys.*, 1971, **54**, 301.
[117] W. Stempfle and E. G. Hoffman, *Z. Naturforsch.*, 1970, **25a**, 2000.

ments may be used to distinguish between relaxation due to intra- and intermolecular dipolar interactions in two spin systems.

The potential, in the application of the Fourier transform techniques described above for selective measurement of T_1 and T_2 of the individual lines in high-resolution n.m.r. spectra, is very promising.

4 Spin-echo Experiments in Liquids

A. Self-diffusion Measurements.—The spin-echo technique using either continuous[118] or pulsed[119] magnetic field gradients provides a more convenient experimental method for the determination of D in fluids than the older classical methods,[120] but its precision is not yet as good, and further development of the method is needed. During the past year, this technique has been increasingly used for measurement of D in liquids and gases, but no further developments in technique have been published. We report here the results of measurements of D in pure liquids not referred to elsewhere.

Sandhu[121] reports measurement, using the continuous-gradient method, of the temperature dependence of D in oxygen-free samples of CH_2Cl_2, CH_2Br_2, and CH_2I_2. Plots of log D vs. $1/T$ are linear and the experimental values of D are in good agreement with values calculated from the significant structure theory of liquids developed by Eyring and co-workers.[122] Less satisfactory agreement is obtained with the Stokes–Einstein and Li–Change ($D = kT/4\pi a\eta$) expression, but the latter is more appropriate than the former. D has also been measured[123] in liquid Na, H_2O, and D_2O by the pulsed-gradient method. At 295 K, $D_{D_2O}/D_{H_2O} = 0.79 \pm 0.02$, which is similar to the value 0·84 for the ratio D_{HDO} (in nearly pure D_2O)/D_{HDO} (in nearly pure H_2O) obtained by measurement of mutual diffusion coefficients in H_2O–D_2O mixtures using optical techniques. This result is consistent with the Cohen–Turnbull diffusion model[124] which predicts that self-diffusion of a trace amount of solute is determined by self-diffusion of solvent molecules.

Both Murday and Cotts[125] and Krüger et al.[125] have used the pulsed-gradient technique to measure D values of ^{6}Li and ^{7}Li in liquid lithium; the former authors critically examine the limitations of the technique for the measurement of D in liquid metals. The objective of these studies was to test the theoretical prediction[126] that in isotopically different classical liquids $D \propto m^{-\frac{1}{2}}$. The experimental ratios (D_6/D_7) measured by these authors for the nearly isotopically pure liquids are 1.18 ± 0.07 at 453 K and 1.09 ± 0.06 at

[118] E. L. Hahn, *Phys. Rev.*, 1950, **80**, 580; H. Y. Carr and E. M. Purcell, *ibid.*, 1954, **94**, 630; H. C. Torrey, *ibid.*, 1956, **104**, 563.
[119] E. O. Stejskal and J. E. Tanner, *J. Chem. Phys.*, 1965, **42**, 288.
[120] P. A. Johnson and A. L. Babb, *Chem. Rev.*, 1956, **56**, 387.
[121] H. S. Sandhu, *Canad. J. Phys.*, 1971, **49**, 1069.
[122] H. Eyring and M. S. Jhon, 'Significant Liquid Structures', Wiley, New York, 1969, p. 88.
[123] J. S. Murday and R. M. Cotts, *J. Chem. Phys.*, 1970, **53**, 4724.
[124] M. H. Cohen and D. Turnbull, *J. Chem. Phys.*, 1959, **31**, 1164.
[125] J. S. Murday and R. M. Cotts, *Z. Naturforsch.*, 1971, **26a**, 85; G. J. Krüger, W. Müller-Warmuth, and A. Klemm, *ibid.*, 1971, **26a**, 94.
[126] R. C. Brown and N. H. March, *Phys. and Chem. Liquids*, 1968, **1**, 191.

463 K; they are close to the square root of the mass ratio $(m_6/m_7)^{\frac{1}{2}} = 1·08$. The dependence of D upon isotopic alloy concentration is found to be relatively weak, in accordance with classical theory, and in this respect is similar to the behaviour recently observed[127] in gaseous H_2–D_2 mixtures. If a Stokes–Einstein-type relationship is applicable, $D\eta$ = const. $\times T$, $\eta \propto m^{\frac{1}{2}}$, and we expect $D_6/D_7 = \eta_7/\eta_6 = 1·08$; however, at the melting point, it is found[128] that $\eta_7/\eta_6 = 1·44 \pm 0·02$.

Measurement of D is frequently useful in separating the contributions to the relaxation rates of spin-$\frac{1}{2}$ nuclei in liquids; this is demonstrated in the study of proton relaxation in liquid benzene by Kosfeld and co-workers.[129] If spin–rotational relaxation is important we would expect D/T_1 to increase with temperature whereas, in fact, in benzene this ratio is observed to decrease over the temperature range 273–373 K, corresponding with the change in viscosity. It is therefore concluded that the spin–rotational interaction has very little influence on T_1^H in this temperature range. From the measured T_1, D, and density, the inter- and intra-molecular contribution to the relaxation are calculated. Kosfeld et al.[130] have also described methods for correcting values of the equilibrium magnetization, as measured by the free precession or spin-echo signal, for losses due to the components of magnetic field instabilities at the Larmor frequency of the nucleus.

Finally, we mention that Hertz[131] has given a brief review of current theories which relate D in liquids to molecular and liquid properties. Also discussed is the possible inter-relation between D and the structural properties of a liquid.

B. Spin-echo Studies of Fluid Flow.—Bulk motion of the sample in the direction of an applied field gradient will affect the spin-echo shape in a way which is dependent upon the dynamic properties of the fluid flow. Deville and Landesman[132] have demonstrated how the spin-echo decay in both single and Carr–Purcell spin-echo experiments differs for laminar and turbulent flow; they also describe the conditions required to check the velocity correlation law in the turbulent flow. Grover and Stringer[133] have also described how the velocity distribution function can be measured and they present experimental data for blood flow in human fingers.

C. Carr–Purcell Spin-echo Experiments.—The Carr–Purcell spin-echo experiment,[113] apart from its conventional use for measurement of transverse relaxation rates in liquids, provides an attractive technique for studying chemical exchange processes in solution.[134] It is somewhat surprising, in view

---

[127] K. P. Müller and A. Klemm, *Z. Naturforsch.*, 1970, **25a**, 243.
[128] N. T. Ban, C. M. Randall, and D. J. Montgomery, *Phys. Rev.*, 1962, **128**, 6.
[129] W. Kietrich, B. Gross, and R. Kosfeld, *Z. Naturforsch.*, 1970, **25a**, 40.
[130] R. Kosfeld and J. Schlegel, *Z. Naturforsch.*, 1970, **25a**, 1743.
[131] H. G. Hertz, *Ber. Bunsengesellschaft Phys. Chem.*, 1971, **75**, 183.
[132] G. Deville and A. Landesman, *J. Phys. (Paris)*, 1971, **32**, 67.
[133] T. Grover and J. R. Stringer, *J. Appl. Phys.*, 1971, **42**, 938.
[134] H. S. Gutowsky, R. L. Vold, and E. J. Wells, *J. Chem. Phys.*, 1965, **43**, 4107; A. Allerhand, *ibid.*, 1966, **44**, 1.

of the importance of such studies, that only one paper has appeared during the year reporting use of this technique. This is by Vold et al.[135] who have used it in conjunction with steady-state lineshape methods to measure protolysis rates in acidic and basic urea solutions from 1—10 molar in urea; good agreement between the two methods is obtained. The model used to analyse the experimental data assumes that the protons undergo exchange between two unequally populated sites, and are scalar coupled in one site to a relaxing nucleus of spin-1. Echo decay rates are analysed by comparison with values calculated by numerical diagonalization of the recursion matrix.[134] Protonation occurs at urea nitrogen rather than oxygen. Below a strength of about 2M urea, the rate constant is independent of urea concentration, but above about 2- or 3-M urea, formation of aggregated species of urea causes the rate constant to depend on urea concentration.

Vold and Chan[136] have suggested that the Carr–Purcell spin-echo experiment can be used, with some attractive advantages over the steady-state method, to study spectra of liquid crystals. This is because, for sufficiently large pulse-repetition rates, the Fourier transform of the echo envelope is the spectrum of an 'average Hamiltonian'

$$\mathcal{H} = \sum_{i<j} \left[D_{ij} I_{zi} I_{zj} + \left(A_{ij} - \frac{1}{3} D_{ij} \right) I_i \cdot I_j \right]$$

where A_{ij} and D_{ij} are the scalar and dipolar coupling constants and the double sum is restricted to the interactions between resonant nuclei only. We note that there are fewer interactions than are observed in the usual steady-state spectrum; field inhomogeneity effects will also be eliminated.

5 Nuclear Spin Relaxation and Diffusion in Liquid Crystals

In view of the current interest in thermotropic and lyotropic liquid crystals and their somewhat unusual physical properties, we have collected together in this section recent relaxation studies in these systems (see also Chapter 10).

A. Thermotropic Liquid Crystals.—Recent measurements[137–142] of T_1^H in nematic liquid crystals show very clearly the unusual nature of relaxation as compared to that observed in normal organic liquids and solids. The most striking observation is the dependence of T_1^H on frequency; it is, over a wide range of ω_0, according to $T_1 = A + B\,\omega_0^{\frac{1}{2}}$. A theory of Pincus,[143] based on

[135] R. L. Vold, E. S. Daniel, and S. O. Chan, *J. Amer. Chem. Soc.*, 1970, **92**, 6771.
[136] R. L. Vold and S. O. Chan, *J. Chem. Phys.*, 1970, **53**, 449.
[137] R. Blinc, D. L. Hogenboom, D. E. O'Reilly, and E. M. Peterson, *Phys. Rev. Letters*, 1969, **23**, 969.
[138] J. W. Deane and J. J. Visintainer, *Phys. Rev. Letters*, 1969, **23**, 1421.
[139] M. Weger and B. Carbane, *J. Phys. (Paris) Colloq.*, 1969, **30**, 72.
[140] R. Y. Dong and C. F. Schwerdtfeger, *Solid State Comm.*, 1970, **8**, 707.
[141] J. J. Visintainer and J. W. Doane, *Mol. Crystals Liquid Crystals*, 1971, **13**, 69.
[142] R. Y. Dong, M. Marusic, and C. F. Schwerdtfeger, *Mol. Crystals Liquid Crystals*, 1970, **8**, 1577.
[143] P. Pincus, *Solid State Comm.*, 1969, **8**, 707.

modulation of the intramolecular dipolar interaction by collective orientational order fluctuations, predicts $T_1 \propto \omega_0^{\frac{1}{2}}$. The frequency dependence predicted is reasonable, but the temperature dependence is not in agreement with experiment. Doane and Johnson,[144] by taking into consideration short-range molecular reorientation as well as collective order fluctuations, have obtained an expression for T_1 similar to the Pincus one, but containing a factor S^2 rather than S. (S is the nematic order parameter.) This theory is in good agreement with the very weak temperature-dependence observed. A similar result is obtained by Lubensky.[145] Theory has, thus far, failed to account for the frequency-independent term A. There are two possible causes for this discrepancy. First, T_1^H measurements are sensitive to those fluctuation modes near their short-wavelength limit; the quadratic dispersion relation which describes these modes would then not be expected to be valid. This cannot, however, be the explanation since $T_{1\rho}^H$ measured in p-methoxybenzylidene-p'-n-butylaniline[141] and in p-azoxyanisole[146] shows little variation with B_1 for values greater than the dipolar linewidth. Secondly, translational diffusion may also relax nuclear spins in the nematic phase.[141,146,147] Indeed, comparison of measured T_1^H and T_1^D in selectively deuteriated liquid crystals[141] shows that intermolecular dipolar interactions are important in proton relaxation. It should, however, be noted that if, as has been suggested, experimental data follow $T_1 = A + B\,\omega_0^{\frac{1}{2}}$, not $T_1^{-1} = A + B\,\omega_0^{-\frac{1}{2}}$, the A term does not simply arise from an additive frequency-independent relaxation mechanism. On the other hand, some authors[146,147] argue that the behaviour of T_1^H is according to $T_1^{-1} = A + B\,\omega_0^{-\frac{1}{2}} + T_1^{-1}{}_{\text{diffusion}}$. The discrepancy between theory and experiment could perhaps be resolved by quadrupole relaxation measurements.

$T^H\rho$ measurements[141] in p-methoxybenzylidene-p'-n-butylaniline at constant B_1 but variable B_0 show that any magnetic field effects on the collective modes do not affect spin–lattice relaxation.

Dong et al.[146] point out that the spectral density functions $J_1(\omega_0)$ and $J_2(2\,\omega_0)$ have a maximum around 800 kHz, implying that, in contrast to normal liquids, the time auto-correlation functions of the order fluctuations decay non-exponentially in nematic phases. In the same paper it is reported that a new low-frequency collective mode, not previously observed, relaxes the spins in the rotating frame at low B_1 values.

Measurements[148] of the ^{14}N linewidth in p-azoxyanisole reveal that the nematic short-range order persists into the isotropic phase.

It has been predicted by Pincus[143] that T_1 should be anisotropic in nematic liquid crystals. By applying an electric field E to change the angle θ_{EB}^0 between B_0 and n_0 (the latter is a unit vector describing the equilibrium

---

[144] J. W. Doane and D. L. Johnson, *Chem. Phys., Letters*, 1970, **6**, 291.
[145] T. C. Lubensky, *Phys. Rev.*, 1970, **A2**, 2497.
[146] R. Y. Dong, W. F. Forbes, and M. M. Pintar, *Solid State Comm.*, 1971, **9**, 151.
[147] R. Y. Dong, *Chem. Phys. Letters*, 1971, **9**, 600.
[148] B. Carbane and W. G. Clark, *Phys. Rev. Letters*, 1970, **25**, 91.

orientation of the molecular axis in the nematic phase), Tarr et al.[149] have shown that the benzene ring proton T_1, measured in p-methoxybenzylidene-p'-cyanoaniline at 383 K, is a function of θ_{EB_0} : T_1 has a minimum at $\theta_{EB_0} = 90°$ and a maximum near the 'magic angle' $\theta_{EB_0} \approx 55°$, where the order parameter is zero. Dong[147] has offered an explanation for this behaviour in terms of collective order fluctuations and translational diffusion.

Several studies have been made using solute molecules to probe the structure of liquid crystals. For example, the anisotropy of molecular diffusion in nematic liquid-crystalline phases is of considerable interest, but its direct measurement by the spin-echo technique is difficult as $T_2^H \approx 10^{-4}$ s in these systems. Murphy and Doane[150] argue that the diffusion of a solute probe molecule should reflect that of the solvent molecules, as is the case for trace amounts of solute in normal liquids. Tetramethylsilane was used as a probe in p-methoxybenzylidene-p'-n-butylaniline: they obtained $D_{\|} = 3·2 \pm 0·3 \times 10^{-9}$ m² s⁻¹ and $D_{\perp} = 0·8 \pm 0·03 \times 10^{-9}$ m² s⁻¹ (the temperature is not specified). The ratio $D_{\|}/D_{\perp} = 4$ is considered reasonable in the light of previous viscosity measurements. Note, D measured[150a] in nematic p-azoxyanisole by the cold neutron scattering method is $0·75 \times 10^{-9}$ m² s⁻¹ and does not depend on temperature. We also expect T_1 and T_2 of solute molecules to reflect the order in nematic liquid crystals. Egozy et al.[151] report measurement of T_1^D and T_2^D in C_6D_6 dissolved in the nematic phase of 4,4'-n-hexylazoxybenzene. Using expressions for T_1 and T_2, calculated in terms of a single correlation time τ_2 for a partially ordered molecule, the authors find $\tau_2 \approx 1 \times 10^{-9}$ s at 348 K. This value is to be compared with the one (5×10^{-11} s) observed in the isotropic phase at 363 K. $T_1^D/T_2^D \approx 4$, demonstrating the preferred average orientation of the benzene molecules.

B. Lyotropic Liquid Crystals.—These systems, which are formed by addition of water to various amphiphilic materials, are of current interest because of their biological associations. So far, research has largely been toward a better understanding of the spin relaxation mechanisms in these systems. McLachlan et al.[152] report T_1^H measurements as a function of temperature and frequency for the micellar liquid-crystalline system comprising 50% D_2O, 36% sodium decyl sulphate, 7% n-decanol, and 7% Na_2SO_4. Surprisingly, there is no apparent change in T_1^H at the phase transition temperatures; similar behaviour has been previously observed by Hansen and Lawson[153] in the system D_2O–dimethyl-dodecylamine oxide. The T_1 data do not appear to be very sensitive to the details of the molecular motion or the structure of the system, nor do fluctuations in long-range order appear to influence relaxation. The results are

---

[149] C. E. Tarr, M. A. Nickerson, and C. W. Smith, *Appl. Phys. Letters*, 1970, **17**, 318.
[150] J. A. Murphy and J. W. Doane, *Mol. Crystals Liquid Crystals*, 1971, **13**, 93.
[150a] R. Blinc and V. Dimic, *Phys. Letters*, 1970, **31A**, 531.
[151] Y. Egozy, A. Lowenstein, and B. L. Silver, *Mol. Phys.*, 1970, **19**, 177.
[152] L. A. McLachlan, D. F. S. Natusch, and R. H. Newman, *J. Magn. Resonance*, 1971, **4**, 358.
[153] J. R. Hansen and K. D. Lawson, *Nature*, 1970, **225**, 542.

Nuclear Spin Relaxation

qualitatively interpreted in terms of a frequency-independent intramolecular dipolar interaction modulated by rotations around the longitudinal chain axis and a frequency-dependent diffusion-modulated interaction. Charvolin and Rigny[154] have shown, by the spin-echo method, that the self-diffusion of potassium laurate molecules in the cubic mesophase of the similar system D_2O–potassium laurate is relatively rapid; $D = 2 \times 10^{-10}$ m² s⁻¹ at 263 K.

Blinc and co-workers[155] have attempted to investigate the flow properties and molecular order in these systems by measuring the temperature dependence of D and T_1 for water in the water channels of both hexagonal 'middle' and lamellar 'neat' sodium palmitate (Napal) liquid crystals. In both cases D increases exponentially with temperature, but decreases abruptly on going from the mesophase to the isotropic liquid. This result is explained in terms of the expression $D = D_{\text{Napal}} p + D_{\text{H}_2\text{O}} (1-p)$, where p is the fraction of time a water molecule is associated with a Napal group. In the isotropic liquid $p \approx 1$ and $D \approx D_{\text{Napal}}$, i.e. diffusion is controlled by the mobility of the Napal groups. On the other hand, in the mesophase $D_{\text{Napal}} \ll D_{\text{H}_2\text{O}}$ and $D \approx (1-p) D_{\text{H}_2\text{O}}$. The abrupt change in D thus reflects a change in p. T_1^H of the water molecules in the mesophase is diffusion controlled; $(T_1/T_2)_{\text{H}_2\text{O}} \approx 2$, demonstrating a preferred orientation of water molecules in the water channels. In contrast, $(T_1/T_2)_{\text{Napal}} \approx 10^2$—$10^4$, demonstrating a much smaller freedom of motion of the hydrocarbon chains compared to the relatively free H_2O molecules.

From the results reported thus far, it is clear that measurements of D can be useful for probing the structure in lyotropic systems, but it is less obvious that T_1^H measurements will be useful, and further work is required to elucidate the spin relaxation mechanisms. The usefulness of non-selective T_1^H measurements is somewhat limited by the presence of many non-equivalent protons in each molecule. On the other hand, there can be little doubt that selective measurement of T_1^C by Fourier transform methods, as outlined in Section 3F, will provide a powerful tool for investigating the structure and molecular dynamics in these complex systems.

6 Spin Relaxation in Solids

A. N.M.R. Saturation in Solids.—Owing to the strong dipolar interactions between nuclear spins, magnetic resonance in solids is a good deal more complicated than in liquids. Before 1962, only two situations could be treated theoretically, namely the cases of very weak or very strong r.f. irradiations of a spin system in a high external magnetic field. The basic theory for the weak r.f. field case was given by Bloembergen, Purcell, and Pound.[156] The interpretation of the behaviour of spin systems in strong r.f. fields was developed by Redfield,[157] who introduced the hypothesis of a spin temperature for the

---

[154] J. Charvolin and P. Rigny, *J. Magn. Resonance*, 1971, **4**, 40.
[155] R. Blinc, K. Easwaran, J. Pirs, M. Wolfan, and I. Zupancic, *Phys. Rev. Letters*, 1970, **25**, 1327.
[156] N. Bloembergen, E. M. Purcell, and R. V. Pound, *Phys. Rev.*, 1948, **73**, 679.
[157] A. G. Redfield, *Phys. Rev.*, 1955, **98**, 1787.

whole rotating-frame Hamiltonian. These theories have been verified by experiment. In 1962, Provotorov introduced a theory to cover the intermediate, partial-saturation region, applicable when B_1 is small compared to the local field. Goldman gives an excellent account of the Provotorov theory in his recently published book.[3] Although this theory is frequently used to explain experimental data for solids, there are many aspects of the theory which have not been experimentally verified. Two papers[158,159] report experimental work concerning the effect of r.f. irradiation on the ^{19}F n.m.r. line in paramagnetic CaF_2 which shows that the behaviour of Zeeman and dipolar energies obeys the theory, at least in the region of extreme narrowing. Similar results were obtained[158] in polycrystalline benzene, in which relaxation is caused by molecular reorientation, in contrast to the previous system where impurity relaxation *via* spin diffusion is dominant. In addition, several new theoretical treatments of n.m.r. saturation in solids have been published.[160]

B. Nuclear Dipole Relaxation.—Jeener and co-workers[161] have developed techniques which enable the spin–lattice relaxation time, T_{1D}, of the dipolar subsystem in a solid to be measured. In diamagnetic solids, the time-independent part of the dipolar interaction sets up the dipolar bath, and the time-dependent part caused by molecular motions provides the spin–lattice relaxation mechanism. T_{1D} is, therefore, related to the molecular motions, but in a way which may differ from the Zeeman relaxation times T_1 and $T_{1\rho}$. In the weak-interaction limit $\tau_c \ll T_2$, van Steenwinkel[162] has shown that:

$$\frac{1}{T_{1D}} = C_1 \tau_c + C_2 \frac{\tau_c}{(1+\omega_o^2 \tau_c^2)} + C_3 \frac{\tau_c}{(1+4\omega_o^2 \tau_c^2)}$$

where the C's are constants containing the strength of the dipolar interactions and the traces of the spin operators; the evaluation of these constants is very difficult. When $\omega_o \tau_c \gg 1$, the first term, corresponding to $m = 0$, will dominate the dipolar relaxation rate. Haeberlen[163] points out that the relaxation steps in the first term always involve three spins, but, as in T_1 processes, the $m = \pm 1$ and ± 2 terms are determined by two-spin systems. He has developed an expression for the $m = 0$ term for a single three-spin system and has used it to calculate T_{1D} due to both intra- and inter-molecular interactions in solid benzene, assuming the molecules make random reorientational jumps about their six-fold symmetry axis. The theoretical results are in close agreement with the measurements of van Steenwinkel.[162] The latter author has published measurements of T_1 and T_{1D} in solid benzene, hexamethylbenzene,

---

[158] J. Haupt and R. van Steenwinkel, *Z. Naturforsch.*, 1971, **26a**, 260.
[159] A. E. Mefed and M. I. Rodak, *Zhur. eksp. teor. Fiz.*, 1970, **59**, 404. [*Sov. Phys. JETP*, 1971, **32**, 220].
[160] T. Shimizu, *J. Phys. Soc. Japan*, 1970, **28**, 811: 1970, **29**, 74; H. Betsuyaku, *ibid.*, 1971, **30**, 641.
[161] J. Jeener, R. Du Bois, and P. Broekart, *Phys. Rev.*, 1965, **139**, A1959; J. Jeener, *Adv. Magn. Resonance*, 1968, **3**, 205.
[162] R. van Steenwinkel, *Z. Naturforsch.*, 1969, **24a**, 1526.
[163] U. Haeberlen, *Z. Naturforsch.*, 1970, **25a**, 1459.

and cyclohexane. A minimum in T_{1D} is observed for τ_c of the order of the rigid lattice linewidth, *i.e.* in the temperature region where the line narrows. Clearly, T_{1D}, like $T_{1\rho}$, is sensitive to low-frequency molecular motions, but the theory is more complicated than for the latter.

C. **Relaxation by Rotational Motion.**—In most cases, relaxation of spin-$\frac{1}{2}$ nuclei in molecular solids is due to the modulation of the dipole–dipole interactions by the rotation of the molecule, and we expect T_1 to be sensitive to the dynamics of the rotational process. In solids, the rotational motion is hindered; the only known example of practically free rotation is in the high-temperature phase of solid H_2. The molecules are considered to undergo coupled librational motion in discrete equilibrium orientations which are determined by the symmetry elements of the crystal lattice. T_1 is not, however, sensitive to these librations, but to the angular displacement between equilibrium orientations. Two classical models are usually assumed for this process when calculating T_1: either fast uncorrelated rotational jumps or slow continuous diffusive motions. Both models give the same expression for T_1; the only rotational parameter involved is the rotational correlation time τ_2, which is related to the jump frequency or the rotational diffusion constant. This theory is found to fit the experimental results in many cases and it would appear that T_1 is insensitive to the rotation mechanism. However, Ivanov[164] has applied his 'random rotational travelling' model to the calculation of the intermolecular T_1 in molecular crystals and shows that the experimental T_1 ought to be sensitive to the number of equilibrium orientations and therefore to the mechanism of rotation. A similar treatment of the intramolecular relaxation rate has been given by Lotfullin and Semin.[165] Clearly, accurate experimental data are required to check these theories; the results should be interesting.

We do not expect the above classical descriptions of the rotation to be reliable at low temperatures where quantum-mechanical rotation may occur. Wallach and Steele[166] report a quantum-mechanical treatment of relaxation for the case of intramolecular dipolar interaction due to independent molecular reorientation about a two-fold symmetry axis and at very low temperatures, where all the torsional transition rates are much smaller than ω_0. The relaxation rate is shown to be directly proportional to the transition rates among the torsional levels and to the extent to which these transitions alter the expectation value of the dipolar Hamiltonian. At very low temperatures, where only transitions between the two lowest torsional levels are important, the apparent activation energy for T_1 should be comparable with the separation of these levels, which is much smaller than the barrier to classical rotation. In this treatment only nuclear-spin-symmetry-retention transitions are taken

---

[164] E. N. Ivanov, *Phys. Stat. Sol.*, 1970, **B42**, 453.
[165] R. Sh. Lotfullin and G. K. Semin, *Doklady Akad. Nauk S.S.S.R.*, 1970, **193**, 1044. [*Sov. Phys. Doklady*, 1971, **15**, 705].
[166] D. Wallach and W. A. Steele, *J. Chem. Phys.*, 1970, **52**, 2534.

into account. In a later paper, however, Wallach[167] considers the role of spin-symmetry-conversion transitions in spin–lattice relaxation for the case of molecules which are able to rotate relatively freely about only one axis in the solid and where the barrier is of the order of 10 kJ mol^{-1}, or more. The symmetry restrictions on the overall wavefunction require that symmetry-conversion transitions must always be accompanied by a simultaneous torsional transition which alters the symmetry of the torsional state. It is shown that for molecules with greater than two-fold symmetry about the rotation axis the intramolecular dipolar interactions can cause only symmetry-conversion nuclear transitions. For molecules with two-fold symmetry, and for which the angle between the rotation axis and the internuclear vector is 90°, the intramolecular interactions cause only symmetry retention transitions. For angles other than 90°, both types of transition are allowed. For the case of intermolecular interactions, both symmetry-retention and -conversion transitions can occur. For a rotational barrier of 10 kJ mol^{-1}, or more, the lowest-lying torsional states of different symmetry are nearly degenerate and it is therefore argued that the rates of symmetry-conversion and symmetry-retention nuclear transitions, caused by either intra- or inter-molecular interactions, should be of comparable magnitude. This explains why a single value for T_1 is always observed in such solids.

D. Spin–Lattice Relaxation by Spin-rotational and Anisotropic Chemical Shielding Interactions.

—Recently, there has been considerable interest in the possibility that the spin–rotational interaction may be important in the relaxation of spin-$\frac{1}{2}$ nuclei in solids as well as in liquids and gases. Evidence is provided by T_1^H measurements in HBr[168] and NH$_4$I[169] and measurements of T_1^F in the plastic crystalline phases of the hexafluorides MF$_6$ (M = S, Se, Te, Mo, and W).[170] However, in none of these studies was it possible to separate the spin–rotational relaxation rate from the other contributions. The assignment is based on the anomalous decrease in T_1 with increasing temperature. Clearly, it is important to compare experimentally derived relaxation rates with values calculated in terms of the various models available for relaxation by this interaction. Sharp and Pintar[171] report an attempt to determine the proton spin–rotational relaxation rate in the cubic phase of NH$_4$I. They assume that the only other contribution to the relaxation is from the intramolecular dipolar interaction; this contribution is calculated using a value for τ_2 estimated from T_1^D measured in ND$_4$I. The derived relaxation rate is frequency-independent, and increases exponentially with temperature with the apparent activation energy 6·3 kJ mol^{-1}, which is to be compared with the

---

[167] D. Wallach, *J. Chem. Phys.*, 1971, **54**, 4044.
[168] M. O. Norris, J. H. Strange, J. G. Powles, K. Marsden, and K. Krynicki, *J. Phys.* (C), 1968, **2**, 422.
[169] M. M. Pintar, A. R. Sharp, and S. Vrscaj, *Phys. Letters*, 1968, **27A**, 1969.
[170] R. Blinc and G. Lahajnar, *Phys. Rev. Letters*, 1967, **19**, 685; P. Rigny and J. Virlet, *J. Chem. Phys.*, 1969, **51**, 3807.
[171] A. R. Sharp and M. M. Pintar, *J. Chem. Phys.*, 1970, **53**, 2428.

value 3·3 kJ mol⁻¹ obtained for $(T_1^D)^{-1}$. This rate was found to be several orders of magnitude greater than values calculated from the rotational-diffusion and rotational-jump models developed for liquids.[28] Either these models are inapplicable to spin–rotational relaxation in solid NH_4I, or the observed relaxation originates from some other interaction. Clearly, further work is required in this area.

Blicharski and Noble[172] have found that T_1^F, measured in solid CF_2Cl_2 at 77 K, shows a sigmoidal dependence on magnetic field strength between $6·0 \times 10^{-3}$ T and 6·4 T. This behaviour is explained in terms of anisotropic chemical shielding and intramolecular dipole–dipole interactions between the two ^{19}F nuclei. Assuming the molecules reorientate by way of 90° reorientational jumps about the C_{2v} axis, the energy difference between the two geometrically non-equivalent equilibrium orientations is 5·9 kJ mol⁻¹. $\sigma_\parallel - \sigma_\perp = 200 \pm 40$ p.p.m., as determined from the ^{19}F second moment.

E. Spin–Lattice Relaxation due to Paramagnetic Impurities.—Relaxation of spin-½ nuclei in non-conducting solids is, when nuclear motion is ineffective, often *via* spin diffusion to paramagentic impurities (ref. 1, pp. 378—389). Bloembergen[173] first showed that there are two steps in the relaxation process, namely diffusion of nuclear Zeeman energy to or from the vicinity of the impurity and the transfer of this energy to or from the lattice by direct interaction of the nuclei and electron spin in the paramagnetic centre. The evolution of the magnetization is governed by the differential equation:

$$\frac{\partial M_z}{\partial t} = D\nabla^2 M_z - (M_z - M_0)\sum_i C_i r_i^{-6}$$

The sum is taken over all paramagnetic impurities, D is the nuclear-spin-diffusion coefficient, r is the distance of the nuclear spin from the impurity, and C is a coefficient that describes the effect of direct relaxation. A numerical solution of this equation has been obtained by Kaplan.[174] Bloembergen introduced the boundary condition that spin diffusion vanishes inside a critical radius r_c. This radius is where the static field of the impurity spin splits adjacent nuclear spin levels by an amount greater than their linewidth, thereby making the spin-diffusion process non-energy-conserving, and causing it to vanish. Horvitz[175] now shows, however, that the fluctuating field created by the impurity spin has components which can induce mutual spin flips between adjacent nuclear spins, thereby producing a spin-diffusion process inside the critical radius. This means that Bloembergen's equation must be evaluated with the boundary condition that there is one spin-diffusion coefficient outside the critical radius and another inside. The importance of including in C electron-pair flips is demonstrated by the experiments of Leifson and Vogel.[176]

[172] J. S. Blicharski and J. D. Noble, *Acta Phys. Polon.*, 1970, **A38**, 295.
[173] N. Bloembergen, *Physica*, 1949, **25**, 386.
[174] J. I. Kaplan, *Phys. Rev.*, 1971, **B3**, 604.
[175] E. P. Horvitz, *Phys. Rev.*, 1971, **B3**, 2868.
[176] O. S. Leifson and E. Vogel, *Phys. Rev.*, 1970, **B2**, 4626.

They have investigated the dependence of the proton relaxation rate in Nd^{3+}-doped $La_2Mg_3(NO_3)_{12},24H_2O$ at 2 K on the angle between B_0 and the crystal symmetry axis; it exhibits a complicated structure which they have shown to be consistent with a nuclear relaxation process in which a proton gives up its Zeeman energy to a pair of electrons such that the Zeeman energy for the three-spin system is conserved; this electron Zeeman energy is then transferred relatively rapidly to the lattice.

Until now, attention has been focused on understanding the impurity relaxation phenomenon itself; however, it ought to be possible to use existing theories to obtain information about the nature of electron traps in solids produced by irradiation. This possibility is demonstrated by the measurements of Lavrencic[177] which show that the influence of impurity concentration, temperature, and frequency on T_1^H in γ-irradiated $NaH_3(SeO_3)_2$ are well described by the theory: $T_1^{-1} \propto N$, $T_1 \propto (\omega_0)^{-\frac{1}{2}}$, implying $\omega_0 \tau_c \gg 1$; τ_c has an activation energy 1·54 kJ mol^{-1}.

F. Spin–Lattice Relaxation by Nuclear Quadrupole Interaction.—First, we consider nuclear quadrupole relaxation as observed in pure nuclear quadrupole resonance experiments. Armstrong and Jeffrey[178] have considered the mechanisms for the quadrupole spin–lattice relaxation of ^{35}Cl nuclei in the $PtCl_6$ complexes K_2PtCl_6, Rb_2PtCl_6, and Cs_2PtCl_6. They calculate the contribution to T_1 due to first-order Raman spin-phonon processes resulting from the internal modes of the $PtCl_6$ octahedra. The values obtained are several orders of magnitude larger than the experimental ones. It is therefore concluded that the internal modes of the $PtCl_6$ complex are relatively unimportant for spin–lattice relaxation of the ^{35}Cl nuclei and that the experimental results primarily reflect the contribution of the rotary lattice mode to the relaxation process. Good agreement is obtained between experimental and calculated rotational mode frequencies. Tzalmona[179] also shows that the ^{14}N nuclear spin–lattice relaxation in CH_3CN is dominated by the rotation of the CH_3 group. T_1 was measured between 77 K and the melting point at 227 K; a minimum of 0·1 s occurs at 88·5 K. The T_1 at this minimum is much too short to be explained in terms of N–H intramolecular dipolar interactions. It is therefore proposed that relaxation is due to the rotation of charge on the methyl hydrogens as a result of hyperconjugation.

Cross-relaxation experiments provide a convenient method for the indirect detection of quadrupole resonances. A novel way of observing cross-relaxation between a spin-$\frac{1}{2}$ and a quadrupolar nucleus is by means of the adiabatic field-cycling procedure, as described by Jones and Daycock.[180] In this experiment the sample is transferred adiabatically from a high field to a low one (a compressed air engine is used to move the sample between the two

---

[177] B. Lavrencic, *Phys. Stat. Sol*, 1971, **A5**, K133.
[178] R. L. Armstrong and K. J. Jeffrey, *Canad. J. Phys.*, 1971, **49**, 49.
[179] T. Tzalmona, *Phys. Letters*, 1971, **34A**, 289.
[180] G. P. Jones and J. T. Daycock, *J. Phys. (C)*, 1971, **4**, 765.

magnetic fields) and any cross-coupling between the magnetic and quadrupolar systems is detected by observing the change in magnetization on return to the high field. Of course, cross-coupling will only considerably affect the observed magnetization provided the cross-relaxation time is shorter than the field-switching time, a condition easily fulfilled experimentally: switching times of the order of 100 ms are used while cross-relaxation times are about 10 to 100 μs. A spin-temperature treatment of this experiment is given and its conclusions are shown to be in agreement with experimental data obtained for imidazole.

G. Spin–Lattice Relaxation Time Measurements in Miscellaneous Solids.—

Molecular Solids. Studies of relaxation in the solid phases of elemental fluorine, boron, and deuterium have been reported. In solid F_2, O'Reilly *et al.*[181] have measured T^F, $T^F_{1\rho}$, and T^F_2 in the β-phase from 53·5 to 45·5 K and in the α-phase down to 4·2 K. The β-phase is a plastic crystal with relatively rapid translational diffusion and very rapid, but anisotropic, rotational diffusion. T_1 is essentially unchanged in going from the liquid phase to the β-phase, as is often observed in plastic crystals, indicating that the molecular rotation exhibits the same characteristics in the liquid and solid. On the other hand, $T_{1\rho}$ and T_2 are substantially changed and are determined by translational diffusion. Using Torrey's theory, τ_D was calculated from T_2 (E_D = 6·2 kJ mol^{-1}); using this value, $T_{1\rho}$ was calculated but found to be a factor of two longer than the experimental value. This discrepancy is due to a temperature-independent contribution to $T_{1\rho}$ from the anisotropic rotation of the F_2 molecules. In the α-phase, a minimum in $T_{1\rho}$ is observed and assigned to a co-operative 15° 'tilt' motion of the molecules. A shallow minimum in T_1 is assigned to spin diffusion to molecular oxygen impurities and occurs owing to the proximity of the oxygen electronic spin–lattice relaxation rate to the ^{19}F Larmor frequency. In powdered β-rhombohedral boron, T_1 of ^{11}B is shown, by Hynes and Alexander,[182] to be determined by the quadrupolar interaction; relaxation *via* spin diffusion to paramagnetic impurities is not important. The recovery of the longitudinal magnetization following a saturation pulse sequence was fitted by the sum of two exponentials having time constants 5 and 52 ms (the central transition was observed); this behaviour is not satisfactorily explained. However, the magnitudes of these relaxation times are interesting: when corrected to eliminate the effect of the magnitude of nuclear spin and nuclear quadrupole moment, so that only the contribution of the lattice vibrations to the relaxation is considered, the strength of the phonon contribution is found to be unusually large compared to the alkali-metal halides. The T_1 values in solid H_2, HD, and D_2 have long been the subject of theoretical study; the experimental literature on solid D_2 is, however, sparse compared with that for H_2 and HD. A detailed study of the relaxation mechan-

---

[181] D. E. O'Reilly, E. M. Peterson, D. L. Hogenboom, and C. E. Scheie, *J. Chem. Phys.*, 1971, **54**, 4194.
[182] T. V. Hynes and M. N. Alexander, *J. Chem. Phys.*, 1971, **54**, 5296.

isms in hexagonal close-packed D_2 from 0·4 K to the triple point at 18·7 K and over a range of concentration of paradeuterium molecules has been reported by Weinhaus et al.[183] T_1^D has also been measured in solid n-D_2 between 0·15 and 5 K by Constable et al.[184]

Proton T_1 and $T_{1\rho}$ have been measured by O'Reilly et al.[185] in the solid V phase (stable above 202 K) of thiourea between 240 and 420 K. T_1 reaches a minimum near 400 K and $T_{1\rho}$ a corresponding minimum near 270 K; these are assigned to the 180° flip motion of the molecule: $\tau_{180° \text{flip}}/s = 5 \times 10^{-15}$ exp(43·1 kJ mol$^{-1}/RT$). Near 330 K, a shallow, secondary minimum in $T_{1\rho}$ is observed and is assigned to a 'ferroelectric mode' process in which the molecules oscillate between the positions they assume in the ferroelectric solid I (stable below 169 K) for the two polar structures of the crystal: for this process $\tau_2/s = 2 \times 10^{-13}$ exp(43·5 kJ mol$^{-1}/RT$).

The importance of methyl group reorientation for proton relaxation in organic solids has long been recognized; they act as relaxation sinks for protons in more rigid parts of the molecule. By comparing the spin–lattice relaxation efficiency of a terminally deuteriated and an undeuteriated C_{17} lithium soap, van Putte and Egmond[186] have shown that, below room temperature, 98% of the relaxation can be accounted for in terms of methyl group relaxation.

T_1^H measurements have been employed by Anderson[187] to study the molecular motions in solid solutions of benzene in thiophen. In C_4H_4S–C_6D_6 and C_6H_6–C_4D_4S samples, it was assumed that the T_1^H data reflect the motions of the proton-bearing molecules alone. On this basis, the temperature dependence of the T_1 minima on composition suggests that as the benzene concentration increases, the motion of the C_6H_6 molecule slows down while that of the C_4H_4S molecule speeds up. This is a surprising result and one wonders to what extent the T_1 data, which were obtained by the null method, reflect the coupling between the proton and deuteron spin systems.

Finally, van Hecke et al.[188] have monitored T_1^H during the conversion of the methane spin modification (ortho and meta→ para) following cooling to 4·2 K in samples doped with oxygen to catalyse the conversion. A quite peculiar time-dependence of T_1 is observed and is indicative of three 'parallel' relaxation mechanisms being involved. The relaxation rate due to mechanism I increases almost linearly with time to an equilibrium value, reached after the conversion time so that, most likely, this mechanism is associated with conversion. The rate due to mechanism II is not a function of time, but shows a slight dependence on O_2 concentration. The rate due to mechanism III is independent of O_2 concentration and conversion time, but decays exponentially with time; it is assigned to methane being suspended for some time

[183] F. Weinhaus, S. M. Meyers, B. Maraviglia, and H. Meyer, *Phys. Rev.*, 1971, **B3**, 626.
[184] J. H. Constable and J. R. Gaines, *Phys. Rev.*, 1971, **B3**, 1556.
[185] D. E. O'Reilly, E. M. Peterson, and Z. M. El Saffer, *J. Chem. Phys.*, 1971, **54**, 1304.
[186] K. van Putte and G. J. N. Egmond, *J. Magn. Resonance*, 1971, **4**, 236.
[187] J. E. Anderson, *Mol. Crystals Liquid Crystals*, 1970, **11**, 343.
[188] P. van Hecke, P. Grobet, and L. van Gerven, *Phys. Letters*, 1970, **33A**, 379.

between two phases. Exactly how spin conversion influences relaxation mechanism I is not clear; all that is shown by these results is that both converting spin modifications have an almost identical relaxation rate, either because of fast exchange between both or because they have a common relaxation mechanism. An attempt has been made by Jones et al.[189] to observe the existence of nuclear spin states in solid SiH_4; the proton magnetization was measured from 30—4·2 K and found to be consistent with the Curie law, indicating, within experimental error, no conversion among nuclear spin states. T_1^H varies from about 30 s at 30 K to more than 10^4 s at 4·2 K.

Ionic Solids. O'Reilly et al.[190] have employed T_1^H measurements to study the change in the motion of the hydronium ion at the monoclinic to orthorhombic transition at 243 K in crystalline perchloric acid monohydrate, H_3O^+ ClO_4^-. In the monoclinic phase below 243 K, second-moment measurements show that the H_3O^+ ion rotates about the three-fold axis of the molecule. T_1^H measurements are consistent with this model and give $\tau_2/s = 2\cdot 8 \times 10^{-13}$ $\exp(20\cdot 2 \text{ kJ mol}^{-1}/RT)$; the pre-exponential factor is consistent with a classical rotor in a three-fold potential. In the orthorhombic phase, $\tau_2/s = 6\cdot 7 \times 10^{-15}$ $\exp(17\cdot 5 \text{ kJ mol}^{-1}/RT)$, as calculated from T_1^D and T_1^H data, and is typical of values in the liquid state; the absence of structure in the deuteron resonance line at room temperature demonstrates that the motion in this phase is essentially isotropic.

It is shown, by Pedersen and Clark,[191] that in $K_2C_2O_4, D_2O$ the 180° flip of the D_2O molecule is responsible for the relaxation of the ^2H nuclei. Of course, this interpretation requires that the electric field gradients at the two deuteron sites of the D_2O molecules be different, a fact which is confirmed by c.w. measurements. The relaxation is described by a set of rate equations which include both spins of a molecule and additional terms corresponding to exchange jumps between the two equilibrium positions. The interval between the 180° flips is $\tau/s = 3\cdot 6 \times 10^{-17} \exp(68\cdot 2 \text{ kJ mol}^{-1}/RT)$; this barrier is in good agreement with a value calculated from a point-charge model.

The temperature dependence of $T_{1\rho}^H$ in powdered lithium hydrazinium sulphate [$Li(N_2H_5)SO_4$] has been measured by Knispel et al.[192] between 140 and 495 K. The $N_2H_5^+$ ions and their interconnections are known to play an important role in the electrical properties of this solid; the $N_2H_5^+$ groups form hydrogen-bonded chains along the c axis of an orthorhombic structure. These $T_{1\rho}^H$ data, in conjunction with previous results[193] for T_1^H and for proton second moment, show that the $—NH_3^+$ reorientation occurs with $\tau_2/s = 5\cdot 0 \times 10^{-14}$ $\exp(18\cdot 0 \text{ kJ mol}^{-1}/RT)$, and the $—NH_2$ part of the $N_2H_5^+$ ion executes 180° flips about the bisectrix of the H—N—H angle with $\tau_2/s = 2\cdot 2 \times 10^{-14} \exp(44\cdot 7$

---

[189] E. P. Jones and L. P. Montgomery, *Phys. Letters*, 1971, **35A**, 229.
[190] D. E. O'Reilly, E. M. Peterson, and J. M. Williams, *J. Chem. Phys.*, 1971, **54**, 96.
[191] B. Pedersen and W. G. Clark, *J. Chem. Phys.*, 1970, **53**, 1024.
[192] R. R. Knispel and H. E. Petch, *Canad. J. Phys.*, 1971, **49**, 870.
[193] J. D. Cuthbert and H. E. Petch, *Canad. J. Phys.*, 1963, **41**, 1629; W. D. MacClement, M. M. Pintar, and H. E. Petch, *ibid.*, 1967, **45**, 3257.

kJ mol^{-1}/RT). Evidence for the motion of the entire N$_2$H$_5^+$ ion with activation energy 73 kJ mol^{-1} is also obtained but the effects of reorientational and diffusional motions are not differentiated.

Antiferromagnetic Solids. In Cu$_2$(CH$_3$CO$_2$)$_4$,2H$_2$O, the Cu^{2+} ions are arranged in pairs: T_1^H measurements[194] between 57 and 130 K show that T_1^{-1} is proportional to the fraction of these Cu^{2+} pairs in the paramagnetic state and that the singlet–triplet separation is $\Delta = (310 \pm 15)$ cm^{-1}, a value which is more reliable than those from previous e.s.r. and susceptibility studies. The angular dependence of T_1^H has been measured[195] in the linear chain antiferromagnetic FeCl$_2$,2H$_2$O and MnCl$_2$,2H$_2$O at, respectively, 4·2 and 1·34 K. In MnCl$_2$,2H$_2$O the experimental data agree with a calculation based on a two-magnon process, but in FeCl$_2$,2H$_2$O an anomalous field effect is found.

Metallic Hydrides. In PuH$_{2\cdot 62}$, the temperature dependence of T_1^H, measured[196] between 116 and 294 K, changes from $T_1 T = A(T-\theta)$ at high temperatures to $T_1 T = B(T-\theta)^2$ at low temperatures, and indicates a change from antiferromagnetic to ferromagnetic behaviour as the temperature is lowered. From the temperature dependence of T_2^H, the activation energy for hydrogen diffusion is found to be 11·5 ± 1·7 kJ mol^{-1}. T^H and T_2^H measurements are also reported[197] for the Y–Ce–H system.

Hexagonal Ice. $T_{1\rho}^H$ has been measured by Valic *et al.*[198] in a single crystal of hexagonal ice from 249·9 to 271·6 K and in spin-locking fields between 0·3 and 2·4 mT. The $T_{1\rho}$ vs T^{-1} plots exhibit minima near the melting point, indicating that in this temperature region relaxation in the rotating frame is in the weak-collision limit. For the proton motion $\tau_2/s = 6 \times 10^{-18}$ exp (59·8 kJ mol^{-1}/RT): this value is claimed to be more accurate than the one determined in a previous experiment.[199] At 250 K and below, the proton motion becomes so slow that the relaxation it causes in the rotating frame is in the strong-collision limit. At this temperature, $T_{1\rho}$ is found to be anisotropic; the experimental data are found to be in an order of magnitude agreement with values estimated using an isotropic order parameter calculated for a simple two proton model of ice.

Molecules Bound to Surfaces. Nagel *et al.*[200] have made an interesting study by way of T_1^H measurements of the effects of π-bonding on the mobility of benzene, cyclohexene, and cyclohexane adsorbed on to Na-Y-zeolites. In all

[194] I. Svare and D. P. Tunstall, *Phys. Letters*, 1971, **35A**, 123.
[195] T. Goto, A. Hirai, and T. Haseda, *Phys. Letters*, 1970, **33A**, 185.
[196] G. Cinader, D. Zamir, U. El-Hanany, Z. Hadari, and D. Degani, *Solid State Comm.*, 1970, **8**, 1703.
[197] E. F. Khodosov and I. I. Khodos, *Fiz. Tverd. Tela*, 1970, **12**, 2745. [*Sov. Phys. Solid State*, 1971, **12**, 2213].
[198] M. I. Valic, S. Gornastansky, and M. M. Pintar, *Chem. Phys. Letters*, 1971, **9**, 362.
[199] R. Blinc, G. Lahajnar, I. Zupancic, and H. Gränicher, *Chem. Phys. Letters*, 1969, **4**, 363.
[200] M. Nagel, D. Michel, and D. Geschke, *J. Colloid Interface Sci.*, 1971, **36**, 254.

cases T_1^H is due to interaction with paramagnetic impurities, but the temperature of the T_1 minimum increases strikingly with the number of π-electrons in the adsorbate: cyclohexane (173 K), cyclohexene (263 K), and benzene (370 K at $\theta = 0.85$). Furthermore, the effect of increasing coverage is to broaden the T_1 minimum for cyclohexane, indicating a distribution of correlation times, while for benzene the T_1 minimum is shifted to higher temperatures, but is always consistent with a single correlation time. The difference in behaviour of benzene and cyclohexane is attributed to their different π-bonding abilities. The benzene molecules are localized on the surface owing to specific interaction between their π-electrons and adsorption centres; the effect of increasing coverage is then to hinder the reorientational motion of the molecules. In contrast, cyclohexane molecules do not interact specifically with the surface and are free to diffuse; relaxation is by translational diffusion.

Woessner et al.[201] have demonstrated that the self-diffusion of water in agar–water gel is essentially unaffected by the presence of the agar macromolecules. However, large proton and deuteron relaxation effects are observed and attributed to the behaviour of a very small fraction of the water molecules which interact with the agar macromolecules. This fraction of 'bound water' molecules is less than 1% in a 10% agar gel. The spin–lattice relaxation of both ^1H and ^2H takes place in the exchangeable hydroxy-groups on the agar macromolecule. The spin–spin relaxation is determined by the spin interactions within the bound water molecules. The correlation time describing the motion of the bound water molecules must therefore be at least a factor of 10^2 longer than that for the hydroxy-groups.

H. Self-diffusion in Molecular Crystals.—In the case of rare-gas solids, theoretical models leading to reasonable quantitative interpretation of diffusion data obtained by both radiotracer and n.m.r. techniques have been formulated.[202] On the other hand, no reliable theories are as yet available for diffusion in molecular solids; moreover, there are considerable discrepancies between diffusion data obtained by n.m.r. and radiotracer techniques in plastic organic solids. Bladon et al.[203] have obtained results which suggest that these discrepancies might well reflect a basic difference in the molecular motions to which these two methods are sensitive. Using a conventional high-resolution n.m.r. spectrometer and very carefully purified samples, they measured the activation energies $E_{n.m.r.}$ for the line-narrowing process in a series of organic solids covering a range of entropies of fusion, ΔS_f; their results are compared in the Table with values of E_D, obtained by radiotracer or plastic deformation techniques. Also included in this table is the value of the activation energy for diffusion in hexamethylethane, as determined by Chezeau et al.[204] from $T_{1\rho}$ measurements, and the values for sulphur and

[201] D. E. Woessner, B. S. Snowden, and Y.-C. Chiu, *J. Colloid Interface Sci.*, 1970, **34**, 283; D. E. Woessner and B. S. Snowden, *ibid.*, p. 290.
[202] H. R. Glydes and J. A. Venables, *J. Phys. and Chem. Solids*, 1968, **29**, 1093.
[203] P. Bladon, N. C. Lockhart, and J. N. Sherwood, *Mol. Phys.*, 1971, **20**, 577.
[204] J. M. Chezeau, J. Dufourcq, and J. H. Strange, *Mol. Phys.*, 1971, **20**, 305.

selenium hexafluoride obtained by Virlet and Rigny[205] using T_{1D} measurements. It is seen that, for solids with high entropies of fusion ($\Delta S_f > 19.2$ J K^{-1} mol^{-1}), $E_{n.m.r.} \approx E_D$, whereas for those solids with low entropies of fusion ($\Delta S_f < 13.8$ J K^{-1}mol^{-1}), $E_{n.m.r.} < E_D$. Solids such as perfluorocyclohexane and triethylenediamine, with entropies in the intermediate region, show, unexpectedly, the former behaviour at low temperatures and the latter at high temperatures. In the high ΔS_f solids, $E_D \approx 2\Delta H_s$ and it is reasonable to

Table Comparison of the apparent activation energies $E_{n.m.r.}$ and E_D for diffusion as measured respectively by n.m.r. and radiotracer or plastic deformation techniques

Compound	Crystal habit	ΔS_f/ J K^{-1}mol^{-1}	$E_{n.m.r.}$/ kJ mol^{-1}	E_D/ kJ mol^{-1}	ΔH_s/ kJ mol^{-1}
Pivalic acid	f.c.c.	6.7	55.4±1	91.2±0.4	59.0
t-Butyl chloride	f.c.c.	7.9	38.9±1	—	31.4
Cyclohexane	f.c.c.	9.2	46.0±2	68.2±1	41.8
Camphene	b.c.c.	9.6	49.7±8	97.9±2	51.0
Sulphur hexafluoride[a]	b.c.c.	10.5	37.7±8	—	23.3
Selenium hexafluoride[a]	b.c.c.	10.5	37.6±8	—	24.9
Neopentane	f.c.c.	12.1	37.0±2	—	32.2
2,2-Dichloropropane	f.c.c.	13.8	26.8±1	—	25.9
Triethylenediamine	f.c.c.	17.2	90.4±2(L)*	—	52.3
		—	56.1±2(H)*	—	—
1,1,1-Trichoroethane	f.c.c.	18.8	64.4±2	—	35.1
Perfluorocyclohexane	f.c.c.	19.2	64.0±2(L)*	61.6±4	33.9
		—	33.1±1(H)*	—	—
Hexamethylethane	b.c.c.	20.1	87.5±2	85.8±2	43.5
Hexamethylethane[b]		—	82±2	—	—
Adamantane	f.c.c.	20.9	153.5±1	146.8±16	66.9

* (L) and (H) denote low- and high-temperature behaviour, respectively. Unless otherwise specified data are taken from: P. Bladon, N. C. Lockhart, and J. N. Sherwood, *Mol. Phys.*, 1971, **20**, 577; [a] J. Virlet and P. Rigny, *Chem. Phys. Letters*, 1970, **6**, 377; [b] J. M. Chezeau, J. Dufourcq, and J. H. Strange, *Mol. Phys.*, 1971, **20**, 305.

assume diffusion proceeds by discrete vacancy migration. Bladon et al.[203] speculate that the transition in the behaviour of $E_{n.m.r.}$ from the high to the low ΔS_f solids reflects a gradual increase in the relaxation around the vacancy; the discrepancy between $E_{n.m.r.}$ and E_D is then attributed to the fact that the tracer technique is sensitive to motion over large distances and long times and detects the overall migration of the disordered defect, which is energetically similar to normal vacancy motion (this explains why $E_D \approx 2\Delta H_s$ irrespective of ΔS_f), whereas the n.m.r. experiment is sensitive to the microscopic motions within the relaxed vacancy which occur on a much shorter time-scale. It is also argued that both the macroscopic and microscopic motions should exhibit normal Arrhenius temperature-dependences and that the different processes will dominate the n.m.r. data in different temperature regions, thereby explaining the partial behaviour observed for the intermediate ΔS_f com-

[205] J. Virlet and P. Rigny, *Chem. Phys. Letters*, 1970, **6**, 377.

pounds. The possibility that this behaviour might result from an experimental artifact cannot be excluded and measurements by other n.m.r. methods are called for. Thus, although there is little doubt that the variation in $E_{n.m.r.}$ (relative to $2\Delta H_s$) with ΔS_f reflects a difference in the physical nature of the defect, we are still far from a definitive model for the diffusion process in plastic solids. Clearly, more radiotracer and n.m.r. studies are required on reliable materials; direct measurement of D by the pulsed-gradient spin-echo method would be useful, too.

Bladon et al.[203] made measurements on both single-crystal and polycrystalline specimens of the same compound and obtained the same results, demonstrating that the physical form of the sample does not affect the n.m.r. data as it does in the case of the radiotracer experiment. The importance of sample purity for n.m.r. measurements in plastic crystals is not clear; the values for $E_{n.m.r.}$ measured by Bladon et al.[203] are in general slightly higher than previously reported values, suggesting that impurities do affect the n.m.r. measurements, whereas Blum and Sherwood[206] show that $E_{n.m.r.}$, as obtained from linewidth measurements in camphene, is unaffected by impurity doping whilst values obtained by radiotracer and creep techniques in the same samples change by 100%. Clearly, further work on the effects of impurity doping are required.

Finally, we mention the interesting temperature dependence of D as measured in succinonitrile by Strange and Terenzi[207] using $T_{1\rho}$ data. Below 260 K, molecular self-diffusion is described by $D/m^2\ s^{-1} = 2.0 \times 10^{-3}$ exp $(-52.7\ \text{kJ mol}^{-1}/RT)$; above 260 K, there is an apparent increase in the activation energy to 62.8 kJ mol^{-1}. It is suggested that this behaviour might be related to the existence of a temperature-dependent equilibrium between *trans* and *gauche* forms of the molecule in the solid. Measurement of the dependence of $T_{1\rho}$ on B_1 shows that $T_{1\rho} \propto \omega_1^2$, and not $\omega_1^{3/2}$ as suggested in a previous study.[208]

I. Ferroelectric Phase Transitions.—An anomalous peak has been observed[209-212] in the nuclear dipolar and quadrupolar spin–lattice relaxation rate vs. temperature plots in the vicinity of the ferroelectric Curie point, T_c. This peak is believed to arise from the coupling of the spins to an unstable lattice mode whose frequency approaches zero at T_c, while the frequencies of the other lattice modes remain characteristically high. Blinc et al.[209] have observed such a peak in the proton relaxation rate measured in powdered $Ca_2Sr(CH_3CH_2CO_2)_6$. In the ferroelectric phase, two out of the six propionate groups are rapidly flipping between two equilibrium positions. The

---

[206] H. Blum and J. N. Sherwood, *Mol. Crystals Liquid Crystals*, 1970, **10**, 381.
[207] J. H. Strange and M. Terenzi, *Mol. Phys.*, 1970, **19**, 275.
[208] J. G. Powles, B. Afronzi, and M. O. Norris, *Mol. Phys.*, 1969, **17**, 489.
[209] R. Blinc, S. Zumer, and G. Lahajner, *Phys. Rev.*, 1970, **B1**, 4456.
[210] G. Bonera, F. Borsa, and A. Rigamonti, *Phys. Rev.*, 1970, **B2**, 2784.
[211] A. Avogadro, E. Cavelius, D. Müller, and J. Peterson, *Phys. Stat. Sol.* 1971, **B44**, 639.
[212] R. Blinc, J. Stepisnik, M. J. Vilfar, and S. Zumer, *J. Chem. Phys.*, 1971, **54**, 187.

correlated fluctuations of the electric dipoles produce fluctuations in the electric polarization which, due to interactions between the dipoles, undergo a critical slowing down as T_c is approached. The anomaly in T_1^H is believed to be due to the coupling of the proton spins to this 'ferroelectric' mode, which is treated by Blinc et al.[209] in terms of an overdamped quasi-spin-wave. In a classical spectral density calculation of T_1^{-1}, they used the fluctuation dissipation theorem to relate the spectral density of the Fourier components of the polarization fluctuations to the imaginary part of the wavenumber-dependent generalized dielectric susceptibility; the predicted temperature dependence of T_1^{-1}, near T_c, is as $|T-T_c|^{-n}$, where $0.5 \leqslant n \leqslant 2$, depending on the details of the interaction between the ferroelectric dipoles and the Brillouin-zone size. Good agreement is obtained between the predicted and experimentally observed temperature dependence of T_1^H.

Measurements[210,211] of T_1 of ^{23}Na in NaNO$_2$ single crystals in the paraelectric and ferroelectric phases show the existence of a logarithmic singularity of the relaxation rate on both sides of T_c. Bonera et al.[210] explain this behaviour by assuming a direct relaxation mechanism caused by the interaction of the quadrupole moment with the electric-field-gradient fluctuations associated with the correlated motions of the electrical dipoles. Using an approach similar to that of Blinc et al.,[209] they show that $T_1^{-1} \propto \ln|T-T_c|$ when the anisotropic character of the electric dipole fluctuations is taken into account. These authors also show that the 'soft' mode can be identified as the flipping motion along the ferroelectric c axis of the electric dipoles associated with the NO$_2^-$ groups. Avogadro et al.[211] show that the critical behaviour of the ^{23}Na relaxation rate can be explained theoretically in terms of the thermodynamics of the ferroelectric phase transition.

The temperature-dependence of the quadrupole splitting in the deuteron magnetic resonance spectra of KD$_2$PO$_4$, KD$_2$AsO$_4$, and CsD$_2$AsO$_4$ shows that the deuterons are jumping between the two off-centre positions in the O—D $\cdots$ O bonds at temperatures above T_c, but are effectively frozen in at one of the two possible sites below T_c.[212] The deuteron relaxation rate, measured by Blinc et al.,[212] shows an anomalous increase on approaching T_c in the paraelectric phase; below T_c it undergoes a sharp decrease followed by a gradual levelling off. For $T > T_c$, $T_1 \propto (T-T_c)^n$, showing the existence of a 'soft' deuteron mode which condenses on approaching T_c. It should be noted that T_1 measurements do not distinguish between a pure relaxational mode and a strongly damped phonon mode.[210] The alternative model for the deuteron motion in terms of independent thermally activated motion of H$_3$PO$_4$ and HPO$_4$ defects, as in ice, predicts an exponential temperature dependence of the relaxation rate.

It is clear from these studies that T_1 measurements can provide valuable and unique information about the behaviour of the unstable lattice modes near the ferroelectric transition in dielectric materials.

7 Pulsed N.M.R. in Solids

During recent years a number of interesting pulse experiments have been developed which are potentially useful for investigating spin–lattice relaxation and the other nuclear spin interactions in solid materials. One of these, the rotating-frame spin–lattice relaxation experiment, was discussed in the previous section. In this section we shall review the two other main areas of current interest. First, we shall discuss multiple-pulse experiments and their application to line-narrowing and the resolution of isotropic and anisotropic chemical shifts, Knight shifts, and electron-coupled spin–spin interactions. An excellent theoretical account of these experiments is given in a recent review article by Mansfield,[213] one of the principal pioneers of these experiments. Secondly, rotating-frame pulsed nuclear double-resonance experiments and their application to the detection of weakly resonant spin systems will be described.

A. Application of Multiple-pulse Experiments to Line-narrowing in Solids.—
The Fourier transform of the free induction decay signal following a single 90° r.f. pulse applied to a spin system at resonance yields the steady-state absorption spectrum. Much of the useful information in this spectrum, such as chemical shifts, Knight shifts, and electron-coupled spin–spin interactions, is obscured by the strong dipolar interactions between the spins. During recent years a variety of multiple-pulse experiments have been developed which are intended to remove or reduce the dipolar interactions but at the same time retaining these other, often more useful, interactions.[214–223] In these experiments cycles of pulses are applied repeatedly; Fourier transformation of the signal obtained at one point in each cycle yields the frequency-domain high-resolution spectrum of the solid. These cycles have the property that $\mathcal{H}_d = 0$ over a full cycle.[219] Of the many possible cycles having this property, the only one which has so far been successfully used to measure chemical shifts in solids is the four-pulse cycle proposed by Waugh, Huber, and Haeberlen;[220] this cycle, which reduces each component of the chemical shift by the factor $(1/3)^{\frac{1}{2}}$ is denoted P_{-y}—$(\tau$—P_x—2τ—P_{-x}—τ—P_y—2τ—$P_{-y})_n$ where P_α represents a 90° r.f. pulse applied along the α axis in the rotating frame. The utility of this cycle was first demonstrated by the measurement of the chemical shift of the ^{19}F nuclei in a single crystal of CaF_2 with respect

[213] P. Mansfield, *Progr. N.M.R. Spectroscopy*, 1971, **8(1)**, 43.
[214] P. Mansfield and D. Ware, *Phys. Letters*, 1966, **22**, 133.
[215] J. S. Waugh and L. M. Huber, *J. Chem. Phys.*, 1967, **47**, 1862.
[216] D. Ware and P. Mansfield, *Phys. Letters*, 1967, **25A**, 651.
[217] J. S. Waugh, L. M. Huber, and E. D. Ostroff, *Phys. Letters*, 1968, **26A**, 211.
[218] P. Mansfield and D. Ware, *Phys. Letters*, 1968, **27A**, 160.
[219] J. S. Waugh, C. H. Wang, L. M. Huber, and R. L. Vold, *J. Chem. Phys.*, 1968, **48**, 662.
[220] J. S. Waugh, L. M. Huber, and U. Haeberlen, *Phys. Rev. Letters*, 1968, **20**, 180.
[221] U. Haeberlen and J. S. Waugh, *Phys. Rev.*, 1968, **175**, 453.
[222] W. A. B. Evans, *Ann. Phys.*, 1968, **48**, 72.
[223] P. Mansfield and K. H. B. Richards, *Chem. Phys. Letters*, 1969, **3**, 169.

to the resonance in liquid $C_6H_5CF_3$ in a mixture of the two:[220] the shift is 43 p.p.m. in the direction of increased diamagnetic shielding, a value a little smaller than the shift of the aqueous F^- ion from $C_6H_5CF_3$ (53 p.p.m.). It has subsequently been applied to several more interesting systems. The ^{19}F absorption spectrum obtained in a solid mixture of tetrafluoroethylene–perfluoromethylvinyl copolymer, at room temperature, has two absorption lines separated by 73 ± 4 p.p.m.; one is symmetrical and attributed to the OCF_3 group, the other is distinctly asymmetrical and is tentatively attributed to an axially symmetric chemical-shift anisotropy in the —CF_2 group; $\sigma_\perp - \sigma_\parallel \approx 50$ p.p.m.[224] In solid perfluorocyclohexane, at 200 K, the AB quartet is resolved and gives $\delta = 17.5 \pm 1.5$ p.p.m. and $J = 310 \pm 40$ Hz, in good agreement with liquid-state data.[225] The absence of asymmetry in the spectrum due to chemical-shift anisotropy is caused, suggests Andrew,[226] by molecular motion. One of the most promising applications of the multiple-pulse technique is the direct measurement of the components of the chemical-shift tensor. Mehring et al.[227] have measured the principal values of the ^{19}F chemical-shift tensor in powdered fluoranil, $C_6F_4O_2$; relative to liquid C_6F_6 they are -101, -62, and $+86$ p.p.m. The components of the ^{19}F chemical-shift tensor in a single crystal of MgF_2 have also been measured;[228] relative to HF they are -26, -11, and $+4$ p.p.m. These results demonstrate quite clearly that the simple four-pulse cycle, despite its theoretical limitations, is practicable and can provide important information about chemical shifts in solids. A six-pulse modification of the four-pulse cycle of Waugh et al. has been proposed by Mansfield,[229] $P_{-y}-(\tau-P_xP_{-y}-\tau-P_y-2\tau-P_{-y}-\tau-P_yP_{-x}-\tau)_n$; this is a reflection-symmetry cycle and eliminates the second-order as well as the first-order dipolar terms, and should therefore be more efficient at line-narrowing. At present the resolution obtainable by the four-pulse cycle is about 150 Hz and is determined by instrumental effects such as r.f. field inhomogeneity over the sample, finite pulsewidths, receiver recovery limitations on τ, and phase errors. Although it should be possible to design more complex reflection-symmetry cycles to compensate for these effects, it is questionable whether or not their application will give a much greater improvement in resolution until pulsed n.m.r. instrumentation is significantly improved. It is, however, now clear that linewidth resolution down to about 50 Hz should be attainable by the multiple-pulse technique. In passing, we mention that the multiple-pulse experiment compares favourably with the specimen rotation technique for line-narrowing. The lower linewidth limit of the latter is determined by the speed of rotation, which is at best 10 kHz. The ^1H and ^{19}F linewidths in solids are often well in excess of 10 kHz, putting them well beyond the present experimental limit. On the other hand, linewidths

[224] J. D. Ellett, U. Haeberlen, and J. S. Waugh, *Polymer Letters*, 1969, **7**, 71.
[225] J. D. Ellett, U. Haeberlen, and J. S. Waugh, *J. Amer. Chem. Soc.*, 1970, **92**, 411.
[226] E. R. Andrew, *Phys. Letters*, 1970, **32A**, 520.
[227] M. Mehring, R. G. Griffin, and J. S. Waugh, *J. Amer. Chem. Soc.*, 1970, **92**, 7222.
[228] L. M. Stacey, R. W. Vaughan, and D. D. Elleman, *Phys. Rev. Letters*, 1971, **26**, 1153.
[229] P. Mansfield, *Phys. Letters*, 1970, **32A**, 485.

of 200 kHz are easily accessible by the multiple-pulse experiment. Furthermore, the specimen-rotation technique leaves only the isotropic part of the chemical-shift tensor.

All of the experiments described in the previous paragraph were made on compounds containing only one abundant nuclear species with non-zero magnetic moment. The reason for this is that in the four-pulse experiment while the pulse train succeeds in enormously reducing the effects of dipolar interactions between like spins, only a reduction by a factor of $\sqrt{3}$ is acquired for the secular interactions between unlike spins.[230] Mehring et al.[230] have shown how the effects of the remaining dipolar broadening can be removed by simultaneously irradiating the non-resonant spins. The ^{19}F spectrum of a mixture of CaF_2 (powdered), NaF (single crystal), and C_6F_6 (liquid) obtained using the four-pulse cycle shows two lines, one from C_6F_6 and the other from CaF_2. The absorption peak of the ^{19}F spin in NaF was too broad to be observed due to coupling to the ^{23}Na spins. In an identical experiment, except with the ^{23}Na spins irradiated with a 180° resonant r.f. pulse once per four-pulse cycle, an additional line due to NaF was observed. The ^{19}F chemical shift of NaF is 114 ± 6 p.p.m. relative to CaF_2 in the direction of increased diamagnetic shielding; this is a surprisingly large value considering both solids are ionic.

An interesting pulse sequence has been described by Rhim, Pines, and Waugh[231] which can be used to make a system of dipolar coupled nuclear spins behave as though the sign of the dipolar Hamiltonian has been reversed. This pulse sequence, referred to as the time-reversal sequence $B(t_B)$, consists of a pair of 90° pulses along the $+y$ and $-y$ axes of the rotating frame, enclosing an even number of contiguous 180°, or n 180°, pulses which alternate between the $+x$ and $-x$ axes. Its effect is to convert the effective dipolar Hamiltonian from $\mathcal{H}_d$ to $-\tfrac{1}{2}\mathcal{H}_d$. A variety of experiments to demonstrate the time-reversal properties of this sequence are described. One of them $[90°_x—\tau—B(t_B = 4\tau)—\tau$, echo] is easily understood and it is interesting to note its analogy to the more familiar inhomogeneous spin-echo experiment, (90°—τ—180°—2τ, echo), in liquids. In CaF_2, echoes for τ substantially larger than T_2 were observed, demonstrating that the spin system does not approach internal thermodynamic equilibrium in a time T_2, contrary to the assumption of the spin-temperature approximation. A similar contradiction has been observed[232] for the transverse relaxation of ^{19}F nuclei in Teflon in the rotating frame at exact resonance and in r.f. fields large compared to the local field: the magnetization which had dephased under the action of the dipolar coupling could, using various pulse sequences, be reversed independently of the length of the r.f. pulse. The apparent conflict between these experiments and the spin-temperature hypothesis is explained by the fact that the time

---

[230] M. Mehring, A. Pines, W.-K. Rhim, and J. S. Waugh, *J. Chem. Phys.*, 1971, **54**, 3239.
[231] W.-K. Rhim, A. Pines, and J. S. Waugh, *Phys. Rev. Letters*, 1970, **25**, 218; *Phys. Rev.*, 1971, **B3**, 684.
[232] W.-K. Rhim and H. Kessemeier, *Phys. Rev.*, 1971, **B3**, 3655.

development of the system is dynamically reversible in a microscopic sense, whereas it may behave irreversibly on a thermodynamic scale. Apart from the phenomenological interest in these experiments, they are potentially useful for removing the effects of dipolar broadening in solids. For example, the sequence 90°_x—$[\tau$—$B(t_B = 2\tau)]_n$ produces a prolonged multiple-echo train; since the effective Hamiltonian during the interval τ is $\mathcal{H}_d$ and during the pulse burst is $-\frac{1}{2}\mathcal{H}_d$, then $\bar{\mathcal{H}}_d = 0$ over each cycle of duration $3t_B/2$. It is claimed that this multiple-echo experiment has been used to observe chemical shifts in solids but no results are given. This new method has a number of apparent advantages over previously described line-narrowing techniques, the most important being that while t_B may be of the order T_2 or more, the line-narrowing efficiency is determined by the subcycle time inside the burst, which can be made extremely small since no observation of the magnetization is necessary during this time.

Hanabusa[233] has used the multiple-pulse technique for sensitivity enhancement in experiments employing the pulse method to trace out broad absorption lines in metals and ferromagnets. Since the multiple-pulse experiment prolongs the decay of the transverse magnetization, one can accumulate more signals per single experiment. Under optimum conditions the enhancement factor is $(T_{2e}/T_2)^{\frac{1}{2}}$ for a given measuring time.

B. Application of Multiple-pulse Experiments to the Study of Molecular Motion.

—In the cyclic multiple-pulse experiments, the effective relaxation time of the transverse magnetization (T_{2e}) is controlled by spin–lattice relaxation effects provided τ is short enough to eliminate the dipolar contribution. Clearly, relaxation effects limit the resolution attainable and are a nuisance when high-resolution spectra are the prime objective; in this case, it is necessary to cool the sample to stop the molecular motion. On the other hand, these experiments do provide a very convenient method for observing slow thermal motion when only one spin species is present. Haeberlen and Waugh,[234] and more recently Mansfield[213] and Schmiedel et al.,[235] have analysed the effects of spin–lattice relaxation in cyclic multiple-pulse experiments and have obtained expressions for T_{2e}, in the case of the two-pulse and four-pulse cycle experiments. T_2 is shown to depend on the Fourier components of the thermal motion in the neighbourhood of the pulse modulation frequency. We expect, therefore, that T_{2e} measured as a function of temperature but at fixed τ will pass through a minimum when $\tau_c \approx \tau$; Schmiedel et al.[235] have shown that this minimum occurs at $\tau_c = \tau/2$ in the four-pulse-cycle experiment. Thus, the correlation time and its temperature dependence can be directly measured. The range of τ_c accessible by this technique is about the same as by the $T_{1\rho}$ method; the former method is, however, more convenient since the relaxation time can be measured in a single shot.

[233] M. Hanabusa, *J. Appl. Phys.*, 1971, **42**, 1077.
[234] U. Haeberlen and J. S. Waugh, *Phys. Rev.*, 1969, **185**, 420.
[235] H. Schmiedel, D. Freude, and W. Gründer, *Phys. Letters*, 1971, **34A**, 162.

C. Pulsed Double Nuclear Magnetic Resonance Experiments.—The spin-locked double-resonance experiment was first proposed by Hartman and Hahn[236] and developed by Lurie and Slichter[237] for the indirect detection of very weak nuclear resonance signals in solids. In this experiment the abundant I spin system is spin-locked along the resonant field B_1I. During spin-locking the I spins are virtually decoupled from the rare S spins. If, however, a second r.f. field is applied at a frequency $\omega_{0S} = \gamma_S B_0$ and with amplitude such that $\gamma_I B_{1I} = \gamma_S B_{1S}$, the Zeeman energy levels for both spin systems are equal in their respective rotating frames. Under these conditions, provided there is strong dipolar coupling between the I and S spins, the two spin systems exchange energy by mutual spin flips and approach thermodynamic equilibrium. When the thermal capacity of the S spin system is small relative to the I spins, the locked I spin system is not significantly demagnetized. If, on the other hand, a train of coherent pulses is applied to the S spins such that each pulse is on for a time $t > T_{CR}$, where T_{CR} is the cross-relaxation time, and is off for a time $\tau > T_{2S}$ so that the S spin system is maintained saturated, there will be a cumulative demagnetization of the I spin system. The I magnetization destruction spectrum observed by monitoring the attenuation of the free induction decay signal, following termination of the spin-locking pulse, as a function of ω_S, then shows a line corresponding to the S resonance. Blinc and co-workers[238] have used this technique to determine the ^{14}N quadrupole coupling tensor in a single crystal of glycine at 413 K. The quadrupole coupling constant is found to be very small, $e^2qQ/h = 745 \pm 20$ kHz, whereas the asymmetry parameter is relatively large, $\eta = 0.61$. Surprisingly, the direction of the largest principal axis of the electric field gradient tensor does not coincide with the C—N direction but makes an angle of 60° with it. As the increased sensitivity over pure ^{14}N quadrupole resonance spectroscopy is of the order 10^3—10^4, this technique should be valuable for the determination of ^{14}N quadrupole coupling constants in biologically important molecules. Hartland[239] has previously used the pulsed double-resonance technique to determine the quadrupole interaction of ^{14}N in a single crystal of $NH_4H_2PO_4$.

In the experiment as described in the previous paragraph, the resolution of the S resonance is generally poor because the I–I dipolar interaction dominates the width of the I magnetization destruction spectrum. Mansfield and Grannell[240] describe an important experimental modification which gives two orders of magnitude improvement in resolution. Their method is closely related to the experiment of McArthur et al.[241] which shows that when $\omega_S = \omega_{0S}$, the S spin free induction decay signal can be mapped out by varying τ from zero to a few times T_{2S}. When the S spins are slightly off

---

[236] S. R. Hartman and E. L. Hahn, *Phys. Rev.*, 1962, **128**, 2042.
[237] F. M. Lurie and C. P. Slichter, *Phys. Rev.*, 1964, **133**, A1108.
[238] R. Blinc, M. Mali, R. Osredkar, A. Prelesnik, I. Zupancic, and L. Ehrenberg, *Chem. Phys. Letters*, 1971, **9**, 85.
[239] A. Hartland, *J. Phys. (C)*, 1969, **2**, 264.
[240] P. Mansfield and P. K. Grannell, *J. Phys. (C)*, 1971, **4**, L197.
[241] D. A. McArthur, E. L. Hahn, and R. E. Walstedt, *Phys. Rev.*, 1969, **188**, 609.

resonance, the transverse S magnetization, which precesses freely during the off-time (τ), can be made to reverse its direction with respect to the S channel r.f. field, creating positive or negative S spin-temperatures when B_{1S} is restored if $\tau \approx T_{2S}$. Since a negative spin-temperature is more effective in destroying the I spins, this leads, for fixed τ, to a modulation of the I spin destruction spectrum at a frequency $1/\tau$ and so improves the resolution of the centre of the S resonance. The centre frequency of the spectrum is easily determined since it is independent of τ. The utility of this technique is demonstrated by application to the detection of the natural abundance ^{13}C resonance in amorphous polytetrafluoroethylene. With $\tau > T_{2S}$, the ^{19}F destruction spectrum is a 50 kHz broad unmodulated line, but, with $\tau < T_{2S}$, a modulated spectrum is obtained with a central peak width of 600 Hz; the effective ^{13}C gyromagnetic ratio ($\gamma/2\pi$) for ^{13}C$_2$F$_4$ is $10{\cdot}7068 \pm 0{\cdot}0003$ MHz T^{-1}. By varying τ for fixed ω_S it should be possible to faithfully map out the free induction decay of the S spins; Fourier transformation would then give the S spin lineshape, enabling chemical shielding anisotropies and other useful information to be obtained. It should be possible, in favourable cases, to resolve the resonances of chemically shifted groups of spins in a molecule.

When the nuclear double-resonance experiment is used to detect very small concentrations of spins, the question arises as to whether or not the sensitivity is limited by the spin diffusion rate within the I spin system. Slusher and Hahn[242] have suggested that the net energy transfer from a single saturated S spin centre is increasingly suppressed by spin diffusion effects as the S spin concentration is reduced below a certain critical value. This behaviour has not been confirmed by subsequent experiments.[243] Moran and Lang[244] have presented a new theoretical treatment of this problem: they develop a diffusion kernel solution to the driven spin diffusion equation and calculate the I magnetization decay rate induced by a dilute concentration of S spin centres. Their results show that for a uniform distribution, spin diffusion suppression of the induced decay rate only depends upon local effects and is independent of the concentration of S spins. When the S spins are aggregated into clusters, spin diffusion suppression could severely limit sensitivity.

In laboratory-frame cross-relaxation experiments, the effects of multiple-quantum processes are not resolved from single-quantum spin-flip transitions. This is not, however, the case in the rotating frame, where for certain off-resonance conditions destruction signals due to multiple-quantum transition may be observed. Mansfield and Cant,[245] using the ^{19}F spins to monitor the resonances of the ^{23}Na spins in a single crystal of NaF, have observed destruction signals caused by two fluorine spin-flips accompanied by one sodium spin-flip. Kunitomo[246] has made similar observations in the ^{23}Na–^{35}Cl

---

[242] R. E. Slusher and E. L. Hahn, *Phys. Rev.*, 1968, **166**, 332.
[243] D. V. Lang and P. R. Moran, *Phys. Rev.*, 1970, **B1**, 53; P. R. Spencer, H. D. Schmid, and C. P. Slichter, *ibid.*, p. 2989.
[244] P. R. Moran and D. V. Lang, *Phys. Rev.*, 1970, **B2**, 2360.
[245] P. Mansfield and G. P. Cant, *Phys. Letters*, 1970, **33A**, 130.
[246] M. Kunitomo, *J. Phys. Soc. Japan*, 1971, **30**, 1059.

Nuclear Spin Relaxation

system. Since the amplitude of a destruction line depends on the spin-locking and cross-relaxation times as well as on the thermal capacity of the two spin systems, cross-relaxation times for both single- and double-quantum transitions can be measured. The cross-relaxation time for a double I flip single S flip is calculated theoretically by Mansfield and Cant.[245]

4
Experimental Techniques

BY D. G. GILLIES

1 Introduction

The scope of application of the n.m.r. technique to chemical problems has increased remarkably during the period covered by this report. Commercial developments continue apace, ranging from those designed for versatile multinuclear operation with Fourier transform facilities and an on-line mini-computer to low-cost machines specifically designed for easy operation on a routine basis by non-specialist users. Perhaps the most dramatic development has been the application of Fourier techniques to ^{13}C spectroscopy.

The basic functions of the spectrometer will be considered in turn, mainly in terms of the continuous wave (c.w.) method, the traditional hunting ground for the chemist. The additional requirements of pulsed n.m.r. spectrometers will also be considered. It is not intended to give a full account of Fourier methods at this time since the subject has recently been reviewed elsewhere.[1] Since this is the first Report in the series, it will be necessary to refer back beyond the relevant time period to trace the context of a particular development. Clearly such reference will have to be selective.

The general advance in electronic technology has contributed markedly to the design and performance of n.m.r. spectrometers. Thus the electronic components of modern spectrometers are very frequently entirely solid-state. Wide use is made of both linear and digital integrated circuits. The development of voltage-controlled oscillators, phase-lock loop techniques, and solid-state devices (such as mixers, multipliers, dividers, and hybrid networks) has allowed a great increase in the sophistication of the r.f. techniques which are employed in n.m.r. The potential of the on-line computer is now being utilized in Fourier transform spectrometers, but it also has great possibilities for direct control of more general experiments.

Existing accounts of experimental methods may be found in standard text books.[2,3] For pulse methods, one may refer to the book by Farrar and

---

[1] D. G. Gillies and D. Shaw, in 'Annual Review of N.M.R. Spectroscopy,' ed. E. F. Mooney, Academic Press, vol. 5 to be published.
[2] J. A. Pople, W. G. Schneider, and H. J. Bernstein, 'High Resolution Nuclear Magnetic Resonance,' McGraw-Hill, London and New York, 1959.
[3] J. W. Emsley, J. Feeney, and L. H. Sutcliffe, 'High Resolution Nuclear Magnetic Resonance Spectroscopy,' Pergamon Press, Oxford, 1965, vol. 1, 1966, vol. 2.

Becker.[4] Two review articles have appeared, the second of which concerns itself solely with commercial spectrometers.[5]

2 The Magnet

A. Permanent Magnets.—The inherent stability of a temperature-controlled permanent magnet and its minimal power and cooling requirements have led to its incorporation in simple machines at the lower end of the price range. There has been a big improvement in design with the advent of the barrel-yoke magnet, which enables a more compact construction to be made with a much reduced stray field; the latter fact minimizes interference from external magnetic fields. Both Varian and Perkin-Elmer incorporate this principle in their smaller systems, which provide proton resonance at 60 MHz in a 5 mm o.d. tube. Resolution attainable is typically about 0.3 Hz at 60 MHz (5 parts in 10^9).

B. Electromagnets.—These are now universally of the low-impedance type, drawing a high current at low voltage from a solid-state power supply. Both the magnet and the power supply are kept at constant temperature by the circulation of water at a controlled temperature. Large magnets are presently available which enable use to be made of larger field gaps at 90 and 100 MHz. This leaves room for larger samples and a flexible probe design. Spinning 12 or 13 mm sample tubes with variable-temperature facilities are provided, with a resolution of about 5 parts in 10^9.

Better performance is now available from the small magnets, with increased facilities; for instance, the Varian NV-14 provides for an 8 mm o.d. spinning sample tube with variable-temperature facility at a resolution of 0.5 Hz at 60 MHz from a magnet with six-inch diameter pole-pieces. Smaller still is the Varian EM 300, giving 0.4 Hz resolution at 30 MHz.

C. Superconducting Magnets.—These presently provide for proton resonance at up to 300 MHz. The magnet is in the form of a solenoid, so that the sample spinning axis and the magnetic field are collinear. This is particularly convenient for studies in liquid-crystal solvents as the spinning does not disturb the orientation. Resolution in apparatus using these magnets is typically 0.6 Hz (2 parts in 10^9) for a 5 mm sample, but developments which are as yet unpublished suggest that the use of larger samples at higher resolution is feasible.[6]

D. Control of Homogeneity.—Optimum homogeneity is held for long periods in systems with field–frequency lock. The superposition of a slow modulation (typically 1 Hz) on to the shim coils controlling the y gradient, and its subse-

---

[4] T. C. Farrar and E. D. Becker, 'Pulse and Fourier Transform N.M.R.,' Academic Press, London and New York, 1971.
[5] D. G. Howery, *J. Chem. Educ.*, 1971, **48**, A327; *ibid.* p. A389.
[6] R. E. Richards, personal communication.

quent phase detection in the control signal output, allows the d.c. shim current to be adjusted automatically for maximum control signal. A digital approach whereby both y gradient and curvature are computer-controlled has been demonstrated.[7]

E. **Stability.**—Permanent and superconducting magnets are inherently sufficiently stable for many purposes, but both may with advantage be fitted with a field–frequency lock system for ultimate reproducibility and stability. Electromagnets are fitted with a flux stabilizer, a device which senses changes in magnetic flux through search coils and generates an opposing flux both directly by means of another pair of buck-out coils and indirectly by feeding an error signal to the magnet power supply. The modern stabilizer circuitry is entirely solid-state and inherently much more reliable than the older types.

Lower field stability is achieved when a Hall-effect probe is used, as in the Varian Fieldial system, but a method of improving the stability by locking to an n.m.r. signal has been described recently.[8]

3 Sensitivity

Inherent sensitivity has been improved since it became possible to use larger samples without serious loss in resolution. Approximate figures for signal-to-noise ratio based on the rather unsatisfactory standard of 1% ethylbenzene are as follows:
60 MHz with 5 mm tube — 30:1; 90 or 100 MHz and 5 mm tube — 60:1; 90 or 100 MHz and 12 or 13 mm tube — 200:1; and 300 MHz with 5 mm tube — 100:1. The higher inherent sensitivity of superconducting systems is obviously important when the amount of sample is limited.

Improvement of signal-to-noise ratio by the method of time averaging is now a routine method, the enhancement being proportional to the square root of the number of scans. The Fourier transform method has further improved the picture. By summing successive decays one can typically expect a factor of ten improvement over c.w. methods for the same expenditure of time. This means that a given signal-to-noise ratio may be achieved in one hundredth of the time required for the c.w. method. Of course, this has already had far-reaching implications in the field of ^{13}C n.m.r. One night's run in f.t. mode is equivalent to 50 days in c.w. mode, enabling studies to be made of nuclear species which are at low concentration in large molecules.

4 Frequency Generation

A. **Single Resonance.**—In the early days of high-resolution n.m.r. it was normal to sweep the magnetic field. The constant radiofrequency was derived by frequency multiplication from a stable quartz-crystal-controlled oscillator housed at constant temperature. It was usual to have several r.f. units to cover

---

[7] R. R. Ernst, *Rev. Sci. Instr.*, 1968, **39**, 998.
[8] W. H. Wing, E. R. Carlson, and R. J. Blume, *Rev. Sci. Instr.*, 1970, **41**, 1303.

a few nuclei of interest. This latter disadvantage was overcome by the Varian V4300B wide-line spectrometer, whose operating frequency was continuously variable from 2—16 MHz. Since the radiofrequency signal was not from a crystal-stabilized source, it was not inherently suitable for high-resolution work. Richards and co-workers[9-11] performed frequency-sweep experiments by varying the frequency over a small range by a motor-driven potentiometer. Spectra could thus be conveniently calibrated by means of a frequency counter, paying suitable heed, of course, to the residual field fluctuations not eliminated by the flux stabilizer. Bothner-By[12] improved on this system by driving the frequency of the spectrometer with a frequency synthesizer whilst the field was held constant by locking to a separate proton sample whose resonance frequency was also locked to the synthesizer system. The technique of using the sidebands generated as a result of audiofrequency field modulation in field–frequency lock spectrometers such as the Varian HA-100 will be discussed here as a method of frequency generation. Modulation may be carried out at several frequencies, and the sidebands produced may be considered as sub-carriers. More details are given in Section 8. By varying an audiofrequency, one has effectively obtained a variable-frequency carrier.

The increasing interest in nuclei other than ^{1}H or ^{19}F was at first catered for by simple extension of the above methods. The increased range of shifts required the use of higher modulation frequencies with wider ranges of sweep. This in turn led to problems which are discussed in Section 8. These sorts of arrangement were clearly stop-gap, and awaited a truly multinuclear approach. Fortunately, the necessary electronic developments were being made so that it became possible to synthesize any desired frequency. The modern method is to operate at a fixed, stable, and preferably high, modulation frequency, and to generate all the requisite r.f. carrier frequencies (variable, if this is necessary for sweep purposes) by frequency synthesis. One or more commercial frequency synthesizers may be used directly or 'custom' synthesis of particular frequencies for particular nuclei may be carried out. One may multiply, divide, add, and subtract frequencies with ever-increasing ease. Division by any number may be achieved very simply with integrated circuits, and for addition and subtraction one uses the wide-band balanced-diode modulators such as the Hewlett-Packard 10514A or Hatfield MD4. These modern developments and conveniences emphasize the validity of the pioneering work of Baker and Burd.[13,14]

Roberts[15] has used a digital frequency sweep spectrometer for some years for ^{13}C and ^{15}N work. The Hewlett-Packard synthesizer is programmed to produce a sweep consisting of small digital steps. The programmer controls both

---

[9] O. W. Howarth, R. E. Richards, and L. M. Venanzi, *J. Chem. Soc.*, 1964, 3325.
[10] D. Herbison-Evans and R. E. Richards, *Mol. Phys.*, 1964, **7**, 515.
[11] D. Herbison-Evans and R. E. Richards, *Mol. Phys.*, 1964, **8**, 19.
[12] A. A. Bothner-By, personal communication.
[13] E. B. Baker and L. W. Burd, *Rev. Sci. Instr.*, 1957, **28**, 313.
[14] E. B. Baker and L. W. Burd, *Rev. Sci. Instr.*, 1963, **34**, 238.
[15] F. J. Weigert and J. D. Roberts, *J. Amer. Chem. Soc.*, 1967, **89**, 2967.

Experimental Techniques

sweep width and sweep rate. Maciel and co-workers[16] have modified a Varian HA-100 and produced a ^{13}C spectrometer with ^{19}F field–frequency lock. Three radiofrequencies are derived from a synthesizer. The two V4311 r.f. units are driven by 15·68 MHz (^{19}F) and 6·28 MHz (^{13}C) signals, giving output frequencies at 94·08 MHz for the ^{19}F field–frequency lock channel and at 25·1 MHz for observation of ^{13}C resonances. The latter frequency may be driven by a voltage derived from a c.a.t. (time-averaging computer) ramp. The third frequency is produced at 100 MHz for use in the proton noise decoupler.

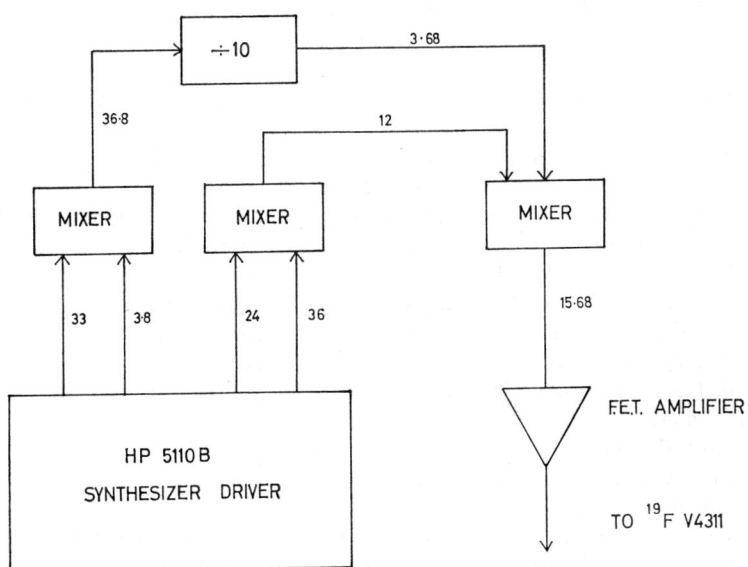

Figure 1 *Example of the use of double balanced mixers in frequency synthesis. (All frequencies in* MHz)
(Reproduced by permission from *Rev. Sci. Instr.*, 1970, **41**, 1458)

The spectrum may be swept either as in the usual HA-100 mode, *i.e.* with a variable audiofrequency modulation, or by sweeping the ^{13}C carrier frequency, as mentioned above. Figure 1 illustrates the generation of the 15·68 MHz signal, as an example of the use of double balanced mixers.

Recently, Manatt and co-workers[17] have used a Hewlett-Packard frequency synthesizer (type 5100A) as the variable-frequency audio oscillator in an HA-100 system. Digital sweep was provided by a commercial linear-sweep controller. A synthesizer-based system built around a permanent magnet has been described.[18] The basic proton frequency was driven by a fixed frequency

---

[16] V. J. Bartuska, T. T. Nakushima, and G. E. Maciel, *Rev. Sci. Instr.*, 1970, **41**, 1458.
[17] M. A. Cooper, H. E. Weber, and S. L. Manatt, *J. Amer. Chem. Soc.*, 1971, **93**, 2369.
[18] N. Boden, J. Capert, W. Derbyshire, H. S. Gutowsky, and J. R. Hanson, *Rev. Sci. Instr.*, 1968, **39**, 805.

derived from a Schomandl synthesizer, as was the 4 kHz modulation for the stabilization channel. Frequency sweep was achieved in the observation channel by deriving a 3—5 kHz signal from a Vidar voltage-controlled oscillator and using this as the input to the single-sideband modulator unit. This left the variable-frequency synthesizer output available for generating precise frequencies for double-resonance experiments.

Gillies[19] uses a Schlumberger FSX 3006S with provision for sweeping any decade (maximum range 1 MHz) in a modified HA-60-IL system. The proton frequency is derived from the master crystal, as is the 10 kHz audiofrequency modulation used in the lock channel. The recorder movement drives a voltage which is used to provide frequency sweep. For observation of proton resonances, the synthesizer is used as a variable-frequency audio oscillator; for observation of other nuclei, the resonance frequency is produced directly by the synthesizer, with the same sweep system. The system enables very accurate, highly expanded spectra to be achieved. By monitoring the interpolation frequency on the normal counter and driving the latter with a clock frequency derived from the master crystal, calibration to an accuracy of 1 part in 10^4 of the swept decade is achieved in one second.

Commercial instruments tend to use a 'custom' synthesis system. A frequency ω' near that of interest may be generated by appropriate multiplication and/or division of a master-crystal frequency. A narrow-range, digitally swept frequency synthesizer may then be used to produce a frequency that when added to ω' yields the desired resonance frequency; this may be swept through the region of interest. The latter function could also be performed by a voltage-controlled oscillator. Under this heading it should be mentioned that a suitable frequency or frequencies must also be generated for use in the receiver for demodulation purposes.

To sum up, in a modern frequency-swept spectrometer design, all fixed frequencies should be related, leaving one variable frequency for sweep purposes. The process of calibration is then reduced to the measurement of this one frequency. If a digital sweep is incorporated, one has an immediate crystal-accurate calibration at any point at which the sweep is stopped. In analogue sweep systems one achieves the most accurate calibration by counting at a high frequency, which is then effectively divided down before manifesting itself as a change in spectrometer frequency.

B. Double Resonance.—*Homonuclear.* The standard method for homonuclear decoupling experiments in most spectrometers is to apply an additional modulation frequency. The r.f. power has to be increased in order to allow sufficient intensity in the perturbing sideband, and in consequence the amplitudes of the audiofrequency modulations for the lock channel (if present) and for the observing channel have to be reduced. The method suffers from the fact that a beat is observed when the observing frequency passes through the irradiating frequency, making observations difficult in the immediate area.

[19] D. G. Gillies, unpublished results.

Experimental Techniques

Secondly, the presence of the strong component disturbs the lock signal, and one cannot irradiate with a frequency too near that of the lock resonance.

A time-division double-resonance scheme was described by Hewitt[20] in 1968. It allowed the decoupling of spectra in liquid-crystal solvents without a trace of the usual beats. Both the measuring and the irradiating radio-frequencies were obtained as sidebands by audio modulating a carrier frequency of 60 MHz, derived from the normal 60 MHz transmitter. Both audiofrequencies, f_{fixed} and f_{variable}, were supplied by frequency synthesizers. In each channel the audiofrequency was first passed through a Schmitt trigger before being presented at an AND gate which was enabled by a pulse. The enabled square wave passed to the modulation input of a Hewlett-Packard mixer (HP 10514A), whose r.f. input was a 60 MHz signal from the transmitter. This produced output frequencies at 60 MHz $\pm f$. The two channels were combined in a tee before being presented to a Hewlett-Packard Type 230A r.f. power amplifier. Pulses at the AND gates ensured that the two audiofrequencies were never present simultaneously; furthermore, the transmitter was gated on and the receiver was gated off during the pulses. The repetition rate was about 10 kHz, a frequency sufficiently high that the switching sidebands were outside the region of interest. It is felt that there is much scope for this type of arrangement in standard spectrometers and that the lock channel might be included as well. A simple control circuit could be made using standard logic modules, eliminating the requirement for several pulse generators.

Heteronuclear. Little more need be said, except that the irradiating frequency should be related directly to the observing frequency by means of a frequency synthesizer. Heteronuclear i.n.d.o.r. experiments are readily performed by sweeping the synthesizer frequency and may be time-averaged if all the frequencies involved are sufficiently stable.[19,21] The Hewlett-Packard power amplifier type 230A is a popular device for increasing the milliwatt power levels of synthesizer outputs to about 5 W. A wide-band amplifier, Marconi type TF 2167, delivers 10 W up to 80 MHz without tuning, and appears to be even more appropriate than the HP230A for most work.

Noise. The concept of noise decoupling was introduced by Ernst[22] and is achieved by noise modulation of a carrier frequency followed by power amplification. A typical arrangement is shown in Figure 2. Several papers featuring noise decoupling have appeared recently.[16,17,23] Maciel and co-workers[16] described a method whereby the r.f. carrier frequency of the Varian noise decoupler could be set to a fixed offset from a standard synthesizer-derived frequency of 100 MHz. One might add here that proton noise decouplers are now standard equipment with ^{13}C spectrometers, where

---

[20] R. C. Hewitt, *Rev. Sci. Instr.*, 1968, **39**, 1066.
[21] G. A. Olah, R. D. Porter, and D. P. Kelly *J. Amer. Chem. Soc.*, 1971, **93**, 464.
[22] R. R. Ernst, *J. Chem. Phys.*, 1966, **45**, 3845.
[23] R. Burton and L. D. Hall, *Canad. J. Chem.*, 1970, **48**, 59.

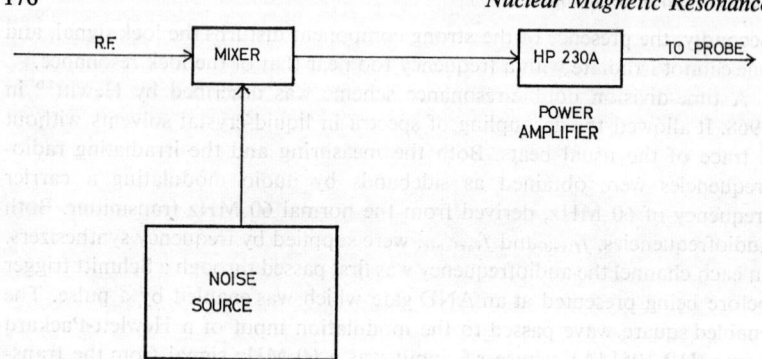

Figure 2 *A typical arrangement for noise decoupling*

the increased signal-to-noise ratio resulting from multiplet collapse and Overhauser enhancement is most welcome.

Universal Shift Reference. It is convenient to point out here that the act of performing a double-resonance experiment, or indeed the use of heteronuclear locking techniques with related frequencies, enables resonance frequencies of all nuclei to be reported at a field which corresponds to proton resonance in tetramethylsilane at a standard frequency of 60 or 100 MHz. The merits of this technique have been discussed.[24]

Freeman and Gestblom[25] performed some very precise experiments involving saturation of parts of resonance lines. The importance of an exact dispersion mode for the lock signal was emphasized. In the short term, the lock condition was good to ±0.001 Hz, whereas long-term stability was about ±0.02 Hz. Very slow sweeps were achieved (15 mHz min⁻¹), the sweep frequency being the divided (by 1000) output of a frequency synthesizer. For most purposes the long-term stability of the lock condition is quite adequate!

5 The Probe

Probes are inevitably becoming more complex with the increasing demands made by the requirements for multinuclear operation, double-resonance capability, and various field–frequency lock modes. Provision for variable-temperature operation with large spinning samples must now also be made. Probes are narrow-band devices in the sense that input and output circuitry must be tuned to the appropriate frequency. The functions outlined above are typically met by the provision in the probe of several coils. A further requirement is that the receiver be isolated from the transmitter, since the nuclear signals are weak compared to the transmitter power level.

A. Single-coil Systems.—Here one coil acts both as transmitter and receiver.

---

[24] R. J. Chuck, D. G. Gillies, and E. W. Randall, *Mol. Phys.*, 1969, **16**, 121.
[25] R. Freeman and B. Gestblom, *J. Chem. Phys.*, 1968, **48**, 5008.

It is tuned to the frequency of interest and typically forms one arm of a twin-tee bridge. The bridge is balanced in the off-resonance condition so that the signal observed at the receiver is essentially only the nuclear resonance signal. Maciel and co-workers[16] wound an extra single coil for ^{19}F lock at 94·1 MHz around the normal 25·1 MHz receiver coil in an HA-100 probe. The coil was one arm of a magic or hybrid tee, a device which performs the same function as the more usual twin-tee bridge. A filter was inserted between the coil and the tee to ensure that the strong decoupling frequency present in the probe did not appear in the lock channel. The use of hybrid tees in n.m.r. was first described by Klein and Phelps.[26] A typical device will work over a wide range of frequencies, although once set up in a particular circuit it suffers from the same frequency sensitivity of its balance as the normal twin-tee network. They are excellent for home-made heteronuclear internal or external lock channels. In both the Bruker and the Varian XL-100 systems the receiver coil is doubly tuned. It is used simultaneously in the crossed-coil mode for the observation channel and as a single coil for the locking channel.

Single-coil systems are used extensively in pulsed spectrometers. They produce a stronger r.f. field for a given input voltage, although with inferior homogeneity compared to other arrangements, *e.g.* crossed coils.

B. Crossed-coil Systems.—Here, the separate and mutually orthogonal transmitter and receiver coils may be tuned independently and are typically separated by a Faraday shield to minimize electrostatic coupling. The insert geometry is arranged such that the coils are nearly electrically orthogonal. The leakage to the receiver coil is minimized by the adjustment of two mechanical paddles which provide capacitative and resistive coupling. Additional electrical balance is provided in the Bruker system. Geils[27] has discussed the advantages (particularly at low temperatures) of electrical probe balancing in crossed-coil systems and has constructed a probe with no flux paddles. By the use of a Faraday shield he was able to reduce the leakage level to 10^{-3} times the transmitter level. The balancing circuitry then matched this residual leakage in amplitude and phase. Another technique which eliminates probe-balancing problems is the use of a time-shared modulation (see Section 8). Automatic probe-balancing techniques have been discussed[28] but there is no widespread use of such schemes.

Richards and co-workers[29] have described crossed-coil-type probes for use in a superconducting solenoid at frequencies of 10—100 MHz. The dimensions of the bore and the fact that the magnetic field is along the axis of the probe make construction awkward. One arrangement consisted of two pairs of Helmholtz coils embedded in epoxy resin. There was insufficient room for paddles, and receiver isolation from the transmitter was 50—60 dB. A less

---

[26] M. P. Klein and D. E. Phelps, *Rev. Sci. Instr.*, 1967, **38**, 1545.
[27] R. H. Geils, *Rev. Sci. Instr.*, 1971, **42**, 265.
[28] M. Birkle and G. Schulz, *J. Phys.* (*E*), 1969, **2**, 846 and references therein.
[29] J. D. Halliday, H. D. W. Hill, and R. E. Richards, *J. Phys.* (*E*), 1969, **2**, 29.

conventional approach, although still of crossed-coil type, was also described which was a development of a design by Redfield.[30] This allowed for a much more compact construction with an intrinsic isolation of 70 dB which, with the addition of paddles, improved to 100 dB.

A crossed-coil design by Mock,[31] designed for operation at 30—60 MHz, allows for rotation of the receiver coil to optimize receiver isolation at 60 dB over the temperature range 1—300 K.

C. Double Resonance.

—An extra transmitter coil tuned to the appropriate frequency is usually provided for heteronuclear decoupling. A less versatile method is to double-tune the transmitter coil. If only a low-level double-resonance experiment is necessary, then it is quite adequate to connect the second frequency directly to the transmitter coil through a blocking circuit tuned to the observing frequency.[19] However, for noise decoupling purposes, where the average power dissipation is high, the coil is made of suitably heavy wire or foil.

Burton and Hall[23] have described modifications of a Varian HA-100 probe which allow for heteronuclear double resonance. Detailed circuit descriptions are given for plug-in modules which provide for double tuning of the probe. Also described is the winding of an additional transmitter coil on to a 40·5 MHz ^{31}P probe for proton decoupling at 100 MHz.

Long and Moritz[32] wound extra separate coils for power decoupling of ^{14}N, ^{2}H, and ^{31}P in an HA-60-IL system. All circuit details were given.

D. Field–Frequency Lock.

—The operation of a single-sample lock requires the simultaneous presence and detection of two frequencies using arrangements such as those described above.

The two-sample technique is especially convenient, and offers ease of operation since the lock is not lost when the experimental sample is withdrawn. The second (permanent) sample is placed as close to the experimental sample as possible so that they experience as nearly equal magnetic fields as possible. The control sample has its own coil system and r.f. circuitry which may be optimized independently of the settings for the observation channel.

E. General Construction.

—The ability to make the flexible probe required by a truly multinuclear approach is governed to a large extent by the field gap of the magnet. If the gap is small, one is usually forced to make a single-purpose probe. The first essential flexibility is the provision for changing the receiver insert. A given insert will typically cover about an octave frequency range e.g. 6—12, 12—25, 25—60, 50—100 MHz. The design is complicated by the requirement for operation over a range of temperatures. This necessitates provision for a flow of nitrogen gas and for thermal shielding by

---

[30] A. G. Redfield, *Rev. Sci. Instr.*, 1956, **27**, 230.
[31] J. B. Mock, *Rev. Sci. Instr.*, 1970, **41**, 129.
[32] G. J. Long and A. G. Moritz, *Mol. Phys.*, 1968, **15**, 439.

Experimental Techniques

means of a dewar. The insert may be placed inside the dewar (XL-100 system) or include the dewar assembly and the proton noise decoupling coil (Bruker system). The transmitter and double-resonance coils may be tuned and matched to their respective transmitters by the addition of a plug-in module. The receiver coil is tuned and matched to the observing and lock preamplifiers in a similar manner.

F. Field Modulation.—Coils are provided for this purpose. The greater range of chemical shifts of nuclei other than hydrogen has raised the frequency required for modulation. Early probes such as the Varian V4331 were of massive construction, and the modulation frequency was limited to a few kilohertz by eddy-current losses in the metal. Modern probes, though still of strong construction, are 'thin sided' to minimize these losses. By mounting coils in silicone rubber to reduce microphonics, leaving slots in the sides to minimize eddy currents, the modulation frequency which is attainable has been raised to about 40 kHz.

G. Variable-temperature Operation. Normal high-resolution probes are limited to the range -150 to $+200$ °C. The traditional method for maintaining a constant temperature in the probe employs the flow of nitrogen gas at a controlled temperature over the sample. The body of the probe is protected from temperature changes by the inclusion of a dewar vessel around it. However, small changes in coil geometry necessitate adjustment of probe balance and detector phase as the temperature changes. The resolution of the probe will also change with temperature. The sensitivity of permanent magnets to changes in sample temperature has been minimized in the Perkin-Elmer systems by circulating water which is at the same temperature as the magnet through the walls of the probe.

The accuracy of temperature measurement and the absence of temperature gradient depend on the probe design. In the simple system the control heater and sensor are immediately below the sample tube, around which the gas stream passes upwards. The temperature of the probe is usually precalibrated by measuring internal shifts in neat methanol (low-temperature range) and in neat glycol (high-temperature range). An apparatus for measuring temperature directly in a spinning sample tube by means of a thermistor has been described.[33] This was mounted on a thin brass tube which could if necessary be raised during the spectral run so as not to degrade the resolution of the n.m.r. spectrum.

Better control with smaller temperature gradients may be anticipated from systems employing counter-flow techniques, in which the exit gas is in thermal contact with the inlet gas. The original design of Shoolery and Roberts[34] incorporated this principle, as does the latest probe for the Perkin-Elmer R12. The latter device also eliminates the use of refrigerant baths for low-

[33] A. L. Van Geet, *Rev. Sci. Instr.*, 1969, **40**, 177.
[34] J. N. Shoolery and J. D. Roberts, *Rev. Sci. Instr.*, 1957, **28**, 61.

temperature operation by expanding high-pressure gas from the cylinder through a nozzle, cooling being achieved by the Joule–Thomson effect. The design also eliminates the need for dry air to be used to drive the spinner since the temperature-control gas leaves through the bottom of the probe.

Norris and Strange[35] have described a probe for use in pulse studies with non-spinning samples, in which liquid nitrogen is introduced directly into the probe, resulting in much improved thermal efficiency and temperatures right down to the boiling point (78 K). Temperature stability of ± 0.5 K could be maintained indefinitely and was not disturbed by replenishing the coolant.

For temperatures higher than 200—300 °C special probes are required. Temperatures in excess of 1100 °C[36] and up to 1000 °C[37] have been attained by placing the electrical heater windings immediately around the sample area.

H. Rapid Sample Spinning.—This is a technique which has been developed mainly by Andrew and which he has recently reviewed.[38] The fast spinning is performed along an axis which is at 54°44′ to the magnetic field, and the result is that dipole–dipole interactions in solids tend to be averaged to zero. Rates of spin of up to 800 Hz have been achieved in a high-resolution probe.[39] Higher rates, up to *ca.* 10 kHz, require special design and utilize either a gas bearing or an axle supported system.

Recently, the measurement of ^{31}P chemical shifts in stationary solid samples has been reported[40,41] using a conventional high-resolution spectrometer.

6 The Preamplifier

Preamplifiers are normally placed as close to the receiver coil as possible, because the signal-to-noise ratio of the instrument is governed by the noise present at the preamplifier input. The advent of solid-state devices has enabled compact and convenient designs to be made. In a typical multinuclear arrangement one still chooses a preamplifier specific to a narrow frequency range with a typical gain of 40—60 dB. For optimum operation the receiver coil is tuned to resonance and its impedance transformed and exactly matched to the input impedance of the preamplifier. Field-effect transistors are commonly used in this situation, often in cascode arrangement. In some systems the preamplifier also includes a mixer so that the output signal is at an intermediate frequency. This situation obtains in the Bruker system and in the Varian 220 MHz spectrometer.

7 The Receiver

Typically, this unit uses a superheterodyne principle. A schematic diagram of

---

[35] M. O. Norris and J. H. Strange, *J. Phys. (E)*, 1969, **2**, 1106.
[36] R. L. Odle and C. P. Flynn, *Rev. Sci. Instr.*, 1964, **35**, 1611.
[37] S. Hafner and N. H. Nachtrieb, *Rev. Sci. Instr.*, 1964, **35**, 680.
[38] E. R. Andrew, 'Progress in N.M.R. Spectroscopy', Pergamon Press, 1971, vol. 8, p. 1.
[39] M. Cohn, A. Kowalsky, J. S. Leigh, and S. Maricic, 'Magnetic Resonance in Biological Systems', Pergamon Press, 1967, p. 45.
[40] K. B. Dillon and T. C. Waddington, *Spectrochim. Acta*, 1971, **27A**, 1381.
[41] K. B. Dillon and T. C. Waddington, *Nature Phys. Sci.*, 1971, **230**, 158.

Experimental Techniques 181

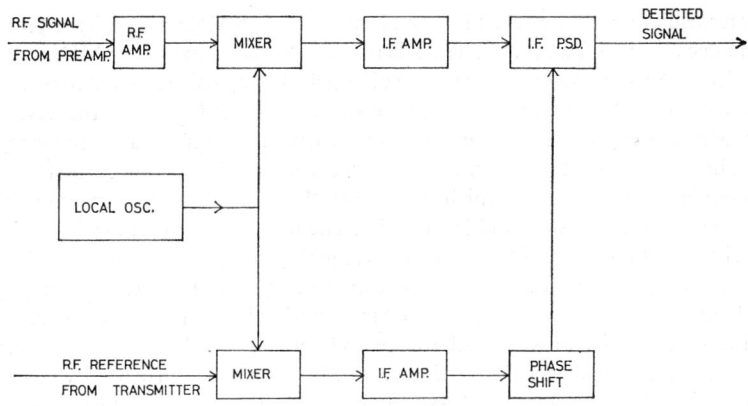

Figure 3 *Schematic diagram of Varian V4311 receiver*

a typical unit, such as the Varian V4311, is shown in Figure 3. The local oscillator is crystal-controlled and chosen so that the intermediate frequency is close to 5 MHz. Any modulation information on the signal is presented at the output as detected audiofrequencies, which are then processed through an audio amplifier and phase-sensitive detectors (p.s.d.'s). This design does not lend itself to multinuclear operation. Although a given unit may be operated over a small range of frequencies, one has to change the quartz crystal each time and align several tuned circuits. A more flexible means of generating a local oscillator frequency is required.

Gillies[19] has used a receiver constructed by Decca, in which the local oscillator is phase-locked to the (swept) radiofrequency (see Figure 4). The

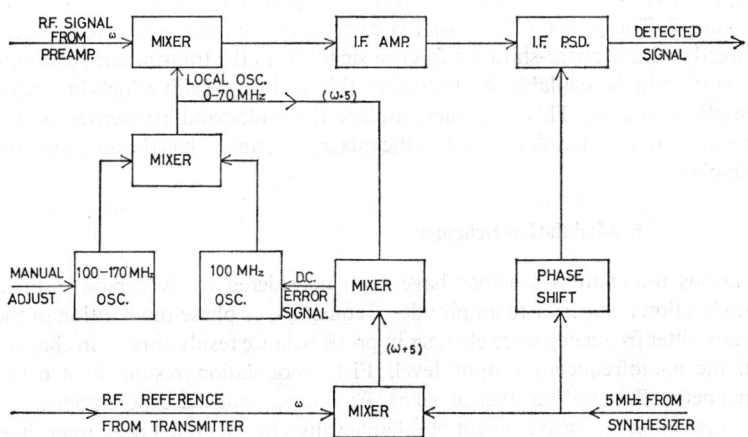

Figure 4 *Schematic diagram[19] of Decca receiver for* 1–70 MHz

intermediate frequency (i.f.) is then always exactly 5 MHz, minimizing phase changes in the i.f. amplifier. The receiver covers the range 1—70 MHz.

In the XL-100 system, the same result is achieved by using as local oscillator a voltage-controlled crystal oscillator which is phase-locked to the system master-crystal. The local oscillator tracks with the swept signal frequency, maintaining a constant intermediate frequency. The i.f., of course, carries the modulation information which constitutes the n.m.r. signals corresponding to centreband and sideband frequencies. The use of a crystal filter in the i.f. strip allows selection of the particular frequency component of interest. This constitutes a single-sideband detection scheme. In the normal audiofrequency phase-detection process, n.m.r. signal information is obtained by phase selection of a particular sideband (or centreband). The other audiofrequency components not detected are still present, but contain information which is unusable. Their presence in the signal applied to the i.f. phase detector does contribute to system noise, and it is this factor that is eliminated in the single-sideband system.

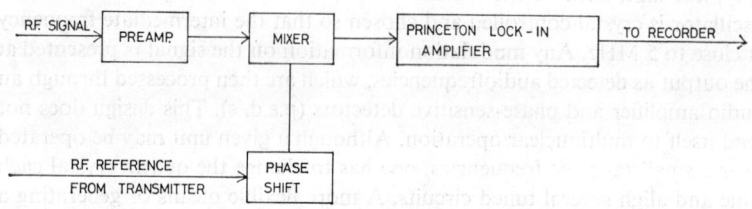

Figure 5 *Schematic diagram of receiver of Richards et al.* [Reproduced by permission from *J. Phys.* (*E*), 1969, **2**, 29]

Richards and co-workers[28] have described a system which does not use the superheterodyne technique. The schematic diagram for this system is shown in Figure 5. The preamplifier signal is applied to a mixer, where it is mixed directly with a phase-shifted reference signal from the transmitter. Adequate overall gain is available by following this with a sensitive lock-in audio-amplifier system. This approach utilizes the wide-band properties of the balanced-diode modulator used as the mixer. It certainly has the advantage of simplicity.

8 Modulation Schemes

Various modulation schemes have been considered by Anderson.[42] Field modulation is superior to amplitude-, frequency-, or phase-modulation of the transmitter frequency since changes in probe balance result directly in changes in the audiofrequency output level. Field modulation results in a direct magnetic effect on the nuclear spins, producing good baseline stabilization and enabling the measurement of reliable values of integrated peak intensities

[42] W. A. Anderson, *Rev. Sci. Instr.*, 1962, **33**, 1160.

Experimental Techniques

when audiofrequency phase-sensitive detection is used. However, it is still possible to produce spurious signals by the modulation coils acting as loudspeakers and producing mechanical vibration of the probe. This effect is minimized by good probe design.

By selecting the phase of the audiofrequency component in the output from the r.f. receiver, operation on centreband or sidebands in absorption or dispersion mode may be achieved. The modulation requirements are different,[27,43,44] centreband operation requiring low r.f. power with a high modulation index and sideband operation requiring the opposite for optimum signal-to-noise ratio. Sideband operation requires less-critical setting-up. The sideband frequency should clearly be larger than the total range of chemical shifts and the amplitude of the modulation greater than the linewidths. This is in contrast to broad-line techniques, where one observes the derivative spectrum and where both the amplitude and frequency of the modulation must be much less than the linewidth to obviate line distortion.

The presence of two audiofrequency modulations with simultaneous phase-sensitive detection allows one output in the dispersion mode to be used for field–frequency control, leaving the other for observation of the spectrum. Sweeping one frequency with respect to the other provides for field or frequency sweep.

Changing an audiofrequency modulation, particularly over the wide ranges demanded by nuclei other than hydrogen, produces large phase shifts which result in spectral distortion in frequency-sweep mode and deterioration of the field–frequency lock in field-sweep mode. There are several ways of compensating for this difficulty. One method is to design a circuit which produces a frequency-dependent phase shift which compensates for the deviation.[45] The range of this method is rather limited. Jenkins and Phillips[46] found that the major cause of the problem in the HA-100 system was a signal delay of 35 μs in the detection system. By building the same delay into the reference channel of each audio-frequency phase-detector, good phase compensation over a 20 kHz range was achieved.

Frequency-dependent effects have also been observed in c.w. ^{13}C spectra employing audiofrequency sweep techniques. These effects were eliminated by observing spectra at a fixed audiofrequency whilst the carrier frequency was swept. During this r.f. sweep the locking audiofrequency was swept in the opposite sense, so as to provide a constant field–frequency lock condition.

Anderson[42] also described a modulation scheme which was first suggested by Arnold in 1955. Several groups around the world use this system to advantage but its appearance on the commercial scene is eagerly awaited. The transmitter is gated on for a short time whilst the receiver is gated

---

[43] W. A. Anderson, 'N.M.R. and E.P.R. Spectroscopy', Pergamon Press, New York, 1960.
[44] O. Haworth and R. E. Richards, 'Progress in N.M.R. Spectroscopy', ed. J. W. Emsley, J. Feeney, and L. M. Sutcliffe, Pergamon Press, Oxford, 1966. vol. 1.
[45] P. Ackerman and F. W. van Denason, *J. Phys. (E)*, 1970, **3**, 811.
[46] P. N. Jenkins and L. Phillips, *J. Phys. (E)*, 1971, **4**, 530.

off. The nuclear signal induced by this burst of r.f. energy is observed by the receiver (gated on), during which time the transmitter is gated off. The method was subsequently demonstrated by Baker, Burd, and Root[47] and given the name 'time-sharing modulation'. The modulation rate is high compared to $1/T_2$, and sidebands are produced at multiples of the modulation frequency. The method has two great assets. The first is that the receiver never 'sees' the transmitter directly, so the output is not affected by changes in probe balance and good base-line stabilization ensues. The other point is that the repetition rate may be high, 50 or 100 kHz, so that one may observe either narrow-line signals spaced over a wide range or broad lines, using the same technique. Clearly, one has to increase the transmitter power to allow for the lower duty cycle, and also the fact that the receiver is not open for the whole time reduces the sensitivity slightly. The experiment may be set up simply by using solid-state mixers as gates and logic modules to generate the appropriate timing sequence. There are two other examples of this mode of operation.[18,48]

9 Computer Techniques

The first example of the use of computers in n.m.r. spectrometry was the so-called computer of average transients c.a.t. The machine is a special-purpose computer, typically with 1024 memory channels each with a capacity of 16 bits. Successive spectra are digitized and coherently added into the memory, producing a signal-to-noise enhancement proportional to the square root of the number of scans. Ernst[49] has discussed sensitivity enhancement in general. The spectra must be added coherently, which implies synchronization of the c.a.t. and spectral sweeps. For those spectrometers which do not have long-term stability of magnetic field the c.a.t. is arranged to search for a peak from which it derives a reference point before each scan. Any type of spectrum may be treated using a c.a.t. so long as it is reproducible. For instance, long-term time-averaged i.n.d.o.r. experiments are perfectly feasible with field–frequency lock and stable frequencies.[19,21]

More sophisticated use may be made of a computer when it is used not only for data processing but for controlling experimental parameters, monitoring machine performance, and general decision-making with respect to the conduct of the experiment. Ernst[7] used an on-line computer to control the magnet homogeneity by monitoring a control signal whilst optimizing the y gradient and curvature. Freeman and co-workers[50] used a computer to generate precise audiofrequencies for programming a divider working from a 25 MHz clock. Thus with computer control over both the frequency and the

---

[47] E. B. Baker, L. W. Burd, and G. N. Root, *Rev. Sci. Instr.*, 1965, **36**, 1495.
[48] E. Lippmaa, J. Past, A. Olivson, and T. Saluvere, *Eesti N.S.V. Teaduste Akad. Toimetised, Fuus.-Mat.*, 1966, **58**, 15.
[49] R. R. Ernst, 'Advances in Magnetic Resonance', ed. J. S. Waugh, Academic Press, 1966, vol. 2, p. 131.
[50] R. Freeman, S. Wittekoek, and R. R. Ernst, *J. Chem. Phys.*, 1970, **52**, 1529.

Experimental Techniques

timing of pulses, experiments can be performed which are not possible manually.

In Fourier transform n.m.r. the small computer has now become part of the apparatus. At the very least it is used to perform the Fourier transformation on free induction decay signals that have been accumulated in a c.a.t. The ability to perform these transformations quickly and easily in the laboratory is in the process of transforming n.m.r. spectroscopy itself! Allowing the computer a more sophisticated role by controlling and monitoring the spectrometer reaps rich rewards in that different types of experiment, such as relaxation measurements, may be performed at will. The subject has recently been reviewed elsewhere.[1]

10 Pulse Spectrometers

The instrumental requirements have recently been reviewed in detail by Farrar and Becker.[4] At first sight the requirements may appear similar to those of the c.w. mode but the detail will often be quite different. The first difference arises from the fact that the transmitter must be capable of producing short (1—100 μs) bursts of r.f. power with peak ratings of 100–10 000 W. In contrast, c.w. transmitters typically provide only one watt or less.

The probe must be able to cope with the consequent high r.f. voltages and the power must be quickly dissipated after the pulse, since it is then that the weak nuclear induction signal is measured. The probe may be of the single- or crossed-coil type. The former are easier to construct, and produce a higher but less uniform r.f. field for a given input voltage. In the crossed-coil type the receiver and transmitter coil adjustments may be optimized separately. This is very convenient since the requirements of fast transmitter recovery (low-Q circuit) and high sensitivity detection (high-Q) are mutually exclusive.

The usual way to overcome this problem in single-coil systems is shown in Figure 6, where the tuned sample circuit consists of L_1 and C_1, and the quarter-wave sections of r.f. cable and crossed diodes allow the circuit to effectively function with two Q values. Essentially the diode-short operates (producing a low-Q situation) only in the presence of the large transmitter pulse. As usual, the signal-to-noise ratio for the whole system is determined by the receiver coil–preamplifier interface, and clearly the preamplifier should have a low noise figure. An additional requirement is a fast recovery from the overload (or cut-off) that occurs during the pulse. During detection of the weak n.m.r. signal it is essential that there be no signal present arising from leakage of r.f. energy from the transmitter. Typical on/off ratios for transmitter gates are 10^9 or greater.

The linearity requirements for the preamplifier and main amplifier are much higher than for c.w. work since the system has to respond faithfully to the high voltages at the start of a decay and the progressively lower voltages during the decay. Typical input range is 1μV to several millivolts, giving output voltages of about 1mV to about one volt. The bandwidth of the receiver system should be wide enough to accommodate all frequency components of interest. In

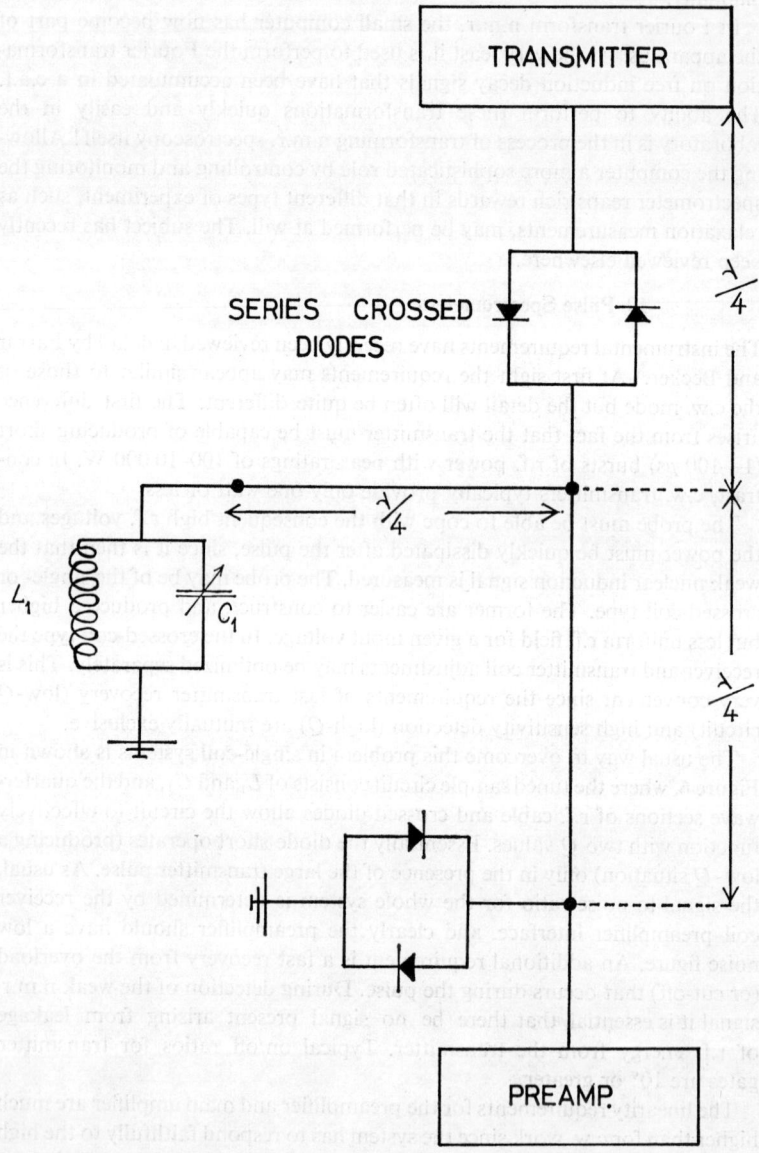

Figure 6 *Typical single-coil pulse arrangement.*
(Reproduced by permission from 'Pulse and Fourier Transform N.M.R.', Academic Press, London and New York, 1971)

Experimental Techniques

the high-resolution situation this produces no additional problem since the requirements for pulse and c.w. operation are about the same. In practice only minor modifications to the detection system are required. Phase-sensitive detection is to be preferrred, and the requisite high stability for the magnetic field is typically provided by a separate c.w. field–frequency lock system.

Although gating techniques have been used in c.w. spectroscopy, pulse work requires more sophisticated gating since the power levels are that much higher. Typically, an r.f. gate also provides r.f. power amplification.[51] Recently a gate with low power requirements using MOSFET devices has been described.[52] Working at 200 MHz it produced 100 dB suppression by providing 28 dB gain in the ON state and 72 dB of attenuation in the OFF state.

The remaining requirement is for a pulse programmer. The basic function of the device is to control the r.f. gate so as to provide pulses of r.f. energy of adjustable length at appropriate times. The programmer should also provide a trigger pulse to initiate the recording of data at the appropriate time. In some experiments the r.f. phase of the pulses in a sequence is changed by the programmer, which activates other circuits in the gate. Arrangement has also to be made to send the appropriate phase reference signal to the receiver.

The required pulsewidths may range from $1\mu s$ to 10 ms or more, and the intervals between pulses may be short (10 μs) or very long (1000 s), depending on the experiment. Suitable devices may be readily constructed using standard digital integrated circuitry. The various experimental sequences can then be selected at a flick of a switch. The time-base circuitry for the programmer should be crystal-controlled.

Interfacing a computer to the r.f. gate provides great flexibility of operation in that not only can one select the experiment by the choice of suitable software, but one can also modify the times within a sequence as the experiment progresses.

In comparing pulse and c.w. spectrometers, the simplifications in the apparatus should not be forgotten. There is no requirement for field modulation in pulse systems since, as in time-shared modulation schemes (see Section 8), the receiver is not on during the pulse. Provision of time-sharing for the lock channel would banish modulation coils from the probe. No sweep facility is required. The inverse pulsewidth determines the maximum bandwidth, and the carrier frequency, which may be derived from a synthesizer, is the offset.

The output from the pulse spectrometer is a time domain signal which, if simple, may be analysed manually. In the complex situation that pertains in high-resolution n.m.r. the free induction decay is also complex. The availability of small computers for laboratory use, together with the advent of the Fast Fourier Transform (f.f.t.) algorithm of Cooley and Tukey,[53] has enabled

---

[51] L. J. Burnett and J. F. Harman, *Rev. Sci. Instr.*, 1968, **39**, 1226.
[52] B. M. Moores, R. H. Munnings, and R. L. Armstrong, *Rev. Sci. Instr.*, 1970, **41**, 1096.
[53] J. W. Cooley and J. W. Tukey, *Math. of Comput.*, 1965, **19**, 296.

the processing of these signals to provide intelligible information in the frequency domain. The method can also readily be used to measure relaxation times of individual lines in a spectrum. The use of the Fourier transform in high-resolution n.m.r. spectroscopy, the detailed instrumental requirements, and applications have recently been reviewed,[1] and are not dealt with specifically in this volume. Some pulse experiments may be performed on standard high-resolution instruments with little modification. They are of necessity low-power experiments, but advantage is taken of this fact to measure the relaxation behaviour of individual lines in the spectrum. Freeman and co-workers[50] produced pulses by gating an audio field modulation frequency. Experiments have also been performed using an HA-100[54] and an A56/60.[55] In the latter system, r.f. gating was employed to produce 90° pulses in the usual way but a magnetic field pulse was used to produce 180° pulses about the z direction. An earlier paper[56] described modifications to a Varian A60 which enabled a variety of methods for measuring T_1 and T_2 to be used.

11 Measurement of Relaxation Times

A. Values of T_1.—C.W. Methods. There are several c.w. methods available and most of these are detailed in the book written by Pople, Schneider, and Bernstein.[2] Some of the methods in current use are cited below; the papers quoted contain all the relevant back references.

Brownstein and Bystrov[57] made a careful study of relaxation in some amides by the saturation recovery technique. The spectrometer was adjusted so that the peak of interest was at resonance; a weak (non-saturating) observing r.f. field was used. On suddenly increasing this field (B_1) to saturation level, a Torrey oscillation was produced[58] before the pen of the recorder settled at a lower value of the peak amplitude. The frequency of the oscillation (Ω) is a measure of the saturation field, as shown in the relation:

$$\Omega = \gamma B_1 \text{ (sat.)}$$

where γ is the magnetogyric ratio. The relation enables the frequency to be used as a measure of the effective r.f. field in these experiments. Some time later the r.f. level was returned to its original low value. The apparent recovery time is equal to T_1 if the observing field is low enough and the other listed conditions are met.

Loewenstein et al.[59] used the progressive saturation method to measure the deuterium relaxation time in C_6D_6.

Adiabatic fast-passage conditions are met when $(B_1/T_2) \ll (\partial B_0/\partial t) \ll \gamma B_1^2$. Under these conditions a passage through resonance inverts the mag-

---

[54] A. Ginsburg, A. Lipman, and G. Navon, *J. Phys. (E)*, 1970, **3**, 699.
[55] E. J. Wells and K. H. Abramson, *J. Magn. Resonance*, 1969, **1**, 378.
[56] J. E. Anderson, J. M. Steele, and A. Warwick, *Rev. Sci. Instr.*, 1967, **38**, 1139.
[57] S. Brownstein and V. Bystrov, *Canad. J. Chem.*, 1970, **48**, 243.
[58] H. C. Torrey, *Phys. Rev.*, 1949, **76**, 1059.
[59] Y. Egozy, A. Loewenstein, and B. L. Silver, *Mol. Phys.*, 1970, **19**, 177.

Experimental Techniques

netization. The T_1 method involves sweeping through resonance τ seconds later in the reverse direction with a weak observing field. The peak intensity reflects the magnetization present and so leads to a value for T_1. Parker and Jonas[60] analysed the method closely and found that the logarithmic plot of intensity versus τ was accurate. (See also ref. 61.) The method has been used[62] for ^{15}N.

If the adiabatic passage is stopped exactly on resonance, then the magnetization precesses around B_1 in the rotating frame and decays according to $T_{1\rho}$, which may or may not be equal to T_1. Deverell et al.[63] used the method to measure exchange rates in the chair–chair interconversion of cyclohexane and outlined the merits of the method for fast processes. Sykes[64] has also used the method for biological applications.

Pulse Methods. These are reviewed in reference 4 and are more flexible in terms of time-scale, the c.w. methods of necessity being appropriate for T_1 values of the order of seconds.

The 180°–τ–90° sequence is standard and simple for a single resonance line. For instance, T_1 values for ^{7}Li have been measured in this way.[65] In complex situations, Fourier transformation of the decay yields a frequency spectrum, which in the case of ^{13}C gives values of T_1 for individual carbon nuclei directly. The response of a spin system to a series of equally-spaced pulses of equal length, or indeed c.w. passages through resonance,[66] is governed by the relaxation times. The method has been applied in ^{13}C Fourier n.m.r. for measuring T_1 values.[67]

The selective pulse experiments of Freeman et al.,[50] mentioned in Section 10, yielded relaxation parameters for individual lines from which actual T_1 values could be deduced when the mechanisms of cross-relaxation had been elucidated. The power level was set to irradiate one line of width 0·1 Hz with a field equivalent to $\gamma B_1/2\pi = 0\cdot 5$ Hz, corresponding to a 180° pulse of 0·25 s. The timing of the pulses and their frequency were computer controlled, so that an arbitrary set of lines could be pulsed in an arbitrary order at arbitrary times. The output signal was not the usual decay but a series of single-cycle sine waves whose gradually reducing amplitude contained the relaxation information. A correction had to be applied to allow for the fact that during the 360° pulse the magnetization spent some time along the transverse axis and hence was also affected by T_2 processes.

The other selective pulse experiments mentioned in Section 10[54,55] produced rotary spin echoes giving $T_{1\rho}$ values. Fourier transform methods have been applied to measure $T_{1\rho}$ in o-dichlorobenzene.[68]

[60] R. G. Parker and J. Jonas, *Rev. Sci. Instr.*, 1970, **41**, 319.
[61] J. A. Glasel, *J. Phys. (E)*, 1968, **1**, 963.
[62] T. Saluvere and E. Lippmaa, *Chem. Phys. Letters*, 1970, **7**, 545.
[63] C. Deverell, R. E. Morgan, and J. H. Strange, *Mol. Phys.*, 1970, **18**, 553.
[64] B. D. Sykes, *J. Amer. Chem. Soc.*, 1969, **91**, 949.
[65] G. E. Hartwell and A. Allerhand, *J. Amer. Chem. Soc.*, 1971, **93**, 4415.
[66] D. C. Look and D. R. Locker, *Rev. Sci. Instr.*, 1970, **41**, 250.
[67] R. Freeman and H. D. W. Hill, *J. Chem. Phys.*, 1970, **53**, 4103.
[68] R. Freeman and H. D. W. Hill, *J. Chem. Phys.*, 1971, **55**, 1985.

B. Values of T_2.—*C.W. Methods.* In the case of quadrupolar nuclei, where the transverse relaxation is dominated by quadrupolar relaxation, the linewidth can be used to give T_2 directly. Thus Loewenstein *et al.*[59] used this method to measure the deuterium T_2 value in C_6D_6 in a nematic solvent. For narrow lines the linewidth is often inhomogeneity-dominated, and so the method is not as useful.

Pulse Methods. The effects of inhomogeneity may be minimized by the use of 180° pulses. The Carr–Purcell sequence is 90°–τ–180°–τ[echo]–180°–τ[echo]... and the true T_2 value is deduced from the echo signal amplitudes. The method has been applied to high-resolution spectra.[69] The echo for a single line was studied by using narrow-band detection with an audio-frequency phase-sensitive detector.

The values of T_2 for the ^{13}C nuclei in *o*-dichlorobenzene were deduced[68] by equating T_2 with $T_{1\rho}$ (see above).

12 Miscellaneous Topics

A. Stochastic Resonance.—This method has been described by Ernst[70] and Kaiser[71] and is discussed in reference 1. The experiment is best performed with pseudo-random noise generated by a computer. This noise is used to modulate the r.f. carrier and the output noise from the spectrometer is digitized and stored. Several runs may be accumulated. The output noise differs from the input noise in that it has been modified by the spin system. The Fourier transform of the input noise, when multiplied by that of the output noise, gives the frequency spectrum. The method may have some practical advantages over conventional Fourier analysis, especially for fluorine.[70]

B. Fourier Difference Spectroscopy.—If the presence of a strong resonance dominates the free induction decay, then after suitable processing this resonance may be used to act as a reference frequency for the other nuclei present, eliminating the need for a field–frequency lock.[72]

[69] R. Freeman and H. D. W. Hill, *J. Chem. Phys.*, 1971, **54**, 301
[70] R. R. Ernst, *J. Magn. Resonance* 1970, 3, 10.
[71] R. Kaiser, *J. Magn. Resonance* 1970, 3, 28.
[72] R. R. Ernst, *J. Magn. Resonance*, 1971, 4, 281.

5
Spectral Analysis

BY R. G. JONES

The trends in spectral analysis over the past twelve months have continued in the increasing, sometimes indiscriminate, use made of the growing number of computer programs available. This application is reflected in the large predominance of papers dealing with analysis of known systems sometimes repeated for the same molecule with better precision and sometimes correcting past errors. This is not an unexpected development since the theory of systems easily handled in the personal approach has been the subject of previous papers, and the computer is often preferred for its convenience, speed, and ability to tackle problems well beyond the limits of the more fundamental methods. This is not to say that the blanket computer approach should always be applied. One paper dealt with later in this report (Section 2B) presents a rational plea for the application of the fundamental approach whenever possible. This appeal, it is hoped, will not fall on unsympathetic ears, especially since the lessons to be learned from the fundamental approach can provide a useful background in rationalizing computerized results.

This report contains a summary of the papers on spectral analysis divided into three main categories: (i) New Spin Systems Studied (ii) New (Original) Methods of Tackling Known Systems (iii) Known Systems Studied. It is heavily weighted toward Section 3 in numbers of papers, but Sections 1 and 2 have been given a proportional weighting in detail of presentation to maintain some balance in the space occupied.

1 New Spin Systems Studied

Analysis of N.M.R. Spectra of the Type $[AMX_n]_2$[1] with $J(XX') = 0$.—The analysis of the M and X parts of $[AMX_n]_2$ systems is described with the explicit expressions derived for M and X transitions when $J(XX') = 0$. The problem is discussed using parameters conveniently similar to the parameters of the $[AX_n]_2$ systems.[2] The basic twelve parameters defining the system are ν_A, ν_M, ν_X, $J(AA')$ [$\equiv J(A)$], $J(MM')$ [$\equiv J(M)$], $J(XX')$ (assumed to be zero), $J(AM)$, $J(AM')$, $J(AX)$, $J(AX')$, $J(MX)$, $J(MX')$: $N(ij)$ and $L(ij)$ parameters are defined as

[1] B. E. Mann, *J. Chem. Soc.* (*A*), 1970, 3050.
[2] R. K. Harris, *Canad. J. Chem.*, 1964, **42**, 2275.

$N(\mathrm{AM}) = |J(\mathrm{AM}) + J(\mathrm{AM}')|$; $N(\mathrm{AX}) = |J(\mathrm{AX}) + J(\mathrm{AX}')|$;
$$N(\mathrm{MX}) = |J(\mathrm{MX}) + J(\mathrm{MX}')|$$
$L(\mathrm{AM}) = |J(\mathrm{AM}) - J(\mathrm{AM}')|$; $L(\mathrm{AX}) = |J(\mathrm{AX}) - J(\mathrm{AX}')|$;
$$L(\mathrm{MX}) = |J(\mathrm{MX}) - J(\mathrm{MX}')|$$

The total magnetic quantum numbers

$$m_\mathrm{T}(\mathrm{X}) = \sum_1^n m(\mathrm{X}_i) = x \quad \text{and} \quad m_\mathrm{T}(\mathrm{X}') = \sum_1^n m(\mathrm{X}'_i) = y$$

are defined with m the eigenvalue of the spin operator I_z for the nucleus in parentheses.

The $2^{2n+4} \times 2^{2n+4}$ matrix which constitutes the mathematical problem can be factorized extensively using $m_\mathrm{T}[\mathrm{AMX}_n]_2$ and in particular for the case considered here, $J(\mathrm{XX}') = 0$ (coupling between nuclei within the magnetically equivalent groups can be omitted completely).[3]

The basis functions are described in general in the form $m(\mathrm{A})$, $m(\mathrm{A}')$, $m(\mathrm{M})$, $m(\mathrm{M}')$, x, and y; the determinant of the full Hamiltonian matrix can be factorized according to the magnetic quantum numbers $m_\mathrm{A} = m(\mathrm{A}) + m(\mathrm{A}')$, $m_\mathrm{M} = m(\mathrm{M}) + m(\mathrm{M}')$, and $m_\mathrm{X} = m_\mathrm{T}(\mathrm{X}) + m_\mathrm{T}(\mathrm{X}') = x + y$ since $I_{z\mathrm{A}}$, $I_{z\mathrm{M}}$, and $I_{z\mathrm{X}}$ commute with the full Hamiltonian.

The general functions which form the basis for one-, two-, and four-dimensional sub-matrices of the Hamiltonian are as follows:

(i) $\alpha\alpha\alpha\alpha\ (x,y)$
(ii) $\alpha\alpha\beta\beta\ (x,y)$
(iii) $\beta\beta\alpha\alpha\ (x,y)$
(iv) $\beta\beta\beta\beta\ (x,y)$ $\Big\}$ 1×1

(v) $\alpha\beta\alpha\alpha\ (x,y)$; $\beta\alpha\alpha\alpha\ (x,y)$
(vi) $\alpha\beta\beta\beta\ (x,y)$; $\beta\alpha\beta\beta\ (x,y)$
(vii) $\alpha\alpha\alpha\beta\ (x,y)$; $\alpha\alpha\beta\alpha\ (x,y)$
(viii) $\beta\beta\alpha\beta\ (x,y)$; $\beta\beta\beta\alpha\ (x,y)$ $\Big\}$ 2×2

(ix) $\alpha\beta\alpha\beta\ (x,y)$; $\beta\alpha\alpha\beta\ (x,y)$
$\alpha\beta\beta\alpha\ (x,y)$; $\beta\alpha\beta\alpha\ (x,y)$ $\Big\}$ 4×4

The one- and two-dimensional sub-matrices have corresponding determinants with analytical solutions but the 4×4 determinants have no simple solutions in general. However, in cases where $2|J(\mathrm{M})| \ll |L(\mathrm{AM}) \pm (x-y)L(\mathrm{AX})|$, the 4×4 determinants factorize respectively into two 2×2 matrices (though $x = y$ leads to a special solution). In all other cases the problem can be solved using a computer. The whole spectral analysis has been treated by the author without using symmetry explicitly.

The M and X transition frequencies are listed in Tables 1 and 2. The parameter χ introduced in Table 2 is convenient since the selection rule for χ transitions becomes $\Delta\chi = \pm 1$; χ is also used as a subscript to label relevant parameters.

[3] A. Saupe and J. Nehring, *J. Chem. Phys.*, 1967, **47**, 5459.

Spectral Analysis

Table 1 *Transition frequencies and intensities for the M nuclei of an $[AMX_n]_2$ system where $J(X) = 0$. The v_M term is omitted*

Transition	Energy	Intensity
$2 \to 1$	$\frac{1}{2}\{J(M)+N(AM)+(x+y)N(MX)-A\}$	$(1+g_e)Y$
$3 \to 1$	$\frac{1}{2}\{J(M)+N(AM)+(x+y)N(MX)+A\}$	$(1-g_e)Y$
$6 \to 2$	$\frac{1}{2}\{-J(M)+N(AM)+(x+y)N(MX)+A\}$	$(1+g_e)Y$
$6 \to 3$	$\frac{1}{2}\{-J(M)+N(AM)+(x+y)N(MX)-A\}$	$(1-g_e)Y$
$7 \to 4$	$\frac{1}{2}\{J(M)+2yJ(MX)+2xJ(MX')+B-C\}$	$\frac{1}{2}(1+g_f)Y$
$7 \to 5$	$\frac{1}{2}\{J(M)+2yJ(MX)+2xJ(MX')-B-C\}$	$\frac{1}{2}(1-g_f)Y$
$8 \to 4$	$\frac{1}{2}\{J(M)+2yJ(MX)+2xJ(MX')+B+C\}$	$\frac{1}{2}(1-g_f)Y$
$8 \to 5$	$\frac{1}{2}\{J(M)+2yJ(MX)+2xJ(MX')-B+C\}$	$\frac{1}{2}(1+g_f)Y$
$9 \to 4$	$\frac{1}{2}\{J(M)+2xJ(MX)+2yJ(MX')+B-D\}$	$\frac{1}{2}(1+g_g)Y$
$9 \to 5$	$\frac{1}{2}\{J(M)+2xJ(MX)+2yJ(MX')-B-D\}$	$\frac{1}{2}(1-g_g)Y$
$10 \to 4$	$\frac{1}{2}\{J(M)+2xJ(MX)+2yJ(MX')+B+D\}$	$\frac{1}{2}(1-g_g)Y$
$10 \to 5$	$\frac{1}{2}\{J(M)+2xJ(MX)+2yJ(MX')-B+D\}$	$\frac{1}{2}(1+g_g)Y$
$12 \to 7$	$\frac{1}{2}\{-J(M)+2xJ(MX)+2yJ(MX')+C-B\}$	$\frac{1}{2}(1+g_f)Y$
$12 \to 8$	$\frac{1}{2}\{-J(M)+2xJ(MX)+2yJ(MX')-C-B\}$	$\frac{1}{2}(1-g_f)Y$
$12 \to 9$	$\frac{1}{2}\{-J(M)+2yJ(MX)+2xJ(MX')+D-B\}$	$\frac{1}{2}(1+g_g)Y$
$12 \to 10$	$\frac{1}{2}\{-J(M)+2yJ(MX)+2xJ(MX')-D-B\}$	$\frac{1}{2}(1-g_g)Y$
$13 \to 7$	$\frac{1}{2}\{-J(M)+2xJ(MX)+2yJ(MX')+C+B\}$	$\frac{1}{2}(1-g_f)Y$
$13 \to 8$	$\frac{1}{2}\{-J(M)+2xJ(MX)+2yJ(MX')-C+B\}$	$\frac{1}{2}(1+g_f)Y$
$13 \to 9$	$\frac{1}{2}\{J(M)+2yJ(MX)+2xJ(MX')+D+B\}$	$\frac{1}{2}(1-g_g)Y$
$13 \to 10$	$\frac{1}{2}\{-J(M)+2yJ(MX)+2xJ(MX')-D+B\}$	$\frac{1}{2}(1+g_g)Y$
$14 \to 11$	$\frac{1}{2}\{J(M)-N(AM)+(x+y)N(MX)-A\}$	$(1+g_e)Y$
$15 \to 11$	$\frac{1}{2}\{J(M)-N(AM)+(x+y)N(MX)+A\}$	$(1-g_e)Y$
$16 \to 14$	$\frac{1}{2}\{-J(M)-N(AM)+(x+y)N(MX)+A\}$	$(1+g_e)Y$
$16 \to 15$	$\frac{1}{2}\{-J(M)-N(AM)+(x+y)N(MX)-A\}$	$(1-g_e)Y$

$g_e = J(A)/A$

$g_f = \{(x-y)L(AX)[(x-y)L(AX)+L(AM)]+J^2(A)\}/(BC)$

$g_g = \{(x-y)L(AX)[(x-y)L(AX)-L(AM)]+J^2(A)\}/(BD)$

$Y = nC_{(n/2+x)} \times nC_{(n/2+y)}$

Where $A = \sqrt{[(x-y)^2L^2(MX)+J^2(M)]}$

$B = \sqrt{[(x-y)^2L^2(AX)+J^2(A)]}$

$C = \sqrt{\{[(x-y)L(AX)+L(AM)]^2+J^2(A)\}}$

$D = \sqrt{\{[(x-y)L(AX)-L(AM)]^2+J^2(A)\}}$

$\tan 2\theta = J(A)/\{\sqrt{[(x-y)L(AX)]}\}$

$\tan 2\phi = J(A)/\{\sqrt{[(x-y)L(MX)]}\}$

$\tan 2\omega = J(A)/\{\sqrt{[(x-y)L(AX)+L(AM)]}\}$

$\tan 2\delta = J(A)/\{\sqrt{[(x-y)L(AX)-L(AM)]}\}$

It is assumed that $2|J(M)| \ll |L(AM) \pm (x-y)L(AX)|$

The author has commented briefly on the comparison between the X part of the $[AMX_n]_2$ system and the X part of the $[AX_n]_2$ system. The sharp doublet observed in $X[AX_n]_2$ spectra for $m(A) = \pm 1$, with separation $N(AX)$, is replaced in the $X[AMX_n]_2$ spectra by two sharp doublets, separation $N(AX) \pm N(MX)$, for $m(A) + m(M) = \pm 2$ and a series of doublets of separation $N(AX) \pm \{[\chi^2 L^2(MX) + J^2(M)][(\chi-1)^2 L^2(MX) + J^2(M)]\}^{\frac{1}{2}}$ for $m(A) + m(M) = 0$ [from $m(A), m(M) = \pm 1$]. The intensity is equally divided between these $m(A) + m(M) = (\pm 2)$ and 0 states.

Table 2 *Transition frequencies and intensities for the X nuclei of an $[AMX_n]_2$ system where $J(X) = 0$. The $v(X)$ term is omitted*

Sub-matrix	Transition†	Energy†	Intensity†
(i)	$1_{\chi-1} \to 1_\chi$	$\frac{1}{2}[N(AX)+N(MX)]$	$n \times 2^{2n}$
(vii)	$2_{\chi-1} \to 2_\chi$	$\frac{1}{2}[N(AX)+A_\chi-A_{\chi-1}]$	$\frac{1}{2}(1+g_a)Z_{(x)}$
	$2_{\chi-1} \to 3_\chi$	$\frac{1}{2}[N(AX)-A_\chi-A_{\chi-1}]$	$\frac{1}{2}(1-g_a)Z_{(x)}$
	$3_{\chi-1} \to 2_\chi$	$\frac{1}{2}[N(AX)+A_\chi+A_{\chi-1}]$	$\frac{1}{2}(1-g_a)Z_{(x)}$
	$3_{\chi-1} \to 3_\chi$	$\frac{1}{2}[N(AX)-A_\chi+A_{\chi-1}]$	$\frac{1}{2}(1+g_a)Z_{(x)}$
(v)	$4_{\chi-1} \to 4_\chi$	$\frac{1}{2}[N(MX)+B_\chi-B_{\chi-1}]$	$\frac{1}{2}(1+g_b)Z_{(x)}$
	$4_{\chi-1} \to 5_\chi$	$\frac{1}{2}[N(MX)-B_\chi-B_{\chi-1}]$	$\frac{1}{2}(1-g_b)Z_{(x)}$
	$5_{\chi-1} \to 4_\chi$	$\frac{1}{2}[N(MX)+B_\chi+B_{\chi-1}]$	$\frac{1}{2}(1-g_b)Z_{(x)}$
	$5_{\chi-1} \to 5_\chi$	$\frac{1}{2}[N(MX)-B_\chi+B_{\chi-1}]$	$\frac{1}{2}(1+g_b)Z_{(x)}$
(ii)	$6_{\chi-1} \to 6_\chi$	$\frac{1}{2}[N(AX)-N(MX)]$	$n \times 2^{2n}$
(ix)*	$7_{\chi-1} \to 7_\chi$	$\frac{1}{2}\left[\frac{\chi}{|\chi|}L(MX)+C_\chi-C_{\chi-1}\right]$	$\frac{1}{4}(1+g_c)Z_{(x)}$
	$7_{\chi-1} \to 8_\chi$	$\frac{1}{2}\left[\frac{\chi}{|\chi|}L(MX)-C_\chi-C_{\chi-1}\right]$	$\frac{1}{4}(1-g_c)Z_{(x)}$
	$8_{\chi-1} \to 7_\chi$	$\frac{1}{2}\left[\frac{\chi}{|\chi|}L(MX)+C_\chi+C_{\chi-1}\right]$	$\frac{1}{4}(1-g_c)Z_{(x)}$
	$8_{\chi-1} \to 8_\chi$	$\frac{1}{2}\left[\frac{\chi}{|\chi|}L(MX)-C_\chi+C_{\chi-1}\right]$	$\frac{1}{4}(1+g_c)Z_{(x)}$
	$9_{\chi-1} \to 9_\chi$	$\frac{1}{2}\left[-\frac{\chi}{|\chi|}L(MX)+D_\chi-D_{\chi-1}\right]$	$\frac{1}{4}(1+g_d)Z_{(x)}$
	$9_{\chi-1} \to 10_\chi$	$\frac{1}{2}\left[-\frac{\chi}{|\chi|}L(MX)-D_\chi-D_{\chi-1}\right]$	$\frac{1}{4}(1-g_d)Z_{(x)}$
	$10_{\chi-1} \to 9_\chi$	$\frac{1}{2}\left[-\frac{\chi}{|\chi|}L(MX)+D_\chi+D_{\chi-1}\right]$	$\frac{1}{4}(1-g_d)Z_{(x)}$
	$10_{\chi-1} \to 10_\chi$	$\frac{1}{2}\left[-\frac{\chi}{|\chi|}L(MX)-D_\chi+D_{\chi-1}\right]$	$\frac{1}{4}(1+g_d)Z_{(x)}$
(iii)	$11_{\chi-1} \to 11_\chi$	$-\frac{1}{2}[N(AX)+N(MX)]$	$n \times 2^{2n}$
(vi)	$12_{\chi-1} \to 12_\chi$	$\frac{1}{2}[-N(MX)+B_\chi-B_{\chi-1}]$	$\frac{1}{2}(1+g_b)Z_{(x)}$
	$12_{\chi-1} \to 13_\chi$	$\frac{1}{2}[-N(MX)-B_\chi-B_{\chi-1}]$	$\frac{1}{2}(1-g_b)Z_{(x)}$
	$13_{\chi-1} \to 12_\chi$	$\frac{1}{2}[-N(MX)+B_\chi+B_{\chi-1}]$	$\frac{1}{2}(1-g_b)Z_{(x)}$
	$13_{\chi-1} \to 13_\chi$	$\frac{1}{2}[-N(MX)-B_\chi+B_{\chi-1}]$	$\frac{1}{2}(1+g_b)Z_{(x)}$
(viii)	$14_{\chi-1} \to 14_\chi$	$\frac{1}{2}[-N(AX)+A_\chi-A_{\chi-1}]$	$\frac{1}{2}(1+g_a)Z_{(x)}$
	$14_{\chi-1} \to 15_\chi$	$\frac{1}{2}[-N(AX)-A_\chi-A_{\chi-1}]$	$\frac{1}{2}(1-g_a)Z_{(x)}$
	$15_{\chi-1} \to 14_\chi$	$\frac{1}{2}[-N(AX)+A_\chi+A_{\chi-1}]$	$\frac{1}{2}(1-g_a)Z_{(x)}$
	$15_{\chi-1} \to 15_\chi$	$\frac{1}{2}[-N(AX)-A_\chi+A_{\chi-1}]$	$\frac{1}{2}(1+g_a)Z_{(x)}$
(iv)	$16_{\chi-1} \to 16_\chi$	$\frac{1}{2}[-N(AX)-N(MX)]$	$n \times 2^{2n}$

A, B, C, and D have the same values as in Table 1, where the subscript χ refers to the value of $(x-y)$.

$g_a = [\chi(\chi-1)L(MX)^2+J^2(A)]/(A_\chi A_{\chi-1})$
$g_b = [(\chi-1)L(AX)^2+J^2(A)]/(B_\chi B_{\chi-1})$
$g_c = \{[\chi L(AX)+L(AM)][(\chi-1)L(AX)+L(AM)]+J^2(A)\}/(C_\chi C_{\chi-1})$
$g_d = \{[\chi L(AX)-L(AM)][(\chi-1)L(AX)-L(AM)]+J^2(A)\}/(D_\chi D_{\chi-1})$

$$Z_{(x)} = \frac{1}{2n}\sum_{r=|x|}^{n} r \, {}^nC_r \cdot {}^nC_{r-|x|}$$

* Sub-matrix (ix) is solved assuming $2|J(M)| \ll |L(AM) \pm (x-y) L(AX)|$
† For all sub-matrices except (ix) χ varies from 1 to n. In the case of sub-matrix (ix) χ varies from $-n+1$ to n.

Spectral Analysis

The remaining series of doublets from the $X[AX_n]_2$ spectrum of separation $[\chi^2L^2(AX)+J^2(A)]^{\frac{1}{2}} \pm [(\chi-1)^2L^2(AX)+J^2(A)]^{\frac{1}{2}}$ is replaced by two series of doublets in the $X[AMX_n]_2$ spectrum with separations $N(MX) \pm [\chi^2L^2(AX)+J^2(A)]^{\frac{1}{2}} \pm [(\chi-1)^2L^2(AX)+J^2(A)]^{\frac{1}{2}}$ and $L(MX) \pm \{[L(AM)+\chi L(AX)]^2+J^2(A)\}^{\frac{1}{2}} \pm \{[L(AM)+(\chi-1)L(AX)]^2+J^2(A)\}^{\frac{1}{2}}$. Illustrations for nominal parameter values would have been a useful addition at this stage.

The above theory has been applied to bis(di-t-butylphosphine)-metal complexes, which have been analysed as $[AMX_{18}]_2$ systems (where A ≡ phosphorus and M ≡ CH protons). The X approximation is justified because $J(MX)$ and $J(MX')$ have so far been observed to be small compared with $|\nu_M - \nu_X|$ in the compounds studied. The spectra of the complexes are discussed neglecting $J(M)$, $J(MX)$, and $J(MX')$ throughout. The X part of the spectrum under these conditions resembles the $X[AX_n]_2$ case except for the observation of an additional series of doublets of separation

$$\{[L(AM)+\chi L(AX)]^2+J^2(A)\}^{\frac{1}{2}} \pm \{[L(AM)+(\chi-1)L^2(AX)+J^2(A)\}^{\frac{1}{2}}$$

In the case of di-t-butylphosphine, $L(AM) \gg L(AX)$, and the outer lines described by the sum between the two terms are normally too weak to be observed. The inner lines produced by the difference between the two terms will be closely packed whatever the magnitude of $J(A)$ and are unlikely to be resolved. The maximum intensity will occur when $\chi = 1$ and $\chi = 0$, as $Z_{(\chi)}$ (defined in the footnote of Table 2) is then a maximum. Hence there is a broad doublet of separation $\frac{1}{2}(\{[L(AM)+L(AX)]^2+J^2(A)\}^{\frac{1}{2}} - \{[L(AM)-L(AX)]^2+J^2(A)\}^{\frac{1}{2}}) =$ ca. $L(AX)[1+J^2(A)L^{-2}(AM)]^{-1}$.

The result for the *trans*-bis(di-t-butylphosphine)-metal complexes, *e.g.* *trans*-[PdCl$_2$(PHBu$^t_2$)$_2$] where $L(AM) = 340$ Hz and $J(A) = 515$ Hz, is the observation of a five-line pattern (2:1:2:1:2) with only $N(AX)$ directly observable (as the separation of the two outer lines).

2 New Methods of Tackling Known Spin Systems

A. Superoperator Direct Method Least-squares Analysis of N.M.R. Spectra.—
Cohen and Emerson report the use of the superoperator 'direct' method for the analysis of [AB]$_2$ and ABC$_2$ systems.[4] This method provides the formulation necessary for any calculation where the direct manipulation of eigenvalues that are the observable transition frequencies is preferred over the traditional energy eigenvalues. The resonance frequencies and intensities are obtained directly by solving a new eigenvalue problem

$$\hbar^D \mathscr{X}_\alpha = [\mathscr{H}, \mathscr{X}_\alpha] = \omega_\alpha \mathscr{X}_\alpha$$

in which $\mathscr{X}_\alpha$ is an eigenoperator of $\hbar^D$, the derivation superoperator belonging to the high-resolution spin Hamiltonian $\mathscr{H}$, whose eigenvalue ω_α is the difference between two eigenvalues of $\mathscr{H}$ and is thus the transition frequency. A weighted least-squares procedure based on an iteration of the transition matrix,

[4] S. M. Cohen and M. T. Emerson, *J. Magn. Resonance*, 1971, **4**, 54.

$$\hbar_{ji,lk}^{\mathrm{D}} = \delta_{ki}\langle\psi_j|\mathscr{H}|\psi_l\rangle - \delta_{jl}\langle\psi_k|\mathscr{H}|\psi_i\rangle$$

where the ψ_i form an orthogonal basis for the Hamiltonian, has been devised to facilitate the analysis of the systems studied. Fewer steps are required in the direct method formulation of the least-squares procedure, since eigenvalues of $\hbar^{\mathrm{D}}$ correspond directly to the observed spectral lines, and the best least-squares values for the spectral parameters are those that minimize the sum of the squared residuals of the observables. The steps taken by the practising spectrum analyst are essentially the same as when using the familiar indirect methods. Input for each program consists of a set of trial values for the n.m.r. parameters, the numbers of lines to be matched with the observed frequency of each tentatively assigned transition together with a weighting factor ranging from one to zero, and the maximum number of iterations desired. The programs provide for the least-squares adjustment of spectral parameters, calculations of theoretical spectra, and computation of the corresponding standard deviation. Iteration on intensities is possible but normally omitted because the data are less reliable than those for frequencies.

The molecules chosen for illustrations of the application of the method were cyclopropane-*cis*-1,2-dicarboxylic acid (ABC_2) and *trans*-1,2-dimethoxycarbonylcyclopropane $[AB]_2$. Repeated spacings are listed for the ABC_2 system, and the source of the $J(AB)$ value (from two line separations) is given but no mention is made of the sub-spectrum which exists for the C_2 singlet state (DDS) and which provides a $|\nu_A - \nu_B|$ value as well as $J(AB)$. The excellent agreement between observed and calculated results was checked using the simple relationships obtained from those transitions for which frequencies and intensities are given explicitly in terms of the spectral parameters. Double quantum transitions have been used to confirm the final assignment and the results fall in line with previous work.

B. **A Systematic Approach to the $[AB]_2$, $[AB]_2MX$, and Derived Systems.**[5]—This paper reports a systematic and consistent approach to $[AB]_2$ and $[AB]_2MX$ systems with respect to the construction and ordering of wavefunctions and energy levels respectively. The authors advocate a fundamental approach to the analysis problem which makes as much use of the experimental spectra as the simplicity of analytical expressions describing the spectral transitions allows, and maintains contact with algebra as far as is practicable in the search for a first best set of parameters to form the basis of an iterative refinement. A critical survey of proton–proton coupling constants obtained by n.m.r. analysis, which indicated that only 10% of the available data were believed to be based on sufficiently reliable assumptions and method,[6] is quoted in support of this personal approach. It is certainly a method to be recommended to those new to the fields of single resonance, double reson-

---

[5] T. J. Batterham and R. Bramley, *Org. Magn. Resonance*, 1971, **3**, 83.
[6] A. A. Bothner-By, 'Advances in Magnetic Resonance,' Academic Press, New York, 1968, vol. 1, p. 195.

Spectral Analysis

ance and multiple quantum resonance although these authors[5] are by no means the first to have thought this the case!

The literature survey carried out for this work was incomplete in that the authors clearly missed the papers by Grant, Hirst, and Gutowsky[7] and Dischler and Englert[8,9] on $[AB]_2$ systems. These omissions substantially diminish the credibility of their claim to be the first to treat such systems systematically! It is to their credit to have produced an orderly table of symmetrized wavefunctions, matrix elements, explicit energy expressions, and eigenfunctions, with systematic sign patterns throughout, and their discourse on boundary limits of the trigonometric functions used in intensity expressions is also worthy of mention.

A significant contribution to spectral analysis has been made by them in their treatment of the 4 × 4 secular determinant where they have used perturbation theory to derive frequencies correct to second order and intensities

Table 3 $[AB]_2$ *Transition frequencies and intensities; A and B transition frequencies are given with respect to* $\frac{1}{2}(\nu_A + \nu_B)$ *for the case* $\nu_A > \nu_B$

Line	A Transition	B Transition	$+A$ or $-B$ Frequency	Intensity
1	$A_2 \leftarrow 2A_1$	$2A_{-1} \leftarrow A_{-2}$	$N/2 + C_1$	$1 - \sin 2\theta_1$
2	$2A_1 \leftarrow 4A_0$	$4A_0 \leftarrow 2A_{-1}$	$\delta + N/2 - C_1 - \varepsilon_4$	$1 - N/(\delta + N/2)$
3	$1A_{-1} \leftarrow A_{-2}$	$A_2 \leftarrow 1A_1$	$-N/2 + C_1$	$1 + \sin 2\theta_1$
4	$1A_0 \leftarrow 1A_{-1}$	$1A_1 \leftarrow 1A_0$	$\delta - N/2 - C_1 + \varepsilon_1$	$1 + N/(\delta - N/2)$
5	$1A_1 \leftarrow 3A_0$	$3A_0 \leftarrow 1A_{-1}$	$K/2 + C_1 + C_5 - \varepsilon_3$	$\sin^2\theta_s - D_1$
6	$2A_0 \leftarrow 2A_{-1}$	$2A_1 \leftarrow 2A_0$	$-K/2 + C_1 + C_5 + \varepsilon_2$	$\cos^2\theta_s - D_2$
7	$1A_1 \leftarrow 2A_0$	$2A_0 \leftarrow 1A_{-1}$	$K/2 + C_1 - C_5 - \varepsilon_2$	$\cos^2\theta_s + D_3$
8	$3A_0 \leftarrow 2A_{-1}$	$2A_1 \leftarrow 3A_0$	$-K/2 + C_1 - C_5 + \varepsilon_3$	$\sin^2\theta_s + D_4$
9	$1B_1 \leftarrow 2B_0$	$2B_1 \leftarrow 1B_0$	$C_2 + C_3$	$\sin^2(\theta_a - \theta_2)$
10	$1B_0 \leftarrow 2B_{-1}$	$2B_0 \leftarrow 1B_{-1}$	$C_4 + C_3$	$\cos^2(\theta_a + \theta_4)$
11	$1B_1 \leftarrow 1B_0$	$2B_1 \leftarrow 2B_0$	$C_2 - C_3$	$\cos^2(\theta_a - \theta_2)$
12	$2B_0 \leftarrow 2B_{-1}$	$1B_0 \leftarrow 1B_{-1}$	$C_4 - C_3$	$\sin^2(\theta_a + \theta_4)$

$2C_1 = (\delta^2 + N^2)^{\frac{1}{2}}$, $2C_2 = [(\delta + M)^2 + L^2]^{\frac{1}{2}}$, $2C_3 = (M^2 + L^2)^{\frac{1}{2}}$, $2C_4 = [(\delta - M)^2 + L^2]^{\frac{1}{2}}$ $2C_5 = (K^2 + L^2)^{\frac{1}{2}}$,
$\sin 2\theta_s = L/2C_5$, $\sin 2\theta_1 = N/2C_1$, $\sin 2\theta_2 = L/2C_2$, $\sin 2\theta_4 = L/2C_4$, $\sin 2\theta_a = L/2C_3$, $-\pi/4 \leq \theta_1 \leq +\pi/4$.
$\delta = \nu_A - \nu_B$ (positive).
$D_1 = (L \sin 2\theta_s)/(2\delta + K - N + 2C_5)$
$D_2 = (L \sin 2\theta_s)/(2\delta - K + N + 2C_5)$ $\varepsilon_1 = U_{45} + U_{46}$, $\varepsilon_2 = -U_{45} + U_{57}$
$D_3 = (L \sin 2\theta_s)/(2\delta + K - N - 2C_5)$ $\varepsilon_3 = -U_{46} + U_{67}$, $\varepsilon_4 = -U_{57} - U_{67}$
$D_4 = (L \sin 2\theta_s)/(2\delta - K + N - 2C_5)$

$$U_{45} = \frac{(N\cos\theta_s - L\sin\theta_s)^2}{2(2\delta - N + K - 2C_5)} \quad U_{46} = \frac{(N\sin\theta_s + L\cos\theta_s)^2}{2(2\delta - N + K + 2C_5)}$$

$$U_{57} = \frac{(N\cos\theta_s - L\sin\theta_s)^2}{2(2\delta + N - K + 2C_5)} \quad U_{67} = \frac{(N\sin\theta_s + L\cos\theta_s)^2}{2(2\delta + N - K - 2C_5)}$$

[7] D. M. Grant, R. C. Hirst, and H. S. Gutowsky, *J. Chem. Phys.*, 1963, **38**, 470.
[8] V. B. Dischler and G. Englert, *Z. Naturforsch.*, 1961, **16a**, 1180.
[9] V. B. Dischler and W. Maier, *Z. Naturforsch.*, 1961, **16a**, 318.

correct to first order for the six transitions relevant to one half of the spectrum. These expressions are given in Table 3 and an illustration of the agreement with experiment is given in Table 4 in terms of an R parameter = $|\nu_A - \nu_B|/(J + J')$ where it can be seen there is an increasing divergence for values of $R < 2$.

The methods described have been extended to $[AB]_2MX$ systems, and subspectral methods have been applied in illustrations involving *p*-fluorophenyl dichlorophosphine and tri-*p*-fluorophenylphosphine.

Table 4 *Comparison of transition frequencies and intensities calculated by exact numerical diagonalization and by the method of reference 5, for* $R = 3\cdot3, 2\cdot2, 1\cdot1,$ *and* 0.

Line	$R = 3\cdot3$				$R = 2\cdot2$			
	Exact		Ref. 5		Exact		Ref. 5	
	ν/Hz	Int.	ν/Hz	Int.	ν/Hz	Int.	ν/Hz	Int.
2	105·161	0·712	105·13	0·74	105·432	0·586	105·54	0·63
4	95·978	1·320	96·00	1·35	96·190	1·493	96·26	1·58
5	107·623	0·152	107·61	0·14	107·736	0·126	107·67	0·09
6	102·843	0·619	102·84	0·67	103·079	0·547	103·06	0·63
7	98·477	0·820	98·48	0·91	98·853	0·790	98·87	1·02
8	93·698	0·378	93·71	0·36	94·195	0·458	94·26	0·44

Line	$R = 1\cdot1$				$R = 0$			
	Exact		Ref. 5		Exact		Ref. 5	
	ν/Hz	Int.	ν/Hz	Int.	ν/Hz	Int.	ν/Hz	Int.
2	106·393	0·308	107·40	0·38	—	—	88·32	1·00
4	96·808	2·05	97·53	2·64	100·00	3·0	102·68	1·00
5	107·823	0·081	107·03	0·03	—	—	136·15	1·05
6	103·507	0·358	103·01	0·56	—	—	107·89	0·26
7	99·946	0·504	100·45	1·81	—	—	101·11	0·26
8	95·630	0·662	96·42	0·70	100·00	1·0	72·85	1·05

C. Use of Double Quantum Transitions.—*Double Resonance: Perturbation of Double Quantum Transitions.*[10] Experiments are described which involve irradiation of double quantum transitions. This method can be used to determine connected transitions of the energy level diagram in a way similar to that involving irradiation of a single quantum transition.[11] The single quantum irradiation would normally be the easier experiment, but the irradiation of a double quantum transition line is claimed to be a more selective experiment because a small offset from exact centering leads to a large asymmetry in the observed doublet splittings of connected transitions. The splitting is proportional to the square of the amplitude of the irradiating field. This paper[10] provides a clearly written and detailed theoretical section based on the rotating frame approach; it also illustrates the application of the tech-

[10] B. Gestblom and O. Hartmann, *J. Magn. Resonance*, 1971, **4**, 322.
[11] R. A. Hoffmann and S. Forsén, *Progr. N.M.R. Spectroscopy*, 1966, **1**, 34.

nique to *o*-dibromobenzene, for which a double quantum transition occurs at the centre of the [AB]$_2$ spectrum. The method can be used to measure the positions of double quantum transitions very accurately (± 0.02 Hz), but it is necessary that they are clearly resolved.

Effect of Double Irradiation (Tickling) of Single Quantum Transitions on Double Quantum Transitions in Systems Simplified by Double Resonance.[12] The detection of double quantum transitions in the spectra of three-spin systems has been shown to lead to a unique assignment of the energy levels and hence coupling constants and chemical shifts. Tickling experiments can be regarded as an alternative source for obtaining similar information, but the double quantum transition method has the advantage that no external oscillator is needed. However, the double quantum transition method meets with some difficulties when the number of nuclei in the system increases beyond three; the number of double quantum transitions increases rapidly and assignment becomes troublesome. An increasing number of combination lines appear in the single resonance spectra and may dominate at high radio-frequency power levels, complicating the observation and identification of the double quantum transitions. The method can be applied, however, to systems involving ABC weakly coupled to other nuclei by performing double resonance experiments to observe ABC transitions when the other nuclei are irradiated. The examples of molecules provided are allylcyanide, ABMX$_2$, and 3-methylthiophen, ABMX$_3$. Irradiation of X nuclei in these cases leaves the ABM spectra with A, B, and M chemical shifts slightly shifted by the Bloch–Siegert effect. A second Bloch–Siegert shift results from the higher radio-frequency power level needed to observe the double quantum transitions and therefore allowance must be made for these effects in the analysis of the complete single resonance low power B$_1$ spectrum. The relationships between the single and double quantum transitions can be defined with reference to the ABM energy level diagram shown in the Figure.

```
                  a
                 ─── ααα
      b                             d
     ─── αβα                       ─── ααβ
                    c
                   ─── βαα
                        f
                       ─── αββ
      e                             g
     ─── ββα                       ─── βαβ
                                    h
                                   ─── βββ
```

Figure

The transition $e \leftrightarrow a$ is a double quantum transition, and single quantum

[12] K. Schaumburg and H. J. Jakobsen, *J. Magn. Resonance*, 1970, **2**, 1.

transitions with an energy level in common with it can be classified in the following way:

$h \leftrightarrow e$ is a progressive transition $\Delta = 3$

$\left.\begin{array}{c} b \leftrightarrow a, \, e \leftrightarrow b \\ e \leftrightarrow c, \, c \leftrightarrow a \end{array}\right\}$ are labelled generic regressive transitions

and $d \leftrightarrow a, \, e \leftrightarrow d$ are non-generic regressive transitions $\Delta = 1$

The intensity of the double quantum transition, when a connected single quantum transition is irradiated (tickled), is dependent on a transition moment product $\langle e|I^+|\kappa\rangle\langle \kappa|I^+|\alpha\rangle$ where $\kappa = b,c,d$. The product involves normal single quantum transitions for $\kappa = b,c$ while in the case $\kappa = d$, $\langle e|I^+|d\rangle$ is zero or very small because $e \leftrightarrow d$ is a combination transition. An increase in intensity of the double quantum transition is expected when a generic regressive transition is irradiated but, in contrast, a much smaller increase in intensity is expected when non-generic transitions are irradiated.

The method provides new information about the relationship between the observed transitions and the energy level diagram. However, it requires some sophistication in instrumentation (external oscillator and frequency sweep), which may not be available to everyone at the level of reproducibility and stability necessary.

3 Studies of Known Spin Systems

A. Systems involving Protons Only.—*Aliphatic Systems.* (i) Butane $[A_3B_2]$.[13] The spectrum of liquid butane at 27 °C at 60 MHz and 100 MHz has been analysed using the UEANMR 2 program, to give the averaged coupling constants. A value of 0·77 kcal mol^{-1} assumed for the barrier to rotation about the central single bond has been used to calculate ^3J_g and ^3J_t, and comparison with cyclohexane, dioxans, and substituted ethanes used to justify that assumption.

(ii) Mannitol derivatives $[AB]_2$.[14] The spectra of two 2,5-*o*-methylene-D-mannitol derivatives have been analysed as $[AB]_2$ systems using specific deuteriation to reduce coupling to other systems. The results have been interpreted as compatible with twisted chair conformations.

Olefinic Systems. (i) *cis-* and *trans*-1,3-dichloroprop-1-ene, ABX_2.[15] The solvent dependence of the coupling constants has been followed closely in these repeat analyses carried out with better accuracy. The solvent dependence of proton couplings in *trans*-1,3-dichloroprop-1-ene has been interpreted on the basis of two discrete rotamers with the *gauche* form favoured by 0·5 kcal mol^{-1} in the vapour, and 1·0 kcal mol^{-1} in polar solvents. The behaviour of the *cis*-compound has been associated with one flexible conformation approximating to the *gauche* isomer with varying geometry in different solvents.

(ii) Alkyl vinyl sulphides, ABC.[16] The chemical shifts were discussed in terms

[13] D. W. Aksnes and P. Albriktsen, *Acta Chem. Scand.*, 1970, **24**, 3764.
[14] J. F. Stoddart and W. A. Szerek, *J. Chem. Soc. (B)*, 1971, 437.
[15] R. J. Abraham and K. Parry, *J. Chem. Soc. (B)*, 1971, 724.
[16] G. Cecarelli and E. Chiellini, *Org. Magn. Resonance*, 1970, **2**, 409.

Spectral Analysis

of mesomeric structures, and the differences in comparison with the corresponding ethers explained by the ability of sulphur to accommodate ten electrons. A predicted correlation between n.m.r. spectral parameters and chemical reactivity (as observed in vinyl ethers for cationic polymerization) awaits confirmation from polymerization experiments.

(iii) *1-Methoxybuta-1,3-dienes*, $ABCDEX_3$.[17] The spectra of four solutions of the *cis-* and *trans-*isomers have been completely analysed as $ABCDEX_3$ systems using NMRIT and NMREN (adapted to the IBM-7090 computer) and the results extrapolated to infinite dilution. Long-range couplings between the X and ABC protons in the *trans-*isomer complicated the spectrum so that a curve-fitting analysis method termed 'Spectral Decomposition' was developed, using a non-linear least-squares method incorporated in an iterative programme needing trial parameters very close to the true values to avoid convergence to spurious minima. Accurate values obtained for the long-range couplin involving the methoxy-protons in the *trans-*isomer were $^4J = -0.34$, $^5J = -0.29$, and $^6J = \pm 0.12$ Hz (± 0.02 Hz). Relative signs of long-range coupling constants have been determined by selective double resonance. The conformation of the butadienyl system in the two isomers is largely planar *s-trans* as deduced from the magnitudes and signs of the coupling constants. The preferred conformation of the methoxy-group in the *trans-*isomer is the planar *s-cis* whereas that of the *cis-*isomer is probably *s-trans*.

(iv) *Vinyl compounds*, ABCX.[18] The carbon-13 satellites in the proton spectra of eleven vinyl compounds have been analysed as ABC[ABCX] at 60 MHz using LAOCOON 3. Additivity relationships have been used to resolve ambiguous cases where more than one data set fitted the spectra. Degeneracy in the vinyltrichlorosilane spectrum made analysis of the carbon-13 spectrum (ca. 40 scans at 22·6 MHz) necessary to obtain unique values for the proton–proton coupling constants. Marked variations were observed in the sign and magnitude of two-bond interactions $^2J(^{13}CH)$. Substituent effects were rationalized in terms of the Pople and Bothner-By theory of geminal coupling constants. Similar but less pronounced substituent effects have been observed in cyclopropanes.

(v) *2-Substituted propanes*, ABX_3.[19] The spectra of seven 2-substituted propanes have been analysed at 60 MHz to obtain accurate values for the long-range $^4J_{cis}$ and $^4J_{trans}$ coupling constants in an attempt to systematize the effects of substituents for the purpose of molecular structure determination and conformational analysis.

(vi) *2,3-Dihydro-furans, thiophens, and derivatives*, $[AB]_2$, $[AB]_2X$, $[AB]_2$-XY.[20] The proton spectra of a number of these compounds have been re-examined and analysed to correct errors arising in the past because of deceptive simplicity. The coupling constants derived were discussed in terms of the mechanism for $^3J(HH)$ in planar heterocyclic molecules.

[17] E. Diez and M. Rico, *J. Mol. Spectroscopy*, 1971, 37, 131.
[18] K. M. Crecely, R. W. Crecely, and J. H. Goldstein, *J. Mol. Spectroscopy*, 1971, 37, 252.
[19] D. G. de Kowalewski, *J. Magn. Resonance*, 1971, 4, 249.
[20] R. J. Abraham, K. Parry, and W. A. Thomas, *J. Chem. Soc. (B)*, 1971, 446.

Heterocyclic Systems. (i) Seleno-[2,3-*b*]- and -[3,2-*b*]-thiophen and derivatives, ABCD and ABX.[21] The proton spectra of the parent compounds have been analysed as ABCD and the 5-carboxy-derivatives analysed as ABX, using a computer. The coupling constants derived were almost identical with the corresponding thienothiophen parameters, and inter-ring coupling showed the same pattern, *i.e.* large only between protons situated in a planar zig-zag configuration. Relative signs were determined from double resonance experiments. Couplings over five bonds were quoted as 0.02 ± 0.04 Hz. Some ^{77}Se–^{1}H coupling constants were derived from satellite spectra and values given as:

for seleno[2,3-*b*]thiophen $^{2}J(\text{SeH}) = \pm 49.2$ Hz, $^{3}J(\text{SeH}) = \pm 9.4$ Hz
for seleno[3,2-*b*]thiophen $^{2}J(\text{SeH}) = \pm 47.6$ Hz, $^{3}J(\text{SeH}) = \pm 7.8$ Hz.

(ii) Indole derivatives, ABC, ABCX, ABCMN.[22] The proton spectra of fifteen indole derivatives have been obtained at 60 MHz and 100 MHz and analysed to provide values for the parameters, including long-range coupling constants over five and six bonds, which have been used to determine the position of substituents in new indole compounds. A numbered structure of indole would make the otherwise commendably brief but relevant text easier to follow.

(iii) Thiepin-1,1-dioxide, [ABC]$_2$.[23] The spectra were obtained at 100 MHz from 15% w/w solutions in deuteriochloroform and analysed as [ABC]$_2$ using LAOCOON 3. The high value of J_{34}(vicinal) and J_{45}(*cis*), together with evidence from the allylic ^4J coupling constant, have been interpreted as consistent with the flattened boat form suggested by *X*-ray diffraction studies. A comparison between the chemical shifts observed in thiepin-1,1-dioxide, 4,5-dihydrothiepin-1,1-dioxide and cycloheptatriene has been used to suggest that there is some extra π-electron delocalization, in keeping with the planar configuration. The shifts due to this effect were estimated to be about -0.5 p.p.m. and the diamagnetic effect of a 'ring current' was suggested as being at least partly responsible.

(iv) Pyrrole-2-aldehyde, ABCMX, pyrrole-2-carboxylic acid, ABCX.[24] The proton spectra have been measured for anhydrous acetone solutions at 60 MHz using nitrogen-14 decoupling. Triple resonance experiments show that the aldehyde proton–NH coupling constant in pyrrole-2-aldehyde has opposite sign to the others. Coupling constants were quoted to three decimal places with an experimental error of 0.02 Hz. The relative sign alternation between ^4J and ^5J coupling constants involving the CHO proton was interpreted as in accord with main contributions to the long-range coupling arising from the π-electrons.

Aromatic Systems. (i) 9-Substituted fluorenes, ABMX, ABMTX.[25] The spectra of forty-three 9-substituted fluorenes have been obtained at 60 MHz in four common solvents. The line positions measured for at least three solution con-

---

[21] A. Bugge, B. Gestblom, and O. Hartmann, *Acta Chem. Scand.*, 1970, **24**, 1953.
[22] J. Y. Lallemand and T. Berneth, *Bull Soc. chim. France*, 1970, **4**, 4091.
[23] M. P. Williamson, W. L. Mock, and S. M. Castellano, *J. Magn. Resonance*, 1970, **2**, 50.
[24] S. Shimokawa, H. Fukui, and J. Sohma, *Mol. Phys.*, 1970, **19**, 695.
[25] K. D. Bartle, D. W. Jones, and P. M. G. Bavin, *J. Chem. Soc. (B)*, 1971, 388.

centrations were extrapolated to infinite dilution and the spectra analysed as ABMX [taking into account coupling between H(9) and protons in the rings, where this was proven using double resonance] with the assistance of computer calculations. A number of empirical correlations involving chemical shifts have been attempted and some space devoted to explaining the deviations from expected relationships. Spin–spin coupling between H(9) and the 9-substituent was discussed in terms of preferred conformations and the coupling between H(9) and fluorene aromatic protons interpreted in terms of a stereospecific benzylic coupling along a zig-zag path similar to that observed in $\alpha,\alpha,2,6$-tetrachlorotoluene at -80 °C.

(ii) *Phenyl derivatives of phenanthrene, AB, ABC, ABCD, $[AB]_2C$.*[26] Approximate values of chemical shifts and coupling constants have been 'extracted' from overlapping (often deceptively simple) spin systems of 9,9′-biphenanthryl, 9-benzylphenanthrene, the 1-, 4-, and 9-phenylphenanthrenes, 9-methyl-10-phenylphenanthrene, and 1-phenylfluoranthrene, in carbon disulphide solutions. The chemical shift data have been used to determine the angles between the planes of the ring systems and ring substituents, with the aid of chemical shifts calculated for different values of the dihedral angles.

(iii) *Monosubstituted naphthalenes, ABCD, ABC.*[27] This paper provides examples of the many techniques which can be brought to bear on analysis problems. Chemical shifts and coupling constants have been obtained with modest accuracy from studies made at 100 MHz and 200 MHz using multiple resonance, including INDOR, and various solvents. Superposition of lines in many of the spectra hampered the use of an iterative LAOCOON program, but LAOCOON was used sometimes to minimize differences between observed and calculated single transition lines. Relative signs were assumed all positive and the study was motivated by the desire to provide more extensive chemical shift data within one molecule to test the theories of proton shielding.

B. Compounds containing Fluorine.—*2,2′-, 3,3′-, and 4,4′-Dihalogenobiphenyl Compounds, ABCD, ABCDX.* The study of substituent effects in aromatic compounds has been made much more reliable by improved spectrometer performance and the high-speed computers available for analysis of the spectra. The spectra of twelve symmetrically substituted dihalogenobiphenyl compounds have been analysed (^{1}H at 60 MHz and ^{19}F at 85·5 MHz) using the LAOCOON II program. The broadening of the low-field multiplet originally observed[29] in 4,4′-difluorobiphenyl has been resolved as fine structure and assigned to an inter-ring H–F coupling which showed a significant solvent dependence (0·16 Hz in C_6D_6 and 0·21 Hz in CCl_4; RMS deviation 0·03 Hz). The coupling was not observed in the 2,2′- or 3,3′-fluoro-compounds and this is explained in terms of reduced conjugation (in agreement with u.v. work).

[26] K. D. Bartle, P. M. G. Bavin, and D. W. Jones, *Org. Magn. Resonance*, 1970, **2**, 259.
[27] J. W. Emsley, S. R. Salmon, and R. A. Storey, *J. Chem. Soc. (B)*, 1970, 1513.
[28] A. R. Tarpley, jun., and I. H. Goldstein, *J. Phys. Chem.*, 1971, **75**, 421.
[29] G. Gestblom and S. Rodmar, *Acta Chem. Scand.*, 1964, **18**, 1767.

Pyridine Compounds.[30] The spectra of fluorinated pyridines obtained at 60 and 100 MHz have been analysed using the least-squares, iterative program LAME in a search for a magnified effect of polarization of bonds and alternation of charge transmission along the molecular framework, arising from the hetero atom, and manifested in a wider range of H–F and F–F coupling constants.

Pentafluorophenol, $(AM)_2X$.[31] The spectrum of pentafluorophenol in different solvents has been studied in an attempt to eliminate or minimize contributions to the changes in the ^{19}F parameters from anisotropic shielding and electric field effects. The effects observed were attributed to hydroxy-group–solvent interaction as a result of (*a*) direct change of π-donor ability of the interacting OH groups on going from one solvent to another and (*b*) polarization of the π-electron cloud caused by a change in the electronegativity of the hydroxy-substituent. LAOCOON 3 was used for the analysis of this $(AM)_2X$ system.

Benzene, Naphthalene, Phenanthrene, and Biphenyl Compounds.[32] The solvent dependence of fluorine–fluorine coupling constants has been studied in a series of mono-, di-, tri-, and tetra-fluoro-benzenes and -naphthalenes using heteronuclear noise-modulated decoupling. The three-bond [$^3J(FF)$] coupling constant showed no correlation with solvent dielectric constant, and this has been interpreted as meaning that the reaction field is not as important in that case. Longer range coupling constants $^4J(FF)$ and $^5J(FF)$ correlate reasonably well with the dielectric constant and therefore it is intimated that, for these couplings, the reaction field is more important. Long-range coupling constants for a number of difluoronaphthalenes, biphenyls, and phenanthrenes are presented for the first time.

Tetrafluorodichlorocyclopropanes, $[AX]_2$, A_4.[33] The ^{19}F spectra of three tetrafluorodichlorocyclopropanes [(*i*) 1,1,2,3(*cis*), (*ii*) 1,1,2,3(*trans*), and (*iii*) 1,1,2,2] obtained at 94·1 MHz have been studied in a number of solvents. The spectra have been analysed using the carbon-13 satellites to lift deceptive simplicity of the $[AX]_2$ system and degeneracy of the A_4 system. The isotope shifts have been measured for the X of ABX_2 in (*i*) to be 0·152 p.p.m., for X of $[AX]_2$ in (*ii*) to be 0·059 p.p.m., and for (*iii*) to be 0·117 p.p.m. The appropriate $^{13}C–F$ coupling constants, $-315·2$, $-309·2$, and $-323·2$ Hz respectively, show a significant variation. The vicinal *cis* (F–F) coupling cannot be distinguished from *trans* (F–F) coupling in (*iii*), and variable-temperature studies on the neat liquids have been used to assist the assignment, since the *cis* (F–F) coupling constants have been observed to be more sensitive to temperature change.

The absolute values of *cis* and *trans* coupling constants have been observed to decrease with temperature, *i.e.* to become more negative or less positive.

[30] W. A. Thomas and G. E. Griffin, *Org. Magn. Resonance*, 1970, **2**, 503.
[31] A. J. Dale, *Spectrochim. Acta*, 1971, **27A**, 81.
[32] M. A. Cooper, H. E. Weber, and S. L. Manatt, *J. Amer. Chem. Soc.*, 1971, **93**, 2369.
[33] L. Cavalli, *Org. Magn. Resonance*, 1970, **2**, 233.

Spectral Analysis

This has been explained in terms of low-lying torsional or vibrational states of the cyclopropane ring and suggested as a means for defining the signs in related compounds.

1,1,2,2-Tetrafluoroethane $[AMX]_2$.[34] The solvent dependence of the coupling constants in 1,1,2,2-tetrafluoroethane has been used to obtain the rotamer energies. The variation of the coupling constants in different solvents was observed to be small, so that analysis in different solvents presented no difficulty and was accomplished by identification of explicit transitions followed by iterative calculations using LAOCOON 3.

C. Compounds containing Phosphorus.—*Bis(dialkylphosphines)*, $[A_3M_3X]_2$, $[A_3X]_2$.[35]

The ^{19}F (94 MHz) and ^{31}P (40·5 MHz) spectra of bis(methyltrifluoromethylphosphine) and thio-bis(methyltrifluoromethylphosphine) have been interpreted as being compatible with a reversible equilibrium between diastereoisomers (*dd*, *ll*, and meso). The temperature-dependent reversible interconversions of the isomers have been attributed to the inversion at the phosphorus atoms. The spectrum of $[Me(CF_3)P]_2$ has been partially analysed as an $[A_3M_3X]_2$ system and the deuteriated derivative $[CD_3(CF_3)P]_2$

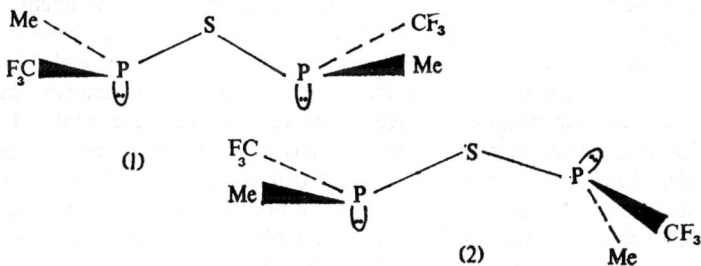

analysed as $[A_3X]_2$ assuming initial values $J(FF) = 0$ and proceeding by a trial and error method using NMRIT without iteration. The analysis of the spectra of the thiobisphosphines (CH_3 and CD_3) indicates that the two stereoisomers have nearly identical $J(PP')$, $J(FP)$, $J(FP')$, and $J(HP) + J(HP')$, but different values of $J(FF')$ (0·3 and 2·0 Hz respectively). The models proposed (1) and (2) were influenced by the X-ray diffraction studies of the nickel carbonyl complex of $(CF_3)_2PSP(CF_3)_2$. The isomer with its ^{19}F spectrum at the highest frequency has been assigned the meso stereochemistry on the basis of the higher $[J(FF')]$ value if the conformation is as (1). The equilibrium constants for the isomerization reactions have been measured at various temperatures from the ^{19}F spectra. The spectrum of $(CF_3)_2POP(CF_3)_2$ has been analysed as $[AX_6]_2$ and the ^{19}F spectrum of $(CF_3)_2PP(CF_3)_2$ has been partially analysed to give $|J(FP) + J(FP')|$. Variable-temperature studies of these compounds revealed no evidence of rotational isomerism.

---

[34] L. Cavalli and R. J. Abraham, *Mol. Phys.*, 1970, **19**, 265.
[35] D. K. Kang, K. L. Servers, and A. B. Bury, *Org. Magn. Resonance*, 1971, **3**, 101.

The $[AMX_{18}]_2$ *System* (see Section 1). Examples of the analysis of this system have been provided in a study[36] of organometallic compounds with two di-t-butyl phosphine groups per molecule. The proton spectra were measured at 90 MHz and analysed, assuming $J(XX') = 0$ and that $^3J(HCH_3)$ is very small in order to simplify the treatment. The results constitute the first reliable measurements of $^2J(trans)$(P–M–P) for Ru^I, Ru^{II}, Ru^{III}, Ir^I, Ir^{III}, and Ni^{II}.

Tri-(3-furyl)phosphine, ABCX.[37] Single (60 MHz) and homonuclear double (100 MHz) resonance proton spectra of tri-(3-furyl)phosphine derivatives have been analysed using sub-spectral methods, and the magnitudes and signs of the ^1H–^{31}P spin–spin coupling constants obtained. The ^1H–^{31}P as well as the ^1H–^1H coupling constants are all of the same sign (*i.e.* positive). For all derivatives the magnitude of the ortho coupling $^3J(P–H_4)$ exceeds the value for $^3J(P–H_2)$ in accordance with the observations for the corresponding ^1H–^1H coupling constants, $^3J_{34}$ and $^3J_{23}$, in furan. Data for tri-(3-thienyl)phosphine oxide, selenide, and sulphide indicated a dependence of $^3J(PH)$ and $^4J(PH)$ on the electronegativity of the atom attached to phosphorus. The known large effect of four-co-ordination of phosphorus with MeI allowed a value of 2·2 Hz to be predicted for $^4J(PH_5)$ in the MeI complex.

1-Phospha-2,6-dioxacyclohexanes.[38] The spectra of three 1-R-4,4-dimethyl-1-phospha-2,6-dioxacyclohexanes obtained at 60 and 100 MHz have been analysed as [AB]$_2$X systems, when the methyl protons were decoupled from the methylene protons at 60 MHz, using the LAOCOON 3 iterative least-squares computer program. Three other compounds were analysed as [AB]$_2$X and references given to the analysis of other closely related compounds. The ring conformations have been deduced using the vicinal $^3J(HH)$ coupling constants and also a tentative dihedral angular dependence of $J(POCH)$ similar to that of $^3J(HH)$. An intermolecular exchange of 1-substituents in compounds such as 1-chloro-3,5-dimethyl-1-phospha-2,6-dioxacyclohexane has been used to account for the temperature and concentration dependence of the proton spectra. The spectra often involved superposition arising from geometric isomers.

The conformational and configurational properties of 4-t-butyl-1-chloro-1-phospha-2,6-dioxacyclohexane and a geometric pair, *cis*- and *trans*-4-t-butyl-1-methoxy-1-phospha-2,6-dioxacyclohexane, have been studied at 60 MHz.[39] Isomer mixtures containing at least 90% of one isomer allowed each analysis to be completed. The spectra were first approximated as being ABXY (4-H ≡ X and P ≡ Y) rather than [AB]$_2$XY and an initial set of coupling constants and chemical shifts derived by hand calculation for use in the LAOCOON 3 iterative computer program. Overlap of many of the lines led to estimated

---

[36] A. Bright, B. E. Mann, C. Masters, B. L. Shaw, R. M. Slade, and R. E. Stainbank, *J. Chem. Soc.(A)*, 1971, 1826.
[37] H. J. Jakobsen, *J. Mol. Spectroscopy*, 1971, **38**, 243.
[38] D. W. White, R. D. Bertrand, G. K. McEwen, and J. G. Verkade, *J. Amer. Chem. Soc.*, 1970, **92**, 7125.
[39] W. G. Bertrude and J. W. Hargis, *J. Amer. Chem. Soc.*, 1970, **92**, 71.

errors of 0·2 Hz but the quoted J values were given to two decimal places. These latter were presumably the computerized values but the authors themselves suggested that this precision was unrealistic. Further, no attempt was made to determine precise values for long-range coupling constants involving the methylene protons on either side of the ring although a tentative value of <2 Hz was mentioned; and a spectrum was computed for apparently arbitrary values of $J(AB)$, $J(AB')$, and $J(BB')$ though these were not claimed to be unique. The emphasis was on the use of the coupling constants in conformational analysis where high precision is not necessary. The *cis*-isomer was shown to be a chair-form with an equatorial t-butyl group and an axial methoxy-group. The *trans*-isomer, also in the chair form, has both substituents axial.

A further study of the 1-phospha-2,6-dioxacyclohexanes[40] included twenty-one 1-R-1-oxo-4-R^1-4-R^2-1-phospha-2,6-dioxacyclohexanes analysed as ABX, AMX, and some as [AB]$_2$X, systems. It is a paper of overwhelming tables because of the diligence of the authors in including temperature, concentration, and solvent effects as well as the basic analysis. There is one system worth mentioning and that is the 3,9-dimethoxy-2,4,8,10-tetraoxa-3,9-diphosphaspiro[5,5]undecane. The proton spectra were obtained at 60 MHz and 100 MHz and divided into four parts labelled CC', GG', MM', and QQ', each of which corresponds to the resonance of two protons. The authors commented that second-order effects were noticeable indicating that $J(H_CH_G)$ and $J(H_MH_Q)$ are not negligible compared to $\Delta\nu(H_CH_G)$ and $\Delta\nu(H_MH_Q)$ respectively. They go on to pick out apparent first-order splittings due to $J(POCH)$ and $J(HC_3C_4C_5H)$ analogous to those found in the monocyclic compounds. A reference is given to evidence for an inter-ring coupling analogous to $J(H_GH_G')$ and then the authors comment that 'the reason for the unexpected features of the C and G resonances is obscure'. It is clear that, more often than not, in such cases approximate (low precision) values of coupling constants are sought as diagnostic parameters in determining the conformation of the molecules, but here it appears that the authors have initially accepted the evidence of second-order effects in theory, and related systems, and then made no attempt to use them to explain features which could not be explained in terms of first-order splittings.

1-Phospha-2,5-dioxa(diaza,dithia)cyclopentanes.[41] The five-membered heterocyclic phosphorus-containing compounds have received attention in the form of 1-fluoro-1-phospha-2,5-dioxacyclopentane, [AB]$_2$MX, 1-fluoro-2,5-dimethyl-1-phospha-2,5-diazacyclopentane, [ABM$_3$]TX, and 1-fluoro-1-phospha-2,5-dithiacyclopentane, [AB]$_2$MX. The proton spectra were obtained at 100 MHz, fluorine at 94·1 MHz, and the phosphorus at 24·3 MHz. The spectra were analysed using heteronuclear decoupling and sub-spectral methods.

---

[40] D. W. White, G. K. McEwen, R. D. Bertrand, and J. G. Verkade, *J. Magn. Resonance*, 1971, **4**, 123.
[41] J. P. Albrand, A. Cogne, D. Gagnaire, J. Martin, J. B. Robert, and J. Verrier, *Org. Magn. Resonance*, 1971, **3**, 75.

The absolute signs of the $^3J(\text{P--H})$ and $^4J(\text{F--H})$ coupling constants have been determined by reference to the known negative sign of the $^1J(\text{P--F})$ coupling constant from selective heteronuclear double resonance experiments. The $^3J(\text{POCH})$ and $^3J(\text{PNCH})$ coupling constants are positive. The small values observed for $^3J(\text{PSCH})$ have negative signs, the larger values have positive signs. All $^4J(\text{FPXCH})$ coupling constants are positive and show a lack of stereospecificity.

Bis-(5,5-dimethyl-1-phospha-2-oxo-6-thionocyclohexane).[42] Tickling and partial decoupling of phosphorus irradiated at ca. 40 MHz has allowed the relative sign of $^1J(\text{PP})$ to be deduced [positive relative to $^3J(\text{HP})$ and $^4J(\text{HP})$] using sub-spectral principles in 1,1-bis-(5,5-dimethyl-1-phospha-2-oxo-6-thionocyclohexane). The importance of $^1J(\text{PP})$ is emphasized as unique in the second row of elements because it is the only homonuclear coupling that can be easily studied in detail. The result supports other evidence that $^1J(\text{P}^{\text{V}}\text{P}^{\text{V}})$ is usually positive, while $^1J(\text{P}^{\text{III}}\text{P}^{\text{III}})$ is negative, except when small as in $\text{Me}_2\text{P}(:\text{S})\text{P}(:\text{S})\,\text{Me}_2$ where it could be either sign.

D. Organometallic Compounds.—*Triphenyl Group IV Element Lithium Compounds, $[AB]_2C$*.[43] The spectra of the aromatic rings have been analysed at 100 MHz for triphenyl(C,Si,Ge,Sn, or Pb)Li compounds in tetrahydrofuran solutions using LAOCOON 3. Conductance studies indicated that the compounds were completely dissociated in tetrahydrofuran. The phenyl proton chemical shifts have been analysed as consistent with electron donation from the Group IV atom into the rings. The electron delocalization was observed not to follow previously established trends for the interaction of M^{IV} atoms with the π-electron system but instead appears to be related to the degree of association between the lithium and Group IV atoms. The mechanism for electron delocalization, *i.e.* $d\pi$–$p\pi$ or $p\pi$–$p\pi$, could not be elucidated from the experimental data. Lithium chemical shifts were measured with respect to external 1M-LiCl (aqueous) at 38·8 MHz.

Allenic Derivatives of Tin, Lead, and Mercury, AB_2X.[44] Satellite spectra arising from metal–proton coupling constants over two and four bonds have been observed in the proton spectra of tin, lead, and mercury allenic compounds. The relative signs of $^2J(\text{MH})$ and $^4J(\text{MH})$ coupling constants have been deduced directly from analysis of the ABX_2 spectra and confirmed using LAOCOON II. The probable signs of the reduced coupling constants were discussed in relation to published data.

π-Cycloheptadienyl Manganese Tricarbonyl Compounds, $A[BCXY]_2$, $ABCDEXYZ$.[45] The spectrum of tricarbonyl-π-cycloheptadienylmanganese has been partially analysed, and a fuller analysis has been accomplished for the 6-*exo*-

---

[42] R. K. Harris and J. R. Woplin, *Chem. Comm.*, 1970, 1391.
[43] R. H. Cox, E. G. Janzen, and W. G. Harrison, *J. Magn. Resonance*, 1971, **4**, 274.
[44] M. P. Simonnin, M. Le Quan Minh, and M.-J. Lecourt, *Org. Magn. Resonance*, 1970, **2**, 369.
[45] M. I. Foreman and F. Haque, *J. Chem. Soc. (B)*, 1971, 418.

methoxy-compound with the help of homonuclear decoupling. The vicinal coupling constants obtained from computer analysis have been interpreted in terms of a rigid skew structure which remains unchanged between $-90\,^\circ\text{C}$ and $145\,^\circ\text{C}$ on the n.m.r. time scale.

Aryl Derivatives of Mercury, $[AB]_2C$, $[AB]_2X$, $[AB]_2$.[46] Advantage has been taken of the large mercury–proton coupling constants to detect the effects of substituent electronegativity changes. Heteronuclear decoupling experiments have been used to confirm the assignment of the mercury–proton coupling constants. The method of analysis of the proton spectra was not described, except for *p*-fluorophenyl mercury chloride which was analysed as ABX at 60 MHz. The comment passed by the authors was that the spectrum was complex and no mention made of approximating to the true $[AB]_2X$ system. *m*-Substituted compounds were not analysed because the spectra were said to be complex.

4 General Comments

This chapter has emphasized the application of n.m.r. to structural problems, in the latter section. This has been a purposeful phenomenon-orientation because it is not possible to check every calculation made in the papers reviewed, but it does provide a cross-section of the problems where the detailed spectral parameters have been used beneficially.

There is a tendency for some authors to overstate their case by providing an excess of n.m.r. data derived from analysis of the spectra in support of their hypotheses or conclusions. Others apparently seek to perpetuate the misnomer 'virtual coupling' (which is an abbreviated misleading way of saying that second-order effects can result in first-order appearance for a spectrum when a particular coupling constant is very small or zero)[47]. A few authors, perhaps recently introduced to the advantages of n.m.r., need to correct their ideas of when nuclei are, and are not, magnetically equivalent as compared to chemically equivalent.[48] In this connection too, it seems that unity will never be achieved in system notation. The most recent notation suggested by Haigh[49] has the distinct merit of saving space, providing a more compact notation, such as $[AX]_4$ for AA′A″A‴ X X′X″X‴.

Books and Reviews.—Two books and one review have appeared which are directly related to analysis problems.

'The Analysis of High Resolution N.M.R. Spectra', R. J. Abraham, Elsevier, Amsterdam, 1971.

'Analysis of N.M.R. Spectra', R. A. Hoffmann, S. Forsén, and B. Géstblom, in 'N.M.R., Basic Principles and Progress', Springer-Verlag, Berlin, Heidelberg, and New York, 1971, Vol. 5.

[46] P. J. Banney and P. R. Wells, *Austral. J. Chem.*, 1971, **24**, 317.
[47] P. Diehl, R. K. Harris, and R. G. Jones, *Progr. N.M.R. Spectroscopy*, 1967, **3**, 1.
[48] J. A. Pople, W. G. Schneider, and H. J. Bernstein, 'High Resolution N.M.R.', McGraw-Hill, New York, 1959.
[49] C. W. Haigh, *J. Chem. Soc.(A)*, 1970, 1682.

'^{13}CH Satellite N.M.R. Spectra', J. H. Goldstein, U. S. Watts, and L. S. Rattet, in 'Progress in N.M.R. Spectroscopy', Pergamon Press, London, 1971, Vol. 8.

Computer Programs.—The most comprehensive collection of computer programs for analysis of n.m.r. spectra is issued by the Atlas Computer Laboratory, Chilton.[50] It includes general information on the library of programs and on the use of the Atlas computer. Further details can be obtained from Dr. R. K. Harris at the University of East Anglia.

[50] R. K. Harris and J. Stokes, 'A Library of Computer Programs for N.M.R. Spectroscopy', Atlas Computer Laboratory, Chilton, Didcot, Berks., 1971.

6
Bandshape Phenomena for Fluids

BY R. K. HARRIS

1 General Introduction

A. Current Developments.—The quantitative information available from continuous-wave n.m.r. spectra of fluids may be classified into two groups:
(a) Chemical shifts and coupling constants — determined from transition frequencies and intensities.
(b) Relaxation times and lifetimes — obtained from linewidths (or, in general, bandshapes).

Of course this is not a rigid division since bandshapes in general also depend on chemical shifts and coupling constants, but it has proved to be an operational one. In fact, in the past the emphasis has been on group (a) investigations, for a number of reasons. Firstly, relaxation times have been of less importance for structure determinations than chemical shifts or coupling constants. Secondly, pulse n.m.r. methods have been of more practical use for determining T_1 and T_2 values than steady-state techniques. This fact, in conjunction with the traditional difference of approach to understanding continuous-wave and pulse n.m.r. spectroscopy (based on the high-resolution spin Hamiltonian and on the semi-classical 'precession' formulations respectively), has led in general to a division of n.m.r. spectroscopists into two groups also! Those concerned with relaxation-time data have been physicists (or physical chemists) using spin-echo equipment, whereas organic chemists and inorganic chemists have determined chemical shifts and coupling constants at ever-increasing rates using high-resolution equipment. It is true that there has been a considerable volume of work on high-resolution bandshapes affected by exchange processes, but this has been mostly directed to rather simple cases treated in a relatively superficial way — and has led to a number of inconsistencies and errors in the literature.

The situation is now changing drastically, and much more attention is being paid to bandshape phenomena. It is worthwhile discussing the reasons for this change:*
(a) The advent of Fourier Transform n.m.r. spectroscopy is tending to close the division between pulse n.m.r. and high-resolution n.m.r. work, thus making the continuous-wave spectroscopist more aware of and interested

---
* See Chapter 4 for a full discussion of the experimental advances mentioned here.

in the effects of relaxation times. Other developments such as selective pulse studies and nuclear Overhauser measurements have contributed to this change.

(b) The knowledge of the theory (particularly density matrix theory) relevant to describing bandshapes is becoming more widespread. In addition, there have been impressive theoretical developments, particularly the use of the Liouville representation, which have extended the area accessible to study.

(c) There have been significant improvements in spectrometer performance. Increases in resolution naturally lead to linewidth differences (arising from T_2 differences) becoming more apparent. Improvements in sensitivity make it easier to study gases, where relaxation linewidths are usually considerable.

(d) Increasing use and greater sophistication of computers led to an easier manipulation of data. This has resulted in the general availability of powerful computer programs.†

The advent of digitization of n.m.r. spectra has greatly accelerated this process and, in particular, has made it feasible to treat bandshapes quantitatively rather than using a 'by eye' comparison with theoretical spectra. The development of computers on-line to n.m.r. spectrometers has given still further impulse in the same direction.

It is anticipated, therefore, that there will be a sharp increase in the amount of research relevant to this chapter in the next few years, and this process has already commenced, as should be clear from the present Report.

B. Coverage of the Report.—This chapter is intended to describe most bandshape n.m.r. work on fluids which has been published during the relevant period (July 1970 to June 1971). In addition, most of the research reported during the first six months of 1970 will be discussed, with some reference to important papers in previous years. However, any bandshape work for oriented molecules will be found in Chapter 10, and there is some discussion of bandshapes, not repeated here, in Chapters 8 and 9. It is, of course, not possible to make an absolute division between measurements dependent on intensities, and those depending on bandshapes, but intensity variations such as those involved in nuclear Overhauser or CIDNP studies are excluded from consideration in this chapter. Adiabatic rapid-passage and other saturation-recovery experiments using continuous-wave equipment are considered in Chapter 3.

Not all chemical applications involving bandshapes will be discussed, e.g. where the n.m.r. content is reckoned as trivial in the sense of containing no novelty. In particular, cases where information about rate constants for chemi-

---

† The existence of at least two libraries of computer programs should be noted.[1,2]

[1] Quantum Chemistry Program Exchange, Chemistry Dept., Indiana Univ., Bloomington, Indiana, U.S.A.
[2] 'A Library of Computer Programs for N.M.R. Spectroscopy', R. K. Harris and J. Stokes, Science Research Council (Atlas Computer Laboratory), Didcot, Berks., 1971.

cal exchange has been obtained solely from observation of coalescence temperatures will only rarely be mentioned. However, more attempt has been made to keep the sections other than Section 3 comprehensive, since much less work is being carried out in these areas.

The division of this chapter into sections is again to some extent arbitrary. Some research papers are clearly applicable to more than one section. There will be cross-references in important cases only. The sections are not expected to correspond to equal amounts of work (the section on chemical exchange is by far the largest). The important case of intermolecular exchange, as treated by the Swift–Connick[3] approach, is considered in Section 3 rather than in Section 5, but other cases where bandshapes are affected both by exchange and by relaxation are dealt with under the relaxation heading.

C. Nomenclature, Notation, and Units.—Since the studies reported in this chapter are increasing rapidly in importance, it is natural that new phraseology should arise, part of which may be adopted while part may be deemed inappropriate. The first point to be made concerns the title of this chapter. The term 'bandshape analysis' would seem to be preferable to 'lineshape analysis' on grounds of logic, but the two expressions appear to be used interchangeably in n.m.r. studies of dynamic equilibria.

The term 'thermal decoupling' appears to be coming into use[4-7] to denote the loss of splitting in spin-$\frac{1}{2}$ spectra coupled to spin-$\frac{1}{2}$ nuclei as the temperature is lowered, due to increased efficiency of quadrupolar relaxation. It is, perhaps, not entirely appropriate since similar effects could occur through solvent changes (especially those affecting viscosity considerably) at a single temperature. Most cases reported recently have been for boron-containing compounds, and since these often involve chemical exchange they are discussed in Section 4D.

The question of notation for exchanging spin systems has received little explicit attention. It would seem useful when intramolecular exchange is between identical spin systems to indicate low- and high-temperature limits with use of an arrow[8] (*e.g.* AB→A_2), though in the Reporter's view AB→C_2 may be preferable. However, this notation becomes difficult when the exchanging systems are not identical. Such systems may be described with a double arrow, for example AB⇌A'B' (ref. 9), though again AB⇌CD may be preferred. The fast-exchange limit then also needs to be specified, and the use of a bar has been suggested,[9] *e.g.* the limit for AB⇌A'B' may be

[3] T. J. Swift and R. E. Connick, *J. Chem. Phys.*, 1962, **37**, 307; 1964, **41**, 2553.
[4] H. Beall, C. H. Bushweller, W. J. Dewkett, and M. Grace, *J. Amer. Chem. Soc.*, 1970, **92**, 3484.
[5] M. Grace, H. Beall, and C. H. Bushweller, *Chem. Comm.*, 1970, 701.
[6] W. J. Dewkett, H. Beall, and C. H. Bushweller, *Inorg. Nuclear Chem. Letters*, 1971, **7**, 633.
[7] C. H. Bushweller, H. Beall, M. Grace, W. J. Dewkett, and H. S. Bilofsky, *J. Amer. Chem. Soc.*, 1971, **93**, 2145.
[8] K.-I. Dahlqvist, S. Forsén, and T. Alm, *Acta Chem. Scand.*, 1970, **24**, 651.
[9] C. J. Creswell and R. K. Harris, *J. Magn. Resonance*, 1971, **4**, 99.

written $\overline{AB}$. However, it is clear that some systematically-considered proposals are required. The case of intermolecular exchange also needs to be explicitly discussed.

Pyper and Harris[10-12] have suggested a notation for spin systems containing quadrupolar nuclei. A prefix superscript to the letter denoting the nucleus is used to indicate the multiplicity of the spin, $(2I+1)$, as for electron spins in atomic spectroscopy. Thus, for instance, a nitrogen nucleus might be denoted ^3A and an NH_2 group would be $^3A^2X_2$.

Reporting of thermodynamic data for exchange suffers from the problem of units. The use of SI (kJ mol^{-1}) is increasing, particularly in European journals (*e.g.* ref. 13) but many results are still in kcal mol^{-1} (*e.g.* ref. 14). Since in other areas of spectroscopy it has in the past been common to quote barriers in yet different units (*e.g.* cm^{-1}), it seems desirable to standardise information. At present much seems to depend on editorial policy. In this chapter kJ mol^{-1} will be used; the conversion factor to kcal mol^{-1} is 0·2390.

D. Reviews and Symposia.—Much of the present rapid development concerning n.m.r. bandshapes may be traced to the impetus given by several timely reviews during the last six years. In particular, those of Redfield[15] on general relaxation theory and Lynden-Bell[16] on density matrix calculations should be mentioned. In the area of chemical exchange effects, the reviews of Binsch[17] and of Johnson[18] have proved to be important.

The period 1970 to mid-1971 has also seen the appearance of several relevant review articles. Since this Specialist Periodical Report is not intended as a review of reviews, these will not be discussed but are merely listed in the references.[19-32] However, in order for the reader to be able to turn readily to those of interest to him, the titles of the recent reviews are included in the reference information, as are the article lengths. It is fitting

---

[10] R. K. Harris, N.A.T.O. Summer School, Coimbra, Portugal, 1968.
[11] R. K. Harris, N. C. Pyper, R. E. Richards, and G. W. Schulz, *Mol. Phys.*, 1970, **19**, 145.
[12] N. C. Pyper, *Mol. Phys.*, 1970, **19**, 161.
[13] J. A. Ladd and J. Parker, *J. Organometallic Chem.*, 1971, **28**, 1.
[14] D. L. Griffith, B. L. Colson, and J. D. Roberts, *J. Amer. Chem. Soc.*, 1971, **93**, 1648.
[15] A. G. Redfield, *Adv. Magn. Resonance*, 1965, **1**, 1.
[16] R. M. Lynden-Bell, *Progr. N.M.R. Spectroscopy*, 1967, **2**, 163.
[17] G. Binsch, *Topics Stereochem.*, 1968, **3**, 97.
[18] C. S. Johnson, jun., *Adv. Magn. Resonance*, 1965, **1**, 33.
[19] H. J. Keller, 'N.M.R. Untersuchungen an Komplexverbindungen', *N.M.R. Basic Principles and Progr.*, 1970, **2**, 1—88.
[20] E. W. Randall and D. G. Gillies, 'Nitrogen N.M.R.', *Progr. N.M.R. Spectroscopy*, 1971, **6**, 119—174.
[21] K. E. Schwarzhans, 'Magnetic Resonance of Paramagnetic Complexes', *Angew. Chem. Internat. Edn.*, 1970, **9**, 196—205.
[22] H. Kessler, 'Detection of Hindered Rotation and Inversion by N.M.R. Spectroscopy', *Angew. Chem. Internat. Edn.*, 1970, **9**, 219—235.
[23] W. E. Stewart and T. H. Siddall, tert., 'N.M.R. Studies of Amides', *Chem. Rev.*, 1970, **70**, 517—551.
[24] V. K. Pogorelyi and I. P. Gragerov, 'Investigation of Rapid Proton Exchange by P.M.R.', *Russ. Chem. Rev.*, 1970, **39**, 875—890.

to mention at this point that outstanding among the 1970—71 review articles relevant to this chapter is that by the late Ragnar A. Hoffman.[29] This publication provides an excellent summary of the theory of n.m.r. bandshapes and includes some original research work unpublished elsewhere. It provides a good working basis for those interested in future developments in this area.

Proceedings of symposia and conferences, as reported in their full form in established scientific journals for 1970—71, are treated here in the same way as reviews: they are listed (including titles) in the references,[33-36] but are not discussed further. Books and chapters of books on relevant topics will not be dealt with here at all.

2 Experimental Features

The processes of digitization, computation, and iteration are fundamental to the advances being made in bandshape analysis of all types. Of course, many studies are still made by eyeball comparison of theoretical and experimental bandshapes (this process is, in fact, remarkably efficient), and also by use of such characteristic parameters as bandwidths, peak separations, *etc.* These methods often rely on the use of appropriate equations describing the bandshape. Drakenberg, Dahlqvist, and Forsén[37] have highlighted the possible dangers in such methods. They find that in the case of the simple two-site exchange problem, use of approximate equations can give rise to considerable errors (up to, for example $\Delta H^{\ddagger} = 77.8$ kJ mol^{-1}, where the most accurate treatment gives 63·2 kJ mol^{-1}).

Hand digitization of spectra followed by computer treatment is, of course, feasible (see, for example, refs. 38 and 11), but is extremely tedious. The growth

---

[25] T. M. Ivanova and G. P. Kugatova-Shemyakina, 'The Application of N.M.R. in the Study of the Conformational Equilibria of Cyclic Compounds', *Russ. Chem. Rev.*, 1970, **39**, 510—528.
[26] G. A. Webb, 'N.M.R. Spectroscopy of Paramagnetic Species', *Ann. Reports N.M.R. Spectroscopy*, 1970, **3**, 211—259.
[27] W. A. Thomas, 'N.M.R. Spectroscopy as an Aid in Conformational Analysis', *Ann. Reports N.M.R. Spectroscopy*, 1970, **3**, 91—147.
[28] W. T. Huntress, 'The Study of Anisotropic Rotation of Molecules in Liquids by N.M.R. Quadrupolar Relaxation', *Adv. Magn. Resonance*, 1970, **4**, 1—37.
[29] R. A. Hoffman, 'Lineshapes in High-Resolution N.M.R.', *Adv. Magn. Resonance*, 1970, **4**, 87—200.
[30] J. C. Davis jun., and K. K. Deb, 'Analysis of Hydrogen Bonding and Related Association Equilibria by N.M.R.', *Adv. Magn. Resonance*, 1970, **4**, 201—270.
[31] B. D. N. Rao, 'Nuclear Spin Relaxation by Double Resonance', *Adv. Magn. Resonance*, 1970, **4**, 271—332.
[32] C. Hall, 'The Application of Chlorine, Bromine, and Iodine N.M.R. Spectroscopy to the Study of Physico-chemical Processes in Liquids', *Quart. Rev.*, 1971, **25**, 87.
[33] *J. Phys. Paris Colloq.*, 1971, **71**, 72—1170.
[34] Symposium on Dynamic N.M.R. Spectroscopy, Abstracts of Papers, 161st National Meeting of American Chemical Society, Los Angeles, 1971.
[35] Symposium: Wide-Line N.M.R., [American Oil Chemists' Society 43rd Fall Meeting, Minneapolis, 1969], *J. Amer. Oil Chemists' Soc.*, 1971, **48**, 1—17 and 47—69.
[36] Berichtsheft: Molekulare Bewegungen in Flüssigkeiten (Herrenalb 1970), *Ber. Bunsengesellschaft Phys. Chem.*, 1971, **75**, 183—283.
[37] T. Drakenberg, K.-I. Dahlqvist, and S. Forsén, *Acta Chem. Scand.*, 1970, **24**, 694.

in automatic digitization procedures (*e.g.* refs. 39 and 40) is therefore not surprising. The stage of home-built digitization has now been partially superseded by the production of commercial equipment by the n.m.r. spectrometer manufacturers (see Chapter 4). This process is aided by the common use of digital chart recorders. However, the logical conclusion of this process of development is not the explicit output of digitized spectra on paper tape or cards, for later treatment by a remote off-line computer, but rather the direct feeding of the digitized data into a local on-line computer (see again Chapter 4). There remains the debatable question of the relative efficiency of small dedicated computers and large multi-access computers. Many of the advantages and disadvantages (accessibility and control, size of problem that can be tackled, speed, *etc.*) are reasonably obvious.

Cox, Riddell, and Williams[40] emphasise the advantage of computer comparison of observed and calculated spectra of exchanging systems, and give several examples of applications (mostly of the simple equal-population $A \rightleftharpoons B$ type). Allowance was made for a sloping baseline, and the area of the computed bandshape was then normalised to that of the experimental spectrum. A least-squares fitting was then performed, with allowance for different values of T_2^{eff} at the two sites. The final calculated spectrum was output to the spectrometer recorder.

Gillen and Noggle[41] examine in detail the common broad-line n.m.r. problem of linewidth errors arising from too high a frequency or amplitude of the audio magnetic field modulation. They give equations for the necessary corrections in both the absorption and dispersion modes, and also present their results in a series of convenient graphs. They conclude that the derivative of the dispersion mode is frequently the more accurate for determinations of T_2 (the derivative of the absorption mode is actually usually used).

Rotaru[42] has published graphs which allow the determination of the true amplitude and width ratios for dispersion-mode n.m.r. spectra which consist of the superposition of a narrow line and a broad line. Both Lorentzian and Gaussian shapes are considered. Although the article is directed towards solid-state spectra, the graphs may occasionally be useful for spectra of fluids.

Mochel and Claxton[43] have taken this process a stage further. They present details of work in which an analog computer is used to resolve composite n.m.r. curves arising from up to seven lines (of either Gaussian or Lorentzian shape). Refinements such as treatment of slanted baselines may be introduced. The methods are applied with success to spectra of butadiene–styrene copolymers. However, the general use of such methods is open to some criticism

---

[38] K.-I. Dahlqvist and S. Forsén, *Acta Chem. Scand.*, 1970, **24**, 797.
[39] R. E. Carter, J. Marton, and K.-I. Dahlqvist, *Acta Chem. Scand.*, 1970, **24**, 195.
[40] B. G. Cox, F. G. Riddell, and D. A. R. Williams, *J. Chem. Soc. (B)*, 1970, 859.
[41] K. T. Gillen and J. H. Noggle, *J. Magn. Resonance*, 1970, **3**, 240.
[42] M. Rotaru, *Rev. Roumaine Phys.*, 1970, **15**, 663.
[43] V. D. Mochel and W. E. Claxton, *J. Polymer. Sci., Part A-1, Polymer Chem.*, 1971, **9**, 345.

since it is possible to obtain a spurious number of lines if they are allowed to be very close, and also knowledge of the shape of individual lines is important but almost impossible to determine *a priori*. Despite these drawbacks, useful information can be obtained when the techniques are used judicially.

Pyper,[12] in calculating spectra of spin systems containing quadrupolar nuclei using density matrix theory, has commented that inclusion of B_0-inhomogeneity effects as effective spin–spin relaxation times on the diagonal elements of the relaxation equations is satisfactory if the inhomogeneities are Lorentzian, but not if they are Gaussian. It is clearly preferable, in principle, to account for inhomogeneities by convoluting calculated lineshapes with an experimental broadening function. Obtaining a valid broadening function is, however, not a trivial matter.[44] Fortunately, results are only sensitive to such instrumental broadening when the linewidth attributable to the process of interest (exchange, *etc.*) is small; caution should clearly be exercised in such cases.

Harris *et al.*[11] have stressed the problems of estimating true baselines for broad experimental spectra and have suggested the use of intensity differences between the arbitrary frequency positions and a 'standard position' for comparison with theoretical spectra. However, these authors also discuss methods of obtaining true experimental baselines graphically in the case of Lorentzian bands. They also stress that under certain conditions spectra may be ill-conditioned to some of the required parameters.

One of the difficulties encountered in comparing thermodynamic activation parameters from different literature sources lies in the choice between using the Arrhenius equation, $k_r = A \exp(-E_a/RT)$, or the Eyring equation, $k_r = (\kappa kT/h) \exp(-\Delta H^{\ddagger}/RT) \exp(\Delta S^{\ddagger}/R)$, and hence in the form of plotting used [log k_r *vs.* $1/T$ or $\log(k_r/T)$ *vs.* $1/T$]. Of course, it is feasible to plot in one way and then to algebraically convert the results to the other type of data using standard formulae.[45] However, since assumptions of linearity are always made, and such simultaneous assumptions for log k_r and $\log(k_r/T)$ are in principle mutually incompatible, it would seem preferable, if both Arrhenius and Eyring parameters are to be quoted, to derive them from independent plots. Otherwise, inconsistencies can arise — for instance Peeling, Schaefer, and Wong[46] give (without comment) values of $\Delta S^{\ddagger}$ for a particular system which differ by a factor of two, according the the procedure used. It certainly seems inconsistent to derive $\Delta H^{\ddagger}$ directly from plots of log k *vs.* $1/T$, as is still sometimes done.[37] A second feature which perhaps needs to be standardised is the use of transmission coefficients in the Eyring equation. The

---

[44] N. C. Pyper, *Mol. Phys.*, 1971, **20**, 449.
[45] See, for example, F. Daniels and R. A. Alberty, 'Physical Chemistry', John Wiley and Sons, New York, 1966, p. 613.
[46] J. Peeling, T. Schaefer, and C. M. Wong, *Canad. J. Chem.*, 1970, **48**, 2839.
[47] B. J. Fuhr, B. W. Goodwin, H. M. Hutton, and T. Schaefer, *Canad. J. Chem.*, 1970, **48**, 1558.
[48] F. Loustalot, M. Loudet, S. Gromb, F. Metras, and J. Petrissans, *Tetrahedron Letters*, 1970, 4195.

use of a factor ½ to take account of every metastable intermediate is common (see, for example, ref. 47), but it is not always feasible to gauge the existence of such intermediates.

Miscellaneous.—Use of 'linewidths' dominated by unresolved splittings due to spin–spin coupling continues to be popular (see, for example, ref. 48), especially for methyl resonances in cyclic systems such as steroids,[49] where the relevant small coupling constants (and hence the linewidths) depend on molecular conformation or configuration. This technique has been used[50] to determine the position of equilibrium in the ring inversion of thiacyclohexanes. For some reason it proved convenient to measure the bandwidth at one-fifth height (this was between 23 and 31 Hz); t-butyl derivatives were used as model compounds. Such 'bandwidth' studies have also been[51] valuable in ^{19}F resonance for fluorocyclohexanes.

3 Exchange of Magnetic Sites

A. Theoretical Work.—The important theoretical developments made in recent years regarding exchange effects on n.m.r. spectra have been consolidated during the year under review. Thus Kleier and Binsch[52] have continued theoretical work[53] on n.m.r. exchange lineshapes using the Liouville representation by examining the possibility of factorising the problem for cases involving intramolecular exchange. They propose a series of theorems regarding general invariance properties in Liouville space. The important case of mutual exchange, where the Hamiltonians for all nuclear configurations are identical (and hence all populations are equal) is then discussed, with the ABC⇌BCA ⇌CAB three-site, mutual exchange system taken as an example. Magnetic equivalence is then considered and illustrations are given for $A_3B_3 \rightleftharpoons C_3D_3$ and mutual exchange of the type $A_3B_3 \rightleftharpoons B_3A_3$. Finally, permutation symmetry is explored, with the example $[AB]_2 \rightleftharpoons [CD]_2$. It is abundantly clear that the methods discussed are extremely powerful and will result in a great increase in the number of types of exchange process studied by n.m.r. in the future. It should be added that Kleier and Binsch have incorporated the symmetry features into their computer program,[54] available through the Quantum Chemistry Program Exchange.[1] Binsch and co-workers[55] also present an experimental example to justify the claim that improved accuracy for

---

[49] H. Schick, E. Grundemann, and G. Hilgetag, *J. prakt. Chem.*, 1970, **312**, 1074.
[50] R. Borsdorf, P. F. Matzen, H. Remane, and A. Zschunke, *Z. Chem.*, 1971, **11**, 21.
[51] J. Cantacuzene and R. Jantzen, *Tetrahedron Letters*, 1970, 3281.
[52] D. A. Kleier and G. Binsch, *J. Magn. Resonance*, 1970, **3**, 146.
[53] G. Binsch, *J. Amer. Chem. Soc.*, 1969, **91**, 1304.
[54] (a) G. Binsch and D. A. Kleier, 'DNMR2 — The Computation of Complex Exchange-broadened N.M.R. spectra', Q.C.P.E. Program 140, 1969. (b) D. A. Kleier and G. Binsch, 'DNMR3 — A Computer Program for the Calculation of Complex Exchange-broadened N.M.R. spectra (modified version for spin systems exhibiting magnetic equivalence or symmetry)', Q.C.P.E. Program 165, 1970.
[55] D. A. Kleier, G. Binsch, A. Steigel, and J. Sauer, *J. Amer. Chem. Soc.*, 1970, **92**, 3787.

n.m.r. studies of exchange processes can often be obtained by deliberately choosing to study complicated spectra. The case they examined was that of the cyclopropyl protons of 2,5-dimethoxycarbonyl-3,4-diazanorcaradiene (1), which give an A_2BC spectrum at -25 °C. As exchange of the type

<pre>
         MeO₂C   H_C
            \   ├─H_B
             N   ⋮H_A
             ‖
             N   ⋮H_A
            /
         MeO₂C

             (1)
</pre>

$A_2BC \rightleftharpoons A_2CB$ becomes increasingly rapid with rise in temperature, the spectra degenerate into the A_2B_2 type. The authors discuss the advantages of the complex nature of the spectra, and the Reporter cannot resist quoting one of their sentences in full:

'The various line separations in the spectra may be regarded as a collection of individual clocks, running on different time scales, that become successively tuned to the changing rate of the molecular process, whereas for a simple spectrum with only a single clock, reliable rate determinations may be limited to a narrow range around the coalescence point.'

They further point out that measurements were feasible over a range of 121 K, spanning four powers of ten in the rate constant. Computation of a single spectrum, using the program DNMR3,[54b] for the system studied took only 21 seconds on a Univac 1107 computer.

The Binsch theory for exchange-broadened n.m.r. lineshapes has also been developed by Anderson and Lee,[56] and by Anderson.[57] In the earlier paper[56] Anderson and Lee apply the theory to three problems:

(a) The mutual exchange $AB \rightleftharpoons BA$ case in which the nuclei have different relaxation times. As expected, the linewidth of the slowly relaxing nucleus depends on the relaxation rate of the rapidly relaxing nucleus. However, this is also true in general in the absence of exchange; Anderson and Lee have not taken this into account since their way of incorporating different relaxation times is to use two different values of T_2 for the diagonal elements of the relaxation operator. This presumably corresponds to assuming that relaxation is by the uncorrelated random fields mechanism.

(b) The mutual exchange $AB \rightleftharpoons BA$ case in which there is direct coupling between the spins (i.e. for oriented molecules). It is apparent that the

[56] J. M. Anderson and A. C.-F. Lee, *J. Magn. Resonance*, 1970, **3**, 427.
[57] J. M. Anderson, *J. Magn. Resonance*, 1971, **4**, 184.

inner lines of the quartet broaden more than the outer lines as the exchange rate increases (the opposite of the case for indirect coupling).

(c) The n.m.r. spectrum of partially oriented dimethyl[$^2H_3$]acetamide was examined over a range of exchange rates. Only theoretical spectra are presented; some experimental spectra are illustrated in a later paper,[58] but it was only possible to obtain one approximate value for the rate constant (see also Chapter 10).

The second paper[57] deals with a double-resonance case (see Section 7).

Extensive use of full n.m.r. bandshape analysis for exchanging systems has been made by Swedish workers, especially Forsén and Dahlqvist, in a series of papers.[8,37-39,59,60,63-66] Several of these articles discuss theoretical or practical aspects of considerable importance. The problems arising from the fact that the bandshape depends on several parameters are extensively discussed.[37,59] In a detailed evaluation of the experimental errors influencing exchange rates, Drakenberg, Dahlqvist, and Forsén[37] discuss the simple two-site, equal-population, uncoupled case. The spectra then depend on three parameters — the lifetime, τ; the effective relaxation time (assumed equal for the two sites), T_2^{eff}; and the chemical shift difference, δv. The authors include inhomogeneity effects in T_2^{eff}; however, variations in homogeneity with temperature were minimised by adjusting the spectrometer to obtain a constant linewidth for the tetramethylsilane signal (as opposed to the normal process of always optimising resolution). The value of T_2^{eff} was estimated by iterative curve fitting in the limits of fast and slow exchange, with linear interpolation between these extremes. The values of δv were obtained by extrapolation from the measured variation under conditions of slow exchange. The authors estimate that an assumption of constant T_2^{eff} would introduce an error of 2·9 kJ mol^{-1} in $\Delta H^{\ddagger}$ in their case (where T_2^{eff} actually varied from 0·24—0·30 s). The possible error arising from extrapolation of δv was thought to be ±1·3 kJ mol^{-1}. Such errors are well in excess of the statistical errors arising from Arrhenius or Eyring plots. In fact, these authors[37] suggest that their result ($\Delta H^{\ddagger}$ = 63·2 kJ mol^{-1}) for the restricted internal rotation of NN-dimethyltrichloroacetamide, using interpolated values for T_2^{eff}, is accurate only to ±2·5 kJ mol^{-1}. They also include a detailed discussion of their method of temperature measurement at the sample, including a table of the temperature gradient above the receiver coil in their Varian A-60A probe.

Dahlqvist[59] discusses in detail the problem of evaluating both rate para-

---

[58] J. M. Anderson and A. C.-F. Lee, *J. Magn. Resonance*, 1971, **4**, 160.
[59] K.-I. Dahlqvist, *Acta Chem. Scand.*, 1970, **24**, 683.
[60] K.-I. Dahlqvist and S. Forsén, *J. Magn. Resonance*, 1970, **2**, 61.
[61] J. Kaplan, *J. Chem. Phys.*, 1958, **28**, 278; 1958, **29**, 462.
[62] S. Alexander, *J. Chem. Phys.*, 1962, **37**, 574, 967.
[63] L. Arlinger, K.-I. Dahlqvist, and S. Forsén, *Acta Chem. Scand.*, 1970, **24**, 662.
[64] L. Arlinger, K.-I. Dahlqvist, and S. Forsén, *Acta Chem. Scand.*, 1970, **24**, 672.
[65] T. Drakenberg and S. Forsén, *J. Phys. Chem.*, 1970, **74**, 1.
[66] P. Stilbs and S. Forsén, *J. Phys. Chem.*, 1971, **75**, 1901.

meters and the equilibrium constant in the case that is one stage more complex than that discussed above — namely the two-site, unequal-population, uncoupled case (the actual example is AX⇌BY but the effect of the X, Y protons on the A, B spectrum is taken into account by the X approximation). Dahlqvist[59] convincingly demonstrates that, even when T_2^{eff} and δv are fixed, spectra can often be fitted by very different values of the lifetime, τ, and population, P (for example, illustrated spectra with $\tau = 0.0021$ s, $P = 0.739$ and $\tau = 0.30$ s, $P = 0.95$ look very similar). He develops a rather complicated method of extracting the correct values of ΔH and $\Delta H^{\ddagger}$ from the experimental spectra together with theoretical three-dimensional plots of τ vs. K vs. a shape function, F. Although he applies this method with success to the internal rotation of the aldehyde group in furan-2-aldehyde, it would seem that it is unlikely to come into general practice. The principal value of the paper is, perhaps, the emphasis on possible ambiguities.

Dahlqvist and Forsén[8,60] have discussed the application of bandshape analysis to complex n.m.r. spectra, and have developed a computer program to tackle many four-spin cases, using the Kaplan[61]–Alexander[62] density matrix methods. They stress that in the ABC⇌BAC case, with only weak or moderate coupling to C, two of the C lines (corresponding approximately to molecules with $\alpha\alpha$ and $\beta\beta$ spin functions for A and B) remain sharp throughout the exchange region. They suggest that these lines may be useful in estimating natural linewidths and in distinguishing between inter- and intra-molecular exchange, though these uses could be dangerous since for strongly coupled three-spin systems all lines are affected by the exchange. Dahlqvist, Forsén, and Alm[8] also discuss the [AB]$_2$⇌[BA]$_2$ and ABCD⇌BADC cases (high-temperature limits A$_4$ and [AB]$_2$ respectively). In the former case the spectra are always symmetrical about their mid-point, but of course this is not true in the latter case. The authors stress two features of exchange studies of complex spin systems: (*a*) as noted by Binsch and co-workers[55] (see above), the complexity of the spectra is often an advantage, and (*b*) the amount of computer time required increases very rapidly as the number of nuclei involved increases, thus placing a limitation on the size of problem that can be tackled.

Dahlqvist and Forsén[60] have used the above theory to study exchange of the type ABCD⇌BADC for benzofuroxan (2). Several spectra are illustrated

(2)

in comparison with theoretical spectra; the variations with temperature are striking. The exchange, indeed, presents a very favourable case for obtaining rate data, since the spectra are complex but the number of parameters is low — only one lifetime is needed to determine the spectrum. The principal experimental problem lies in obtaining information on the variation of chemical shifts, coupling constants, and 'natural' linewidths with temperature. However, Dahlqvist and Forsén[60] comment that the richness of detail in the spectrum allows the determination of two of the relevant chemical shift differences even down to lifetimes of the order of 10^{-3} s (these shifts do not, in fact, show any significant temperature dependence). The third shift difference of importance could be determined at all temperatures investigated.

The Swedish workers have applied their methods to a number of chemical cases, in addition to those mentioned above. They are summarised below:

(a) Internal rotation in substituted neopentylbenzenes.[39] These are examples of ABX$\rightleftharpoons$BAX exchange, where the AB region was studied and the effect of X taken into account by use of the 'X approximation'.

(b) Internal rotation in 2-acetylfuran,[63] an AMX$\rightleftharpoons$BNY case. The modified Bloch equations were used, but line intensities in the slow-exchange limit were corrected for first-order effects. Differences (by a factor of ca. 1·5) in rates measured using 100 MHz and 60 MHz equipment were attributed to systematic errors in temperature determinations at 60 MHz by somewhat less than 4 °C.

(c) Internal rotation in N-methylpyrrole-2-aldehyde.[64] This is a similar case to (b) above but the isomers were present in very unequal amounts [for (b) the ratio was found to be $1 \cdot 0 \pm 0 \cdot 1$]. The principal effects are therefore pronounced broadening of some lines (separate signals for the low-temperature limit could only be detected for some of the resonances). The maximum broadening of the signal chosen for evaluation (CHO) was only 50%. The method of Dahlqvist[59] (see above) was used to simultaneously evaluate ΔH and $\Delta H^{\neq}$, but the accuracy claimed is not high.

(d) Internal rotation about the C$\equiv$N bond[65] for [^{15}N]formamide. The exchange problem in this case was of the ABCX$\rightleftharpoons$BACX type, and was treated using the 'X approximation'. The spectrum of the C region (aldehyde proton) enabled a distinction to be drawn between internal rotation and intermolecular exchange of the NH_2 protons. The ^{15}N-substituted compound was used in order to avoid the complications due to quadrupolar effects. There is no appreciable effect of solvent on the barrier parameters, but the n.m.r. parameters are altered by solvent, so the bandshapes change.

(e) Internal rotation in urea.[66] In this case only a mean value of $\Delta G^{\neq}$ was obtained because there were considerable uncertainties in the chemical shifts and 'natural' linewidths.

(f) Intermolecular exchange of enolic protons in 1-formylphenylacetone.[38] This involves the collapse of doublet splittings, and was treated on the simple two-site, equal population, uncoupled basis since the two nuclei involved were well separated in chemical shift. The McConnell equation[67] was therefore used (except in the limit of very fast exchange, where it is inapplicable) and the data extracted by iterative computer fitting of the bandshape. The authors studied several concentrations but were unable to decide if the activation parameters are concentration-dependent. Because of the catalytic effect of impurities, they took their highest E_a values as most reliable, and concluded that since the existence of an intramolecular hydrogen-bond in this system does not appear to substantially affect exchange rates, the exchange mechanism must largely retain hydrogen-bond energy.

Creswell and Harris[9] have examined intramolecular exchange cases of the simplest type to incorporate both unequal populations and second-order effects in the spin system. This they refer to as AB⇌A′B′ exchange. They describe the density matrix theory briefly (an earlier theoretical discussion had appeared by Johnson,[68] but coupled to a rather poorly-defined experimental example) and show experimental spectra, which were fitted by comparison with the appearance of theoretical spectra by eye and by measurement of peak-height-to-trough ratios, *etc.* The effects of population and chemical-shift variation with temperature are discussed. The authors emphasise the advantage given by spectral complexity since this results in a wide temperature range over which spectral variations occur. The temperature range is further extended for one of the two examples because the molecule also involves a simple A⇌X exchange.

Jouanne and Heidberg[69] have reported complete lineshape studies of AB⇌BA exchanging systems, though the computational basis of their work has still to be published. The exchange studied was the internal rotation about the ring-carbon–nitrogen bond of certain N-n-alkyl-2,4,6-trinitroanilines, the ring protons forming the AB system. The resonance of the NH proton also occurred in the same spectral region, and its effects had to be subtracted. The authors discuss the importance of variable 'natural linewidths' on their results; they sacrifice their data obtained for the extremes of fast and slow exchange in order to minimise this problem.

Leigh[70] has discussed 'effective relaxation rates' $\overline{T_1}^{-1}$ and $\overline{T_2}^{-1}$ for systems involving chemical exchange, using modified Bloch equations,[67] and has given explicit expressions for these parameters. These equations are simplified by Leigh so as to be applicable to the effects of dilute solutions of paramagnetic spins on nuclear relaxation times; the results are identical with those previously published in the literature (*e.g.* those of Swift and Connick[3] for $\overline{T_2}^{-1}$), but the derivations are simpler and have fewer restrictions. Strictly, Leigh's deductions are made ignoring the radiofrequency ω_1 and are therefore directly applicable to pulse experiments rather than to continuous-wave

---

[67] H. M. McConnell, *J. Chem. Phys.*, 1958, **28**, 430.
[68] C. S. Johnson, jun., *J. Magn. Resonance*, 1969, **1**, 98.
[69] J. von Jouanne and J. Heidberg, *Ber. Bunsengesellschaft Phys. Chem.*, 1971, **75**, 261.
[70] J. S. Leigh, jun., *J. Magn. Resonance*, 1971, **4**, 308.

work. Except when saturation occurs, the steady-state result will presumably be the same, however.

Reeves and Shaw[71] have published a comprehensive treatment of the effects of chemical exchange on steady-state spectra in cases where the modified Bloch equations[67] apply. They incorporate many-site exchange, site-dependent relaxation times, differing site populations, and saturation effects. Particular attention is paid to the simple two-site, uncoupled exchange system, and many computed bandshapes (both adsorption and dispersion mode) are illustrated. Alternative forms of the basic bandshape equation are described, corresponding to specific assumptions about the relaxation processes involved. The authors stress the necessity for complete bandshape treatments in n.m.r., the necessity of computation, and the convenience of their equations for computer calculations. No observed lineshapes are reported in this publication,[71] but more details are promised later.

The Alexander approach[62] has been used to simulate n.m.r. spectra involving intermolecular exchange in cases where second-order effects appear in the spectra. Yamamoto and Kamezawa[72] investigated the spectra of a number of alcohols in which the OH protons are exchanging (promoted by the presence of a small amount of H_2O). The slow-exchange spectra were AB_2X, AB_2X_3, and AB_2KX, where A is the exchanging proton. The spectra of the X protons, which are not directly coupled to A, are affected markedly by the exchange. When $\nu_A - \nu_B \neq 0$ the X spectra become unsymmetrical at intermediate exchange rates. This asymmetry allows the relative signs of coupling constants to be determined. A sharp line is always obtained at $\nu_X = 0$ for 2,2-dichloroethanol (AB_2X case), arising from the states in which the spins of the CH_2 protons have a singlet wavefunction. The spectra illustrated show the elegance of the density matrix approach. The authors compare their results at coalescence with those obtained using the Gutowsky, McCall, and Slichter[73] approach, but they do not discuss the kinetics of the exchange in any detail.

The growth of 'total bandshape analysis' for treating n.m.r. spectra of exchanging systems has led to a decline in the use of specific spectral parameters, e.g. peak separations and ratios of maximum to minimum intensities. Indeed the latter tendency has occurred partly as a result of criticism of some of the earlier work of this type. However, Ramey and co-workers[74] have reported extensive investigations of the linewidth method for a variety of amides, usually giving good agreement with results from total bandshape work. The cases examined were of the two-site, equal-population exchange type; the Gutowsky–Holm[75] equations were used. Ramey et al.[74] present a family of curves of half-width versus lifetime for a range of chemical-shift

[71] L. W. Reeves and K. N. Shaw, *Canad. J. Chem.*, 1970, **48**, 3641.
[72] O. Yamamoto and N. Kamezawa, *J. Magn. Resonance*, 1970, **3**, 269.
[73] H. S. Gutowsky, D. W. McCall, and C. P. Slichter, *J. Chem. Phys.*, 1953, **21**, 279.
[74] K. C. Ramey, D. J. Louick, P. W. Whitehurst, W. B. Wise, R. Mukherjee, and R. M. Moriarty, *Org. Magn. Resonance*, 1971, **3**, 201.
[75] H. S. Gutowsky and C. H. Holm, *J. Chem. Phys.*, 1956, **25**, 1228.

differences, and stress the need for accurate correction to allow for inhomogeneities and couplings. The authors discuss the significance of their results in detail; in the case of dimethylformamide itself they disagree with previous findings by about 17 kJ mol^{-1} for the value of E_a. They believe this may be due to the existence of more than one rate process. The paper as a whole appears to represent an attempt at rationalising discrepancies in the literature rather than a proposal that the linewidth method is to be preferred to total bandshape analysis. In principle, the total bandshape method should be best.

A novel single-parameter method has been suggested by Heuring and Brey,[76] who investigated the moment method for studying chemical exchange. They apply their theory to the simple A$\rightleftharpoons$X equal-population, two-site exchange, with zero linewidths in the absence of exchange. Only the second moment was discussed and, although the authors show the method to be feasible, there does not seem to be any reason why it should be preferred to total bandshape methods. The extension of the second moment to cover unequal populations, finite non-exchange linewidths, and spin–spin coupling is discussed but no experimental examples are given. Errors are given considerable attention.

Delpuech[77] has discussed in detail how the nitrogen inversion rate of an amine may be calculated from the mean lifetime of the corresponding ammonium cation, as determined by n.m.r. (variation of pH). In particular, he shows the limitations of the validity for the Saunders and Yamada[78] approximate formulae.

Dreeskamp and Hildenbrand[79] have illustrated a convincing example of the determination of relative signs of coupling constants from a qualitative examination of line-broadening effects, due to exchange, for complex spin systems. The molecule concerned is Me_2PF_3, which undergoes an intramolecular 'pseudorotation' exchange process, rendering the ^{19}F nuclei equivalent at high temperatures. As the temperature is raised from -20 °C, some lines in the ^1H spectrum broaden while others remain sharp. The latter arise from molecules in which the fluorine nuclei all have α spin or all have β spin. Since the sharp lines include those at the extremes of the spectrum, the two F–H coupling constants must have the same sign.

Similar retention of sharp lines has been noted[80,81] for the intramolecular exchange process of $H_2Fe[PhP(OCHMe_2)_2]_4$ and similar compounds, as studied by ^1H n.m.r. A doublet of triplets at -16 °C becomes a quintet at $+51$ °C (a second rate process affects the spectrum at temperatures below -16 °C); the outermost lines are unaffected by the exchange. This contrasts with the case of $H_2Ru(PhPEt_2)_4$, where it is the innermost lines which remain

[76] V. P. Heuring and W. S. Brey, *J. Magn. Resonance*, 1970, **3**, 468.
[77] J.-J. Delpuech, *Org. Magn. Resonance*, 1970, **2**, 91.
[78] M. Saunders and F. Yamada, *J. Amer. Chem. Soc.*, 1963, **85**, 1882.
[79] H. Dreeskamp and K. Hildenbrand, *Z. Naturforsch.*, 1971, **26b**, 269.
[80] E. L. Muetterties, *Accounts Chem. Res.*, 1970, **3**, 266.
[81] P. Meakin, L. J. Guggenberger, J. P. Jesson, D. H. Gerlack, F. N. Tebbs, W. G. Peet, and E. L. Muetterties, *J. Amer. Chem. Soc.*, 1970, **92**, 3482.

sharp.[80] For the related compound $H_2Fe[PhP(OEt)_2]_4$ no sharp features are retained through the exchange-sensitive region.[80,81] Retention of sharp features for the C region of the ABCX (at slow exchange) system of [^{15}N]formamide enabled Drakenberg and Forsén[65] to conclude that the effects were due to internal rotation, not intermolecular exchange.

B. Examples of Intramolecular Exchange.—Four techniques for extending the range or accuracy of n.m.r. exchange studies merit mention at this point. These techniques, which have been in use during the period covered by this review (as discussed later), are:

(a) Increase of available chemical-shift range by studies of the ^{19}F resonance of CF_2 groups instead of the ^{1}H resonance of CH_2 groups.[82-84]

(b) Increase of available chemical-shift range by studies of ^{13}C resonance instead of ^{1}H resonance.[82]

(c) Extension of the chemical-shift range by the use of paramagnetic species.[85]

(d) Increase of feasible temperature range by combining n.m.r. bandshape data with direct kinetic studies (by n.m.r. or other techniques) for slow exchange ($k \leqslant 0\cdot1$ s^{-1}).[86-90]

For most of the reported n.m.r. work the above techniques were either not available or not feasible, but they are becoming increasingly used. However, techniques (a) and (c) suffer from the disadvantage that chemical modifications are made to the system, and that therefore the data may not be relevant to the parent compounds.

Hindered Internal Rotation. In a series of papers [46,47,91-93] Schaefer and his co-workers have studied the n.m.r. bandshapes of several toluenes and xylenes, substituted both in the ring and in the side-chain. The exchange process involved is internal rotation about the side-chain-to-ring C—C bond, as in (3). In this case[47] the exchanging system is ABX⇌BAX, and the authors wrote their own computer program, based on the density matrix method, to

---

[82] D. Doddress, C. Charrier, B. L. Hawkins, W. O. Crain, L. Harris, and J. D. Roberts, *Proc. Nat. Acad. Sci. U.S.A.*, 1970, **67**, 1588.
[83] H. J. Reich, E. Ciganek, and J. D. Roberts, *J. Amer. Chem. Soc.*, 1970, **92**, 5166.
[84] D. T. Hill and K. C. Ramey, *Helv. Chim. Acta.*, 1970, **53**, 1184.
[85] R. von Ammon, R. D. Fischer, and B. Kanellakopulos, *Chem. Ber.*, 1971, **104**, 1072.
[86] C. H. Bushweller, J. W. O'Neil, M. H. Halford, and F. H. Bissett, *J. Amer. Chem. Soc.*, 1971, **93**, 1471.
[87] C. H. Bushweller, J. Grolini, G. U. Rao, and J. W. O'Neill, *J. Amer. Chem. Soc.* 1970, **92**, 3055.
[88] C. H. Bushweller, *Chem. Comm.*, 1970, 51.
[89] K. von Bredow, A. Jaeschke, H. G. Schmid, H. Friebolin, and S. Kabuss, *Org. Magn. Resonance*, 1970, **2**, 543.
[90] C. H. Bushweller, J. W. O'Neil, M. H. Halford, and F. H. Bissett, *Chem. Comm.*, 1970, 1251.
[91] B. H. Barber and T. Schaefer, *Canad. J. Chem.*, 1971, **49**, 789.
[92] H. G. Gylulai, B. J. Fuhr, H. M. Hutton, and T. Schaefer, *Canad. J. Chem.*, 1970, **48**, 3877.
[93] J. Peeling, B. W. Goodwin, T. Schaefer, and C. Wong, *Canad. J. Chem.*, 1971, **49**, 1489.

calculate the total bandshape. Relatively speaking, the X resonances remain sharp because the X proton has only a single type of site. The computer program calculates peak positions and half-widths for the AB region, and provision is made for overlapping peaks (though it is not entirely clear how). Comparison between experimental and theoretical spectra can be made visually or by consideration of the half-widths. The authors discuss and make allowance for the effect of chemical-shift variation with temperature. They also give careful consideration to the errors, finally quoting E_a to $\pm 5\cdot4$ kJ mol^{-1}. An effective $T_2 = 1\cdot38$ s (to account for magnetic field inhomogeneity in perdeuteriotoluene solution) was estimated from the resonance width of a small amount of impurity α,α,α,2,4,6-hexachlorotoluene; it would seem that this T_2 is used in the diagonal elements of the relaxation matrix rather than in a final convolution. One other minor feature is worthy of comment. The authors[47] try to draw a distinction between 'exchange of Larmor frequencies'

(3)

and 'exchange of sites' for the ring protons. Such a distinction appears meaningless since it depends on assuming that the ring is fixed while the side-chain rotates. One of the later papers[92] in this series deals with the exchange (3a)⇌(3b) in CS$_2$ solution, where it becomes ABC⇌BAC. In this case the program DNMR[53,54] was used for the computations. The Canadian workers endeavoured to check whether undetected systematic errors occurred for their work by studying[93] a very different exchange system of the same chemical type, that of α,α,α',α',2,4,5,6-octachloro-*m*-xylene. The exchange in this case is of the four-site, negligibly-coupled type, for which modified Bloch equations[67] could be used. The molecule exists as a mixture of three conformers on the n.m.r. timescale at low temperatures. Thus the problem involves two equilibrium constants, but in fact there are in addition only two independent exchange lifetimes. The equilibrium constants were measured at low temperatures, and extrapolated to high temperatures using theoretical values of entropy differences (symmetry contributions only). The rest of the spectral analysis proceeded as in the earlier cases. The authors emphasise that because there are a number of different coalescence features, and the spectral changes occur over a range of 110 K, the thermodynamic parameters are rather accurate. Mark and Pattison[94] have also examined (by ^{1}H n.m.r.) hindered internal rotation in molecules containing $CHCl_2$ groups bonded to a benzene ring (in their cases *ortho* pairs of $CHCl_2$ groups occurred), but no detailed bandshape work has yet been reported. Coalescence temperatures and values of $\Delta G_c^{\ddagger}$ were given for two compounds.

Complete bandshape studies have been carried out,[95] using ^{19}F resonance, of internal rotation in a number of substituted ethanes, namely $ClBr_2C \cdot CClBrF$, $ClBr_2C \cdot CCl_2F$, $Cl_2BrC \cdot CCl_2F$, $ClBr_2C \cdot CBr_2F$, $Cl_2BrC \cdot CClBrF$, $ClBr_2C \cdot CHBrF$, $Cl_2BrC \cdot CHClF$, and one propane, $Cl_2BrC \cdot CBrCl \cdot CF_3$. The authors used Alexander's approach[62] for intermolecular exchange (except for the CF_3 rotation in the propane) because the exchanging species have different populations. The first five ethanes were examples of three-site, unequally-populated $A \rightleftharpoons B \rightleftharpoons C$ or of two-site, unequally-populated $A \rightleftharpoons B$ exchange. The protonated cases were of the type $AX \rightleftharpoons BY \rightleftharpoons CZ$ (which was treated using 'X-approximation' subspectral techniques); the ^{1}H bandshapes were also studied in these cases. Theoretical and experimental spectra were compared visually, and the authors note that above the coalescence temperature 'many combinations of $\Delta G^{\ddagger}$ values could be found to reproduce the observed spectra' — the uniqueness of the $\Delta G^{\ddagger}$ values was not rigorously established. In general, Arrhenius (or Eyring) rate plots were not used, it being assumed that $\Delta S^{\ddagger}$ is negligible. However, this was palpably not so for two of the compounds, and investigations on one of these, $ClBr_2C \cdot CClBrF$, gave an average $\Delta S^{\ddagger} = -25 \cdot 9$ J mol^{-1} deg^{-1}. Since all three barriers in general influence the spectral shapes, the authors conclude that each rotational interconversion is followed by immediate deactivation and therefore that the 'free-rotor' concept is not valid for these compounds. For $Cl_2BrC \cdot CBrCl \cdot CF_3$, rotations about the two C—C bonds could be separately investigated — a nice demonstration of the use of n.m.r. for a double rate process. At -62 °C rotation about the C—CF_3 bond was fast but that about the other bond was slow on the n.m.r. timescale. The slower process was investigated by matching experimental and theoretical spectra on an $A_3 \rightleftharpoons B_3 \rightleftharpoons C_3$ basis. Some semiquantitative information about the faster process was obtained by treating the fast-exchange limit only and neglecting geminal F–F coupling.

Attempts to re-investigate internal rotation in amides by n.m.r. continue, in order to resolve the discrepancies in earlier literature. A study of *NN*-dimethyl-formamide and -acetamide in a variety of solvents has been reported;[96] careful attention was paid to obtaining an appropriate chemical-shift difference in each case. Four different algebraic 'lineshape parameters' describing the exchange were used to evaluate the rate constants from the spectra, and there is some discussion of their relative merits. Arrhenius and Eyring activation parameters are derived and discussed. There are significant variations in the

---

[94] V. Mark and V. A. Pattison, *Chem. Comm.*, 1971, 553.
[95] F. J. Weigert, M. B. Winstead, J. I. Garrels, and J. D. Roberts, *J. Amer. Chem. Soc.*, 1970, **92**, 7359.
[96] A. Calzolari, F. Conti, and C. Franconi, *J. Chem. Soc.*, (*B*), 1970, 555.

Arrhenius parameters with amide concentration and with some solvents, but $\Delta G^{\ddagger}$ seems to be sensibly constant(!). Curiously, values of $\Delta H^{\ddagger}$ are not reported, though those of $\Delta S^{\ddagger}$ appear. Ng[97] has re-examined the spectrum of NN-dimethyltrifluoroacetamide using $^1H-\{^{19}F\}$ double resonance and complete bandshape analysis, with visual fitting, to the Gutowsky–Holm[75] equations for exchange between two negligibly-coupled, equally-populated sites. The activation parameters agree with those from spin-echo work, but not with earlier high-resolution results, obtained using the intensity-ratio method.

Filleux-Blanchard and Durand[98] have observed effects of temperature variation on the n.m.r. spectra of $MeNHC(S)NH_2$ due to two rate processes. Internal rotation about the $C-NH_2$ bond is considerably faster than that about the $C-NHMe$ bond. Activation parameters were obtained by comparing theoretical and experimental spectra, apparently visually. The theoretical spectra were generated using modified Bloch equations,[67] 'adapted to include coupling', but little detail is given of the method. Results from some related molecules are also discussed.

Activation parameters have been reported[99] for the $s\text{-}cis \rightleftharpoons s\text{-}trans$ (C—C) isomerism of a vinylogous amide, $Me_2NCH=CHCOMe$, studied using modified Bloch equations.[67] The shape of the vinylic protons was examined with 'X approximation' allowance for the coupling, i.e. the system is treated as $AX \rightleftharpoons BY$ (unequal populations). The analysis appears to be of a standard type but little detail is given. It is unfortunate that this is so since it is not possible to judge the accuracy of the results without knowing, for instance, how the problems of chemical shifts and 'natural' linewidths varying with temperature were tackled (similar comments could be made about many of the papers discussed in this review). The same compound was also studied (in a different solvent) by Dabrowski and Kozerski,[100] who used line broadening of the C-methyl resonance near the fast- and slow-exchange limits to determine exchange lifetimes. They avoided the intermediate region of exchange but relied on consistency of the separate data for high and low temperatures. Although this procedure appears satisfactory, it is clearly preferable to have data over the full range where possible. The two papers[99,100] give very different values for ΔH^* and ΔS^* (though the results for ΔG^* do not differ greatly); this emphasises that there is still considerable need for caution in making and reporting n.m.r. exchange studies.

Activation enthalpies for internal rotation in an N-alkyl-N-nitrosoaniline, obtained by the n.m.r. bandshape method, have been reported,[101] but no details were given of the procedure.

Bushweller et al.[86,90] have obtained (the two publications seem to contain

---

[97] S. Ng, *J. Chem. Soc.*, (A), 1971, 1586.
[98] M.-L. Filleux-Blanchard and A. Durand, *Org. Magn. Resonance*, 1971, 3, 187.
[99] M.-L. Filleux-Blanchard, H. Durand, and G. J. Martin, *Org. Magn. Resonance*, 1970, 2, 539.
[100] J. Dabrowski and L. Kozerski, *J. Chem. Soc.* (B), 1971, 345.

substantially the same material) rate data for the conformational isomerism of an *N*-acetylpyrrolidine derivative by both the n.m.r. total bandshape and the thermal stereomutation methods. They stress the added accuracy inherent in the use of both methods, since a wide temperature range may be studied. The n.m.r. data were obtained from the bandshape due to A⇌X unequal-population, two-site exchange, with allowance made both for the variation of the chemical shift with temperature and for unequal non-exchange linewidths (1·8 Hz and 1·4 Hz).

The interconversion of two conformational variants of 9-(α-naphthyl)-fluorene (4) has been studied[102] by three n.m.r. methods, namely (i) two

(4)

(5)

different linewidth approaches, and (ii) a complete bandshape approach (references are cited but few details given). The activation parameters from (i) and (ii) differ by up to 4·2 kJ mol^{-1} but this fact is not discussed. The nature of the exchange process is discussed. The same paper[102] also gives some approximate results for a conformational change of t-butyl-9,9'-bifluorenyl (5). The n.m.r. spectrum of this molecule gives evidence for a second exchange process at lower temperature.

Nitrogen Inversion. Lehn and Wagner[103] report briefly on exchange processes

---

[101] J. T. D'Agostino and H. H. Jaffé, *J. Org. Chem.*, 1971, **36**, 992.
[102] K. D. Bartle, P. M. G. Bavin, D. W. Jones, and R. L'Amie, *Tetrahedron*, 1970, **26**, 911.

in medium-ring monocyclic and dicyclic systems involving nitrogen inversion (in related compounds there has been some ambiguity as to whether the process is ring inversion or nitrogen inversion). The preliminary communication,[103] however, reports $\Delta G^{\ddagger}$ values only at coalescence temperatures ($\Delta G_c^{\ddagger}$). The same authors[104] also list $\Delta G_c^{\ddagger}$ values for nitrogen inversion in a number of five-membered cyclic amines. In these cases they deliberately assume that estimation of $\Delta H^{\ddagger}$ and $\Delta S^{\ddagger}$ from bandshapes would be inaccurate (a somewhat defeatist attitude, though perhaps justified in some cases). They therefore calculate $\Delta H^{\ddagger}$ from $\Delta G_c^{\ddagger}$, assuming $\Delta S^{\ddagger}$ to be ± 20.9 J mol^{-1} deg^{-1}; they suggest that this procedure is useful in order to compare barriers for different molecules. In using literature data they carry through the same procedure, since they are, probably rightly, sceptical of reported values of $\Delta S^{\ddagger}$ as high as 71 J mol^{-1} deg^{-1} for nitrogen inversion.

The exchange processes for compounds of the type PhCH$_2$NMeX [X = OMe, OH, or Cl] have been re-examined using n.m.r. spectra and computer-simulated bandshapes by Griffith, Olson, and Roberts.[14] They conclude (in distinction from earlier work) that the exchange involved must be inversion at nitrogen.

Ring Inversion. With the increasing availability of instruments capable of ^{13}C resonance, it is not surprising that exchange processes are now being studied by this technique. There are normally advantages over the use of ^{1}H resonance in (a) the larger chemical shift range for ^{13}C, and (b) the fact that when ^{1}H noise decoupling is used there are few complications in ^{13}C resonance arising from spin–spin coupling. Thus Doddrell *et al.*[82] have studied the ring inversion of 1,1,3,3-tetramethylcyclohexane using total bandshape analysis (of which few details are given) for the exchanging methyl carbon nuclei. The results (E_a is 17.6 kJ mol^{-1} smaller than the value for cyclohexane itself) indicate that there is very little increase in strain in the transition state over that in cyclohexane. The same authors[82] also made a complete bandshape study in ^{19}F resonance of the AB⇌BA exchange for two *gem*-difluoro-1,1,3,3-tetramethylcyclohexanes. However, it is shown that this 'fluorine labelling' technique leads to substantially different activation parameters from those for the analogous unfluorinated compounds.

The 'fluorine conformational probe' method (^{19}F resonance) has also been employed by Jefford, Hill, and Ramey[84] for the AB⇌BA systems of several cyclohexanes containing the CF$_2$ group. It is claimed that the total bandshape method was used but this is a little misleading since it applies only to the theoretical calculations; the fitting was done using a corrected half-width as the sole parameter. Moreover, the 'corrections' applied were very large (up to 66 Hz). Both Arrhenius and Eyring rate parameters were obtained (from several plots) and are discussed. In one case $\Delta S^{\ddagger}$ is negative; the value is therefore somewhat suspect.

[103] J. M. Lehn and J. Wagner, *Chem. Comm.*, 1970, 414.
[104] J. M. Lehn and J. Wagner, *Tetrahedron*, 1970, **26**, 4227.

Von Ammon, Fischer, and Kanellakopulos[85] have reported the use of paramagnetic ions for studies of exchange processes. The enhanced chemical-shift differences make it possible to interpret variable-temperature spectra even for quite complex spin systems (the authors used the technique for examining inversion of a cyclohexyl ring). Although no detailed bandshape work was carried out, the method may prove to be a generally powerful one, though interaction with paramagnetic ions clearly modifies the thermodynamic activation parameters themselves.

The Gutowsky–Holm[75] method of analysis (measuring peak-separations) was used by Wulz, Brune, and Hetz[105] to obtain thermodynamic parameters for the ring inversion of 3,3,6,6-tetramethyl-1,2,4,5-tetraoxan (6) (a simple two-site exchange case) in a variety of solvents. Although the results are reasonable, the method is not to be recommended; the authors do not discuss their errors in any detail and some of the final plots appear distinctly curved.

(6)   (7)

However, it is certain that the solvent affects the barrier significantly ($\Delta H^{\neq}$ varies from 49·0 to 68·2 kJ mol^{-1}), and it is concluded that repulsive forces between the non-bonding electron pairs of the oxygen atoms make a substantial contribution to the barrier.

Bushweller et al.[87,88] have used a mixture of total n.m.r. bandshape and direct thermal transformation to study the chair-to-twist process for 3,3,6,6-tetramethyl-1,2,4,5-tetrathian (7), for which the twist form is more stable than the chair. It is not clear on what basis the bandshape analysis was performed, but the process appears to have been treated as a case of three-site exchange (chair$\rightleftharpoons C_{2v}$ twist). At all events it was found that the best fit of experimental and theoretical spectra was obtained when the direct chair$\rightleftharpoons$chair interconversion rate was set at zero. This provides an interesting case where n.m.r. bandshape analysis gives evidence of the mechanism of an equilibrium.

Ring inversion in dispiro[2,2,2,2]decane (8), studied by n.m.r., has been the subject of a brief report.[106] Complete bandshape analysis of the cyclohexyl proton resonance ($[AX]_2 \rightleftharpoons [XA]_2$) was used.

Jones and Ladd[107] have treated the ring inversion of some 5,5-dimethyl-2-spiro-1,3-dioxans using the Gutowsky–Holm[67] equations for exchange between two uncoupled sites. They extend the Gutowsky–Holm treatment (unequally-populated sites) to the situation where the two sites have unequal values of T_2^{eff}. However, the cases studied have equally-populated sites and

---

[105] K. Wulz, H. A. Brune, and W. Hetz, *Tetrahedron*, 1970, **26**, 3.
[106] J. B. Lambert, J. L. Gosnell, jun., D. S. Bailey, and L. G. Greifenstein, *Chem. Comm.*, 1970, 1004.
[107] V. I. P. Jones and J. A. Ladd, *Trans. Faraday Soc.*, 1970, **66**, 2948.

Bandshape Phenomena for Fluids

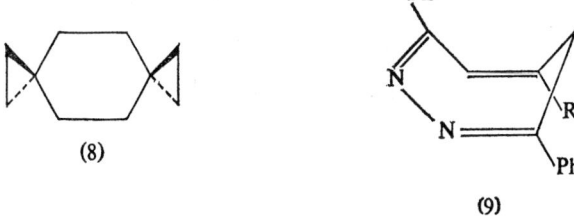

(8)   (9)

are taken to have equal T_2^{eff} values. The results show a considerable range of $\Delta S^{\ddagger}$ values, and the reasons for this situation are discussed by the authors.

Ring-inversion activation parameters have been reported[108] for piperazine and NN-dimethylpiperazine using bandshape methods. In the second case the exchange was treated as of the AB⇌BA type (rather a crude assumption). In neither case do the authors claim great accuracy in the results.

Ring inversion of $4H$-1,2-diazepines (9) has been studied[109] by the n.m.r. method. Although the case is actually ABX⇌BAX, it was treated as AB⇌BA, the small long-range values of $^4J(\text{AX})$ and $^4J(\text{BX})$ being taken into account widths by using the TMS signal as a monitor. Gerig and Ortiz[110] have studied ring inversion in 2-methylene-cis-decalins by n.m.r. In the case of 1,1,3,3-tetradeuterio-2-methylene-cis-decalin the exchange was of the type AB⇌CD, in which one of the low-temperature parts showed nearly degenerate chemical shifts. Exchange rates were obtained by computer fitting but little detail is given, which is unfortunate because the bandshape changes are rather subtle. The paper also reports[110] studies by ^{19}F n.m.r. of 2-fluoromethylene-cis-decalin, which is a more favourable bandshape case of the same type. It was necessary to correct for chemical-shift variations (by an extrapolation procedure). Unfortunately, the exchange parameters obtained for the two molecules are very different; the authors suspect the data for the fluorinated compound because of the large value of $\Delta S^{\ddagger}$ (84 J mol^{-1} deg^{-1}). Variation of the equilibrium constant with temperature was apparently not taken into account and may not have been considered as a source of error. Stress is laid on the fact that the way in which the bandshapes change leaves no ambiguity about which components of the two AB-type, low-temperature part-spectra are exchange-related.

Von Bredow et al.[89] have studied a system of intramolecular chemical exchange of some interest, namely that given by 4,5,6-trithia-1,2-benzocyclohept-1-ene and some of its 3'6'-derivatives (10). These compounds give evidence of two rate processes. The slower one is assigned to inversion of the seven-membered ring, and the faster one to its pseudorotation. The two low-temperature conformers are presumed to have a chair form and a boat form

---

[108] R. G. Lett, L. Petrakis, A. F. Ellis, and R. K. Jensen, *J. Phys. Chem.*, 1970, **74**, 2816.
[109] U. Svanholm, *Acta Chem. Scand.*, 1971, **25**, 640.
[110] J. T. Gerig and C. E. Ortis, *J. Amer. Chem. Soc.*, 1970, **92**, 7121.

(10)

respectively. Crystallization isolated one form, so the n.m.r. bandshape method and the direct equilibration method were both used for the inversion process; the results are in good agreement.

The group working with Roberts[83] has continued to exploit the large chemical shift range for ^{19}F in studies of intramolecular exchange processes

(11)

by n.m.r. The inversion of 7,7-dicyano-2,5-bis(difluoromethyl)norcaradiene (11) was examined, using the non-equivalence of the *gem*-fluorine nuclei. The ABX⇌BAX system (AB part) was computer-fitted to yield rate constants and hence thermodynamic parameters for the exchange (E_a, A, and $\Delta G^{\ddagger}$ only are quoted). The nature of the intermediates in the exchange process is discussed.

Phosphorus-containing Compounds. Consecutive papers in *J. Amer. Chem. Soc.* by Gorenstein and Westheimer[111] and by Gorenstein[112] discuss n.m.r. exchange processes for alkyloxyphosphoranes containing a five-membered P―O―C=C―C ring. The second paper treats the bandshape analyses in detail. The systems treated were of several types:
(i) Uncoupled two-site exchange, with equal populations.
(ii) $A_3B_3X \rightleftharpoons B_3A_3X$, with $J(AB) = 0$. This case (AB region) was treated by subspectral methods, *i.e.* as the superposition of two independent systems of type (i), differing as a result of the spin of X ($\equiv {}^{31}P$).

[111] D. Gorenstein and F. H. Westheimer, *J. Amer. Chem. Soc.*, 1970, **92**, 634.
[112] D. Gorenstein, *J. Amer. Chem. Soc.*, 1970, **92**, 644.

(iii) $A_3B_3X \rightleftharpoons C_3D_3X$, with unequal populations but $J(AB) = J(CD) = 0$. This case was again treated by subspectral methods, and it was only necessary to use the equations for uncoupled two-site exchange, with superposition.

(iv) Uncoupled three-site exchange, with equal populations, but complicated by the presence of ^{31}P such that superposition of two subspectra was again required. The equations required for this type of subspectrum are explicitly given.

(v) $ABX \rightleftharpoons BAX$, with $J(AB) \neq 0$.

For the first three types the Gutowsky–McCall–Slichter[73] approach sufficed, and this was extended to type (iv). However, a density matrix treatment was necessary for (v). Type (iii) produced an example of interest, where the exchange mechanism could be deduced from the spectra. This concerns the exchange found to be $(12) \rightleftharpoons (13)$. The bandshape of the methoxyl protons

(12) (*trans*)   (13) (*cis*)

was studied, and agreement with theoretical shapes could only be obtained when the process exchanging the starred groups (and simultaneously exchanging the unstarred groups) was considered. Calculations based on exchange between the equatorial groups of the two forms (and simultaneously between the axial groups), for example, conspicuously failed to reproduce the observed spectra. Comparison of experimental and theoretical spectra was apparently by eye. The attention paid to various experimental problems, such as variation of chemical shifts, effective spin–spin relaxation times, and, in some cases, relative populations, was not great, and Gorenstein[112] places generous error limits (± 8—16 kJ mol^{-1} for $\Delta H^{\neq}$) on the activation parameters he reports. Curiously, values of $\Delta S^{\neq}$ do not appear to be quoted. Further interest in the work attaches to the fact that in some cases several rate processes were detected. The pseudorotation processes are discussed in considerable detail, together with their n.m.r. implications, in the first paper.[111]

Intramolecular rate processes in aminophosphines have been studied by Cowley *et al.*[113,114] using complete bandshape simulation, normally with visual

---

[113] A. H. Cowley, M. J. S. Dewar, W. R. Jackson, and W. B. Jennings, *J. Amer. Chem. Soc.*, 1970, **92**, 1085.

[114] A. H. Cowley, M. J. S. Dewar, W. R. Jackson, and W. B. Jennings, *J. Amer. Chem. Soc.*, 1970, **92**, 5206.

comparison of computed and observed spectra (though in two cases iterative fitting to digitized spectra was employed). The examples gave first-order slow-exchange spectra, and were mostly computed using a many-site exchange program ascribed to Saunders,[115] though in actual fact the problems could have been treated by superposition of separate two-site exchange subspectra. For most of the compounds spectral fitting over a large temperature range was not carried out, and the quoted results are mostly for coalescence temperatures, on the grounds that $\Delta S^{\ddagger}$ is likely to be small. One of the two compounds studied in detail gave $\Delta S^{\ddagger} = +12 \cdot 6$ J mol^{-1} deg^{-1}; the other (which is not identified unambiguously but is probably the arsenic compound PhAsClNMe$_2$) gave $\Delta S^{\ddagger} = 79$ J mol^{-1} deg^{-1} — a result that is plausibly explained away in terms of broadening due to some other process which occurs for several of the compounds at very low temperatures, < -100 °C (necessitating a 'natural linewidth' of 8 Hz at -108 °C for the study of the principal process). The authors discuss the exchange process in some detail, and conclude that P—N nternal rotation is probably the rate-determining process in these molecules.

Metal Complexes. Kalck et al.[116] have studied n.m.r. bandshapes for the cyclopentadienyl ligands in the *cis*⇌*trans* intramolecular rearrangement of compounds of the type $(\pi\text{-}C_5H_5)M(CO)_2LH$ [M = W; L = PMe$_3$, PEt$_3$, PPh$_3$; and M = Mo, L = PMe$_3$]. Their calculations follow the Kubo–Sack[117,118] method for the two-site uncoupled case. Their estimation of the variation of [*cis*]/[*trans*] is of some interest — they use the average high-temperature values of *J*(PMH) together with the individual values (assumed temperature-independent) for the *cis*- and *trans*-forms from low-temperature measurements. They conclude that variations of ΔH correlate with those of $\Delta H^{\ddagger}$, in accordance with Polanyi's relation. The expected increase in a barrier when one of the potential minima (or a metastable intermediate) is destabilised is also mentioned by Furtsch, Dierdorf, and Cowley[119] in their work on exchange in phosphorus(v) fluorides. They imply that the intramolecular interconversion route for Me$_2$PF$_3$ is relatively unfavoured because the methyl groups are more reluctant to be axial than, say, the chlorine atoms in Cl$_2$PF$_3$.

Two exchange processes have been shown[120] to affect the n.m.r. spectra of a series of bis(diphenylarsino)methane derivatives of Mo^{II} and W^{II}. The high-temperature process is of the simple, uncoupled, two-site exchange type (actually the systems are in principle coupled but the coupling constants are

---

[115] (a) M. Saunders, *Tetrahedron Letters*, 1963, 1699; (b) M. Saunders in 'Magnetic Resonance in Biological Systems', ed. A. Ehrenberg, B. G. Malmström, and T. Vänngörd, Pergamon Press, Oxford, 1967, p. 85.
[116] P. Kalck, R. Pince, R. Poilblanc, and J. Roussel, *J. Organometallic Chem.*, 1970, **24**, 445.
[117] R. Kubo, *Nuovo Cimento, Suppl.*, 1957, **6**, 1063.
[118] R. S. Sack, *Mol. Phys.*, 1958, **1**, 163.
[119] T. A. Furtsch, D. S. Dierdorf, and A. H. Cowley, *J. Amer. Chem. Soc.*, 1970, **92**, 5759.
[120] M. W. Anker, R. Colton, and C. J. Rix, *Austral. J. Chem.*, 1971, **24**, 1157.

negligibly small), corresponding to intramolecular exchange between the non-equivalent ligands, one of which is chelated while the other is unidentate. The low-temperature process is of AB⇌BA nature for each type of ligand, due to a second 'fluxional' intramolecular process which is not well-described. The spectra were simulated using the Kaplan–Alexander approach[61,62] and compared (for the most part visually) with the experimental spectra. In general the spectra were sensitive to the two rate processes over different temperature ranges, so values from the sensitive regions were extrapolated for use in the insensitive regions. Arrhenius activation energies are reported, and in some cases values for the high- and low-temperature processes are very similar (this is a particular point of interest about the work). In all cases the low-temperature processes have very similar activation energies for the two types of ligand.

Bandshape studies[121] have been used to observe the rate of inversion of configuration at Sn for asymmetric tin compounds. The situation was confused by effects originating from population changes of a conformational nature, and quantitative results for the inversion are sparse.

Variable-temperature ^{1}H n.m.r. studies have proved[122] to be valuable for the measurement of rates of optical inversion in 'labile' Co^{II} chelates. As with other dynamic n.m.r. studies of transition-metal compounds, the paramagnetic nature of the species facilitates the study by causing large chemical-shift differences, as the author emphasises. In consequence he was able to measure lifetimes as short as *ca.* 10^{-7} s, compared with the limit of *ca.* 10^{-2} s previously attained for analogous diamagnetic compounds. The system studied was $Co(acac)_2$(4,7-phen) where acac = acetylacetonate and 4,7-phen ≡ 4,7-dimethyl-1,10-phenanthroline; the chemical shift difference between the non-equivalent terminal methyl groups of the acac ligand was ~40 p.p.m. Only data from the slow-exchange region were used, so only approximate thermodynamic parameters for the exchange are quoted. The interpretation of this experiment has, however, been queried by Jurado and Springer,[123] who reported a preliminary n.m.r. study of enantiomerization of tris-(2,6-dimethylheptane-3,5-dionato)aluminium(III) and related compounds.

C. **Examples of Intermolecular Exchange.**—Research reported in this area during the relevant period consists in large part of studies of ligand exchange from metal ions, using the approach of Swift and Connick.[3] Such applications will be discussed in more detail below. Of course, 1970 and 1971 have also witnessed a surge of work using 'shift reagents' (complexes of lanthanide elements) for increasing chemical shifts as an aid to spectral interpretation (see Chapter 9), but this work does not seem to have produced a corresponding increase in research on linewidths in these systems, as yet. However, Ahmad

---

[121] D. V. Stynes and A. L. Allred, *J. Amer. Chem. Soc.*, 1971, **93**, 2666.
[122] G. N. La Mar, *J. Amer. Chem. Soc.*, 1970, **92**, 1806.
[123] B. Jurado and C. S. Springer, jun., *Chem. Comm.*, 1971, 85.

et al.[124] have measured the widths of t-butyl proton signals for tris-(2,2,6,6-tetramethylheptane-3,5-dionato)-complexes of a wide range of lanthanide metals in CCl_4, both alone and in the presence of cyclohexanone and hexan-1-ol. No interpretation has yet been put forward, unfortunately. The 'shift reagent' technique has also been extended to ^{14}N resonance, and ^{14}N linewidths for organic molecules in solutions containing $Yb(dpm)_3$ and $Eu(dpm)_3$ have been reported[125] (dpm≡dipivalomethanato). Again, however, there is no discussion of the widths found.

A further noticeable development recently has been the increasing use of n.m.r. resonance of nuclei other than ^1H and ^{19}F in intermolecular exchange studies of all types, as will be shown below. The advantages of increased chemical shift ranges for 'other nuclei' is found to outweigh the disadvantage of low sensitivity.

One novel, interesting technique for obtaining exchange rate information in cases where separate signals can be seen for free water and water in the first co-ordination sphere of a diamagnetic metal will be mentioned at this point. Neely and Connick[126] examined the ^{17}O n.m.r. of a solution of $MgClO_4$ containing sufficient Mn^{2+} to broaden the bulk water resonance to at least ten times that of the Mg-bound water signal, the width of which could then be measured accurately.

Studies of Ligand Exchange (the Swift–Connick Approach[3]*).* This method continues to be extensively used to obtain information about ligand-exchange kinetics from n.m.r. linewidth (T_2) measurements (plus data on contact shifts and solvation numbers) for solutions of paramagnetic transition-metal complexes containing a large excess of the ligand. Thus Frankel[127] has studied dimethyl sulphoxide (DMSO) complexes of Co and Ni in various solution mixtures, and analysed the results in terms of the various limiting forms of the Swift–Connick equations,[3] correcting for ligand relaxation contributions other than those depending on the metal. They report thermodynamic exchange parameters, and find that exchange rates are virtually independent of bulk medium effects (although [free DMSO]:[complex] was never less than 50:1, the proportion of DMSO in the solvent was sometimes as low as 1:30). Exchange of water from a terdentate Schiff-base complex of Ni^{II} in aqueous solution has been similarly studied[128] using ^1H n.m.r. linewidth measurements. Data from the temperature range 3—86 °C were used to derive thermodynamic parameters for the exchange. It was found that in order to fit results at both 60 MHz and 100 MHz it was necessary to include an outer-

---

[124] N. Ahmad, N. S. Bhacca, J. Selbin, and J. D. Wander, *J. Amer. Chem. Soc.*, 1971, **93**, 2564.
[125] W. Witanowski, L. Stefaniak, H. Januszewski, and Z. W. Wolkowski, *Tetrahedron Letters*, 1971, 1653.
[126] J. Neely and R. Connick, *J. Amer. Chem. Soc.*, 1970, **92**, 3476.
[127] L. S. Frankel, *Inorg. Chem.*, 1971, **10**, 814.
[128] J. E. Letter and R. B. Jordan, *J. Amer. Chem. Soc.*, 1971, **93**, 864.

sphere contribution to the line-broadening. Langford and Tsiang[129] quote ^{1}H n.m.r. widths of *ca.* 1·3 Hz for MeNO$_2$ in MeNO$_2$–DMSO mixtures containing Ni(DMSO)$_6$(NO$_3$)$_2$ as evidence that MeNO$_2$ has access to the second co-ordination sphere of nickel.

Balahura and Jordan[130] have used ^{1}H n.m.r. line-broadening studies of the Swift–Connick[3] type on (H$_2$O)$_5$CrN$_3^{2+}$ in aqueous perchloric acid to show that the exchange occurring is that of water protons and not of protons from co-ordinated HN$_3$. The dependence of linewidth on both temperature and pH was examined, and activation parameters for the exchange reactions and for T_{2M}, the transverse relaxation time of the protons in the co-ordination sphere, were obtained. Leffler[131] has measured the ^{1}H n.m.r. linewidth of CH$_2$Cl$_2$ in aqueous solutions containing Cu^{2+}, Cr^{3+}, Co^{2+}, Fe^{2+}, Mn^{2+}, and Ni^{2+} hexa-aquo-ions as a function of metal ion concentration and, to some extent, temperature. The results were used to discuss the solvent ordering in the second co-ordination sphere.

As stated earlier, the Swift–Connick method,[3] in common with other n.m.r. techniques, is being increasingly used with the resonance of nuclei other than ^{1}H and ^{19}F. For instance, West and Lincoln[132] have used a wide-line ^{14}N system, operating in both the sideband and the first-derivative modes (the latter corrected for modulation broadening), to study exchange of acetonitrile on Fe^{II}. Separate resonances due to co-ordinated MeCN are never seen, but kinetic information is obtained from n.m.r. linewidths. Unfortunately, the paper reports linewidths in field units, a practice for broad-line work which has largely ceased elsewhere in the literature, at least in chemical studies. Studies of ligand exchange from paramagnetic ions by ^{14}N resonance have also been carried out by Purcell and Marianelli,[133] who looked at acetonitrile exchanging from the first co-ordination sphere of Mn^{II}. The electron relaxation times were measured by e.s.r. and the exchange lifetimes obtained, using this information, from n.m.r. linewidths corrected for the values for pure acetonitrile (the latter being dominated by quadrupolar relaxation, with $T_2^{-1} = 265 \pm 12$ s^{-1} at 25 °C). There was some evidence of a contribution from the second co-ordination sphere at low temperatures. The Swift–Connick approach[3] was very satisfactory and yielded the first-order rate constant and thermodynamic parameters for the exchange. The ligand-exchange kinetics for NCS$^-$ and H$_2$O on paramagnetic Co^{II} have been investigated by Zeltmann and Morgan[134] using ^{14}N and ^{17}O resonance (both linewidths and shifts), measured as a function of both temperature and thiocyanate concentration. Both shifts and widths were corrected using data for solutions not containing paramagnetic ions. Widths were obtained either using sidebands or by peak-to-peak measurements for the derivative absorption mode. The data were

[129] C. H. Langford and H. G. Tsiang, *Inorg. Chem.*, 1970, **9**, 2346.
[130] R. J. Balahura and R. B. Jordan, *Inorg. Chem.*, 1970, **9**, 2639.
[131] A. J. Leffler, *J. Phys. Chem.*, 1970, **74**, 2810.
[132] R. J. West and S. F. Lincoln, *Austral. J. Chem.*, 1971, **24**, 1169.
[133] W. L. Purcell and R. S. Marianelli, *Inorg. Chem.*, 1970, **9**, 1724.
[134] A. H. Zeltmann and L. O. Morgan, *Inorg. Chem.*, 1970, **9**, 2522.

fitted using the Swift–Connick approach,³ including a number of approximations designed in general to reduce the number of parameters involved to the minimum. Inclusion of water activity gave improved results. The authors state that 'some degree of non-averaging' was apparent, but separate resonances for the various complexes were never seen. Because of the large number of parameters (chemical shifts of the complexes, their electron transverse relaxation times, equilibrium constants, and rate constants for the exchange) it is difficult to estimate the accuracy or to prove the uniqueness of the results (full details of the computer program used for least-squares analysis of the data are not given). Activation energies both for the exchange and for the relaxation times were derived. The hydration of ferrous and nickelous ions has been studied by Chmelnick and Fiat[135] using ^{17}O resonance (chemical shift and linewidths) of the water molecules, together with the standard formulae. Separate signals were observed for bound and free water. The linewidths for the bound H_2O (measured using the derivative mode, corrected for modulation broadening) are large — between 11 and 65 kHz — but apparently this presented no particular problems. The linewidth is assumed to be dominated by the exchange (the quadrupolar contributions are said to be within experimental error), and the electron relaxation times are derived. The study indicates that both Fe^{II} and Ni^{II} are six-fold co-ordinated by water.

Measurement of ^{35}Cl n.m.r. linewidths has provided[136] information about ligand-exchange kinetics of tetrahedral chlorocobalt(II) species. Specifically, line broadening (with Cl^- ion in excess) is observed in the presence of pyridine but not in its absence. The observations are interpreted in terms of exchange through the $[CoCl_3(pyridine)]^-$ species; rate constants are derived and the mechanism discussed.

Other Studies of Intermolecular Exchange. The (paramagnetic) tetrahedral ⇌ (diamagnetic) planar isomerization of some transition-metal complexes may be conveniently studied by n.m.r. La Mar and Sherman[137] used linewidths to obtain kinetic information about this process for bis-(n-alkyldiphenylphosphine)nickel(II) dihalides. Under slow-exchange conditions the peaks assigned to the paramagnetic species have wide-ranging chemical shifts because of contact and pseudo-contact effects. Linewidths gave rates of isomerization near both slow- and fast-exchange limits (total bandshape fitting was not used), and the authors derive values for the thermodynamic parameters governing exchange. Corrections were made for variation of non-exchange linewidths, but there is little discussion of errors and in several cases spectra in the limit of slow exchange were not attainable. Similar work on the same type of compounds (though only from measurements in the slow-exchange limit) has been reported by Pignolet et al.[138]

---

[135] A. M. Chmelnick and D. Fiat, *J. Amer. Chem. Soc.*, 1971, **93**, 2875.
[136] R. E. Gentzler, T. R. Stengle, and C. H. Langford, *Chem. Comm.*, 1970, 1257.
[137] G. N. La Mar and E. O. Sherman, *J. Amer. Chem. Soc.*, 1970, **92**, 2691.
[138] L. H. Pignolet, W. De W. Horrocks, jun., and R. H. Holm, *J. Amer. Chem. Soc.*, 1970, **92**, 1855.

The collapse of ^{207}Pb satellites, observed in proton resonance of the nitrilotriacetic acid (NTA) complex of lead, due to ligand exchange has been studied by Rabenstein[139] as a function of excess NTA concentration and of pH. The presence of zero-spin lead meant that the problem was a four-site one. Spectra were computer-simulated using modified Bloch equations. Measurement of bandwidths provided the means of matching theoretical and experimental spectra. Clearly, the spectra depend on a substantial number of rate constants, but the author was able to determine the individual values by varying the solution conditions. However, it is not clear whether there are any ambiguities in the spectral fitting, and the errors in the rate constants receive scant attention.

Intermolecular exchange of parts for Grignard reagents and related compounds continues to attract some attention. Thus Ladd and Parker[13] use ^{7}Li resonance lineshapes to study exchange between phenyl-lithium and p-tolyl-lithium. By carefully taking equimolar ratios they reduce the lineshape problem to the equal-population, two-site, uncoupled case. They conclude that the process is probably bimolecular, *i.e.* *LiPh + LiTol ⇌ LiPh + *LiTol.

Dwek and co-workers[140] used ^{17}O resonance to study the variation in the exchange of oxygen between IO_3^- and H_2O with pH. In neutral solution separate resonances are seen; that for IO_3^- has a width of 317 Hz. Detailed measurements could usually only be made on the water signal, which first broadened as the pH decreased, then sharpened (after coalescence); since the fraction of oxygen present as IO_3^- was small, the Swift–Connick[3] approach was used to derive rates of exchange, which are then discussed in relation to plausible mechanisms. The main process appears to be second-order with respect to H_3O^+.

Furtsch, Dierdorf, and Cowley[119] have suggested an intermolecular route to exchange in phosphorus(v) fluorides. This conclusion relies on the usual criterion that if bonds are broken the relevant coupling constants are averaged to zero, whereas intramolecular exchange averages coupling constants to a non-zero value (except by a chance set of relationships between individual values). In the cases of Me_3PF_2 and Me_2PF_3, heating from -110 °C to $+30$ °C causes ^3J(FPCH) to be lost but ^2J(PCH) to be retained. The authors conclude that the F—P bond is ruptured — collapse of the F–P coupling occurs at *ca.* 100 °C for Me_2PF_2. Lineshape studies (using the Saunders[115] computer program for multi-site exchange) result in a very low value of $\Delta H^{\ddagger}$ (15.6 kJ mol^{-1}) and a large negative value of $\Delta S^{\ddagger}$ (-180 J mol^{-1} deg^{-1}).

The kinetics of fluorine-atom exchange between SiF_6^{2-} and HF (aq. solution) were studied by Russian workers[141] using ^{19}F n.m.r. linewidths in the slow-exchange region, but few details are given.

---

[139] D. L. Rabenstein, *J. Amer. Chem. Soc.*, 1971, **93**, 2869.
[140] R. A. Dwek, Z. Luz, S. Peller, and M. Shporer, *J. Amer. Chem. Soc.*, 1971, **93**, 77.
[141] A. G. Kucheryaev, V. A. Lebedev, and I. M. Ovchinnikov, *J. Struct. Chem.*, 1970, **11**, 858.

Lee and Sheldrick[142] report curious bandshape changes with temperature for a solution of dimethylthallium ethoxide (probably dimeric) in deuteriotoluene. As the temperature is lowered each of the methyl doublet peaks [split by J(TlH)] becomes steadily broader until the temperature is -10 °C, when there is virtually a coalescence. On further cooling the two peaks gradually sharpen again, to give the original value of J(TlH). The spectrum is symmetrical throughout the temperature range and the ethoxide resonances remain sharp. The addition of Me$_3$Tl or Me$_2$TlI had no effect and the authors conclude that the phenomenon does not arise from exchange involving ionic species; they advance three other possible explanations:

(a) Exchange of methyl groups bonded to thallium.
(b) Exchange of dimer with a small concentration of monomer [the monomer and dimer would have to have very similar chemical shifts and large values of J(TlH) of opposite sign].
(c) Fast monomer–dimer exchange with (Tl, Tl) dipolar relaxation (the authors proceed, however, to discuss scalar rather than dipolar coupling).

None of these possibilities attracts Lee and Sheldrick's full assent, and the problem remains unsolved.

4 Effects of Quadrupolar Nuclei

A. Theoretical Work.—Pyper[12] has questioned the validity of the view that the n.m.r. spectra of spin-½ nuclei scalar-coupled to a quadrupolar nucleus may be treated as though exchange were taking place between sites differentiated by the z-component of spin of the quadrupolar nucleus, as suggested earlier by Pople.[143] However, he has shown, using a more rigorous approach based on Redfield relaxation theory,[15] that for a spin system consisting of first-order groups of magnetically-equivalent spin-½ nuclei coupled to one another and to a single quadrupolar nucleus (a ^{3}A^{2}K$_n{}^2$X$_x$.... system), the spectrum of the spin-½ nuclei consists of the overlap of bands centred at the normal first-order positions (and with normal intensities), with shapes given by the simple Pople[143] expression:

$$I(x) \propto [45 + \eta^2(5x^2 + 1)]/[225x^2 + \eta^2(34x^4 - 2x^2 + 4) + \eta^4(x^6 - 2x^4 + x^2)]$$

In this expression $x = \Delta v/J$ and $\eta = 10\pi T_1 J$, where Δv is the separation from the band-centre, T_1 is the spin–lattice relaxation time of the quadrupolar nucleus, and J is the appropriate coupling constant between spin-½ and quadrupolar nuclei. The expression above is given for the spin-1 case only but the theory of Pyper[12] is more general. He also concludes that the resonance of the quadrupolar nucleus in such cases is the summation of Lorentzian bands at the positions expected on a first-order basis, with half-widths given by $(\pi T_1)^{-1}$.

---

[142] A. G. Lee and G. M. Sheldrick, *Trans. Faraday Soc.*, 1971, **67**, 7.
[143] J. A. Pople, *Mol. Phys.*, 1958, **1**, 168.

The theory of Pyper[12] has been tested to some extent by Harris et al.[11] for the fast-relaxation limit, spin-1 case. They chose to study chlorofluoropyridines with the spin systems (ignoring the chlorine nuclei) $^3A^2X_2$, $^3A^2K^2X_2$ and $^3A^2K^2P^2RX$. It proved possible by computer-treatment of the spin-$\frac{1}{2}$ bandshapes to obtain information by decomposing the bands into their components. This enabled accurate values of F–F coupling constants to be obtained, and also values of J^2T_1. The parameters J and T_1 cannot be found separately in the fast-relaxation limit without further information, but this was forthcoming from ^{15}N-satellite ^{19}F spectra and from ^{14}N resonance linewidths (assuming $T_{1N} = T_{2N}$). Even in the absence of such extra data, however, ratios of linewidths for different ^{19}F nuclei give ratios of squares of the appropriate coupling constants to nitrogen. Moreover, if the coupling constants are independent of temperature, Arrhenius activation energy plots for T_{2N} and for J^2T_{1N} provide a check of the equality of T_{1N} and T_{2N}. Indeed, data on activation energies may be obtained even in the absence of information on T_{2N} or J.

Pyper[44] has also, however, stressed that application of the Redfield relaxation theory to general second-order systems containing quadrupolar nuclei and spin-$\frac{1}{2}$ nuclei becomes disproportionately complicated for even moderate-sized molecules. He therefore takes the common extreme type of case which occurs when the quadrupolar nuclei are relaxing rapidly, and treats it by considering the spin-$\frac{1}{2}$ nuclei alone as the 'spin system'. The quadrupolar nuclei become the 'lattice', which now has the novelty of itself undergoing dissipative motion (relaxation) by coupling to other modes of motion (the 'bath'). The theory is developed using the Liouville representation and its advantages and limitations are fully discussed. The application is straightforward if there are no overlapping lines in the spin-$\frac{1}{2}$ spectrum. Pyper[44] considers special cases where (i) the lattice contains a single quadrupolar nucleus (the $^2A^2B^3X$ system is taken as an example), (ii) the lattice consists of two isochronous spin-1 nuclei, and (iii) there is only dipolar coupling between spins and lattice.

The $^2A^2B^3X$ case has been further examined by Harris and Pyper.[144] They discuss general lineshape expressions for any rate of X relaxation (a computer program was written to perform such calculations), and also for the fast-relaxation limit. The theoretical variations of the widths of the four AB lines are fascinating. There are in general both 'adiabatic' and 'non-adiabatic' contributions, arising from the diagonal and off-diagonal elements of the relaxation Hamiltonian respectively. These terms may be viewed as due to fluctuations in the transition frequency, and uncertainty broadening, respectively. The two A lines always have a common width, which in general differs from the common width of the two B lines. The detailed effects depend on the relaxation rate of the quadrupolar nucleus compared to the separation (ω_{23} in rad s^{-1}) of the central two energy levels of the AB system. Two extreme

---

[144] R. K. Harris and N. C. Pyper, *Mol. Phys.*, 1971, **20**, 467.

situations may be distinguished. When $\omega_{23}{}^2 T_{1x}{}^2$ is very large the non-adiabatic effect vanishes, and it is possible for there to be no quadrupolar contribution to the B lines [defining B by $|J(BX)| < |J(AX)|$] even when $J(BX)$ is appreciable. At the other extreme ($\omega_{23}{}^2 T_{1x}{}^2$ vanishingly small) — the 'white power spectrum' case — the situation is more as expected, but some of the information content is lost. Away from this extreme, there may be dynamic line shifts from the simple AB line positions and it is possible to obtain the relative signs of $J(AX)$ and $J(BX)$ from the effects of temperature or of change of spectrometer frequency on the linewidths. Unfortunately, only the white power case has been observed so far in practice. Harris and Pyper[144] studied 2-bromothiazole and were able to extract both $|J(BX)/J(AX)|$ and $J(AX)^2 T_{1x}$ from the linewidths. The former was found to vary with solvent, and the temperature-dependence of the latter was used to obtain activation energies to molecular rotation. Earlier in the year under review, Kumar, Krishna, and Rao[145] had also reported a bandshape study of 2-bromothiazole, though they only considered the white power spectrum case. The emphasis of their work, however, was on homonuclear $^1H-\{^1H\}$ double-resonance experiments (see Section 7) and they were able to obtain appreciably more information by this method, including the sign of $J(BX)/J(AX)$. They made use of ^{14}N linewidths and ^1H adiabatic rapid-passage experiments. Their double-resonance work used linewidth data in addition to frequency and intensity information, though the calculations of double-resonance linewidths were carried out as though the spin system were weakly coupled. In fact it is the double-resonance linewidths that allow the sign of $J(BX)/J(AX)$ to be determined. Their comprehensive series of experiments allowed the authors to conclude that external random fields do not play a significant part in the ^1H relaxation for 2-bromothiazole; all the results could be explained on the basis of internal (^1H, ^1H) dipole–dipole effects, plus the contribution of scalar coupling to ^{14}N. The paper[145] contains a similar study of 2,3,4-trichloronitrobenzene, which also forms a $^2A^2B^3X$ spin system.

The case of quadrupolar relaxation has also been discussed by Sykora[146] (see Section 5). He treats the case of molecules containing spin-$\frac{1}{2}$ nuclei together with a single spin $I > \frac{1}{2}$. His theory is, however, only applicable when the spin-$\frac{1}{2}$ resonance consists of well-resolved lines (*i.e.* quadrupolar relaxation is slow). He also describes the case of a $^3A^3B$ spin system, which is intrinsically of greater interest. He lists energy-level widths and gives details of the spectrum for typical cases, emphasising that the linewidth pattern is in general unsymmetrical. In addition, Sykora[146] suggests that for a $^3A^3B^2X$ spin system the energy-level widths may be tested using the X transition linewidths. No experimental examples are given.

---

[145] A. Kumar, N. R. Krishna, and B. D. N. Rao, *Mol. Phys.*, 1970, **18**, 11.
[146] S. Sykora, *J. Chem. Phys.*, 1971, **54**, 2469.

B. Resonances of Spin-½ Nuclei Coupled to Quadrupolar Nuclei.

Lehn and co-workers,[147] following their previous work[148] on chemical applications of n.m.r. studies involving deuterium relaxation, report studies on electron donor–acceptor complexes, using donors labelled with deuterium in CHD groups. The deuterium spin–lattice relaxation times, T_q, were obtained from the ^{1}H lineshape for the CHD group (by computer fitting) as a function of the fraction, α, of donor molecules complexed. The dependence on α is also calculated, using an adaptation of the theory of Anderson and Fryer,[149] as a function of the four rate constants (for exchange and relaxation). Two of these parameters are measured independently, and two obtained by comparison of the experimental and theoretical dependence of T_q on α. The authors stress the utility of the method for studying very fast reactions (rate constants of the order of 10^{10} s^{-1}); they discuss the significance of the relative rates of the four processes. The comments of Marshall[150] (see Section 4D) are, however, relevant to this work.

The effects of counter-ion and solvent on the ^{1}H n.m.r. spectra of tetra-alkylammonium ions have been studied in some detail.[151] The bandshapes were simulated using the Pople equation,[143] and values of the parameter $p = 3(e^2Q_q)^2\tau_c/40\hbar$ derived. However, the authors state that agreement between calculated and observed spectra is not good but is improved when proton natural linewidths are taken into account. Variations in p are discussed in terms of changes in τ_c and eq, arising from solvent interactions and ion pairing. The anomalous broadening of α-proton resonances for some amino-acids as pH increases has been shown[152] to arise in considerable measure from an increase in ^2J(NH) (combined with relatively invariant quadrupolar relaxation). The variations in ^2J(NH) were observed using compounds enriched in ^{15}N.

Dean and Gillespie[153] report linewidths of up to 1700 Hz for ^{19}F resonance of fluorine nuclei bonded to tantalum and niobium in mixtures of TaF$_5$ and NbF$_5$ (respectively) with SbF$_5$ in SO$_2$ClF solution. Tebbe and Muetterties[154] illustrate curious bandshapes for the paramagnetic species Co(S$_2$PF$_2$)$_3$. They believe that these reflect the effects of ^{59}Co–^{19}F coupling, but they offer no detailed reasons for the shapes. There is clearly a large contribution to the linewidth from the paramagnetic nature of the compound.

The commonly accepted view that quadrupolar relaxation of ^{10}B is faster than that of ^{11}B has been challenged by Akitt,[155] who shows that the larger nuclear spin of ^{10}B more than compensates for its larger quadrupole moment. The implications of this conclusion for bandshapes of spin-½ nuclei coupled

---

[147] C. Brevard and J. M. Lehn, *J. Amer. Chem. Soc.*, 1970, **92**, 4987.
[148] C. Brevard, J. P. Kintzinger, and J. M. Lehn, *Chem. Comm.*, 1969, 1193.
[149] J. E. Anderson and P. A. Fryer, *J. Chem. Phys.*, 1969, **50**, 3784.
[150] A. G. Marshall, *J. Chem. Phys.*, 1970, **52**, 2527.
[151] D. W. Larsen, *J. Phys. Chem.*, 1970, **74**, 3380.
[152] R. L. Lichter and J. D. Roberts, *Spectrochim. Acta*, 1970, **26A**, 1813.
[153] P. A. W. Dean and R. J. Gillespie, *Canad. J. Chem.*, 1971, **49**, 1736.
[154] F. N. Tebbe and E. L. Muetterties, *Inorg. Chem.*, 1970, **9**, 629.
[155] J. W. Akitt, *J. Magn. Resonance*, 1970, **3**, 411.

to boron are discussed and the example of BF_4 illustrated. The ^{19}F spectra show that, as the correlation times are increased owing to increasing solution viscosity (glycerol was added), multiplet collapse of the $^{10}BF_4^-$ and $^{11}BF_4^-$ resonances occurs at roughly the same point.

'Thermal decoupling' of ^{11}B from ^{1}H (see section 4D) has been observed for $(C_5H_5)_2Zr(BH_4)_2$ and $(C_5H_5)_3UBH_4$ by von Ammon et al.[156] The ^{1}H spectrum coalesces and sharpens as the temperature is lowered, while the ^{11}B resonance broadens. For the zirconium compound the ^{1}H and ^{11}B half-widths at $-70\ °C$ are 30 Hz and 250 Hz respectively. Relaxation effects in ^{1}H resonance due to the presence of ^{10}B and ^{11}B are also briefly discussed by Leach and Onak[157] for some pentaborane(9) derivatives, but no detailed bandshape measurements were made.

C. N.M.R. Spectra of Quadrupolar Nuclei.—*Nitrogen*.

Jenks[158] has measured ^{14}N spin–lattice relaxation times for a variety of molecules both from ^{14}N linewidths (assuming $T_{1N} = T_{2N}$) and from progressive saturation of ^{14}N resonance (see Section 6). He uses the results, which range from $6·2 \times 10^{-5}$s for aqueous hexamethylenetetramine to $2·7 \times 10^{-2}$s for nitromethane, to deduce ^{14}N nuclear quadrupole coupling constants *via* estimation of an 'apparent molecular volume'. Linewidths were obtained using a sideband technique based on tube-replacement. Determinations[133] of T_{1N} and T_{2N} for acetonitrile by ^{14}N n.m.r. progressive saturation and linewidth measurements respectively demonstrated that $T_{1N} = T_{2N}$ within the considerable experimental error. Addition of paramagnetic Mn^{II} ion was shown to affect the effective T_{2N} but not T_{1N}.

Many ^{14}N n.m.r. linewidths have been recorded[159] for substituted heterocyclic and aromatic compounds. The results were interpreted on an anisotropic reorientation model, and are used (together with viscosities) to obtain ^{14}N quadrupole coupling constants. Similar studies have been made of cyanopyridines.[160] Use has also been made[161] of ^{14}N linewidths (dominated by quadrupolar relaxation) as a function of temperature for $K_4Mo(CN)_8$; the authors find reasonable agreement of correlation-time data with those derived from e.s.r. measurements on $K_3Mo(CN)_8$. The n.m.r. measurements assumed that there were no effects on the ^{14}N spectrum arising from the existence of a variety of Mo isotopes.

^{14}N Linewidth data have been reported[162] for several Group IV isothiocyanates and for HNCS and HNCO. Only in the case of HNCO was any

---

[156] R. von Ammon, B. Kanellakopulos, G. Schmid, and R. D. Fischer, *J. Organometallic Chem.*, 1970, **25**, C1.
[157] J. B. Leach and T. Onak, *J. Magn. Resonance*, 1971, **4**, 30.
[158] G. J. Jenks, *J. Chem. Phys.*, 1971, **54**, 658.
[159] T. Saluvere and E. Lippmaa, *Eesti N.S.V. Teaduste Akad. Keem. Geol.*, 1970, **19**, 275.
[160] T. Saluvere and E. Lippmaa, *Eesti N.S.V. Teaduste Akad. Toimetised Fuus. Mat.*, 1970, **19**, 436.
[161] R. Poupko, H. Gilboa, B. L. Silver, and A. Loewenstein, *Ber. Bunsengesellschaft Phys. Chem.*, 1971, **75**, 279.
[162] K. M. Mackay and S. R. Stobart, *Spectrochim. Acta*, 1971, **27A**, 923.

splitting due to spin–spin coupling observed. Measurements of ^{14}N linewidths have also been reported for azo-compounds,[163] the hyponitrite ion,[163] the nitro(nitryl)-group,[164] the nitro-group,[164] isocyanates,[165] cyanates,[165] and nitrocarbanions,[166] but there is almost no discussion of the results.

Other Nuclei. The ^{11}B spectrum of liquid BF_3 has been studied at a variety of temperatures by Bacon *et al.*[167] The shape was iteratively fitted from theoretical spectra by computer to the two variables $J(BF)$ and T_{2B} (assumed in the discussion to be equal to T_{1B} and to be quadrupolar-dominated). The ^{19}F spectrum of BF_3 with enriched 97% ^{10}B was also fitted (though not iteratively) to yield $J(BF)$ and T_{1B}. This represents one of only very few accurate treatments involving nuclei of spin >1 (^{10}B has $I = 3$) in the literature. The activation energies for molecular rotation of BF_3 obtained by three methods (in ref. 167 and previous work) are in good agreement, though the ratio of ^{10}B and ^{11}B relaxation times at a given temperature shows a surprising discrepancy from the expected value. The authors suggest that there may be systematic errors in spectral fitting, or that non-quadrupolar mechanisms may contribute significantly to T_{1B}. The study also revealed a substantial variation of $^1J(BF)$ with temperature.

Values of T_2 for ^{35}Cl were obtained by Allerhand[168] from bandshape measurements in the absorption derivative mode using wide-line n.m.r. equipment. However, linewidths were not measured as such, but a 'complete bandshape analysis' was carried out by digitizing the spectrum and fitting a Lorentzian shape to it. The author emphasises, correctly, that such procedures are now accepted as giving improved accuracy in exchange problems and should do so for relaxation measurements. The computer program written for the fitting procedure takes the position of the baseline as an adjustable parameter. The accuracy claimed for the measurement of T_2 is $\pm 5\%$ at maximum. Linewidth data for ^{35}Cl n.m.r. of liquid Cl_2 at various temperatures and of an aqueous chlorine solution have been reported.[169] The temperature dependence of T_2^{-1} for liquid chlorine is found to be that of ηT^{-1}, where η is the bulk viscosity.

Gillen and Noggle[170] have obtained spin–spin relaxation times for ^{35}Cl and ^{14}N in $VOCl_3$, CCl_3CN, and BCl_3 by measuring linewidths as a function of temperature, using wide-line n.m.r. techniques. The derivative of the dispersion mode was selected and corrections made for both frequency and modulation broadening; the relaxation times span the range $1 \cdot 10 \times 10^{-5}$—

---

[163] J. Mason and W. von Bronswijk, *J. Chem. Soc.* (A), 1971, 791.
[164] J. Mason and W. von Bronswijk, *J. Chem. Soc.* (A), 1970, 1763.
[165] K. F. Chew, W. Derbyshire, N. Logan, A. H. Norbury, and A. I. P. Sinha, *Chem. Comm.*, 1970, 1708.
[166] M. Witanowski and S. A. Shevelev, *J. Mol. Spectroscopy*, 1970, **33**, 19.
[167] J. Bacon, R. J. Gillespie, J. S. Hartman, and U. R. K. Rao, *Mol. Phys.*, 1970, **18**, 561.
[168] A. Allerhand, *J. Chem. Phys.*, 1970, **52**, 3596.
[169] C. Hall, D. W. Kydon, R. E. Richards, and R. R. Sharp, *Mol. Phys.*, 1970, **18**, 711.
[170] K. T. Gillen and J. H. Noggle, *J. Chem. Phys.*, 1970, **53**, 801.

$24\cdot2 \times 10^{-5}$ s for ^{35}Cl and $3\cdot88 \times 10^{-4}$—$11\cdot3 \times 10^{-4}$s for ^{14}N. The results are used to make important deductions regarding whether molecular rotation proceeds by small-step Brownian diffusion or by an inertial process (see Chapter 3).

'High-resolution' n.m.r. spectra for ^{9}Be ($I = 3/2$) have been reported.[171] In many cases the lines are broad; widths are occasionally quoted but no detail is given.

Lindqvist and Lindman[172] have studied the line broadening of the ^{85}Rb wide-line n.m.r. signal for rubidium ions in the presence of a variety of natural and artificial humic acids. It is noted that neither ^{35}Cl nor ^{133}Cs signals are affected by the acid.

D. Effects of Magnetic Site Exchange on N.M.R. Spectra for Spin Systems containing Quadrupolar Nuclei.

—A detailed theoretical treatment of quadrupolar relaxation in the presence of chemical exchange has been presented by Marshall[150] in terms of a joint conditional probability for changes in site and in electric field gradient direction at the quadrupolar nucleus, for the case $\tau_{ex} \ll T_1, T_2$ ('fast exchange' limit). The n.m.r. linewidth for the quadrupolar nucleus depends both on the exchange lifetimes and on the correlation times for molecular rotation at each site. A surprising feature of the result is that the linewidth is affected by exchanges between equivalent sites (this cannot be accounted for by use of modified Bloch equations). Marshall[150] relates his calculations to two important types of experiment:

(a) Use of the Swift–Connick[3] approach for species exchanging between sites of very different spin–spin relaxation times. Marshall shows that the Swift–Connick formalism will only be valid when the exchange lifetime is much longer than the rotational-correlation time in the individual sites — that is, when negligible relaxation arises from changes in site.

(b) The 'halide-probe' experiment[173] for determining macromolecular rotational-correlation times, τ_{rot}. Such experiments have typically given values of τ_{rot} which are about an order of magnitude too short to correspond to rotation of the macromolecule as a whole. Marshall indicates that this could occur through neglect of the effect of τ_{ex}.

Hall et al.[174] have also given a detailed treatment of multi-site exchange effects in the n.m.r. spectra of quadrupolar nuclei. They derive closed-form expressions for the three-site and four-site cases under certain restrictions, in particular assuming that one species is dominant. The quadrupolar nature of the system was taken into account by allowing each site to have a different value of T_2. Moreover, the situation discussed is solely that where a linewidth increment for the resonance of the dominant species is observed. However,

---

[171] R. A. Kovar and G. L. Morgan, *J. Amer. Chem. Soc.*, 1970, **92**, 5067.
[172] I. Lindqvist and B. Lindman, *Acta Chem. Scand.*, 1970, **24**, 1097.
[173] T. R. Stengle and J. D. Baldeschweiler, *J. Amer. Chem. Soc.*, 1967, **89**, 3045.
[174] C. Hall, D. W. Kydon, R. E. Richards, and R. R. Sharp, *Proc. Roy. Soc.*, 1970, **A318**, 119.

there is a detailed discussion of the effects of concentration and temperature changes on the linewidth increment. Of particular interest is the discussion of effects for two different isotopes of the same species; the ratio of broadenings proves to be very helpful because quadrupolar contributions can be estimated from knowledge of quadrupole moments. The chemical equilibrium $Cl^- + Cl_2 \rightleftharpoons Cl_3^-$ was examined as an example. Values of T_2 for the various sites were estimated from quadrupolar coupling constants whereas the correlation time was derived from the spectrum. It is assumed that chemical shifts are too small to affect the fitting of the spectra (experiments showed that there is no gross field dependence of the linewidths). All spectra were corrected for instrumental effects on broadening, and the value for the rate constant of trichloride ion dissociation is proposed to be $8(\pm 4) \times 10^6$ s^{-1}. It is clear that in spite of an elegant and comprehensive study of the exchange, the derivation of a precise value of the rate constant proved to be difficult.

Spectra of spin-½ nuclei in systems containing a quadrupolar nucleus undergoing exchange vary with temperature in a complicated way. Dewkett, Beall, and Bushweller report[6] results for a relatively simple, but nonetheless interesting, system containing boron. This involves the B_3H_7 protons of $(PhCH_3)_2$-$NMe(B_3H_7)$. The rates of exchange and quadrupolar relaxation are sufficiently different that each process affects the spectrum in a different temperature region. Thus the ^1H resonance of the B_3H_7 group is sharp at -80 °C ('thermal decoupling' of B from H occurs); as the temperature is increased it broadens because quadrupolar relaxation becomes less effective; when the temperature is lowered the resonance also broadens because the intramolecular exchange is being slowed. The effects of both exchange and quadrupolar relaxation on ^1H n.m.r. spectra of boron-containing species was also the subject of an earlier short communication by Beall et al.[4] for several compounds containing the B_3H_8 group. The effects are marked for TlB_3H_8 which gives a single-line spectrum, half-width ~ 10 Hz, at -127 °C. When the temperature is raised the spectrum broadens as quadrupolar relaxation of ^{11}B and ^{10}B grows less efficient. Intramolecular exchange is presumably rapid even at -127 °C. Both processes affect the spectrum of $(Ph_3P)_2CuB_3H_8$. A similar case to that of TlB_3H_8, namely $(Ph_3P)_2CuBH_4$, was reported separately[5] by the same group of workers in the same month but in a different journal. The same group also published[7] an extension of the work on TlB_3H_8 and related carboranes a little later. This new publication repeated much of the earlier paper,[4] but gave some results of bandshape fitting for the exchange process in $(Ph_3P)_2CuB_3H_8$. Full details were not given, but the exchange mechanism assumed was total scrambling of the protons in the B_3H_8 moiety. For several reasons stated by the authors, the kinetic results are subject to large errors. Such 'thermal decoupling' effects were also noted for the ^1H n.m.r. of the $B_3H_8^-$ ion by Marynick and Onak[175] (exchange being rapid at all temperatures studied). These authors found that for $Me_4N^+B_3H_8^-$ at -90 °C irradiation

[175] D. Marynick and T. Onak, J. Chem. Soc. (A), 1970, 1160.

at ^{11}B frequencies produced no change in the ^{1}H spectrum. The ^{11}B spectrum itself broadened ($\Delta\nu_{\frac{1}{2}} \sim 500$ Hz at -100 °C) as the temperature was lowered, as expected. Marynick and Onak[175] also found that the paramagnetic ion Mn^{2+} also caused 'decoupling' of their spectra; both ^{11}B and ^{1}H spectra appear as singlets, the ^{1}H spectrum being considerably broader than the ^{11}B spectrum ($\Delta\nu_{\frac{1}{2}} \sim 400$ Hz and ~ 150 Hz respectively).

Pinkowitz and Swift[176] have studied the effect of 'complex' formation of ammonia with a 'free' electron on the ^{1}H n.m.r. bandshape. Dilute solutions of potassium in liquid ammonia were used. The dipolar ^{14}N–electron spin–lattice relaxation affects the linewidths for the bulk ammonia resonance *via* the exchange process between complexes. The observed bandshape was analysed in terms of three parameters [J(NH), and the lifetimes for quadrupolar relaxation and for exchange relaxation] using a damped least-squares procedure on a Univac 1108 computer. Modified Bloch equations were derived and used for this process. The values of J(NH) and the quadrupolar relaxation lifetime were relatively unaffected by the presence of the potassium. The exchange relaxation lifetime is discussed in terms of the ^{14}N–electron dipolar spin–lattice relaxation time, which, it is concluded, is dominated by the contact term rather than the dipole–dipole term. Since the electron Larmor frequency and the average lifetime of a given NH_3 molecule in a given solvated electron species, $\tau(e^-)$, are both of the order of 10^{-12} s, the n.m.r. lineshape was field-dependent. Measurement of the bandshape at both 60 MHz and 100 MHz therefore enabled both $\tau(e^-)$ and the average solvation number of the electron to be determined independently.

5 Relaxation Effects — Linewidths

A. Theoretical Work.—A series of papers by Sykora[146,177,178] have discussed n.m.r. linewidths for multi-spin systems using a somewhat simplified theoretical approach in terms of energy-level widths, calculated on the basis of transition probabilities. Emphasis is placed on the separation of the functions depending on the nuclear spin coupling structure on the one hand and the spectral density functions on the other. The first paper,[177] in addition to developing the general theory, discusses in detail the case of intramolecular dipolar relaxation for some simple spin systems, *i.e.* AB, A_2, ABX, A_2X, A_3, and A_3X. The notation seems a little cumbersome, and not enough account seems to have been taken of earlier work in this area (the only reference given for relaxation theory is Abragam's book), but the general comments on the spin systems concerned would appear to be useful. The limitations of the theory are discussed; cases of degeneracy present difficulties. The second paper[178] treats the case of n.m.r. line-broadening due to intermolecular relaxation ('random fields'). It is concluded that there will be a roughly equal

---

[176] R. A. Pinkowitz and T. J. Swift, *J. Chem. Phys.*, 1971, **51**, 2858.
[177] S. Sykora, *J. Chem. Phys.*, 1970, **52**, 4818.
[178] S. Sykora, *J. Chem. Phys.*, 1970, **52**, 5949.

broadening of all transitions, in contrast to the effects described in the first paper.[177] This result is particularly emphasised for the AB spin system. Sykora[146] develops his theories considerably in the third paper of the series, where he considers all broadening mechanisms induced by molecular re-orientation. He treats these in two groups: (*a*) those of tensor character — the dipolar, quadrupolar, and chemical shift anisotropy mechanisms, and (*b*) those of vector character — intermolecular relaxation and the spin-rotation mechanism. Expressions are given for calculating linewidths in each case (see also Section 4). Cross terms between pairs of relaxation mechanisms are also discussed. Unfortunately no experimental examples are given in any of the three papers;[146,177,178] it is to be hoped that some applications will be published subsequently, especially since some of Sykora's conclusions appear to differ from those published elsewhere based on more detailed treatments.

Alla and Lippmaa[179] have explored the implications of the Redfield equation for the high-temperature limit. They quickly specialise their equations for the case of a neat liquid with a molecular two-spin system (non-equivalent $I = \frac{1}{2}$ spins). Further simplifications follow when intermolecular dipole–dipole relaxation alone is considered, and equations are derived for formal transition probabilities between the four energy levels of the system. Explicit equations are given for the transition probabilties and for the linewidths; useful graphs are also plotted for these quantities as a function of the degree of mixing in the two-spin system. The implications for general nuclear Overhauser effects in the first-order limit are pointed out, and comparisons drawn with the case of intramolecular dipole–dipole relaxation.

Sillescu[180] considers molecular re-orientation in liquids in some detail. The Debye model (Brownian motion) and the random-jump model for rotation are both extended to take account of time fluctuations in their rate (rendering the treatment applicable to situations described by a distribution of correlation times). The theory is applied to e.s.r. lineshapes for the case of slow re-orientation where the rotational correlation time is comparable to the spin relaxation time. In principle, the treatment is also applicable to n.m.r. Different lineshapes are given by the Debye and jump models.

B. Paramagnetic Effects.—La Mar and Van Hecke report[181] an interesting line-broadening effect for n.m.r. spectra of paramagnetic complexes. Frequently line-broadening in such complexes is caused by chemical exchange (see Section 3). However, any 'exchange' process can in principle cause such effects; in the publication in question,[181] hindered internal rotation is reported to cause such broadening by modulation of the contact shift. The simplified expression for such broadening is given as: $T_2^{-1} = 4(\gamma \Delta B)^2 \tau_m/3$, where ΔB is the observed contact shift and τ_m is the methyl rotational time. Measurements were made on methyl-substituted phenanthroline and bipyridine complexes of Cr^{II}. Variable-temperature studies show that linewidths at high temperature

---

[179] M. Alla and E. Lippmaa, *J. Magn. Resonance*, 1971, **4**, 241.
[180] H. Sillescu, *J. Chem. Phys.*, 1971, **54**, 2110.
[181] G. N. La Mar and G. R. van Hecke, *J. Chem. Phys.*, 1970, **52**, 5676.

are dominated by bulk ligand exchange, but at low temperatures the methyl rotation effect dominates. Arrhenius plots indicate barriers of *ca.* 75 kJ mol^{-1}, in some instances, which, it was suggested, may be traced to severe steric interactions. The linewidths show a strong dependence on the applied magnetic field, as expected. The authors sound a note of caution regarding their theories because there are some serious quantitative discrepancies in the observed effects. In fact, in a later note[182] they produce evidence that the high barriers are primarily electronic in origin rather than steric.

^{1}H N.m.r. studies of paramagnetic transition-metal complexes are often difficult because of the large linewidths involved. However, there should be a reduction in width by a factor $(\gamma_D/\gamma_H)^2$ when deuteron n.m.r. of corresponding deuteriated compounds is used. Johnson and Everett[183] have investigated this possibility for a number of metal acetylacetonates, and report both ^{1}H and ^{2}H linewidths. The reduction factor is generally smaller than theoretically expected; it is not clear why. In some cases the spectra were affected by substantial isotope effects on the isotropic shifts.

A somewhat similar phenomenon to that of 'thermal decoupling' (see Section 4D) is the cause of bandshape changes induced for certain spin systems by the addition of paramagnetic ions. For instance, it has been shown[184] that addition of Co(acetylacetonate)$_2$ to solutions of pyridine and related compounds not only results in shifts and broadening of the α-proton signals but also simplifies the spectrum of the β-protons by removing the influence of $\alpha-\beta$ coupling. This 'decoupling' was attributed to rapid relaxation of the α-protons; the authors[184] did not discuss the bandshape changes in detail.

A similar effect has been observed[185,186] for trialkyl phosphites in solutions containing paramagnetic Co^{II} or Ni^{II} salts. In this case P–H coupling effects are progressively eliminated from the proton spectrum as the concentration of the paramagnetic complex increases.

C. Gas-phase Studies.

The improvements in sensitivity now available with the 'new generation' of commercial spectrometers have not yet resulted in a large increase in the number of gas-phase studies of bandshapes, though this may be expected in the near future. However, Mohanty and Bernstein[187,188] have measured ^{19}F linewidths for CF$_4$ under a variety of conditions of temperature, pressure, and solvent gases. This, of course, gives T_2, but the authors immediately make the assumption that $T_1 = T_2$ and confine themselves to a discussion of T_1. A distinct medium effect is observed for CF$_4$, T_2 varying between 29 ms and 57 ms at 303 K and 20 atm total pressure; this is explained

---

[182] G. N. La Mar and G. R. van Hecke, *Chem. Comm.*, 1971, 274.
[183] A. Johnson and G. W. Everett, *J. Amer. Chem. Soc.*, 1970, **92**, 6705.
[184] E. Gillies and M. C. Baird, *Chem. Phys. Letters*, 1970, **7**, 451.
[185] R. Engel, *Chem. Comm.*, 1970, 133.
[186] R. Engel and A. Jung, *J. Chem. Soc. (C)*, 1971, 1761.
[187] S. Mohanty and H. J. Bernstein, *J. Chem. Phys.*, 1970, **53**, 461.
[188] S. Mohanty and H. J. Bernstein, *J. Chem. Phys.*, 1971, **54**, 3730.

in terms of collision rate and viscosity. The authors comment that linewidths are reproducible to 5% but believe that T_2 is only accurate to 10%. It would seem that a complete bandshape treatment would give higher accuracy. The earlier paper[187] also gives data for SiF_4 and SF_6 (pure gases only).

Dayan and Chiglien[189] have obtained spin–spin relaxation times for protons in mixtures of simple molecules in the gas phase by measuring linewidths, and have extrapolated their results to infinite dilution conditions. Both dipole and quadrupole electric moments of the solvent gases are shown to be important in determining T_2.

D. Miscellaneous Width Studies.

It has been pointed out[190] that many n.m.r. spectra show bandshape features that cannot be accounted for simply in terms of spectral analysis for chemical shifts and coupling constants alone. Some examples of cyclotriphosphazatrienes showing such behaviour were given. Many other systems giving similarly 'anomalous' bandshapes are also symmetrical (having chemical equivalence) and have two or more equivalent phosphorus nuclei. A number of possible reasons were suggested which may apply in particular cases: (i) unrecognised second-order effects, (ii) the effect of many overlapping lines, not apparent from 'stick-plots', (iii) unrecognised intramolecular exchange, (iv) unrecognised long-range coupling, (v) effects of quadrupolar nuclei coupled to nuclei other than those whose resonance is observed, and (vi) relaxation effects specific to certain lines in the spectrum. The paper[190] does not make definite conclusions for particular cases, but emphasises that the phenomena are of common occurrence and that spectral analyses which rely on constancy of linewidths throughout the spectrum may be suspect when there are overlapping lines.

The anomalous ^{19}F linewidths for pentafluorophenyl compounds, previously[191] ascribed possibly to a relaxation phenomenon, are now attributed[192] to long-range F–F coupling, since T_1 measurements failed to give evidence of abnormally short relaxation times.

6 Relaxation Effects — Saturation

Jenks[158] has used the progressive saturation method in ^{14}N resonance for measuring values of T_{1N}. In this case the effect of the audio-frequency magnetic field modulation, ω_m, must be considered. Jenks describes the evaluation of T_{1N} in the two limits, $\omega_m T_{1N} \ll 1$ and $\omega_m T_{1N} \gg 1$. In the former case, determination of the value of the radio-frequency field, B_1, which gives the maximum ^{14}N signal suffices for the derivation of T_{1N}. The value of $B_1(max)$ is obtained from a plot of $\gamma^2 B_1^2$, against $(\gamma B_1/S_{max})^{2/3}$, where S_{max} is the

---

[189] E. Dayan and G. S. Chiglien, *Compt. rend.*, 1970, **270**, B, 631.
[190] E. G. Finer, R. K. Harris, M. R. Bond, R. Keat, and R. A. Shaw, *J. Mol. Spectroscopy*, 1970, **33**, 72.
[191] K. W. Jolley and L. H. Sutcliffe, *Spectrochim. Acta*, 1968, **24A**, 1191.
[192] L. H. Sutcliffe and G. J. T. Tiddy, *Spectrochim. Acta*, 1970, **26A**, 282.

maximum signal strength for a given value of B_1. The intercept is $-1/T_{1N}T_{2N}$, giving T_{1N} since T_{2N} is measured from the linewidth in the absence of saturation. The uncertainty in T_{1N} is said to be 10%. When $\omega_m T_{1N} \gg 1$, T_{1N} is evaluated in a similar way but using a different equation. In this case the uncertainty is 25%. These methods clearly do not rely on assuming $T_{1N} = T_{2N}$.

The progressive saturation method has also been used[133] in ^{14}N resonance for determining spin–lattice relaxation times for acetonitrile, both with and without the presence of paramagnetic Mn^{II} ion.

Reeves and Shaw[71] have considered the effects of saturation on n.m.r. spectra of systems involving exchange of the multi-site, uncoupled, first-order type (see also Section 3). They give details of some calculations for the two-site, equal-population case. Saturation effects are found to be most severe in the limits of very slow and very fast exchange. The effect is considered from the aspect of possible errors in derived rate constants due to partial saturation.

Freeman and Hill[193] have reported the equivalent of a 'progressive saturation' method for determining T_1 in the Fourier Transform pulsed mode. However, although the method involves reaching a steady state between relaxation and the effect of the pulses, the measurements are obviously not taken under continuous conditions of equilibrium between the spins and a radio-frequency field, so that the method has as much analogy with saturation-recovery and adiabatic rapid-passage experiments as with continuous-wave progressive saturation.

7 Bandshapes in Multiple Resonance

Bucci and co-workers[194–196] have used the second-quantization formalism to calculate n.m.r. double-resonance bandshapes. In their first paper,[194] which deals with spin tickling, they show that first-order second-quantization treatment gives the same results as the usual rotating-frame approach, but that second-order terms modify the bandshape appreciably in certain circumstances. Some theoretical bandshapes are illustrated to show the differences between the two approaches, but no experimental data are available for comparison. In the second[195] and third[196] papers the work is extended to deal with saturation, and, in particular, with multiple-quantum transitions. In certain circumstances, when a number of photons of one radio-frequency are absorbed and the same number emitted at the second radio-frequency, a broad dispersion-like signal is predicted when the two frequencies are nearly equal. Such a signal was observed in the case of chloroform.[196] The theory is more successful in predicting intensities and bandshapes than the Yatsiv approach.[197]

[193] R. Freeman and H. D. W. Hill, *J. Chem. Phys.*, 1971, **54**, 3367.
[194] P. Bucci, A. M. Serra, P. Cavaliere, and S. Santucci, *Chem. Phys. Letters*, 1970, **5**, 605.
[195] P. Bucci, P. Cavaliere, and S. Santucci, *J. Chem. Phys.*, 1970, **52**, 4041.
[196] P. Bucci, M. Martinelli, and S. Santucci, *J. Chem. Phys.*, 1970, **53**, 4524.
[197] S. Yatsiv, *Phys. Rev.*, 1959, **113**, 1522, 1538.

The paper by Gestblom and Hartman[198] discussed in Chapter 7 treats the double-resonance case where high radio-frequency powers in v_1 are used, such that double-quantum transitions are observed. In a short section the authors discuss the lineshapes in double-quantum tickling experiments which arise from magnetic field inhomogeneity. As in normal tickling experiments, progressive doublet lines are broader while regressive doublet lines are sharper than in single resonance. Calculated spectra are illustrated using Gaussian field inhomogeneities.

Kumar and Gordon,[199] in their study of Overhauser effects in two-spin systems (see Chapter 7), have commented on the difficulty of obtaining measures of intensity of individual lines in double resonance, both because of small unresolved splittings caused by 'tickling', and because of the variations[200] in the inhomogeneity contribution to linewidths for progressive and regressive lines.

Cases including Exchange.—The effects of double resonance on spectra of exchanging systems may yield useful information regarding both exchange and relaxation. Yang and Gordon[201] have discussed a formalism for such cases which combines the Redfield relaxation treatment[15] with the Alexander exchange formalism;[62] this is developed explicitly for the intermolecular exchange case and applied to the AB_2 system 2,2,2-trichloroethanol. The comparison of experimental and theoretical spectra was made visually, producing exchange rates and relaxation times to an accuracy of *ca.* 10%. Only cases near the slow-exchange limit were studied (indeed the use of the method for such cases is emphasised). Thus, most of the effects distinguished rely rather on saturation transfer rather than on bandshapes in the more limited sense dealt with here (see Chapter 7).

Fung and Olympia[202] have also considered double-resonance effects in the presence of chemical exchange, using the density matrix formalism. The ^{19}F spectrum of $F_2Br\ C\cdot CClBr_2$ was studied, and the exchange parameters determined by single-resonance experiments; the bandshape was fitted to the expected $AB \rightleftharpoons BA \rightleftharpoons X_2$ system. The spectrum is relatively insensitive to the *gauche*⇌*gauche* exchange rate. The double-resonance experiments then yield information about the relaxation mechanisms. The transfer of saturation is emphasised. It is found that the use of single phenomenological values of T_1 for each isomer is not sufficient to describe the bandshapes (especially intensities) and a full density-matrix treatment has to be employed. The authors conclude that the random-field relaxation mechanism is dominant.

Anderson[57] has combined the Binsch theory of exchange with Redfield relaxation theory[15] to produce a description of the nuclear magnetic double-

---

[198] B. Gestblom and O. Hartmann, *J. Magn. Resonance*, 1971, **4**, 322.
[199] A. Kumar and S. L. Gordon, *J. Chem. Phys.*, 1971, **54**, 3207.
[200] R. Freeman and B. Gestblom, *J. Chem. Phys.*, 1967, **47**, 2744.
[201] P. P. Yang and S. L. Gordon, *J. Chem. Phys.*, 1971, **54**, 1779.
[202] B. M. Fung and P. L. Olympia, jun., *Mol. Phys.*, 1970, **19**, 685.

resonance spectrum for a mutually exchanging two-spin system. Both 'tickling' and 'decoupling' situations are discussed. Only theoretical spectra are illustrated, but it is shown that the double-resonance technique can be helpful in obtaining exchange rates, since at intermediate values the rich structure of the double-resonance spectra makes for easy comparison between experiment and theory (cf. the similar point made with regard to complex spin systems in Section 3), while at low exchange rates the strongly decoupled spectrum provides a ready measure of the exchange rate relative to the relaxation rate.

7
Multiple Resonance

BY D. SHAW

1 Introduction

The aim of high-resolution n.m.r. is to gain information, usually of chemical interest, from the interaction of magnetic nuclei with a magnetic field. This interaction is studied by observing the resonant absorption of energy when a precise field to frequency ratio (of the exciting energy) is achieved. The experiment is normally carried out by sweeping either the basic magnetic field or the exciting frequency while keeping the other constant; thus one resonance at a time is studied. In this chapter experiments will be discussed where two or more resonances are excited at the same time. All the resonances will be nuclear resonances, *i.e.* ENDOR (electron–nuclear double resonance) and ELDOR (electron–electron double resonance) will not be considered. Double-resonance experiments were first proposed by Bloch in 1954,[1] and successfully achieved in the same year by Royden, who decoupled the ^{13}C in methyl iodide while observing the proton resonance.[2]

The technique of double resonance was first reviewed, by Baldeschwieler and Randall, in 1962,[3] and subsequently, homonuclear double resonance has been reviewed by Hoffmann and Forsén[4] and McFarlane has reviewed heteronuclear decoupling.[5] Recently, an excellent account of double resonance and its applications in chemistry has appeared.[6] The normal terminology and classifications used by these authors (especially see ref. 6) will be used in this chapter.

The usage of multiple-resonance techniques has dramatically increased following the inclusion of the required electronics to perform these experiments as standard equipment in some commercial spectrometers, firstly for homonuclear experiments and more recently for heteronuclear experiments (see Chapter 4). Today virtually every spectrometer can perform homonuclear decoupling experiments with ease, and a very large proportion of n.m.r. assignments are based on double-resonance results. If it were not for double-resonance experiments in the form of proton noise decoupling,[7] the

---

[1] F. Bloch, *Phys. Rev.*, 1954, **94**, 496.
[2] V. Royden, *Phys. Rev.*, 1954, **96**, 543.
[3] J. D. Baldeschwieler and E. W. Randall, *Chem. Rev.*, 1963, **63**, 81.
[4] R. A. Hoffman and S. Forsén, *Progr. N.M.R. Spectroscopy*, 1966, **1**, 15.
[5] W. McFarlane, *Ann. Rev. N.M.R. Spectroscopy*, 1968, **1**, 135.
[6] W. von Philipsborn, *Angew. Chem. Internat. Edn.*, 1971, **10**, 472.

current emergence of ^{13}C n.m.r. as a technique of enormous power, particularly in organic chemistry, would almost certainly not have occurred. Indeed, under the definition given above, virtually every n.m.r. experiment performed by a modern spectrometer is a multiple-resonance experiment. This occurs by their almost universal use of a separate resonance experiment to 'lock' the field to the frequency, hence permitting the use of precalibrated charts, *etc.*[8,9]

This chapter will be concerned with recent developments in n.m.r. involving multiple resonance. To describe every use of multiple resonance, especially in spectral assignment, would be almost impossible and has specifically not been attempted. New developments in each type of multiple-resonance experiment will be discussed, together with any history that is necessary to place them in perspective.

2 Theory of Multiple Resonance

The basic theory of multiple resonance n.m.r. was formulated by Bloch,[10,11] based on a density matrix approach. Redfield[12,13] later reformulated this theory, and in this form it has been the basis for most subsequent theoretical work on double resonance. Baldeschwieler and Randall[3] in their review gave a concise outline of the basic theory of multiple resonance, and recently Nageswara Rao has given a comprehensive account of the density matrix theory with particular respect to relaxation effects.[14] The reader is referred to these reviews and that of Hoffman and Forsén[4] to obtain a general background to double-resonance theory.

The theory of double as well as single resonance can be approached from two points of view. In the so-called 'indirect' method the spin Hamiltonian operator (or its matrix equivalent) is used to find the eigenvalue of the spin system: line frequencies and intensities are then indirectly calculated from these eigenvalues. In the 'direct' method[15] transition frequencies are calculated directly as eigenvalues of a super-operator (or its matrix equivalent). The latter approach is equivalent, from the point of view of computation, to the use of Liouville-space. This approach has been applied to double-resonance experiments, originally with rather disappointing results,[16] but latterly with more success.[17] The later work was more successful owing to the choice of a better basis set of Hamiltonian operators for use within the super-operator. The method successfully yielded the complete double-resonance spectrum (12 transitions) of an AX system including those due to the B_2 field itself. This

[7] R. R. Ernst, *J. Chem. Phys.*, 1966, **45**, 3845.
[8] R. Freeman and D. H. Whiffen, *Proc. Phys. Soc.*, 1962, **79**, 794.
[9] H. Primas, Fifth European Congress on Molecular Spectroscopy, Amsterdam, 1961.
[10] F. Bloch, *Phys. Rev.*, 1956, **102**, 104.
[11] F. Bloch, *Phys. Rev.*, 1957, **105**, 1206.
[12] A. G. Redfield, *IBM J. Res. Development*, 1957, **1**, 19.
[13] A. G. Redfield, *Adv. Magn. Resonance*, 1965, **1**, 1.
[14] B. D. Nageswara Rao, *Adv. Magn. Resonance*, 1970, **5**, 271.
[15] C. N. Banwell and H. Primas, *Mol. Phys.*, 1963, **6**, 225.
[16] J. M. Anderson, *J. Magn. Resonance*, 1970, **4**, 184.
[17] B. Gestblom, O. Hartmann, and J. M. Anderson, *J. Magn. Resonance*, 1971, **5**, 174.

method can be extended to include relaxation effects *via* the Redfield theory,[12,13] which is in a similar super-matrix format.

The most powerful and general approach to double-resonance theory is *via* the statistical density matrix approach.[14,18] In this approach, which is a direct method and similar in principle to that mentioned above, the density matrix of the spin system and the lattice together is evaluated. The density matrix, σ, is described by the equation:

$$\frac{d\sigma}{dt} = -i[\mathcal{H}(t), \sigma] - \Gamma(\sigma - \sigma_0)$$

$\mathcal{H}(t)$ is the system Hamiltonian including all static and time-dependent (*e.g.* B_1 and B_2) external fields:

$$\mathcal{H}(t) = \mathcal{H}_0 + \mathcal{H}_1(t) + \mathcal{H}_2(t)$$

Γ is the relaxation super-operator (matrix) and σ_0 is the density matrix describing the system at thermal equilibrium. The time dependence of $\mathcal{H}(t)$ on B_2 is normally removed by transforming into a frame of reference rotating about the z-axis at frequency ω_2. The experimentally observed magnetization $M_y(t)$ is obtained as the expectation value defined in terms of σ. This approach has been successfully applied to relaxation studies in double resonance.[14]

Nageswara Rao and co-workers, by inclusion of further terms in $\mathcal{H}(t)$ to allow for the effect of electric fields, have extended this theory to the ABX case where X is a quadrupolar nucleus ($I > \frac{1}{2}$).[19] They have shown that a study of the relaxation processes involved in double-resonance experiments can yield the relative signs of the coupling constants involving the quadrupolar nucleus. The technique was demonstrated by showing that J(AN) (9·7 Hz) has the same sign as J(BN) (2·9 Hz) in 2-bromothiazole.

A slightly different approach to the double-resonance problem has been taken by Bucci *et al.*[20] Their approach, already used for single-resonance experiments,[21-23] involves a second quantization of the r.f. fields, as opposed to the normal rotating-frame notation, in defining $\mathcal{H}(t)$. With a single exciting frequency this approach is equivalent to the normal rotating-frame technique. However, for two or more excitations the second-quantization approach is the simpler since only time-independent Hamiltonians are necessary and hence σ can be obtained under stationary conditions. They applied this technique to the simple case of a spin-$\frac{1}{2}$ nucleus, the proton of chloroform: at low observing power their results were in line with those of previous workers,[14] but they extended the study to higher observing powers and at these powers multiple-quantum transitions were observed. These were of two types: first, if n quanta

[18] F. Bloch, *Phys. Rev.*, 1956, **102**, 104.
[19] A. Kumar, N. R. Krishna, and B. D. Nageswara Rao, *Mol. Phys.*, 1970, **18**, 11.
[20] P. Bucci, M. Martinelli, and S. Santucci, *J. Chem. Phys.*, 1970, **53**, 4524.
[21] A. Di Giacomo and S. Santucci, *Nuovo Cimento*, 1969, **63B**, 407.
[22] P. Bucci, P. Cavaliere, and S. Santucci, *J. Chem. Phys.*, 1970, **52**, 4041.
[23] P. Bucci, P. Cavaliere, S. Santucci, and A. M. Serra, *Chem. Phys. Letters*, 1970, **5**, 605.

were absorbed from B_1 and n quanta emitted at ω_2, or *vice-versa*, dispersion-like signals were produced; secondly, when n quanta were absorbed by B_1 and $n \pm 1$ quanta were emitted at ω_2, or *vice-versa*, spectra with both emission and absorption lines present were observed. The experimental spectra agree well with the theoretical predictions.

The theory of multiple resonance has been extended to triple resonance,[24-27] and some of the theoretical predictions confirmed by experimental results. An AMX system was investigated with both irradiating fields being of 'tickling' power $[(\gamma B_1/2\pi) = 0.7$ Hz].[26] The predicted combination signals[24,25] at $(\nu_1 + \nu_2 - \nu_3)$ and $(\nu_1 - \nu_2 + \nu_3)$ were recorded using a phase-sensitive detector referenced to these frequencies.[26] The AX case was studied using *trans*-chloroacrylic acid at 100 MHz under conditions where one of the irradiating fields was strong and the other weak.[27] The main results of the experiments confirmed the conclusions of the theory previously proposed,[28] *i.e.* splitting of the signals into doublets, the appearance of combination signals for a three energy-level fragment even when the irradiated transitions possess common energy levels, and a sharp resonance nature of the effects.

3 Double Resonance in the Presence of Chemical Exchange

When two nuclei are taking part in chemical exchange and their lifetimes in any of the sites involved are short enough to be comparable with their T_2, then line broadening results (see Chapter 6). This is the basis of the study of kinetics by n.m.r. lineshape analysis.[29] Since T_1 is longer than T_2 for most cases, the applications of n.m.r. can be extended to slower reaction rates by the study of effects dependent on T_1 rather than on T_2. Double-resonance experiment sinvolving transfer of saturation[30,31] are of such a type.

The transfer of saturation method has recently been extended to processes involving exchange between three non-equivalent sites, and the average lifetime of the protons in the oxygen, methylene, and vinyl centres ($\tau = 4.2$, 3.8 and 14 s respectively) of the keto–enol equilibrium of acetylacetone determined.[6] The transfer of saturation method can also be applied to the study of the relaxation mechanisms by following the path of the spin saturation through a reaction. Such an approach has been used[32] to study (*a*) the left–right exchange in an asymmetric π-methallyl system and (*b*) σ, π rearrangement in symmetrical dimeric π-allylpalladium complexes. These reactions are too slow to be followed by linewidth studies[32] but show themselves very amenable to study by the saturation transfer method.

[24] V. F. Bystrov, *J. Mol. Spectroscopy*, 1968, **28**, 81.
[25] V. F. Bystrov, *J. Magn. Resonance*, 1970, **2**, 267; *ibid*., 1970, **3**, 350.
[26] V. A. Afana'ev and V. F. Bystrov, *J. Magn. Resonance*, 1970, **3**, 357.
[27] V. F. Bystrov, *J. Magn. Resonance*, 1970, **3**, 350.
[28] E. Kundla, *Izvest. Akad. Nauk Est. S.S.R., Ser. Tekh. i Fiz.-Mat. Nauk*, 1968, **17**, 475.
[29] H. Kessler, *Angew. Chem.*, 1970, **82**, 237.
[30] S. Forsen and R. A. Hoffman, *J. Chem. Phys.*, 1964, **40**, 1189.
[31] D. S. Kavakoff and E. Namanwirth, *J. Amer. Chem. Soc.*, 1970, **92**, 3234.
[32] P. W. N. M. Van Leenwen and A. P. Praat, *J. Organometallic Chem.*, 1970, **22**, 483.

Advantage has been taken of saturation transfer to reduce the intensity of the HOD line in proton spectra using D_2O as a solvent.[33] Systematic decoupling throughout the spectral region revealed two frequencies at which the intensity of the HOD band dramatically decreased. These frequencies correspond to the chemical shifts of the two types of OH groups present in the isoprenaline solute. This technique could prove useful, not only in removing unwanted water signals, enabling the region close to them to be studied, but also in characterizing chemical shifts of interchangeable protons. It is estimated that such chemical shifts could be measured with an accuracy of ±5 Hz in favourable cases.[33]

When double-resonance n.m.r. is used to study reaction mechanisms the basic assignment of the lines must be certain, as with any form of spectroscopy. A recent study of the mechanism of internal rotation in 5-methyl-dichlorosilylcyclopentadiene,[34] undertaken to distinguish between a 1,2 or a 1,3 shift mechanism, came out in favour of the unusual 1,3 mechanism. A later, more detailed, analysis of the same spectra demonstrated that a better fit to the spectral data could be obtained if the assignment of the AA'BB' system from the cyclopentadiene ring was reversed. With the reversal of this assignment the spectroscopic data now supported the much more widely accepted 1,2 shift mechanism.[35]

The theory of double resonance in the presence of chemical exchange has been approached from two standpoints: firstly, by a combination of the chemical exchange formalism of Alexander[36] with the relaxation formalism of Redfield[12] within a density matrix approach based on a double-resonance basis set of operators;[37,38] secondly, by a combination of the exchange lineshape theory of Binsch[39] with the Redfield relaxation theory.[40] Under the influence of chemical exchange, double-resonance spectra acquire new features: at high irradiating powers decoupled spectra are produced whose intensity varies from positive through zero to negative as the exchange rate passes through the relaxation rate;[40] at lower powers, both positive and negative peaks can occur for a given exchange rate.

Two systems have recently been carefully studied, namely 2,2,2-trichloro-ethanol (AB_2)[37] and the internal rotation of 1,2-dibromo-1,1-dichloro-2,2-difluoroethane $(AB+C_2)$.[38] In the latter case ^{19}F n.m.r., with its larger chemical shifts and coupling constants, was used to enable strong irradiation of one line while the effects on the other lines were observed, without instrumental perturbations from the exciting field. Computer fitting of the experimental spectra yielded chemical exchange correlation times and produced an internal

---

[33] J. Feeney and G. C. K. Roberts, *Chem. Comm.*, 1971, 205.
[34] N. M. Sergeyer, G. I. Arramenko, and Yu. A. Ustynyuk, *J. Organometallic Chem.*, 1970, **22**, 77.
[35] F. A. Cotton and T. J. Marks, *Inorg. Chem.*, 1970, **9**, 2802.
[36] S. Alexander, *J. Chem. Phys.*, 1962, **47**, 967, 974.
[37] P. P. Young and S. L. Gordon, *J. Chem. Phys.*, 1971, **54**, 1779.
[38] B. M. Fung and P. M. Olympia, *Mol. Phys.*, 1970, **19**, 685.
[39] G. Binsch, *J. Amer. Chem. Soc.*, 1969, **91**, 1304.
[40] J. M. Anderson, *J. Magn. Resonance*, 1971, **4**, 184.

rotation barrier for *trans–gauche* interconversion of 39·7 ±0·5 kJ mol⁻¹. Both these studies produced good agreement between the theory and the experiment. The dominant relaxation mechanism in the presence of chemical exchange was found to be that of an uncorrelated external random field.[14]

Comparison of experimental and theoretical double-resonance spectra allows the determination of chemical exchange rates and relaxation parameters even when the exchange rates are slow compared with the linewidths involved. Under these conditions more conventional n.m.r. methods are inapplicable. Further developments in the field should be rewarding because of the rich structure of the spectra obtained, which will permit accurate rate and relaxation measurements to be made by fitting of the experimental results to the calculated data.

4 Chemical Applications of Multiple-resonance Techniques

The applications of multiple resonance are as wide as those of n.m.r. itself, (see ref. 4 for an excellent account and classification of the uses of double resonance). This section will review the latest additions to, and extensions of, these applications.

Spin Decoupling.—Spin decoupling is the oldest[2,41] and the most common (excluding field–frequency locking) multiple-resonance experiment. In this method a strong r.f. field is applied at the resonance frequency of the nucleus under investigation. Under the influence of this strong r.f. field the quantization of the irradiated spin changes from the z-axis to the x-axis and consequently its scalar coupling with the remaining spins is removed from the spectrum; hence the name of the experiment.

The applications of homonuclear spin decoupling are far too numerous to mention. Also, heteronuclear spin decoupling experiments are becoming much more common, both for spectral simplification and for determination of chemical shifts in 'hetero' nuclei. For accurate measurements of chemical shifts the INDOR technique with its lower power, and hence freedom from Bloch–Siegert effects,[42] and greater accuracy is to be preferred. The technique of heteronuclear spin decoupling has in the past year been applied to measuring chemical shifts in a wide range of nuclei, *e.g.* in ^{19}F,[43] in ^{13}C,[44] in ^{31}P,[45] in ^{14}N,[46,47] in ^{15}N,[48] and in ^{117}Sn and ^{119}Sn.[49] Indeed, the double-

[41] W. A. Anderson, *Phys. Rev.*, 1956, **102**, 151.
[42] F. Bloch and A. Siegert, *Phys. Rev.*, 1950, **57**, 522.
[43] A. B. Foster, R. Hems, and L. D. Hall, *Canad. J. Chem.*, 1970, **48**, 3937; C. W. M. Grant and L. D. Hall, *Canad. J. Chem.*, 1970, **48**, 3537.
[44] W. McFarlane, *Chem. Comm.*, 1970, 418.
[45] H. J. Jakobsen, *J. Mol. Spectroscopy*, 1971, **38**, 243.
[46] H. Sioto and K. Nukada, *J. Amer. Chem. Soc.*, 1971, **93**, 1072, 1077.
[47] P. Hampson, A. Mathias, and R. Westhead, *J. Chem. Soc.* (*B*), 1971, 397; F. W. Wehrli, W. Giger, and W. Simon, *Helv. Chim. Acta*, 1971, **54**, 229.
[48] D. Gagnaire, R. Ramasseul, and R. Rassat, *Bull. Soc. chim. France*, 1970, 415; L. Paolillo and E. Becker, *J. Magn. Resonance*, 1970, **3**, 200.
[49] A. Tupciauskas, N. M. Sergeyer, and Yu. A. Ustynyuk, *Mol. Phys.*, 1971, **22**, 179.

resonance method for shift measurements has become so important that in his discussion of possible reference componds for nitrogen Becker includes suitability for double-resonance studies as a major factor in his choice of $\overset{+}{N}Me_4$ as a reference.[50]

As mentioned in the introduction, the technique of proton noise decoupling[7] is assuming an enormous practical importance through its ability to enhance the sensitivity of, and simplify, ^{13}C n.m.r. spectra. The simplification of ^{13}C spectra, brought about by the decoupling of the protons present in the molecules, has greatly aided the interpretation of carbon spectra. However, performing such experiments removes the information present in the multiplicity of the carbon lines, i.e. how many protons are directly bonded to a particular carbon. This information can be regained using the technique of 'off-centre double resonance'. Here the proton decoupler is run in a coherent mode at a frequency which is not the resonant frequency of any of the protons in the molecule under investigation. The carbon lines now demonstrate their multiplicity but the splittings involved are less than their original coupling value, hence simplifying the spectra. Most of their Overhauser effect is maintained. It is possible to obtain the useful Overhauser enhancement, but retain the complete spin–spin decoupling, by making use of the fact that on removal of the decoupling field the axis of quantization on which spin decoupling depends returns to the z-axis instantaneously, while the population changes of a generalized Overhauser effect return slowly (see p. 264). A manual continuous-wave (c.w.) experiment on enriched methyl iodide has shown that, by switching off the noise decoupler just prior to recording the spectrum, the Overhauser enhancement can be utilized while preserving the spin coupling constants at their correct value.[51] This experiment in its described form is impractical for routine use, but may prove to be useful when combined with a Fourier transform spectrometer. Proton noise decoupling can, of course, help with the study of nuclei other than carbon, for example it has been applied to simplify the ^{19}F spectra of fluoroaromatics,[52] with spectacular results. However, the full consequences of noise decoupling are only just becoming recognized (see p. 269).

Spin decoupling experiments can also be used to determine the relative signs of spin coupling constants.[53] The sign of a coupling constant is of theoretical[54] and occasionally diagnostic[55] importance. Following the work of Dreeskamp,[56] McFarlane,[5] and other workers,[6] the signs of all the common coupling constants are now known. However, a few of the more 'exotic'

---

[50] E. D. Becker, *J. Magn. Resonance*, 1971, **4**, 142.
[51] J. Feeney, D. Shaw, and P. J. S. Pauwels, *Chem. Comm.*, 1970, 554.
[52] M. A. Cooper, H. E. Weber, and S. L. Manatt, *J. Amer. Chem. Soc.*, 1970, **93**, 2369.
[53] J. P. Maher and D. F. Evans, *Proc. Chem. Soc.*, 1961, 208.
[54] M. Barfield and D. M. Grant, *Adv. Magn. Resonance*, 1965, **1**, 149; M. Barfield and B. Chakrabarti, *Chem. Rev.*, 1969, **69**, 757.
[55] H. Gunther, *Z. Naturforsch.*, 1969, **24b**, 680.
[56] C. Schumann and H. Dreeskamp, *J. Magn. Resonance*, 1970, **3**, 204, and references therein.

coupling constants have had their signs determined during the past year, e.g. $^1J(^{31}P^{19}F)$,[57] $^1J(^{125}Te^1H)$,[57] and $^3J(^{15}N^{19}F)$[58] being found negative, and $^1J(^{31}P^{15}N)$[58] being positive.

Generalized Overhauser Effect.—The Overhauser effect as originally defined is the polarization of nuclei which occurs as the result of saturating the electron spins in a metal.[59,60] However, common usage has made the term describe the changes in population which result from internuclear polarization based on relaxation mechanisms (dipolar coupling, etc.). Intensity changes which occur in double-resonance experiments due to scalar coupling effects between connected transitions are said to be caused by 'generalized Overhauser effects'.[4] The intensity changes caused by the latter effect are small and must be detected by comparison with transitions unaffected by the second r.f. field. The generalized Overhauser effect is normally utilized *via* the less tedious INDOR experiment. The advent of pulse techniques in high-resolution n.m.r., with their ability to obtain a complete spectrum in a fraction of a second, has permitted the time scale of n.m.r. studies to be decreased. Freeman has taken advantage of this property to monitor the decay of population changes caused by the generalized Overhauser effect in the CHO proton of acetaldehyde.[61] The quartet was measured at a variable time after the removal of the decoupling field from one half of the methyl doublet signal. The spectra obtained showed the intensities of the quartet relaxing to their equilibrium 1:3:3:1 ratio.

INDOR.—The INDOR experiment was first described by Baker[62] and is performed by recording the intensity of one transition with a non-saturating field of fixed frequency, while a second field of lower power is swept through the spectral region of interest. Changes in intensity occur when the frequency of the second field coincides with the frequency of a line which has a common energy level with the monitored transition. So far, surprisingly, little use has been made of INDOR. Perhaps this is because (until the latest generation of n.m.r. spectrometers appeared) the required electronics have not been available as standard equipment on commercial spectrometers. The uses have so far been confined to the measurement of chemical shifts of 'other nuclei' or to the determination of the signs of coupling constants. The INDOR technique has been reviewed.[63] 1970 has at last seen the emergence of homonuclear INDOR as a powerful technique for the solution of complex molecular structures.

An excellent account of the practice and application of the INDOR technique has been published.[64] The instrumental conditions to be used, and

---

[57] W. McFarlane, J. F. Nixon, and J. R. Swain, *Mol. Phys.*, 1970, **19**, 141.
[58] A. H. Cowley, J. R. Schweiger, and S. L. Manatt, *Chem. Comm.*, 1970, 1491.
[59] A. Overhauser, *Phys. Rev.*, 1953, **89**, 689; *ibid.*, 1953, **92**, 411.
[60] A. Abragam, 'The Principles of Nuclear Magnetism,' O.U.P., 1961, Chap. 9.
[61] R. Freeman, *J. Chem. Phys.*, 1970, **53**, 457.
[62] E. B. Baker, *J. Chem. Phys.*, 1962, **37**, 911.
[63] V. J. Kowalewski, *Progr. N.M.R. Spectroscopy*, 1969, **5**, 1.

Multiple Resonance

analyses of the commonly encountered spin systems, are given, as well as examples of the application of the technique to selected problems. The power of extensive INDOR investigations is well illustrated by the study[65] of the structure of the thermal dimerization product of 11,13-dioxo-12-methyl-12-aza[4,4,3]propellane. The spectrum of the product is exceedingly complex; even at 220 MHz only four protons can be resolved and hence conventional double-resonance techniques cannot be conveniently used to assist in the analysis. The assignment was achieved by monitoring one of the resolved protons and using INDOR to locate a coupled proton hidden in the complexity of the rest of the spectrum. The newly-located proton was then itself used as a probe and by INDOR a further coupled proton was located. Thus by systematic use of INDOR eleven of the twelve skeletal protons were detected. Other examples of the use of INDOR have recently appeared.[66,67]

The original applications of INDOR were of heteronuclear nature, use being made of the high sensitivity *etc.* of the proton to study nuclei whose direct observation was difficult.[5] This use of hetero INDOR continues,[68] *e.g.* the work on $^{13}CH_3{}^{12}CN$.[68a] The use of this technique will probably expand with the ease of such experiments on modern spectrometers and the tendency to convert existing spectrometers for the purpose.[67c] The growth of hetero INDOR will, however, be retarded by the increasing use of pulse spectrometers. The increased sensitivity of this type of spectrometer makes direct observation of the less sensitive nuclei a practical possibility.

Nuclear Overhauser Effect (NOE). The 'Nuclear Overhauser Effect' is distinguished from the 'generalized Overhauser effect' in that (i) it is either inter- or intra-molecular in nature (*i.e.* scalar coupling between the nuclei involved is not essential), and (ii) it involves the high power irradiation of the *total* signal due to one type of nucleus while the resonance of a second type of nucleus is monitored. This effect was first demonstrated[69] on the chloroform proton while irradiating the cyclohexane protons in a mixture of the two compounds. The structural significance of the effect was realised by Anet and Bourn,[70] and its use has increased rapidly since then. The more general and theoretical aspects of the Overhauser effect will be dealt with later.

The applications of NOE are usually to structural problems, being based on its dependence on internuclear distance (r^{-6}). Bell and Saunders[71] have

---

[64] F. W. van Deursen, *Org. Magn. Resonance*, 1971, **3**, 221.
[65] O. Sciacovelli, W. V. Philipsborn, C. Amith, and D. Ginsburg, *Tetrahedron*, 1970, **26**, 4589.
[66] D. H. R. Barton, P. N. Jenkins, R. Letcher, and D. A. Widdowson, *Chem. Comm.*, 1970, 391.
[67] (*a*) R. Burton, L. D. Hall, and P. R. Steiner, *Canad. J. Chem.*, 1970, **48**, 2679; (*b*) G. W. M. Grant and L. D. Hall, *ibid.*, 1970, **48**, 3537; (*c*) R. Burton and L. D. Hall, *ibid.*, 1970, **48**, 59.
[68] (*a*) F. W. Wehrli and W. Simon, *Helv. Chim. Acta*, 1969, **52**, 1749; (*b*) P. J. Banney, D. C. McWilliam, and P. R. Wells, *J. Magn. Resonance*, 1970, **2**, 235.
[69] R. Kaiser, *J. Chem. Phys.*, 1965, **42**, 1838.
[70] F. A. L. Anet and A. J. R. Bourn, *J. Amer. Chem. Soc.*, 1965, **87**, 5250.
[71] R. A. Bell and J. K. Saunders, *Canad. J. Chem.*, 1970, **48**, 1114.

measured and collected a large number of NOE values in compounds where the intermolecular distance can be obtained. Their results exhibit the theoretical dependence of NOE enhancement on the sixth power of the internuclear distance. This paper, as does the paper of Schirmer et al.,[72] gives a good account of the basic theory of NOE and the practical and instrumental aspects of performing NOE experiments. The connection between the Overhauser effect and internuclear distance, which allows the distance ratios to be determined for molecules in solution, has been used to study the problem of the conformation of cycloguanosines. The computed distances, from NOE data, were within 5—10% of the distances measured on molecular models constructed using bond lengths and angles known from X-ray diffraction on similar compounds. It was found to be possible to use the distances thus obtained to distinguish between the various possible conformers that the molecule could take up in solution.[72]

Bell and Saunders[73] have used NOE experiments to study the internal rotation of dimethylformamide; they observed the transfer of NOE enhancement by chemical exchange, thus demonstrating the rotation of the methyl groups about the partial double bond. These workers noted a solvent dependence of the NOE effect in dimethylformamide by comparing their results with those obtained by Anet and Bourn.[70] This solvent dependence was also observed in the detailed study of the relaxation and NOE effects in this compound by Brownstein and Bystrov.[74]

Transient Phenomena.—Several studies involving transient effects in connection with ^{13}C have appeared. Ziessow has applied the Torrey oscillation approach to the measurement of ^{13}C chemical shifts, using ^{1}H[75,76] and ^{19}F,[77] with their superior sensitivity, as the observing nuclei. In this method one line of the observing nucleus is saturated by B_1 while B_2 is swept rapidly through the region of interest. When B_2 passes through a transition with an energy level in common with the line monitored by B_1, the sudden spin pumping induces positive or negative 'Torrey oscillations' in the observed line.

Yonemoto[78] has observed ^{13}C spectra, under field sweep conditions, while irradiating protons with a low-power field. Positive and negative intensity changes in the ^{13}C spectrum of benzene and other simple organic compounds were observed owing to transient dynamic polarization effects. The ^{13}C resonance condition, and the field–frequency ratio necessary to affect the proton resonance involved, are approached simultaneously, and traversed in a

---

[72] R. E. Schirmer, J. H. Noggle, J. P. Davis, and P. A. Hart, *J. Amer. Chem. Soc.*, 1970, **92**, 3266.
[73] J. K. Saunders and R. A. Bell, *Canad. J. Chem.*, 1970, **48**, 512.
[74] S. Brownstein and V. Bystrov, *Canad. J. Chem.*, 1970, **48**, 243.
[75] D. Ziessow and E. Lippert, *Ber. Bunsengesellschaft Phys. Chem.*, 1970, **74**, 568.
[76] D. Ziessow, *J. Chem. Phys.*, 1971, **55**, 984.
[77] D. Ziessow, *Chem. Comm.*, 1971, 463.
[78] T. Yonemoto, *J. Chem. Phys.*, 1971, **54**, 3234.

time short compared with the relevant relaxation times. Rules for predicting whether absorption or emission of energy will occur are given; these depend on the relationship between the two transitions involved.

5 Relaxation Effects in Multiple Resonance

This section will deal with multiple-resonance studies where the main interest or consequence of the study has involved relaxation effects. Such studies fall under two main classifications: those concerned with investigations of, or into, the Overhauser effect, and those concerned with relaxation for its own sake. Not surprisingly, this type of study has become increasingly important, and hence popular, with the upsurge of ^{13}C n.m.r. and the revolution in relaxation measurements brought about by Fourier transform spectrometers.[79]

Relaxation in the Presence of Incoherent Decoupling.—Probably the most important type of multiple-resonance experiment is noise decoupling. The consequences of using decoupling, especially of protons while observing ^{13}C, has come under study in the last year. Noise decoupling has three major effects on the resulting ^{13}C spectrum: first, and most obviously, decoupling takes place and all the lines with multiple structure due to proton–carbon couplings are caused to collapse to singlets; secondly, Overhauser effects causing enhancements in intensity up to a factor of 2 can occur (*i.e.* intensity in the decoupled case is up to three times as great as that in the single-resonance case); lastly, and largely unexplored, changes in T_2^* occur.

The use of proton noise decoupling is normal in ^{13}C n.m.r. and drastically affects the transverse relaxation time T_2^* of any ^{13}C coupled to protons. This effect has been largely ignored to date, since in c.w. experiments the result is simply line broadening. Such additional line broadening goes unnoticed amid all the other line broadening effects which are present under experimental conditions normally used to record ^{13}C spectra. Specifically, the sweep broadening has been tolerated in order to obtain increased sensitivity.

In Fourier spectroscopy the changes of T_2^* produced by this mechanism have very important consequences: hence with the upsurge of Fourier transform spectroscopy they have come in for serious study. The presence of noise decoupling adds a further mechanism to those available for transverse relaxation.[80,81] This effect was observed in an investigation[80] into methods of measuring T_1 by the Fourier transform technique, and was demonstrated experimentally with the spin echoes produced by applying a 180°–τ–90° pulse sequence to a ^{13}C-enriched sample of methyl iodide. Good echoes were obtained when the decoupler was switched off or was in a coherent mode, whereas only negligible refocussing took place when the decoupler was operated in an incoherent mode. The phenomenon was also observed in later

---

[79] D. G. Gillies and D. Shaw, *Ann. Rev. N.M.R. Spectroscopy*, 1972, to be published.
[80] R. Freeman and H. D. W. Hill, *J. Chem. Phys.*, 1971, **54**, 3367.
[81] R. R. Shoup and D. L. Vanderhart, *J. Amer. Chem. Soc.*, 1971, **93**, 2053.

work on $T_{1\rho}$ by Fourier transform techniques, where to obtain satisfactory values of $T_{1\rho}$ either no decoupling or coherent decoupling had to be used while obtaining the free induction decay.[82] T_2 has been measured by conventional spin-echo methods for the methyl ^{13}C in CH_3I and $CH_3CO_2CD_3$, and found[81] to be much shorter than T_1 (3·9 vs. 13·4 and 6·1 vs. 19·2 s, respectively). Such behaviour will occur whenever the lifetime for the spin states of a proton coupled to the ^{13}C is shorter than their mutual coupling ($2\pi J \gg \tau_H^{-1}$). The consequence of this property is that the linewidth of a ^{13}C resonance cannot be less than the natural linewidth of any proton coupled to it.

In the presence of noise decoupling, ^{13}C linewidths are a function of the decoupling power.[80] The B_2 field decreases the lifetimes, τ_H, of the proton spin states and at low r.f. power, where the condition $\tau_H^{-1} \gg 2\pi J$ holds, T_2^* is decreased. However, as B_2 is increased and $\tau_H^{-1} \ll 2\pi J$, the relaxation becomes inefficient, since the protons are completely decoupled, and this contribution to T_2^* is absent.

The transverse relaxation induced by the mechanism described above has specially important consequences in Fourier transform spectroscopy. Multi-pulse techniques, such as DEFT[83] and SEFT,[84] have been proposed† and were predicted to achieve better sensitivity than conventional FT. The increased sensitivity was to be achieved by refocussing the inhomogeneity component of the transverse magnetization after one pulse, in time for the next pulse. For this to be advantageous T_1 must equal T_2, and the total T_2^*, defined by the equation:

$$\frac{1}{T_2^*} = \frac{1}{T_2} + \frac{1}{T_2'} + \frac{1}{T_2^s}$$

(where T_2 is the natural relaxation time, T_2' is the contribution from magnetic field inhomogeneity, and T_2^s is the contribution from the scalar relaxation outlined above), must be dominated by T_2', as only T_2' is refocussable. The multiple pulse sequences are not more efficient than the conventional FT pulse sequence for carbon, as under normal conditions T_2^* is not dominated by T_2' (it may be if very high-power decoupling fields were used, when the contribution from T_2^s would decrease) and in fact conventional FT does refocus itself to quite a marked extent[85] (see Chapter 3, and also reference 79 of this chapter).

Overhauser Studies.—The intensity changes of n.m.r. lines produced during double resonance by the Overhauser effect have their origin in relaxation

† DEFT is Driven Equilibrium Fourier Transform, and SEFT is Spin-Echo Fourier Transform.

[82] R. Freeman and H. D. W. Hill, *J. Chem. Phys.*, 1971, **54**, 3367.
[83] E. D. Becker, J. A. Ferretti, and T. C. Farrar, *J. Amer. Chem. Soc.*, 1969, **91**, 7784.
[84] A. Allerhand, *J. Amer. Chem. Soc.*, 1970, **92**, 4482.
[85] R. Freeman and H. D. W. Hill, *J. Magn. Resonance*, 1971, **4**, 366.

effects, hence the study of the Overhauser effect is closely linked with that of relaxation studies. The generalized Overhauser effects in the AX and AB cases of 2,2-dichloroacetaldehyde and 1-bromo-2-chloroethylene, respectively, have been studied.[86] The B_2 power used was low such that only the populations of various levels were affected, and not the spin energies. Under these conditions the eigenstates of the unperturbed Hamiltonian can be used to solve the equation of motion of the appropriate density matrix, and explicit expressions for the intensities of the double-resonance transitions can be found, for different relaxation mechanisms. The fractional changes of intensity produced by the second r.f. field were found to be sensitive to the relaxation mechanisms considered, and relatively insensitive to the power and frequency offset. The relaxation was found to be dominated by dipole–dipole interactions (90%), with a contribution (10%) from random field effects (cf. the situation in the presence of chemical exchange).

Not surprisingly, great interest has been shown in heteronuclear Overhauser enhancements, particularly when observing ^{13}C and ^{15}N with proton decoupling. Kuhlmann et al.[87] have used the density matrix formalism in a detailed study of the ^{13}C-{H} case. They discuss factors affecting the Overhauser enhancement in terms of competition between the dipole–dipole mechanism and other relaxation processes. The maximum Overhauser effect ($\frac{1}{2}\gamma_H/\gamma_{13C} = 1.98$ for ^{13}C-{H}) is obtained when the relaxation of the observed spin is dominated by dipole–dipole relaxation with the decoupled spin. Using the rigid symmetrical adamantane molecule as an experimental test, they confirmed the theoretical predictions that the Overhauser effect is, if dipole–dipole relaxation dominates, *independent* of the number of directly bonded protons. Deviations from this condition should prove a useful probe for studying anisotropic tumbling in solution. In contrast to the Overhauser effect (under similar limitations on the dominant relaxation mechanisms), T_1 is inversely proportional to the number of bonded protons. (This proportionality has been put to good effect by Allerhand[88] using Fourier transform techniques.[75]) After evaluating the possible relaxation processes, Kuhlmann et al.[87] conclude that, as the enhancement on both the CH and CH$_2$ carbons is the same for adamantane within experimental error (1.50 ± 0.08) and $T_1^{CH}/T_1^{CH_2} = 2$ (20.5 ± 2 and 11.4 ± 1 s), direct dipole–dipole relaxation is the dominant process. The same conclusion has been arrived at from a study of formic acid.[89]

The ^{15}N nucleus is, because of its negative magnetogyric ratio, an interesting system for Overhauser studies. Here the maximum enhancement is -3.93-($\frac{1}{2}\gamma_H/\gamma_{15N}$). Lichter and Roberts have studied [^{15}N]ammonium chloride[90] and found, at high pH, a maximum effect of only 50% theoretical. As the pH is raised the signal first broadens, disappears because of exchange with the

---

[86] A. Kumar and S. L. Gordon, *J. Chem. Phys.*, 1971, **54**, 3207.
[87] K. F. Kuhlmann, D. M. Grant, and R. K. Harris, *J. Chem. Phys.*, 1970, **52**, 3439.
[88] A. Allerhand and D. Doddrell, *J. Amer. Chem. Soc.*, 1971, **93**, 2779.
[89] T. D. Alger, S. W. Collins, and D. M. Grant, *J. Chem. Phys.*, 1971, **54**, 2820.
[90] R. L. Lichter and J. D. Roberts, *J. Amer. Chem. Soc.*, 1971, **93**, 3200.

medium, and reappears as a non-inverted signal. These observations are explained in terms of the lack of dominance of dipole–dipole relaxation; important contributions come from spin–rotation and chemical exchange. The importance of spin–rotation relaxation for ^{15}N has also been reported elsewhere.[91]

The ^{15}N case is interesting, as here a decrease of the Overhauser effect causes the signal to go from negative through zero to positive. Hence it is possible to have a case where, given the correct balance of relaxation processes, a signal present without decoupling can be eliminated by decoupling; 1,4-diaminobutane is such a case.[90]

On the ground of sensitivity, the Overhauser effect is an advantage. However it is variable from nucleus to nucleus within a molecule, and without a detailed knowledge of the relaxation processes present cannot be predicted, hence integrals obtained under multiple-resonance conditions are of doubtful value. Competition from relaxation mechanisms other than intramolecular dipolar (or exchange-modulated scalar) coupling can only, however, lower the Overhauser effect and if another process were made to dominate the relaxation, then the Overhauser effect would be reduced to zero. With zero Overhauser effect, quantitative integrals could be obtained from spectra simplified by decoupling.

La Mar[92] and Natusch[93] have both suggested the use of paramagnetic molecules to provide the dominant relaxation path and hence eliminate the Overhauser effect in ^{13}C. The former has demonstrated the feasibility of the method and the preferability of paramagnetic ions over free radicals. Natusch[93] has analysed the system and shown that the paramagnetic ion preferentially affects T_1, such that a limiting condition is reached before T_2^* is appreciably affected (*i.e.* before line broadening occurs), assuming complete decoupling, *i.e.* $2\gamma_H B_2 \gg \pi J(\text{CH})$, and its consequential line broadening (see the previous section). The effect on the carbon nucleus of paramagnetic relaxation of the proton, transmitted *via* the CH scalar coupling (the so-called 3-spin effect[92,93]) was considered and found to be unimportant.

Further developments along the above lines will obviously be forthcoming and will hopefully lead to quantitative ^{13}C spectra. These spectra would be obtained without the Overhauser enhancement, but with the increasing sensitivity of ^{13}C n.m.r. (owing to the use of Fourier transform techniques), this is of less importance than it would have been a few years ago. If there is a specific interaction of the paramagnetic centre with the solute, *e.g.* a spin label in a biomolecule, then a study of the differential suppression of the Overhauser effect could yield valuable information about such an interaction.[92]

---

[91] E. Lippmaa, T. Saluvere, and S. Laisaar, *Chem. Phys. Letters*, 1971, **11**, 120.
[92] (*a*) G. N. La Mar, *Chem. Phys. Letters*, 1971, **10**, 230; (*b*) G. N. La Mar, *J. Amer. Chem. Soc.*, 1971, **93**, 1040.
[93] D. F. S. Natusch, *J. Amer. Chem. Soc.*, 1971, **93**, 2567.

6 Conclusion

The use of multiple-resonance techniques in high-resolution n.m.r. is rapidly increasing; the days when any multiple-resonance experiment was an achievement in its own right are over. With modern spectrometers virtually any multiple-resonance experiment can be performed as a standard operation. The application of spin decoupling in spectral assignments is now commonplace and its use will become more and more a standard procedure. At last the power of INDOR is being realised and a rapid growth of this technique is to be expected in the near future. In the areas where n.m.r. itself is rapidly expanding, particularly ^{13}C, multiple resonance techniques are virtually essential and their use is taken for granted. During the past year chemists have become increasingly interested in relaxation effects and are beginning to use them routinely as diagnostic aids; this is again particularly true of ^{13}C spectroscopy. Here the changes in relaxation mechanisms brought about by proton decoupling effects can give information about the relaxation of protons coupled to the carbons under study, as well as increasing the basic sensitivity of the experiment. The future developments of n.m.r., from the point of view of both technique and applications, will be closely associated with, if not dependent on, multiple-resonance experiments. The developments of next year are awaited with great interest.

8
Macromolecules and Solids

BY E. G. FINER

1 Introduction

The field covered by this chapter is very wide, and therefore only a selection of the published articles can be reviewed. An attempt has been made to include most of the work which appears important to the Reviewer, although there are probably many gaps (unfortunately sometimes caused by non-availability of journals). The bias in favour of biological macromolecules may partly reflect the Reviewer's own interests, but is probably mainly due to the recent dramatic upsurge in the use of n.m.r. in biological studies, caused to a large extent by improved instrumental techniques (especially those leading to higher sensitivity)—see Chapter 4.

Aspects of the n.m.r. of solids which are relevant to physics but not to chemistry (*e.g.* studies of magnetic properties) have not been covered.

2 N.M.R. Studies of Macromolecules

A. Lipid–Water and Soap–Water Systems.—*Pulse and Wide-line Methods.* At the beginning of the review period, there was some confusion in the literature about the nature and origin of the n.m.r. absorptions given by some aqueous lipid and soap systems, especially those giving lamellar phases. Papers published during this period have helped to clarify the situation. Tiddy[1] published a variable-frequency pulsed n.m.r. study of a lyotropic liquid crystal, the system sodium caprylate–decanol–water, which forms both lamellar and reversed hexagonal phases at the temperatures and mole fractions examined. Earlier workers had postulated that relaxation in similar systems was not simply due to dipole–dipole interactions, but that observed linewidths were due to magnetic field gradients or diffusion through such gradients. Part of the reason for these postulates was an observed dependence of the measured spin–spin relaxation time, T_2, on the pulse spacing in the Carr–Purcell–Meiboom–Gill pulse sequence. Tiddy showed that this was an artefact, the observed change being due to the fact that experimental conditions approach those used for measuring $T_{1\rho}$, the spin–lattice relaxation time in the rotating frame, as the pulse spacing decreases. In effect, the 'T_2' measurement becomes a measurement of a time which is dependent on both T_2

---

[1] G. J. T. Tiddy, *Nature Phys. Sci.*, 1971, **230**, 136.

and $T_{1\rho}$. Measurement with a straight Carr–Purcell sequence gives T_2 directly, and Tiddy's results show that in this case there is no dependence on pulse spacing. Further, Tiddy found that his lamellar system gave a maximum value of T_2 if the lamellae were macroscopically aligned and all oriented at $\theta = 55°$ to the magnetic field (3 $\cos^2\theta = 1$), showing that the relaxation mechanism is at least partly dipolar in origin.

Another study of a soap–water system, this one (potassium laurate–D_2O) forming lamellar and cubic phases, also concluded that the linewidth was dipolar in origin. Charvolin and Rigny[2] measured the free induction decay of the lamellar phase, and found that they could analyse it in terms of three regions — one exponential decay, and two 'solid-like' decays, of the form $\exp(-M_2t^2/2)$, where M_2 is the second moment and t is the time. Thus the continuous wave (c.w.) resonance consists of the superposition of two Gaussian and one Lorentzian curves. The explanation is that the end of the hydrocarbon chain has liquid-like motions, the middle is of intermediate fluidity, and the polar end is rigid. Increasing the temperature causes a decrease in the extent of the rigid region and an increase in that of the 'free' region, as does an increase in the amount of water present. The dipolar origin of the linewidth was shown by a spin-echo experiment, where a $\frac{\pi}{2}-\tau-\frac{\pi}{2}$ sequence with a $\frac{\pi}{2}$ phase shift gave an echo, whereas a $\frac{\pi}{2}-\tau-\pi$ sequence with a $\frac{\pi}{2}$ phase shift did not. The authors also point out that the 'dipolar echo' can lead to non-exponential decay in Carr–Purcell trains. From their measurements, which show that $T_2 \ll$ the spin–lattice relaxation time T_1, and also that T_1 is frequency-independent, it follows that there are two different relaxation mechanisms, one involving rapid motion and one slower motion. The authors deduce that the former motion consists of twisting about the C–C bonds of the hydrocarbon chains (the observation of only a single T_1 means that there is spin diffusion); the latter is two-dimensional diffusion of the molecules in the bilayer. Diffusion is faster in the cubic mesophase, giving a decay in a magnetic field gradient which depends on $\exp(-kt^3)$, where k is a constant.

Further relaxation studies of a lamellar system were reported by Chan et al.,[3] who studied lecithin dispersions. They found a non-exponential free induction decay, with an effective T_2 of 135 μs, corresponding to the bulk of the magnetization of the sample (i.e. there were no other significant decays). The T_1 decay was exponential, giving a value for T_1 of 220 ms. The authors present an interesting method of separating fairly sharp signals from broad lines which manifest themselves as disturbances in the baseline: they obtained a Fourier-transformed spectrum where the receiver was switched off for 250 μs after each pulse, thus filtering out lines of $T_2 < 250$ μs (corresponding to lines of linewidth greater than about 1 kHz). Most of the CH_2 protons did not contribute to this spectrum. In order to test the proposition that the linewidth originates in internal magnetic field gradients, the effective T_2 was measured at 3·76, 8·92, and 14·1 kG (0·376, 0·892, and 1·41 T). The respective relaxation rates

[2] J. Charvolin and P. Rigny, J. Magn. Resonance, 1971, **4**, 40.
[3] S. I. Chan, G. W. Feigenson, and C. H. A. Seiter, Nature, 1971, **231**, 110.

were found to be 6900, 6700, and 7400 ± 500 s^{-1}, which the authors state to be independent of field strength within experimental error. However, work to be published by Finer, Flook, and Hauser shows the rise at 1·41 T to be real, values of the relaxation rate becoming much larger at stronger fields. One conclusion[3] from the data presented by Chan *et al.* is that the ends of the chains are more mobile than the rest of the methylenes; this explains the distribution of T_2's shown by the non-exponential free induction decay and the removal of most CH$_2$ resonances by the Fourier transform filtering experiment. It can also be concluded[3] that spin diffusion exists, explaining the observation of only one T_1. The linewidth arises from residual dipole–dipole interactions.

The role of internal magnetic field gradients in these systems was investigated further by Drakenberg *et al.*[4] These authors studied the effects of magnetic susceptibility anisotropies on the water resonance lineshape in a lamellar bilayer system consisting of water, n-octylamine, and n-octylamine hydrochloride. The water signal is split and broadened, the broadening (about 2 p.p.m.) being proportional to the applied field. This is because magnetic inhomogeneities produce the effect that the field experienced by a water molecule between bilayers which are oriented parallel to the external field is slightly different from that experienced by a water molecule between bilayers oriented perpendicular to the external field. The inhomogenities may be caused both by the different magnetic susceptibilities of water and the amphiphile, and by chemical shift anisotropies. The authors successfully predict lineshapes for spinning and non-spinning samples, assuming a spherical distribution of lamellar directions in the sample.

Another study[5] of a lyotropic system used 50% D$_2$O, 36% sodium decyl sulphate, 7% n-decanol, and 7% sodium sulphate; this forms a viscous isotropic phase between −10 and +9 °C, a nematic phase from 9 to 60 °C, and an isotropic phase above 90 °C. Saturation-recovery and direct methods were used to measure T_1 between 2·5 and 60 MHz. It was found that T_1 increases with frequency, and that there are no discontinuities at the phase boundaries (it has been shown previously that there are no discontinuities for other systems). The T_1 curves can be fitted by two models. The first one involves two correlation processes, one fast one (frequency-independent) and one frequency-dependent one. The fast one can be assigned to CH$_2$ rotation, and the slower to some sort of intermolecular rotational diffusion. The second model assumes that there is only one process, *viz.* translational diffusion (between micelles). Both models account for the lack of dependence of T_1 on phase. The authors point out that the correlation function for molecular motion is not a very sensitive indicator of that motion. Linewidth measurements do show sharp changes at phase boundaries, with a decrease in both the viscous isotropic and fluid isotropic phases, again as has been shown before for similar systems. The authors interpret the line broadening in terms

---

[4] T. Drakenberg, A. Johansson, and S. Forsén, *J. Phys. Chem.*, 1970, **74**, 4528.
[5] L. A. McLachlan, D. F. S. Natusch, and R. H. Newman, *J. Magn. Resonance*, 1971, **4**, 358.

of non-spherical averaging of dipolar interactions, and state that this involves ordering by the external magnetic field. However, it seems to the Reviewer that other ordering processes would be expected to produce the same effect.

McDonald and Peel[6] imposed macroscopic order on mixtures of 1-monooctanoin and water (H_2O or D_2O) by smearing the material between cover slips and making sandwiches fifteen thick to produce an oriented stack. This stack was placed in a 10 mm n.m.r. tube, and the lineshape was recorded as a function of orientation, θ, with respect to the applied magnetic field, using two modulation widths. Two doublets were observed, the narrower one being 3 p.p.m. to low field of the wider one; both doublet splittings fitted well to the expression $K(3 \cos^2\theta - 1)$, the two values of K being 19·1 and 186 μT. The narrow line was assigned to OH protons, and the wider line to chain CH_2 protons. Raising the temperature decreased the linewidth of each component of the OH doublet without altering the splitting, whereas the splitting of the CH_2 doublet decreased. These experiments showed that many hydroxy and methylene groups are fixed, or rotate with axes fixed, perpendicular to the sample plates. The degree of order $S = (3 \cos^2\xi - 1)/2$, where ξ is the angle between individual molecular axes and the overall direction of orientation, was calculated by comparing K for the chains with that calculated for simple rotation about the long axis. The degree of orientation is less than is obtained by magnetic orientation of some nematic liquid crystals. The small value of the splitting observed for the hydroxy resonance shows that there is only a small anisotropy in the motion, and this agrees with results obtained[6] by deuterium resonance.

High-resolution Methods. Oldfield and Chapman[7] reported a ^{13}C pulse Fourier transform (FT) n.m.r. study of lecithins, using 1 kHz bandwidth noise decoupling of protons and a ^{19}F lock. Four to five thousand pulses were required for solution studies (using tubes of diameter 11·6 mm), and twenty to thirty thousand for studies of a dispersion. The paper presents both the absorption spectra and the square root of the power spectra, although presumably no extra information is obtainable from the latter. A chloroform solution of dipalmitoyl lecithin gave a sharp spectrum, as it does in the proton resonance, but only some of the lines from an H_2O dispersion of egg-yolk lecithin were sharp. In particular, the signal from the bulk of the chain methylene carbons was broad. Peak assignments were made using model compounds and intensity measurements. The lack of resolution of chain carbon peaks was ascribed to either intrinsically short spin–spin relaxation times or incomplete irradiation of the proton spectrum. The latter is probably at least a contributing factor, since there is considerable intensity in the proton spectrum well outside a bandwidth of 1 kHz. The conclusion drawn in the paper is that the NMe_3 group, which gives a sharp peak, is mobile; this has also been established by earlier proton studies.

[6] M. P. McDonald and W. E. Peel, *Trans. Faraday Soc.*, 1971, **67**, 890.
[7] E. Oldfield and D. Chapman, *Biochem. Biophys. Res. Comm.*, 1971, **43**, 949.

The broadening produced by paramagnetic species has been used[8] to investigate the structures formed when aqueous dispersions of egg-yolk lecithin are subjected to ultrasonic irradiation. Solutions of manganese sulphate were used to differentiate between the internal and external sides of phospholipid bilayers formed into closed spherical vesicles. If sonication was carried out in the presence of the ions (10^{-3} mol l^{-1}), the water linewidth was broadened to 60 Hz and the NMe_3 peak from 6 to 120 Hz, while the hydrocarbon chain methylene peak was only broadened from 20 to 30 Hz. This fits in with the normally accepted picture of the bilayer, with the hydrocarbon chains in the interior and removed from contact with the water. Addition of the manganese after sonication resulted in the broadening of the signal from only about half the NMe_3 residues, since the rest were on the inside surfaces of the closed vesicles and not accessible to the bulk water. A similar experiment with reversed egg-yolk lecithin micelles in deuteriobenzene, where any water present is held in the polar interior of the micelle, also produced broadening of the water and NMe_3 resonances on the addition of manganese. No effect was observed on the linewidth of the benzene proton resonance (from residual C_6D_5H).

Different causes of broadening in the high-resolution spectrum of sonicated aqueous phospholipid dispersions were investigated by Hauser et al.[9] when they studied the cyclic polypeptide antibiotic alamethicin and its interaction with phospholipids. Alamethicin, which contains eighteen amino-acid residues, complexes ions, but no changes in the spectrum were observed when ions were added. This is in contrast to effects observed with other similar antibiotics which transport ions across phospholipid membranes. Aggregation of the peptide with increasing concentration manifested itself by considerable line broadening, the linewidth $\Delta v_{\frac{1}{2}}$ increasing from 10 to 50 Hz for a typical peak. Even greater effects were observed in the spectra of egg-yolk lecithin and phosphatidylserine when alamethicin was added, the high-resolution spectrum being broadened out until it was lost in the baseline at low lipid: peptide molar ratios. Following this process by integrating the high-resolution spectrum with respect to an external reference showed that the process could be treated as a simple equilibrium between 'complexed' and 'uncomplexed' lipid, one molecule of alamethicin putting six hundred phosphatidylserine molecules into the 'complexed' state (i.e. that state not giving a high-resolution spectrum). The line broadening was interpreted as being due to hydrophobic interactions causing a loss of segmental chain mobility. Other techniques were used to rule out effects such as increased viscosity and slower particle tumbling due to an increase in particle size. X-Ray and electron microscope work showed that n.m.r. was proving to be a sensitive method of following a phase change induced by the peptide.

---

[8] L. D. Bergel'son, L. I. Barsukov, N. I. Dubrovnia, and V. F. Bystrov, *Doklady Akad. Nauk S.S.S.R. Biofiz.*, 1970, **194**, 703.

[9] H. Hauser, E. G. Finer, and D. Chapman, *J. Mol. Biol.*, 1970, **53**, 419.

B. **Polyamino-acids, Peptides, and Proteins.**—*Proton Studies.* A controversy which has been current over the last few years involves the use of n.m.r. to study helix–random-coil transitions in polyamino-acids. In mixed solvents where certain polyamino-acids are partly helical, *e.g.* chloroform–TFA (trifluoracetic acid), two peaks (separated by 0·5 p.p.m. or less) have sometimes been observed in the α-CH region and assigned to helical and random-coil regions respectively. This conflicts with kinetic data, which show that exchange between the two forms is much too rapid to allow separate peaks to be observed. One explanation has been that systems studied have been polydisperse, and the longer chains might have a different secondary structure from that of the shorter chains. Now Tam and Klotz[10] have published an investigation of this problem, studying poly-DL-alanine and poly-L-alanine in mixtures of $CDCl_3$ and TFA. Two α-CH peaks are observed, the high frequency one increasing in intensity and moving to higher frequency as the concentration of TFA is increased. This occurs with both peptides. However, the DL peptide does not form a helix if the D- and L-residues are randomly distributed (although the authors did not check this with their particular samples), and therefore the two peaks cannot be assigned to helix and random-coil forms. Instead, the peaks are both due to random-coil forms, one with protonated and one with unprotonated amide groups. The authors state that their experiments also rule out polydispersity as an explanation of the two peaks, although they do not consider the possibility that ease of protonation could be influenced by chain length. Similar results are quoted from the literature for poly(benzyl glutamate) in the DL form (70% random) and the L form (helical).

A typical study of helix–coil transitions using the original interpretation of the two peaks has been published by Warashina *et al.*[11] These authors studied poly-L-leucine, poly-L-alanine, and poly-L-methionine in chloroform–TFA mixtures. They found that high molecular weight chains gave single peaks for the NH and α-CH resonances, while low molecular weight chains gave two peaks (which they assigned to helical and random-coil conformations). The H–D exchange rates of the NH groups were studied by n.m.r., and were found to obey first-order kinetics for the high molecular weight chains, and slow first-order kinetics for the 'helical' part combined with rapid exchange for the 'random coil' part for the smaller peptides. The authors deduced that the helix–coil interconversion of the low molecular weight polypeptides has a longer relaxation time than that of the high molecular weight peptides. The n.m.r. estimates of % helix were in reasonable, but not excellent, agreement with estimates from o.r.d. studies. If the results of ref. 10 were applied to this work, presumably it would be concluded that the relaxation time being measured is for protonation of the amide group in random-coil forms.

Joubert *et al.*[12] have published a detailed and lengthy investigation of

---

[10] J. W. O. Tam and I. R. Klotz, *J. Amer. Chem. Soc.*, 1971, **93**, 1313.
[11] A. Warashina, T. Iio, and T. Isemura, *Biopolymers*, 1970, **9**, 1445.
[12] F. J. Joubert, N. Lotan, and H. A. Scheraga, *Biochemistry*, 1970, **9**, 2197.

motions and interactions in some water-soluble polymers of the N^5-(ω-hydroxyalkyl)-L-glutamines, where alkyl = Et, Pr^n, or Bu^n, *i.e.* H–[HN–CHR^m–CO]$_n$–OH ($n \sim 20$—750), where R^m = CH$_2$CH$_2$CONH(CH$_2$)$_m$OH, (m = 2, 3, or 4). The helix content can be adjusted by varying the amount of methanol in the solvent; in aqueous solution, the helix content (as measured by o.r.d.) depends on m, n, and the temperature. Those solutions with a high helix content give n.m.r. spectra which show that peaks become broader the nearer the resonating protons are to the (rigid) backbone. The signal from helical α-CH protons is too broad to be observed. Random-coil peptides give sharp spectra, especially at elevated temperatures (*e.g.* 70 °C). Chemical shifts of side-chain resonances are not dependent on the helix content. Changes in the chemical shift of the α-CH with temperature have two origins—changes in the helix content, and changes in the average conformation of the random coil. The former effect amounts to up to 0·10 p.p.m. and the latter 0·03 p.p.m. for a temperature change from 10 to 60 °C. This paper summarizes the confusing situation about whether two α-CH resonances are observed for partly helical systems, and whether, if only one peak is seen, a shift is observed for the peak as the equilibrium is displaced. Like ref. 10, this paper concludes that solvation of the random-coil form by strong acid is the cause if two peaks are seen; α-CH peaks in a helical system give resonances too broad to be observed. In this study, the absence of strong acid means that ionization of side-chains and solvation of peptide bonds are not problems.

Another use of the chemical shift of the α-CH in polyamino-acids was demonstrated by Deber *et al.*,[13] who found evidence for *cis* peptide bonds in proline oligomers. Proline is unusual in that its linear peptides can occur in the *cis*-form, and this paper shows that the α-CH then has a chemical shift 0·25 p.p.m. to higher frequency than in the *trans*-form. This property was used to study relative populations and rates of interconversion in a range of proline oligomers.

Peptide studies provide one of the few examples of the possible use of coupling constants in n.m.r. studies of macromolecules. Relevant work has relied mainly on the dependence of ^3J(HNCH) on the dihedral angle, ϕ, between the α-CH and the peptide NH, this angle giving valuable information about the conformation of the peptide backbone. Gibbons *et al.*[14] have shown that a Karplus-type relationship (as proposed earlier by Bystrov) is approximately correct, by predicting values of this coupling constant for various configurations of a peptide chain. Within an accuracy of ± 1 Hz, their values for L-amino-acids are 2, 8, 9, 10, and 7 Hz for right-handed and left-handed α-helices, anti-parallel and parallel β-pleated sheets, and random coil (full rotational freedom) respectively. The first two values are interchanged for D-amino-acids. Experimental values are 6·5, 3, and 8·5 Hz for random coil,

---

[13] C. M. Deber, F. A. Bovey, J. P. Carver, and E. R. Blout, *J. Amer. Chem. Soc.*, 1970, **92**, 6191.
[14] W. A. Gibbons, G. Némethy, A. Stern, and L. C. Craig, *Proc. Nat. Acad. Sci. U.S.A.*, 1970, **67**, 239.

right-handed α-helix (L-amino-acids) and anti-parallel β-pleated sheet structures respectively. Measurements of this coupling constant therefore provide a good guide for distinguishing between these possible structures. Uncertainties can be reduced by finding which conformations are forbidden on a conformational energy map. Tonelli and Bovey[15] published a related study, assuming that $^3J(\text{HNCH}) = A \cos^2 \phi$, where $A = 8.5$ Hz ($0° \leq \phi \leq 90°$) and 9.5 Hz ($90° \leq \phi \leq 180°$). They used calculated energies (as a function of ϕ) to obtain an average value of the coupling constant, weighting the summation according to the population with a given angle. The result predicts $^3J(\text{HNCH})$ for a random coil to be 6.1 Hz, which is slightly lower than measured values. The earlier Karplus-type angular dependence, proposed by Bystrov *et al.*, gives slightly better results. It may be concluded, however, that these calculations are rather insensitive to the exact form of angular dependence used.

By studying a poly-β-peptide, Glickson and Applequist[16] were able to use $^3J(\text{HCCH})$ to give conformational information. They used poly-β-alanine (in aqueous solution), this being the simplest poly-β-amino-acid, with the structure $(\text{NHCH}_2\text{CH}_2\text{CO})_n$. The simple A_2X_2 spectrum showed that there is rapid internal rotation, and therefore the polypeptide is essentially disordered in aqueous solution. The possibility of the spectrum being deceptively simple was ruled out by a variety of methods.

It is of course necessary to assign the spectra of heteropolypeptides before conformational information can be extracted from the coupling constants, and Brewster and Bovey[17] have published details of the methods they used for a particular nonapeptide, cyclolinopeptide A. Techniques used included double resonance, deuterium exchange of NH protons, temperature dependence of chemical shifts, and energy calculations to distinguish between amino-acids giving a small value of $^3J(\text{HNCH})$.

Coupling constants cannot be determined from the spectra of proteins because the intrinsic linewidths are too great. For useful studies of protein interactions to be made, the sequence and preferably the three-dimensional structure should already be known. One example of a study on a protein of which the full structure cannot be determined by X-ray crystallography is by Boublík *et al.*,[18] who published an investigation of the conformational changes undergone by histone F2b when ionic strength, concentration, and pH are changed. Their method was to study the sequence to find where there is a high concentration of amino-acids which give assignable peaks in the n.m.r. spectrum, *e.g.* aromatic residues; fortunately there is an unequal distribution of residues in histone F2b. The authors followed broadening of the n.m.r. lines as the segmental mobility of various residues was reduced by increasing ionic strength, concentration *etc.* They found that, in all cases, one region of the chain is involved in the induced changes, *viz.* that containing peptides 60—102. In

---

[15] A. E. Tonelli and F. A. Bovey, *Macromolecules*, 1970, **3**, 410.
[16] J. D. Glickson and J. Applequist, *J. Amer. Chem. Soc.*, 1971, **93**, 3276.
[17] A. I. Brewster and F. A. Bovey, *Proc. Nat. Acad. Sci. U.S.A.*, 1971, **68**, 1199.
[18] M. Boublík, E. M. Bradbury, C. Crane-Robinson, and E. W. Jones, *European J. Biochem.*, 1970, **17**, 151.

addition, pH-induced changes involved further residues at each end of the chain. O.r.d. was used to identify in more detail what changes were taking place. Another study[19] not depending on a three-dimensional crystal structure used n.m.r. 'difference spectroscopy' to determine the binding sites of cobalt in the random-coil protein gelatin. The spectra of gelatin in the presence and absence of Co^{2+} ions were subtracted to give a difference spectrum corresponding to the resonances which were grossly shifted by the cobalt. Comparison of this spectrum with one computed by summing the resonances expected from all the aspartic and glutamic acid residues (including a bandshape factor) showed that the cobalt was binding to these residues.

Another example of the use of n.m.r. 'difference spectroscopy' was published by King and Bradbury,[20] who examined the well-characterized protein ribonuclease. After obtaining the spectrum, the pH was altered and the spectrum run again, resulting in a shifting of the peaks from histidine residues; subtracting the two spectra gives the positions of peaks due to these residues. Not only the $C(2)$–H resonances, but the normally unobservable $C(4)$–H resonances were picked up by this technique. Plotting the chemical shifts as a function of pH gave the pK values of the histidines. These were also found by Rüterjans and Pongs,[21] who then observed the changes produced by the addition of guanosine-3′-phosphate to obtain evidence about the involvement of the histidine residues in the enzyme active site. The pK values of the histidine residues of staphylococcal nuclease were the subject of another study,[22] in which the authors used curve-fitting to identify the imidazole $C(2)$–H resonances. Peak crossing as the pH is changed produces difficulties in obtaining plots of chemical shift against pH, and so the spectra were fitted to a series of overlapping Lorentzian curves to produce values of peak areas, chemical shifts, and linewidths. The chemical shifts were then fitted to an equation relating change of chemical shift (δ) to pH:

$$\text{pH} = \text{p}K + \log\left[\frac{\delta_{max}-\delta}{\delta-\delta_{min}}\right]$$

Hence precise pK values were obtained. To fit the spectra from the four histidine residues six curves were used, including one for the baseline and one for an additional spurious peak. Results for the peak areas, which should all be the same, were not very impressive, possibly because the assumption of a Lorentzian peak shape may not be valid. The overall results of this analysis differ from those published earlier, this study showing no evidence for a conformational equilibrium involving histidine. This is because the additional spurious peak was not recognized as such in the earlier work, but was thought to derive from a histidine residue in a different environment.

Most detailed work on protein n.m.r. has involved residues giving reson-

[19] P. I. Rose, *Science*, 1971, **171**, 573.
[20] N. L. R. King and J. H. Bradbury, *Nature*, 1971, **229**, 404.
[21] H. Rüterjans and O. Pongs, *European J. Biochem.*, 1971, **18**, 313.
[22] J. S. Cohen, R. I. Shrager, M. McNeel, and A. N. Schechter, *Nature*, 1970, **228**, 642.

ances well shifted from the main envelope of peaks, especially histidine residues (as is clear from the work described above). The presence of a haem group produces paramagnetic shifts which increase the number of residues giving such resonances. An example of a study using this effect has been given by Patel et al.,[23] who examined myoglobin in H_2O. Comparison of the spectrum with that obtained in D_2O enabled exchangeable hydrogens to be identified. Peaks observed between $\delta10$ and $\delta15$ p.p.m. included the NH resonances of two tryptophans, one arginine, and one histidine. Assignments were based on considerations of the X-ray crystal structure, together with results obtained using chemical modifications of the protein.

[13]C *Studies.* This year has seen the beginning of what will undoubtedly become a popular field, *viz.* [13]C studies of polypeptides and proteins. Gibbons et al.[24] published a study of gramicidin S, a decapeptide which has previously been well studied by proton n.m.r., the latter studies probably providing the closest approach yet to determining the conformation of a protein in solution. The ^{13}C spectrum was obtained by c.w. methods at 25·15 MHz, using noise irradiation of the proton spectrum and spectrum accumulation. Concentrated solutions (300—500 mg in 2 ml of methanol or dimethylsulphoxide) could be employed for this peptide. The compound contains 56 carbon atoms, but a two-fold symmetry axis means that 28 resonances should be observed. In fact 22 were found, demonstrating that there are few superpositions. Assignments were made using spectra of the amino-acids, together with published correlation rules. The c.w. [13]C spectrum of a protein, lysozyme, was obtained by Lauterbur[25] using 25% solutions and 6260 accumulations (taking 87 hours). There was general agreement between the observed spectrum and that obtained by summing the resonances expected from the component amino-acids. Better spectra have been obtained more quickly by the use of FT spectroscopy; Chien and Brandts[26] also studied lysozyme, and accumulated 120 000 transients to obtain the spectrum. The relative intensities in the spectrum differed from those calculated on the basis of the amino-acid composition owing to varying ratios of T_1 to T_2.

Ribonuclease A has also been studied by ^{13}C FT spectroscopy. After 65 000 accumulations, Allerhand et al.[27] were able to detect peaks arising from as few as three equivalent carbon nuclei per molecule. The extra information obtainable by using ^{13}C as opposed to ^{1}H spectra is perhaps less than was originally hoped in the case of proteins, since ribonuclease A contains 575 carbon atoms in 124 amino-acid residues, and yet the ^{13}C spectrum showed a

---

[23] D. J. Patel, L. Kampa, R. G. Shulman, T. Yamane, and B. J. Wyluda, *Proc. Nat. Acad. Sci. U.S.A.*, 1970, **67**, 1109.
[24] W. A. Gibbons, J. A. Sogn, A. Stern, L. C. Craig, and L. F. Johnson, *Nature*, 1970, **227**, 840.
[25] P. C. Lauterbur, *Appl. Spectroscopy*, 1970, **24**, 450.
[26] J. C. W. Chien and J. F. Brandts, *Nature New Biol.*, 1971, **230**, 209.
[27] A. Allerhand, D. W. Cochran, and D. Doddrell, *Proc. Nat. Acad. Sci., U.S.A.*, 1970, **67**, 1093.

maximum of perhaps 35 peaks, many of them overlapping. The spectrum could again be simulated fairly well by using the chemical shifts of the constituent amino-acids. Application of the 'partially relaxed Fourier transform' technique to this spectrum[28] produced the first case where a protein ^{13}C spectrum was used to give new molecular information. Values of T_1 for the different types of carbon atom were obtained. For carbonyl groups of the native and denatured protein, T_1 was 416 and 539 ms respectively; corresponding figures for α-carbons, β-carbons, and ε-carbons were 42 and 120 ms, ~40 and 99 ms, and 330 and 306 ms respectively. For those carbons with directly-bonded protons, the T_1 values are determined by dipole–dipole interactions with the decoupled protons. This enables estimates of the rotational correlation times to be made, giving a figure of about 30 ns for the α-carbons in the native state and about 0·4 ns on denaturation. A corresponding increase in motional freedom is calculated for the side-chain β-carbons.

^{19}F *Studies.* Addition of a label containing ^{19}F as a probe not subject to interference from the majority of the nuclei present has many of the advantages of e.s.r. spin-labelling, and does not suffer from the disadvantage that the label added has to be fairly bulky. Huestis and Raftery[29] have exploited this by labelling ribonuclease S with the trifluoroacetyl group, bonded to lysines 1 and 7. Ribonuclease S consists of two peptides formed by cleavage of ribonuclease A between residues 20 and 21, and retains full enzymatic activity. The short peptide was separated, chemically modified, and recombined with the main part of the protein to produce an enzyme which retained full enzymatic activity, indicating that the presence of the label did not produce gross distortions of the enzyme structure. The three ^{19}F resonances of the peptide, assigned to groups on ε-trifluoroacetyl- and αε-bis(trifluoroacetyl)-lysines, showed chemical shift and linewidth changes on recombination with the S-protein. Addition of an inhibitor also produced shifts. A corresponding change in chemical shift with pH enabled the pK of the group producing the shift to be found; using this and the X-ray three-dimensional structure, it was concluded that inhibitors slightly alter the position of histidine 119.

C. Synthetic Polymers.—The use of n.m.r. in the study of molecular motions in polymers is well established. McCall[30] has published a useful small review summarizing some n.m.r. studies of molecular relaxation mechanisms in polymers. An example of the use of the study of such mechanisms to obtain information about molecular motions is provided by Cuniberti,[31] who measured values of T_1 and T_2 for fractionated samples (with different molecular weights) of poly(dimethylsiloxan), both in the amorphous state and in solution in tetrachloroethylene. The low density and large Si–O bond distance

---

[28] A. Allerhand, D. Doddrell, V. Glushko, D. W. Cochran, E. Wenkert, P. J. Lawson, and F. R. N. Gurd, *J. Amer. Chem. Soc.*, 1971, **93**, 544.
[29] W. H. Huestis and M. A. Raftery, *Biochemistry*, 1971, **10**, 1181.
[30] D. W. McCall, *Accounts Chem. Res.*, 1971, **4**, 223.
[31] C. Cuniberti, *J. Polymer Sci., Part A-2, Polymer Phys.*, 1970, **8**, 2051.

means that the most important relaxation is between protons on the same methyl group. Rotation of the methyl is too fast to provide an efficient relaxation mechanism, and relaxation therefore arises from 'pendant motion' of the methyl with respect to the chain. The paper sets out the relaxation behaviour to be expected for different types of molecules containing such a group, *e.g.* for low-viscosity amorphous polymers $T_2 < T_1$ and there is a possibility of non-exponential relaxation behaviour due to a distribution of rotational correlation times. Approximate calculations of the rotational correlation time of the chain segments for poly(dimethylsiloxan) show that it is roughly proportional to the square root of the viscosity of the liquid. The polymer is very flexible.

Much of the literature published on n.m.r. studies of synthetic polymers is concerned with the determination of stereochemical configurations. This subject has been reviewed by Cheradame.[32] The review discusses some of the practical points involved in n.m.r. studies of polymers, and then considers n.m.r. methods of determining their configuration. Different types of diads, triads, and tetrads are considered, together with determination of isomerization of position and degree of polymerization of monomers. Applications of results are discussed. Dombroski *et al.*[33] have shown how spectra of polymers can be simplified by deuteron substitution. They polymerized stereospecifically deuteriated *cis*- and *trans*-vinyl methyl ether monomers to obtain isotactic and atactic polymers. Chloroform solutions of vinyl methyl ether polymers were studied at 100 MHz, and the OMe and β-CHD resonances were fitted by a computer routine to a number of overlapping peaks, the area of each peak being calculated by the routine. Each peak from a different tetrad sequence had a different chemical shift, which could be assigned. Another study using a computer to resolve overlapping peaks[34] used an analog computer program to resolve up to seven Lorentzian or Gaussian curves, including skew and slanting or curved baselines. The authors studied butadiene–styrene copolymers, and resolved highly overlapping peaks in the styrene aromatic region to obtain information about the length of the styrene blocks.

Katritzky and Smith[35] showed that contact shift reagents produced the same effects on the spectra of poly(ethylene oxide) and poly(methyl methacrylate) as they do with non-polymers. In the case of tris(dipivalomethanato)-europium, the metal atoms are associated with lone pairs on the oxygen atoms of the polymers. The authors were able to resolve three initially overlapping peaks from *C*-methyls in isotactic, heterotactic, and syndiotactic triads in poly(methyl methacrylate), as can be done simply by using benzene as a solvent. Ferguson[36] studied the 220 MHz spectra of several polyethylenes,

[32] H. Cheradame, *Bull. Soc. chim. France*, 1971, 2023.
[33] J. R. Dombroski, A. Sarko, and C. Schuerch, *Macromolecules*, 1971, **4**, 93.
[34] V. D. Mochel and W. E. Claxton, *J. Polymer Sci., Part A-1, Polymer Chem.*, 1971, **9**, 345.
[35] A. R. Katritzky and A. Smith, *Tetrahedron Letters*, 1971, 1765.
[36] R. C. Ferguson, *Macromolecules*, 1971, **4**, 324.

polypropylenes with different degrees of stereoregularity, and ethylene–propylene copolymers. The spectra of the polypropylenes were affected by the different possible tetrad and pentad sequences, and some of the relevant resonances were resolved and identified. Polypropylenes and ethylene–propylene copolymers were also studied by ^{13}C magnetic resonance;[37] different chemical shifts were found for syndiotactic and isotactic carbons, especially the methyl carbons. Carbon-13 n.m.r. was also used to study the stereochemical configurations present in poly(vinyl chloride),[38] using 2,4-dichloropentane and 2,3-dichlorobutane as model systems. The meso and racemic forms of these two models were used to find chemical shift differences: for 2,4-dichloropentane, these differences in chemical shifts amounted to 1·13 p.p.m. for >CHCl, 0·20 p.p.m. for >CH$_2$, and 1·00 p.p.m. for –Me. Differences of up to −1·99 p.p.m. were found for 2,3-dichlorobutane. Thus ^{13}C chemical shifts are very much more sensitive than ^{1}H chemical shifts to stereoisomeric forms. The spectrum of commercial poly(vinyl chloride) was then assigned to various proportions of syndiotactic, heterotactic, and isotactic triads on the basis of the model studies. Finally, Schaefer has used ^{13}C FT n.m.r. spectroscopy to study three different types of polymer: polyelectrolytes,[39] polyacrylonitrile,[40] and some cross-linked polymers.[41] For the last group, sharp resonances were obtained even below the glass transition temperatures.

D. Other Macromolecules.—There is a large background of n.m.r. studies on conformational equilibria in mobile oxygen-containing ring systems, providing information about conformational and configurational equilibria in polysaccharides. Studies on individual sugar rings have often been aimed at determining the relative roles of electrostatic and steric interactions in determining configuration. An example of such work has been reported by Durette and Horton,[42] who studied conformational equilibria of some alkyl tri-O-acetyl- and tri-O-benzoyl-β-D-ribopyranosides in [^{2}H$_6$]acetone solution. They used ^{1}H n.m.r. at 100 MHz to examine the effect of variation of the substituent at C-1 on the conformational populations, obtaining average coupling constants and hence populations, using model compounds to estimate the coupling constants $J_{4,5}$(axial) = 11·1 Hz and $J_{4,5}$(equatorial) = 1·5 Hz. Assuming these values apply to the ribopyranosides within ±0·5 Hz, they were able to find free-energy changes for the equilibria by varying the temperature. The advantage of this method was that no reliance was placed on a Karplus-type equation. Such studies should be applicable to long-chain polysaccharides, since linewidths are often narrow enough for coupling constants

[37] W. O. Crain, A. Zambelli, and J. D. Roberts, *Macromolecules*, 1971, **4**, 330.
[38] C. J. Carman, A. R. Tarpley, and J. H. Goldstein, *J. Amer. Chem. Soc.*, 1971, **93**, 2864.
[39] J. Schaefer, *Macromolecules*, 1971, **4**, 98.
[40] J. Schaefer, *Macromolecules*, 1971, **4**, 105.
[41] J. Schaefer, *Macromolecules*, 1971, **4**, 110.
[42] P. L. Durette and D. Horton, *Carbohydrate Res.*, 1971, **18**, 289.

to be resolvable. Keilich et al.[43] used chemical shifts and coupling constants to study chloroform solutions of amylose and dextran derivatives (methylated, acetylated, and benzoylated), together with the corresponding derivatives of α-D-glucose and β-maltose. These parameters gave information about the configuration and conformation of the monomer units and the type of glycosidic bonding. Whyte and Englar[44] have reported a study of cell-wall glucan from the red alga *Rhodymenia pertusa*. After extracting the glucan (molecular weight 33 000) and forming the permethylated derivative, the authors obtained 100 MHz spectra of [$^2H_6$]benzene–CDCl$_3$ (6:1) solutions. Methylated potato amylose (molecular weight 106 000) was also studied. The anomeric protons give signals to higher frequency than other protons on the sugar rings, and resonate at characteristic chemical shifts: α-linked and β-linked rings produce resonances at $\delta 5.68$ and 4.47 p.p.m. respectively. It was thus possible to establish that the cell-wall glucan is β-(1→4) linked.

N.m.r. studies of nucleic acids have not been as common as studies of other biopolymers, but one example has been reported by Kearns et al.[45] These authors examined the hydrogen-bonded protons of tRNA in water. The non-exchangeable protons in tRNA only give poorly resolved resonances, but the use of H_2O (rather than D_2O) as a solvent enables reasonable signals to be observed from the exchangeable NH ring protons (ca. $\delta 13$ p.p.m.). For purified tRNA$^{phe}_{yeast}$ linewidths of about 50 Hz were obtained. Increasing the temperature leads to line broadening, some peaks broadening faster than others, because the sticking time of the hydrogen-bonded protons is reduced, leading to uncertainty effects on the linewidths.

Finally, Feeney and Roberts[46] have reported a novel technique which should prove useful in n.m.r. studies of a range of biologically important macromolecules, where studies are normally carried out on D_2O solutions. Often resonances near the HDO chemical shift are obscured by the relatively intense peak from residual HDO. Furthermore, exchange with the D_2O means that NH and OH groups in the macromolecules do not give measurably intense proton signals — hence the significant body of work using H_2O solutions reported in this chapter. By irradiating at the frequency of the NH or OH protons, Feeney and Roberts transferred saturation to the HDO resonance *via* the exchanging protons, causing a reduction in signal intensity. Thus nearby resonances could be observed, and in addition the chemical shift of the NH or OH protons was identified.

3 N.M.R. Studies of Solids

The study of molecular motions in solids by c.w. n.m.r. was one of the earliest uses of the technique, and still continues to be useful. Arnold and Eastmond[47]

[43] G. Keilich, E. Seifert, and H. Friebolin, *Org. Magn. Resonance*, 1971, **3**, 31.
[44] J. N. C. Whyte and J. R. Englar, *Canad. J. Chem.*, 1971, **49**, 1302.
[45] D. R. Kearns, D. J. Patel, and R. G. Shulman, *Nature*, 1971, **229**, 338.
[46] J. Feeney and G. C. K. Roberts, *Chem. Comm.*, 1971, 205.
[47] B. Arnold and G. C. Eastmond, *Trans. Faraday Soc.*, 1971, **67**, 772.

Macromolecules and Solids

have studied molecular motions in acrylic and methacrylic acids and barium and lithium methacrylates, measuring second moments as a function of temperature. Their aim was to investigate a possible correlation between the extent of molecular motion in the monomer lattice and the rate of the solid-state polymerization reaction. Only the crystal structure of acrylic acid was known (out of the molecules studied), and so only for that compound could values of the second moment be calculated for various possible molecular motions. The second moment is divided into three contributions: intramolecular, intermolecular up to a cut-off radius, and intermolecular beyond the cut-off radius (integrated over space). The sum was calculated to be 10·23 G^2 (0·1023 mT^2) for a rigid lattice, to be compared with an experimental value of 9·23 ± 0·86 G^2 below −80 °C. Thus the lattice is rigid at low temperatures. Consideration of the expected reduction in second moment for rotation about a single axis led to the conclusion that the mode of motion above −80 °C is oscillation about the crystal *a* axis. These results, together with a qualitative study of the other solids, indicate that there is a correlation between the rate of polymerization and the amount of molecular motion.

In principle, it is possible to obtain more information from a broad-line spectrum than is obtainable from the second moment alone. Freude *et al.*[48] have shown that lineshape analysis is more useful for interpreting the proton resonance spectra of hydroxy-groups on the surfaces of zeolites. The second moment involves only an average distance between nuclei, *i.e.* a statistical distribution of spins on the surface, but a better interpretation of the data invokes the presence of two spin systems. Lineshapes are, of course, commonly used in the analysis of the motions of systems containing more than one type of mobile group. A combination of broad-line n.m.r. and T_1 measurement is a powerful way of studying solid systems, and has been employed by Ripmeester and Dunell[49] to examine molecular motions and phase transitions in a number of alkali-metal stearates and oleates between 77 and 460 K. Relaxation times were measured by adiabatic rapid passage. Pulse methods have been used by Waugh and co-workers for a number of years in the investigation of solids, especially using pulse trains designed to eliminate the effects of dipolar interactions. In the past, compounds with only one type of magnetic isotope (*e.g.* ^1H or ^{19}F) have been used, since the four-pulse train employed only reduces by a factor of $\sqrt{3}$ the secular interactions between unlike spins. However, this problem is overcome[50] by a suitable double irradiation experiment. Without the double irradiation, a mixture of crystalline sodium fluoride, powdered calcium fluoride, and liquid perfluorobenzene gave only two components in the fluorine resonance after Fourier transformation. With irradiation of ^{23}Na, all three resonances were visible, revealing a chemical shift of 114 ± 6 p.p.m. between the two ionic fluorides. The pulse-FT technique has also been used to study chemical shielding anisotropy, $\Delta\sigma$, in solid CS_2 and

---

[48] D. Freude, D. Müller, and H. Schmiedel, *Surface Sci.*, 1971, **25**, 289.
[49] J. A. Ripmeester and B. A. Dunell, *Canad. J. Chem.*, 1971, **49**, 731.
[50] M. Mehring, A. Pines, W.-K. Rhim, and J. S. Waugh, *J. Chem. Phys.*, 1971, **54**, 3239.

CaCO$_3$.[51] Since dipolar coupling between ^{13}C nuclei is very small, the ^{13}C nucleus is useful for studying $\Delta\sigma$ in solids. An enriched (60%) sample of ^{13}CS$_2$ was examined at about 100 K, to determine whether spin relaxation *via* $\Delta\sigma$ explains the unusually short T_1 in liquid CS$_2$. The spectrum obtained was a typical powder pattern of the type expected from a compound with axial symmetry, and values of $\sigma_\parallel = 285 \pm 10$ p.p.m. and $\sigma_\perp = 140 \pm 6$ p.p.m. were found (referred to σ_{iso}, the room temperature isotropic chemical shift). Thus $\Delta\sigma$ is 425 p.p.m., agreeing very well with predictions made on the basis of the short T_1 in the liquid. The natural dipolar linewidth is only about 0·6 G(0·06 mT). For the study of CaCO$_3$, the sample used was not enriched. A modified driven-equilibrium FT technique was used, and values of $\sigma_\parallel = 10 \pm 3$ and $\sigma_\perp = -66 \pm 3$ p.p.m. with respect to a benzene reference were obtained, giving $\Delta\sigma = 76$ p.p.m.

Values of T_1 in solids are often determined by spin diffusion to a relaxation sink, which is frequently a reorienting methyl group. Measurement of T_1 over a temperature range then gives the activation barrier for rotation of the methyl group. Van Putte and Egmond[52] have carried out an experimental check that such reorientation is indeed the dominant relaxation mechanism in lithium heptadecanoate, and that other mechanisms (*e.g.* other motions, paramagnetic impurities) are not important. They measured T_1 for MeC$_{15}$-H$_{30}$CO$_2$Li and obtained highly exponential recoveries, tending to rule out intermolecular spin–spin interactions and effects of cross-correlation in the methyl groups, their absence probably being a result of a very efficient spin diffusion process. On replacing the terminal Me by CD$_3$, T_1 recoveries were less exponential because spin-diffusion pathways are longer (only very few Me groups were left). An increase in the value of the T_1 at the minimum (163 K) by a factor of fifteen on deuteriation shows that Me reorientation is the dominant relaxation process, and the authors were able to conclude that other mechanisms contribute by less than 2% to the relaxation process in the undeuteriated compound. At higher temperatures, other mechanisms (CH$_2$ oscillations and relaxation *via* paramagnetic impurities) become important.

Andrew and Jasinki[53] have described the effect of rapid macroscopic rotation of solids about the 'magic angle' when the n.m.r. spectrum of the solid is already partly narrowed by internal motion. There are two cases of such narrowing: (*a*) rapid but restricted anisotropic molecular motion reduces the linewidth to a smaller plateau value, and (*b*) isotropic molecular motion narrows the line incompletely because it is not fast enough. The method developed by Andrew *et al.* involves macroscopic rotation at an angle sec^{-1} $\sqrt{3}$ with the magnetic field and a speed ω_r at least comparable with the spectral width of the static solid. Satellites at $\pm\omega_r$ are removed from the central resonance, and reduced in intensity as ω_r is increased. In case (*a*), macroscopic rotation can accomplish further substantial narrowing, since the narrowing

---

[51] A. Pines, W.-K. Rhim, and J. S. Waugh, *J. Chem. Phys.*, 1971, **54**, 5438.
[52] K. van Putte and G. J. N. Egmond, *J. Magn. Resonance*, 1971, **4**, 236.
[53] E. R. Andrew and A. Jasinki, *J. Phys. (C)*, 1971, **4**, 391.

is effectively divisible into two parts, one produced by molecular motion and one by the macroscopic rotation. In case (b), there is only one narrowing mechanism, and to produce further narrowing requires motion faster than the molecular rotation. Similar conclusions have been reached by Doskočilová and Schneider[54] on the basis of an experimental study. These authors used polymers with a relatively narrow static linewidth. Defining $2\Delta v_m$ as the peak-to-peak distance in the derivative spectrum, samples of polystyrene were spun at speeds v_r having values between 0.45 and $0.69\Delta v_m$ ($2\Delta v_m = 28.6$ kHz). The expected narrowing and production of side-bands was obtained. A similar result was obtained with poly(methyl methacrylate), with $2\Delta v_m = 15.0$ kHz and $v_r = 0.6$ to $1.1\Delta v_m$. However, no narrowing was obtained with a sample of polyisobutene, with $2\Delta v_m = 1.6$ kHz and $v_r = 6\Delta v_m$. In the latter sample, the molecular reorientation process responsible for producing a relatively narrow line is not fast enough to produce complete narrowing, but is presumably isotropic [case (b)]. Samples of liquids absorbed in a solid matrix were also studied; ethanol soaked in diatomaceous earth gave a spectrum of linewidth 300 Hz, and rotation of the sample perpendicular to the applied field ($v_r = 4$ kHz) reduced this to 200 Hz. Magic angle rotation produced a high-resolution spectrum identical with that from pure ethanol.

Neu et al.[55] studied solid t-butanol and t-amyl alcohol by high-resolution n.m.r., finding that some of the molecules gave a high-resolution signal. That of the t-butanol disappeared with time, and may have been due to the presence of impurities. That of the t-amyl alcohol was time-independent, the proportion of molecules contributing to the high-resolution spectrum depending on the temperature (e.g. 0.2% at -50, 3% at -15, 100% at the melting point, -8 °C). Addition of impurities increased this proportion. However, the mobile fraction was not simply due to the presence of impurities, since a high-resolution spectrum was still observed at -55 °C, and a eutectic mixture would not have so low a melting point. No hysteresis was observed on cooling and re-heating, which indicated that the presence of interstitial liquid was not the explanation. The authors finally were unable to explain their observation in terms of a single phenomenon.

4 N.M.R. Studies of Small Molecules Interacting with Macromolecules or Solids

A. Bound Water.—There is some disagreement in the literature about the interpretation of the n.m.r. spectra given by H_2O and D_2O hydrating collagen fibres. This has been discussed by Fung and Trautmann.[56] The spectrum consists of a doublet plus a central singlet. One theory holds that the adsorbed water is undergoing anisotropic motion such that the time-averaged orientation of the interhydrogen vectors has a greater probability of being parallel

---

[54] D. Doskočilová and B. Schneider, *Chem. Phys. Letters*, 1970, **6**, 381.
[55] J.-M. Neu, D. Canet, and P. Granger, *J. Chim. phys.*, 1971, **68**, 475.
[56] B. M. Fung and P. Trautmann, *Biopolymers*, 1971, **10**, 391.

with than perpendicular to the collagen chain axis; another holds that water molecules are diffusing rapidly between collagen fibres, and the motions are slightly anisotropic because some hydrogen-bonded states of water structures in the pores between strands are more likely than others. The difficulty in the former theory lies in explaining why the splittings are only 4% of their rigid lattice values, and why they change when the water content is altered. Fung and Trautmann also find difficulties with the latter theory, because it is difficult to accept as a physical picture. They suggest that the observed phenomena can be explained by postulating that part of the adsorbed water is bound to the collagen triple helix, while the rest is rotating, and that there is rapid exchange between the two forms. An excess of water on the surface of the fibres cannot exchange rapidly with water inside the sample, and gives rise to the central peak. This apparently reasonable postulate is well justified by order-of-magnitude calculations. The authors also study the effect of ions on the system, using both n.m.r. and e.s.r.

Khanagov[57] has also discussed the interpretation of the spectra of water bound to oriented polymers, including collagen, DNA, and cellulose. The doublet splitting is proportional to $(3 \cos^2\theta - 1)$, where θ is the angle between the fibre axis and the applied field. Anisotropic motions of the water molecules which give rise to this splitting can easily be simulated by a model which gives the correct angular dependence, but the model cannot be based on the molecules having fixed centres. This is because the splitting would then be larger, and would not decrease with increasing water content (in agreement with the ideas of Fung and Trautmann[56]). Khanagov gives a model involving diffusion between sites; the more symmetrical the 'figure of movements', the smaller is the n.m.r. splitting. In fact, the observed value of the splitting is so small that only small distortions from 'ideal' symmetry are required. Small oscillations of water molecules in a tetrahedrally symmetrical ice-like structure are sufficient to produce the observed n.m.r. spectra.

Lynch and Haly[58] have also considered this problem, this time by studying water bound to rhinoceros horn keratin, which provides a solid mass of aligned fibrous material (it should be noted that the legends to the figures in their paper have been interchanged). They measured T_2 of the water as a function of angle and found that it varies approximately as $(3 \cos^2\theta - 1)^{-1}$ (which is plotted incorrectly in their paper). The authors concluded that there is a slight tendency for the water molecules absorbed by the keratin to rotate preferentially about axes approximately parallel to the fibre direction, but were not able to distinguish between the detailed models mentioned above. Dehl[59] has further studied bound water in these systems, examining the deuterium spectrum of D_2O in kangaroo tail tendon collagen parallel fibres. He found the normal angular dependence. About 2·6 molecules of D_2O per collagen residue do not freeze, even at -50 °C, an estimate obtained both by

---

[57] A. A. Khanagov, *Biopolymers*, 1971, **10**, 789.
[58] L. J. Lynch and A. R. Haly, *Kolloid-Z.*, 1970, **239**, 581.
[59] R. E. Dehl, *Science*, 1970, **170**, 738.

calorimetry and by assuming that the observed splitting is a weighted average between that of ice and that of mobile D_2O.

Blinc et al.[60] have used the field-gradient spin-echo method to measure translational self-diffusion coefficients of water molecules in the water channels of lyotropic liquid crystals of sodium palmitate. They found that diffusion was rapid, dropping only at the mesomorphic–isotropic phase boundary. This drop is due to all the water being associated with the sodium palmitate in the isotropic phase. The ordering of the water molecules was studied by measuring the deuterium quadrupole splitting for D_2O; this was near zero in the gel and isotropic phases, but increased in the lamellar and hexagonal phases. Even at its largest, about 500 Hz, the splitting is so small compared to the value for ice (about 200 kHz) that the anisotropy of motion can only be slight. Hansen[61] also studied diffusion of water (D_2O), in muscle and brain tissue. By measuring the diffusion coefficient as a function of pulse separation, he demonstrated that the diffusion is restricted, the average distance between water molecules and the constrictive barriers being less than 10 μm. He also measured T_1 and T_2, finding exponential recovery curves and T_2 values of 18 and 33 ms in muscle and brain respectively. Values of T_1 were somewhat higher. Hansen was unable to detect any lines with very short values of T_2. His conclusion was that either the water is in only one type of environment, or there is rapid exchange between sites. Carr–Purcell–Meiboom–Gill measurements of T_2 gave longer values than $\frac{\pi}{2}$-τ-π measurements, and Hansen concluded that there is diffusion across local field gradients (but see the discussion of this point earlier in this Report).

Glasel[62] studied relaxation of water in a suspension of glass beads, mistakenly stating that this is directly relevant to previous papers on relaxation in heterogeneous lipid–water and soap–water systems, these papers having studied the amphiphile molecules rather than the water. He packed an n.m.r. tube with glass spheres down to 15 μm diameter, and measured the proton and deuterium relaxation rates of added H_2O or D_2O. Plotting relaxation rates against sphere diameter showed an increase with decreasing diameter (increasing surface area); D_2O gave much less dependence than H_2O, and T_2 was shorter than T_1 for H_2O. Mechanical trapping of molecules greatly reduces the effective self-diffusion coefficient, resulting in a large intermolecular relaxation for protons. Glasel points out that proton data cannot be relied on for information about intramolecular effects.

Kuntz has published a series of studies in which peptides and proteins are examined *via* their effects on unfreezeable water in aqueous solutions. A study of the effect of urea on protein hydration[63] reports changes in the linewidth and intensity of a narrow signal, present at -25 °C in a frozen protein solution and ascribed to bound water. Addition of urea broadens this line, *e.g.* 8M

---

[60] R. Blinc, K. Easwaran, J. Pirš, M. Volfan, and I. Zupančič, *Phys. Rev. Letters*, 1970, **25**, 1327.
[61] J. R. Hansen, *Biochim. Biophys. Acta*, 1971, **230**, 482.
[62] J. A. Glasel, *Nature*, 1970, **227**, 704.
[63] I. D. Kuntz and T. S. Brassfield, *Arch. Biochem. Biophys.*, 1971, **142**, 660.

urea added to 10% bovine serum albumin in 0·01M-KCl solution and rapidly frozen to -25 °C causes an increase in $\Delta v_{\frac{1}{2}}$ from 550 to 1600 Hz. This is ascribed to increased water binding by the denatured protein. Other possible origins of the n.m.r. signal are discussed and shown to be improbable. Another study on a range of polyamino-acids[64] showed that the amount of bound water is not simply related to the polypeptide conformation. Basic and acidic amino-acid residues, together with some other hydrophilic residues, are hydrated more than the less polar residues at -35 °C. Comparison of measured hydration of some proteins with that calculated by summing the hydrations expected from the constituent residues shows that the observed value is always slightly smaller, because some of the residues are buried in the interior of the protein. The difference is removed by denaturation. Examination of the linewidth of the residual water signal[65] shows that systems which are random coils at room temperature give sharper lines than those which are in a helical form. Changes occur over a pH range similar to that observed at room temperature for the helix → coil transition. Kuntz therefore concludes that the rapid freezing procedure traps the solution conformation of the polypeptides.

B. Bound Ions.—Studies of bound ions have been both direct and indirect, the latter class often involving studies of relaxation of the water in aqueous solutions of paramagnetic ions plus a macromolecule. Cohn and Reuben[66] have reviewed the use of paramagnetic probes in n.m.r. and e.s.r. studies of phosphoryl transfer enzymes. About 100 such enzymes are known; all require a bivalent metal cation, and all can use Mn^{2+}. Typical studies have determined binding constants, the number of binding sites, the environment of the metal ion, and the structure and configuration at the active site. Dwek et al.[67] have published a useful review of the theoretical background to studies of proton relaxation enhancement by lanthanide cations, these ions being useful because e.g. Eu^{II} is chemically very similar to Ca^{II}, an ion which is of great importance in biological systems. The relaxation of the water of hydration of the ion is determined by the electron spin relaxation time, the lifetime of the bound state, and the rotational correlation time of the solvated ion. Two cases can be distinguished, exchange between bound and bulk water being fast in one case and slow in the other. Measuring the relaxation rates of the water in the presence and absence of a macromolecule which binds the ion gives the dissociation constant. The authors studied the interaction of Gd^{III} and lysozyme, using two different frequencies of measurement to aid determination of the rotational correlation time and energy of activation at 25 °C. Rotation provides the dominant relaxation mechanism in this case.

[64] I. D. Kuntz, *J. Amer. Chem. Soc.*, 1971, **93**, 514.
[65] I. D. Kuntz, *J. Amer. Chem. Soc.*, 1971, **93**, 516.
[66] M. Cohn and J. Reuben, *Accounts Chem. Res.*, 1971, **4**, 214.
[67] R. A. Dwek, R. E. Richards, K. G. Morallee, E. Nieboer, R. J. P. Williams, and A. V. Xavier, *European J. Biochem*, 1971, **21**, 204.

An example of work using Mn^{2+} is a paper by Navon,[68] who studied proton relaxation in aqueous solutions of Mn^{2+} with carboxypeptidase A and pyruvate kinase. In each case, there is a direct binding of substrate and inhibitors to the metal ion. Examination of the frequency dependence of the relaxation times T_1 and T_2 enabled the data to be fitted to equations allowing for the different sources of relaxation, and led to hydration numbers of 1.08 ± 0.1 and 2.04 ± 0.2 for manganese–carboxypeptidase A and manganese–pyruvate kinase, respectively.

Another indirect study of metal binding has used ^{35}Cl as a probe.[69] The quadrupolar ion is bound to the metal ion investigated, and studying ^{35}Cl linewidths at 5.88 MHz gave information about the binding of a wide range of metal ions to bovine serum albumin. Since the linewidth from bound chloride is about 10^5 times that of aqueous Cl^-, owing to an increased electric field gradient and a longer rotational correlation time, a high concentration of metal–protein complex is not required. Typical linewidths obtained were of the order of 100 Hz. Replacement of bound Cl^- by CN^- enabled the number of bound chloride ions per metal ion to be found. The number of metal ions bound per protein molecule was found from plots of linewidth against molar ratio of protein to metal ion.

Haynes *et al.*[70] have made a direct study of $^{23}Na^+$ complexing by ionophores which act as mobile carriers of alkali-metal ions in membranes (*e.g.* valinomycin, monactin). They measured linewidths and chemical shifts at 15.9 MHz, determining quadrupolar coupling constants by calculating rotational correlation times from the simple Debye equation. The estimated error in this procedure was $\pm 30\%$, which may be rather optimistic. The order of electric field gradients found agrees with that predicted from space-filling models and published crystal structures. A good correlation was obtained between the chemical shifts and the complex stability constants, indicating that ion–(ionophore oxygen) interactions determine the stabilities of the complexes. Most complexes showed rapid exchange between free and bound sodium.

C. Other Small Molecules.—Hammes and Tallman[71] have followed the interaction of L-epinephrine (adrenalin) with sonicated lecithin particles in D_2O. Measurements of T_2, oddly 'defined' in this paper as $2/\Delta v_{\frac{1}{2}}$ s, were made from linewidths, ignoring possible changes in chemical shifts of protons giving overlapping peaks. By following T_2 as a function of concentration, the changes being due to an increase in correlation time of the molecules bound to the phospholipid particles, an association constant was estimated. Deductions about binding sites, based on the doubtful presence of differential line broadening in the spectrum, are probably suspect. Fung and Sarney[72] have studied

[68] G. Navon, *Chem. Phys. Letters*, 1970, **7**, 390.
[69] J. L. Sudmeier and J. J. Pesek, *Analyt. Biochem.*, 1971, **41**, 39.
[70] D. H. Haynes, B. C. Pressman, and A. Kowalsky, *Biochemistry*, 1971, **10**, 852.
[71] G. G. Hammes and D. E. Tallman, *Biochim. Biophys. Acta*, 1971, **233**, 17.
[72] B. M. Fung and S. G. Sarney, *Biochim. Biophys. Acta*, 1971, **237**, 135.

the interaction of urea and guanidine hydrochloride with albumins by following the ^{14}N linewidths in the spectra of the denaturing agents. A 30% increase in linewidth was found on increasing the concentration of bovine serum albumin to 20 mg ml^{-1}; viscosity effects were ruled out by the lack of a corresponding increase with denaturant concentration or in the presence of increasing concentrations of polylysine. It was concluded that the denaturants bind weakly to the protein.

Several studies of complexes of small molecules (*e.g.* inhibitors) with enzymes have been reported in the past. Gerig and Rimerman[73] have examined the *N*-formyl-L-tryptophan–chymotrypsin complex in D$_2$O solution, monitoring chemical shifts in the formyl tryptophan spectrum as a function of added enzyme, to give the chemical shifts in the complex. Only the aromatic protons showed any significant shift changes, resonating at lower frequency in the complexed state. There was no discrepancy between calculated ring-current shifts, based on the *X*-ray crystal structure of the complex, and the observed shifts; hence the structure of the complex in solution is probably the same as in the solid state. Lanir and Navon[74] have studied the binding of sulphonamides to zinc bovine carbonic anhydrase. Line broadening at two frequencies was followed as a function of temperature and concentration; the binding constants obtained from these data agreed well with those obtained from studies of the enzymic activity. The rotational correlation times of aromatic ring protons in the inhibitors were the same as those of the protein, while those of methyl groups were much faster. Examination of the frequency dependence of T_2 led to the conclusion that the relaxation mechanism of the methyl protons was intramolecular, while that of the aromatic protons mainly involved close protons on the protein molecule.

[73] J. T. Gerig and R. A. Rimerman, *Biochem. Biophys. Res. Comm.*, 1970, **40**, 1149.
[74] A. Lanir and G. Navon, *Biochemistry*, 1971, **10**, 1024.

9
Medium Effects

BY M. I. FOREMAN

1 Introduction

Whilst the literature covered by this article spans primarily the period July 1970 to June 1971, earlier work has also been included in the relatively few cases where this seemed especially desirable.

The term 'Medium Effects' has, for the present purposes, been rather freely interpreted to include the effects of dissolved paramagnetic metal complexes on the nuclear resonance positions of other solute material, a topic which is currently exciting great interest; otherwise the discussion has been restricted to effects due to hydrogen-bonding, ion-pairing, and aromatic solvents.

2 Hydrogen-bonding Effects

A. Proton Shifts.—Protons are generally deshielded to some degree on forming a hydrogen-bond to some other atom. Consequently, protons bound to electronegative atoms such as oxygen or nitrogen may exhibit a marked solvent-, concentration-, or temperature-dependence due to solute–solvent hydrogen-bonding interactions, or to disruption by the solvent of solute–solute self-associated species. The hydroxyl protons of polyhydric alcohols and phenols in a molten eutectic mixture of alkali-metal acetates, for example, are considerably deshielded relative to their normal resonance positions.[1] The effect is probably due to strong solute–solvent hydrogen-bonding, rather than to the electrostatic fields prevailing within the medium, since the resonance positions of protons within the hydrocarbon fragments remain unchanged.[1]

Increased shielding of the N-H proton resonance of N-methylpropionamide has been reported[2] on dilution with water, implying a net reduction in the extent of hydrogen-bonded self-association. Conversely, addition of the amide to water leads to deshielding of the nucleus, due to initial strong reinforcement of the water structure about the amide molecule up to 0.85 mole fraction of water. Thereafter the trend is reversed. Shielding of the CH_3, CH_2, and N-CH_3 protons occurring at 0.85 mole fraction of water is suggested to be

[1] L. L. Burton, S. Sherer, and E. R. VanArtsdalen, *J. Phys. Chem.*, 1971, **75**, 1338.

evidence of possible clathrate hydrate formation.[2] The methyl resonance positions for NN-dimethylamides are also susceptible to the effects of changes in solvent and temperature because of hydrogen-bonding.[3] The behaviour of the chemical shifts is reported in detail for a wide range of solvents.[3] Raynes and Raza believe the observed magnetic non-equivalence of the methyl groups of dimethylformamide to be wholly due to the effect of the solvent; other factors affecting the environment of the methyl groups being mutually self-cancelling.[4]

Graham and Chang[5] have attempted a quantitative approach to the study of amide chain-association in solution, based on measurement of the shift of the N-H proton resonance of N-monosubstituted amides. Chain-associated species are disrupted by dilution with carbon tetrachloride, thereby increasing the shielding of the N-H proton. Dimerization is considered as one equilibrium:

$$\text{monomer} + \text{monomer} \underset{}{\overset{K_{1,2}}{\rightleftharpoons}} \text{dimer}$$

All further chain-lengthening steps are then considered as one generalized process:

$$\text{monomer} + n\text{-mer} \underset{}{\overset{\bar{K}}{\rightleftharpoons}} (n+1)\text{-mer} \quad (n \geqslant 2)$$

Using expression (1), relating the observed shift (δ_0) to the monomer shift (v_m) and hydrogen-bond shift (v_D), it is possible to evaluate $K_{1,2}$ and $\bar{K}$ via the ratio $\bar{Y}/Y$ (see ref. 6). $\bar{K}$ proved to be ten times as great as $K_{1,2}$, suggesting a lower free-energy change for formation of chain aggregates over that for formation of the dimer.[5]

$$\delta_0 = \frac{\bar{Y}}{Y}(v_m - v_D) + v_D \qquad (1)$$

The extent of hydrogen-bonding, or 'structuredness', of pure liquid water can be modified either thermally or by the addition of ionic species. A quantitative measure of the effect of ionic material may be obtained by combining the two phenomena in the notion of a 'structure temperature',[7] being that temperature at which pure water has the same internal hydrogen-bonded structure as does the given solution. For example, at 20 °C a 1·0-molar aqueous solution of $MgSO_4$ has a structure temperature of 11 °C;[8] the structuredness of the solution is therefore greater than for pure water at 20 °C. For KNO_3 the reverse is the case. The position of the water resonance is affected by

---

[2] J. F. Hinton and C. E. Westerman, *Spectrochim. Acta*, 1970, **26A**, 1387.
[3] A. Calzolari, F. Conti, and C. Franconi, *J. Chem. Soc.*, (B), 1970, 555.
[4] W. T. Raynes and M. A. Raza, *Mol. Phys.*, 1971, **20**, 339.
[5] L. L. Graham and C. Y. Chang, *J. Phys. Chem.*, 1971, **75**, 776.
[6] L. A. LaPlanche, H. B. Thompson, and M. T. Rogers, *J. Phys. Chem.*, 1965, **69**, 1482.
[7] R. H. Fowler and J. D. Bernal, *Trans. Faraday Soc.*, 1933, **29**, 1049.
[8] W. Luck, *Ber. Bunsengesellschaft. Phys. Chem.*, 1965, **69**, 69.

temperature and by addition of ionic species, and it has been suggested[9] that the structure temperature of a solution may be simply determined by observing at what temperature the chemical shift of pure liquid water is the same as that of the given solution.

Such an approach seems likely to be valid only if there is a simple direct relationship between the extent of hydrogen-bonding and the proton chemical shift. It is, however, thought that thermal changes in the degree of excitation of the hydrogen-bond stretching frequency can have a marked effect on the observed shift.[10,11] Nonetheless, values obtained by this method agree reasonably well with results from i.r. work.[8]

B. Heteronuclear Shifts.—The factors contributing to the observed chemical shift of a proton in a given hydrogen-bonding situation are generally rather complex.[12] Furthermore, it is known that the hydroxyl proton resonance position of acetic acid is markedly sensitive to the presence of traces of water,[13] as also is the hydroxyl proton resonance of phenol at high dilution in electron-donating solvents.[14] There has therefore been an increased interest in the effect of hydrogen-bond formation on nuclei other than hydrogen. In such cases, shift perturbations should be more indicative of electron redistribution at the appropriate nucleus (and therefore a potentially better probe of the nature of the interaction) and also less sensitive to small amounts of impurities. Nuclei which are particularly amenable to study are those of ^{17}O, ^{14}N, and ^{15}N.

Reuben[15] reports that, relative to the vapour shift, the ^{17}O resonance of liquid $H_2^{17}O$ at 24 °C is deshielded by 36 p.p.m., and at infinite dilution in dioxan by 18 p.p.m. Distinguishing between 'H-bonds', formed by proton donation, and 'O-bonds', formed by donation of the oxygen electron pairs, and considering only the following processes for the dioxan–water system,[16]

$$H—O—H + \text{dioxan} \rightleftharpoons H—O—H \cdots \text{dioxan}$$

$$\text{dioxan} + H—O—H \cdots \text{dioxan} \rightleftharpoons \text{dioxan} \cdots H—O—H \cdots \text{dioxan}$$

Reuben argues that the observed 18 p.p.m. shift results from the formation on average of 3/2 H-bonds per $H_2^{17}O$. Each H-bond therefore contributes 12 p.p.m. to the observed shift. Combining these figures, and taking account of the fact that in liquid water each molecule forms two H-bonds and two

---

[9] N. D. Milovidova, B. M. Moiseev, and L. I. Fedorov, *Zhur. strukt. Khim.*, 1970, **11**, 136.
[10] N. Muller and R. C. Reiter, *J. Chem. Phys.*, 1965, **42**, 3265.
[11] M. Alei, jun., and A. E. Florin, *J. Phys. Chem.*, 1969, **73**, 863.
[12] A. D. Buckingham, T. S. Schaefer, and W. G. Schneider, *J. Chem. Phys.*, 1960, **32**, 1227.
[13] N. Muller and P. I. Rose, *J. Amer. Chem. Soc.*, 1963, **85**, 2173.
[14] N. N. Shapet'ko, L. K. Vasyanina, and D. N. Shigorin, *Zhur. strukt. Khim.*, 1970, **11**, 540.
[15] J. Reuben, *J. Amer. Chem. Soc.*, 1969, **91**, 5725.
[16] N. Muller and P. Simon, *J. Phys. Chem.*, 1967, **71**, 568.

O-bonds, leads to a value of 6 p.p.m. per O-bond. The conclusion is therefore that proton donation contributes most to the observed ^{17}O shift.

Alei and Florin[11] arrive at the opposite view by a slightly more broadly-based argument. The dilution curves for the ^{17}O resonance of $H_2{}^{17}O$ in ammonia and trimethylamine are quite distinct, whereas those for acetone and trimethylamine are similar. In the case of the former solvents the different effect that each has on the ^{17}O shift is attributed to the different extent to which the solvent protons can accept electrons, since each of the solvents has a similar capacity to accept protons. Conversely, the protons of trimethylamine and acetone are more likely to exhibit a similar capacity as electron-acceptors, but will not show similar proton-acceptor behaviour. It is argued therefore that electron-pair donation from the oxygen of $H_2{}^{17}O$ is the factor which most determines the ^{17}O shift.

Neither of the above arguments is wholly convincing. However, Alei and Florin's proposal[11] finds strong support in subsequent studies of the ^{15}N-resonance position in hydrogen-bonded species.[17] Relative to the shift in the vapour state, the ^{15}N shift of $^{15}NH_3$ at infinite dilution (Δ) is expressed empirically as a linear sum of the following interactions:[17]

(a) Nitrogen electron-pair interactions (σ_{Ax} terms) with

Solvent OH proton (σ_{AOH})
Solvent NH proton (σ_{ANH})
Solvent methyl group (σ_{AMe})
Solvent ethyl group (σ_{AEt})

(b) Interactions of solvent oxygen (σ_{BO}) or nitrogen (σ_{BN}) electron pairs with $^{15}NH_3$ protons (σ_{By} terms)

i.e.
$$\Delta = \sum_x C_x \sigma_{A_x} + \sum_y C_y \sigma_{B_y} \quad (2)$$

where $C_y = 0$ or 1, and C_x = relative abundance of the interacting site in the solvent molecule.

Values for the parameters σ_{Ax}, σ_{By} were obtained[17] by regression analysis of Δ values for $^{15}NH_3$ and $^{15}NMe_3$ in a series of solvents (see Table 1).

Table 1[17] *Shift parameter values for $^{15}NH_3$ and $^{15}NMe_3$*

	$^{15}NH_3$	$^{15}NMe_3$
σ_{AOH}	−25·2	−11·9
σ_{ANH}	−19·1	− 5·4
σ_{AH}	−16·5	− 4·7
σ_{AMe}	−13·2	− 4·4
σ_{BO}	+ 3·3	0
σ_{BN}	+ 2·1	0

[17] W. M. Litchman, M. Alei, jun., and A. E. Florin, *J. Amer. Chem. Soc.*, 1969, **91**, 6574; *ibid* 1970, **92**, 4828.

Interactions with the nitrogen electron-pair, including those involving hydrocarbon substituents from the solvent, appear to contribute most to the observed shift. Solute proton-donation seems relatively unimportant, negligibly so for ^{15}NMe$_3$. Such an empirical approach does not, of course, prove the model; there is, however, an impressive degree of consistency about the results. For example, in tetraethylmethane, a solvent not included in the regression analysis and for which σ_{AEt} should be the only contributor to the ^{15}N shifts, Δ for ^{15}NMe$_3$ is indeed equal to σ_{AEt}.[17] Other ^{15}N studies of Paolillo and Becker[18] agree in general terms with those of Alei and Florin. In addition, they report a correlation between $J(^{15}\text{N}-^1\text{H})$ and the hydrogen-bonding capacity of the solvent for [^{15}N]aniline.

Saitô and Nukada[19] show the ^{14}N resonance to be a more sensitive probe to hydrogen-bonding than the proton resonance. In the case of amides the nucleus is deshielded on formation of hydrogen-bonds in both proton-donor and proton-acceptor solvents, implying a reduction in electron density about the nitrogen atom. In proton-donor solvents the bond forms at the carbonyl oxygen, enhancing contributions from the canonical form:

$$\begin{array}{c} R^2 \\ \diagdown \\ C = \overset{+}{N} \\ \diagup \diagdown \\ -O R^1 \end{array}$$

Likewise it is argued that, in proton-acceptor systems, electron density about the nitrogen may be reduced due to repulsion of the nitrogen $2p_z$ electrons towards the carbonyl group by the electrons involved in the hydrogen-bond.

The above authors[19] regard donation of the nitrogen electron-pair to be unlikely, since these electrons are partially involved in the carbonyl π-system. That nitrogen donation can, however, occur is known from the enhanced rate of rotation about the amide C—N bond due to protonation in acidic media.[20] It may therefore prove to be an over-simplification to neglect entirely shift contributions from this source.

The ^{13}C resonance of ^{13}CHCl$_3$ is also deshielded by hydrogen-bond formation, and for a variety of proton-acceptor solvents a linear correlation has been demonstrated between ^{13}C and ^{1}H infinite dilution shifts relative to the shifts in cyclohexane, and also between ^{13}C shifts and the $J(^{13}\text{C}-^1\text{H})$ coupling.[21] One major hydrogen-bonded species is thought to be present in the

---

[18] L. Paolillo and E. D. Becker, *J. Magn. Resonance*, 1970, **2**, 168.
[19] H. Saitô, Y. Tanaka, and K. Nukada, *J. Amer. Chem. Soc.*, 1971, **93**, 1072, 1077.
[20] B. G. Cox, *J. Chem. Soc.*, (B), 1970, 1780.
[21] R. L. Lichter and J. D. Roberts, *J. Phys. Chem.*, 1970, **74**, 912.

medium (S = solvent molecule):

$$S \cdots H - C(Cl)(Cl)(Cl)$$

With benzene as solvent only the ^{13}C and $J(^{13}C-^{1}H)$ shifts show any correlation. Here the bond is thought to be formed with the benzene π-electron system:[21]

$$\text{C}_6\text{H}_6 \cdots H - C(Cl)(Cl)(Cl)$$

(A similar interaction has been proposed[19] between the N-H proton of pyrrole and benzene.) In such a case the proton shift is affected largely by the magnetic anisotropy of the benzene ring, and the correlation between ^{13}C and ^{1}H shifts is no longer evident.[21]

Smith and Ihrig[22] have undertaken a similar study of $^{13}CFCl_3$. Here the $J(^{13}C-^{19}F)$ coupling is not dependent upon solvent and the ^{19}F resonance shows an increased shielding in hydrogen-bonding solvents. The suggestion is put forward that, as a consequence of the increased polarity of the molecule, reaction field effects play an important part in determining chemical shifts, in addition to any hydrogen-bonding which may be present.

3 Ions in Solution

Ionic species generally exhibit resonance positions which are sensitive to changes of solvent and of counter-ion. Such behaviour is often attributed to an equilibrium in solution between 'intimate' ion-pairs and solvent-separated ion-pairs undergoing rapid exchange:

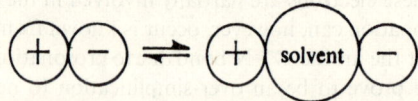

    'intimate' ion-pair   solvent-separated ion-pair

This approach has been used to explain the proton resonance behaviour of fluorenyl salts of alkali metals,[23] in line with earlier work involving absorption spectroscopy.[24,25]

Cox[26] has shown that alkali-metal ions shield the proton resonances of carbazole nitranion in the order $Rb^+ > K^+ > Na^+$, the order being reversed

---

[22] W. B. Smith and A. M. Ihrig, *J. Phys. Chem.*, 1971, **75**, 497.
[23] R. H. Cox, *J. Phys. Chem.*, 1969, **73**, 2649.
[24] T. E. Hogen-Esch and J. Smid, *J. Amer. Chem. Soc.*, 1966, **88**, 307, 318.
[25] L. I. Chan and J. Smid, *J. Amer. Chem. Soc.*, 1966, **90**, 4654.
[26] R. H. Cox, *Canad. J. Chem.*, 1971, **49**, 1377.

in the cation-solvating solvent dimethoxyethane. This suggests appreciable solvent separation of the ions in the latter case. The metal ion is thought to be associated with the π-electron cloud of carbazole nitranion,[26] the carbanions from 4,5-methylenephenanthrene,[26] and fluorene carbanion,[27] with the single exception of Li^+ and carbazole nitranion where the interaction is apparently between Li^+ and the nitrogen atom.[26]

Taylor and Kuntz[28] have investigated the effect of solvent and counter-ion on the methyl resonances of the ions (1) and (2). Relative to $CDCl_3$ the

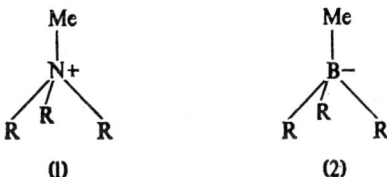

resonance is shielded by aromatic solvents in the case of the cation, deshielded for the anion. The extent of this 'solvent shift' depends both on counter-ion and solvent. Assuming that the solvent shift derives in the main from the solvent-separated ion-pair, the authors propose that the magnitude of the effect is an indication of the extent to which the above equilibrium lies to the right. For a given cation and solvent, a small anion favours intimate ion-pairing and the solvent shift is observed to be small. Conversely, larger solvent shifts are reported for larger anions, which are readily displaced from the intimate ion-pair by solvent molecules, solvent-separated ion-pairs being thereby favoured. For a given solvent there is a similar dependence of the anion solvent shift on cation size.

The concentration- and solvent-dependence of the ^{23}Na resonance in solutions of ^{23}NaI, and to a lesser extent $^{23}NaSCN$, may also be attributed to ion-pairing equilibria.[29] On the other hand, the ^{23}Na resonance of $^{23}NaBPh_4$ and $^{23}NaClO_4$ is insensitive to concentration, which suggests extensive solvent-separation of these ions.[29] Likewise, salicylaldehyde anion is separated by solvent from both Li^+ and Na^+ in D_2O, the anion proton-shifts being independent of the ion concentration.[30] In dimethyl sulphoxide (DMSO), however, the anion shifts do depend on concentration, particularly with Na^+ as counter-ion. Effectively an ion-pairing equilibrium is again invoked:[30]

[27] R. H. Cox, H. W. Terry, jun., and L. W. Harrison, *J. Amer. Chem. Soc.*, 1971, **93**, 3297.
[28] R. P. Taylor and I. D. Kuntz, jun., *J. Amer. Chem. Soc.*, 1970, **92**, 4813.
[29] R. H. Erlich, E. Roach, and A. I. Popov, *J. Amer. Chem. Soc.*, 1970, **92**, 4989.
[30] B. M. Fung and P. Trautmann, *J. Inorg. Nuclear Chem.*, 1970, **32**, 1393.

Cogley, Butler, and Grunwald[31] have analysed the effect of added ions on the proton resonance of water at high dilution in propylene carbonate.[32] For the case of 1:1 salts (MA), under conditions where only one water proton resonance is observed owing to rapid exchange, the monoaquation constant (K_1) is obtained from relation (3),[31]

$$(v - v_0) = C_S \left(\frac{K_1^M(v_1^M - v_0) + K_1^A(v_1^A - v_0)}{1 + C_S(K_1^M + K_1^A)} \right) \qquad (3)$$

where v and v_0 are the water resonance shifts extrapolated to zero water concentration in the presence and absence of salt, C_S is the total salt concentration, and K_1^M and K_1^A are the monoaquation constants for cation and anion respectively. The above expression is a simplified one depending on the fact that $K_1[H_2O]_{free} \ll 1$. Plotting $(v - v_0)/C_S$ versus $(v_0 - v)$ gives a linear plot of gradient $(K_1^M + K_1^A)$ and intercept $[K_1^M(v_1^M - v_0) + K_1^A(v_1^A - v_0)]$. The division of the combined ion parameters into the separate ion contributions is discussed; some of the values so obtained are quoted in Table 2.[31]

Table 2[31]

Ion	$K_1{}^a$/kg mol^{-1}	$(v_1 - v_0)$/Hz^b
Li$^+$	6·5±0·3	102±5
Na$^+$	1·4±0·1	67±7
Cl$^-$	6·2±0·3	100±3

^{a} at 36·4 °C; ^{b} at 60 MHz

The temperature-dependence of the water resonance in aqueous ionic solutions has been used[33] as a means of determining ionic hydration numbers. The total effective hydration number (h) can be related[34] [equation (4)] to m, the stoicheiometric ionic molality; m_1, the molality of exchangeable protons donated by the solvent; $d\delta/dT$, the rate of change of the proton shift of the solution with temperature; and $d\delta_N/dT$, the rate of change of the pure water shift with temperature. The equation requires that the proton shift for water in the ion solvation shell be temperature independent, and neglects contributions to the solution shift from the effects of the ions on the hydrogen-bonded structure of the solvent. A separation of the cation and anion effects is made; the value so obtained for Mg^{2+} agrees with an independently determined result.

$$h = \frac{55 \cdot 5}{m} \left[1 - \frac{(m_1 + 111 \cdot 1)}{111 \cdot 1} \cdot \frac{d\delta/dT}{d\delta_N/dT} \right] \qquad (4)$$

[31] D. R. Cogley, J. N. Butler, and E. Grunwald, *J. Phys. Chem.*, 1971, **75**, 1477.
[32] H. L. Friedman, *J. Phys. Chem.*, 1967, **71**, 1723.
[33] F. J. Vogrin, P. S. Knapp, W. L. Flint, A. Anton, G. Highberger, and E. R. Malinowski, *J. Chem. Phys.*, 1971, **54**, 178.
[34] P. S. Knapp, R. O. Waite, and E. R. Malinowski, *J. Chem. Phys.*, 1968, **49**, 5459.

Medium Effects

A more direct method for determining hydration numbers has been utilized by Fratiello, Schuster, and co-workers.[35] Aqueous ionic solutions diluted with acetone may be cooled to a point (≈ -100 °C) at which rapid site exchange is slowed sufficiently to permit the observation of two water resonances, one due to water in the cation solvation shell, the other to 'free' water (actually an exchange average between 'free' water and water in the anion solvation shell, since such exchange is still rapid.[36]) Hydration numbers may be determined from the peak areas under such conditions. Systems of UO_2^+ and Sn^{4+} ions have recently been studied.[35,37] The estimated lifetime of water molecules in the cation shell is about 5×10^{-3} s and, interestingly, addition of acid does not affect the resonance position of 'solvating' water. Proton exchange must therefore occur preferentially with free water, rather than with water in the cation solvation shell. Symons *et al.*[36] have described methods by which the proton resonance position of methanol in the anion solvation shell may be determined under low-temperature conditions similar to those above, for which the cation solvation shell, but not the anion shell, is resolved.

Clearly, n.m.r. is proving to be a valuable tool in the study of ionic solutions. Equally clearly, however, much work remains to be done to justify or refute the many simplifying assumptions often made in more quantitative studies of these relatively complex systems.

4 Aromatic Solvent Induced Shifts [ASIS Effects]

The ASIS phenomenon continues to be one of the more perplexing n.m.r. problems. A full account (1967) is given in Laszlo's review;[38] briefly, however, proton resonances of solute molecules having π-electron systems and/or polar substituents show a marked solvent dependence, particularly in aromatic solvents. It is usual to define a solvent shift parameter $\Delta_{i,\text{solvent}}$, being the shift of the i-th solute proton in a given solvent relative to the shift in an appropriate 'inert' solvent, usually carbon tetrachloride or cyclohexane, *i.e.* as shown in equation (5).

$$\Delta_{i,\text{benzene}} = \delta_{\text{benzene}} - \delta_{\text{CCl}_4} \qquad (5)$$

It has often been profitable to consider ASIS effects in terms of an association between solvent and solute having a 1:1 stoicheiometry, and for which there exists a definite average relative orientation of the components on the n.m.r. timescale.[38] If benzene is the solvent, the geometry of such a complex may be inferred from the ASIS for the solute protons and a consideration of the Johnson–Bovey mapping of the anisotropic magnetic environment of

---

[35] A. Fratiello, V. Kubo, R. E. Lee, S. Peak, and R. E. Schuster, *J. Inorg. Nuclear Chem.*, 1970, **32**, 3114.
[36] S. Ormondroyd, E. A. Phillpott, and M. C. R. Symons, *Trans. Faraday Soc.*, 1971, **67**, 1253.
[37] A. Fratiello, S. Peak, R. E. Schuster, and D. D. Davis, *J. Phys. Chem.*, 1970, **74**, 3730.
[38] P. Laszlo, *Progr. N.M.R. Spectroscopy*, 1967, **3**, 231.

the benzene molecule.[39] This approach is still used with apparent success,[40] although, as will be seen, it is increasingly being called into question.

Baker and Wilson[41] report ASIS data incompatible with the formation of a 1:1 complex, at least in a general sense. By extrapolating variable-temperature results, a benzene solvent shift of 1·20 p.p.m. is quoted for the acetone resonance at 0 K, a value which the authors believe is too large to be attributed to the anisotropy of a single associated benzene molecule, assuming reasonable distance of approach. Furthermore,[41] an attempt was made to extract values of n for the aggregation process:

$$\text{solute (S)} + n \text{ benzene (B)} \rightleftharpoons \text{complex (SB}_n) \quad (6)$$

from chemical shift data for a series of binary mixtures of benzene. The values reported (Table 3) argue against the generality of a 1:1 stoicheiometry for solute–solvent associations of this type.

Table 3[41] *Aggregation numbers (n) for benzene associated with various solutes*

Solute	n
Chloroform	1·0
Bromoform	1·2
Dichloromethane	1·5
Acetone	2·5
Camphor	1·9
5α-Androstan-11-one	2·8

Conversely, Feeney and Pauwels[42] have studied the association of ketones with benzene in carbon tetrachloride as a diluting solvent, concluding that the association has a 1:1 rather than an exclusively 2:1 stoicheiometry. The 'goodness of fit' between the experimental data and the theoretical curve for 1:1 association was taken to be a good indication that such a stoicheiometry was real. Other work, however, relating to charge-transfer complex systems discussed below, suggests that such an approach can be misleading. Use of Baker and Wilson's method[41] for determining the aggregation number n results in values for the acetone–benzene and 5α-androstan-11-one–benzene systems in carbon tetrachloride of 0·97 and 1·02 respectively.[42] Baker and Wilson[41] observed that dilution of the binary ketone–benzene system with carbon tetrachloride leads to values of n which differ from those for the binary system alone, an observation which they attributed to the possibility of weak complexation between benzene and carbon tetrachloride.[43] The presence of a diluting solvent may account for the differences between the two sets of results, although Feeney and Pauwels[42] subscribe to the view that the higher benzene concentrations attained by Baker and Wilson[41] may be responsible.

[39] C. E. Johnson and F. A. Bovey, *J. Chem. Phys.*, 1958, **29**, 1012.
[40] M. I. Foreman and F. Haque, *J. Chem. Soc. (B)*, 1971, 418.
[41] K. M. Baker and R. G. Wilson, *J. Chem. Soc. (B)*, 1970, 236.
[42] J. Feeney and P. J. S. Pauwels, *J. Chem. Soc. (B)*, 1971, 515.
[43] R. Anderson and J. M. Prausnitz, *J. Chem. Phys.*, 1963, **39**, 1225.

Medium Effects

Clearly, the general acceptance of a 1:1 stoicheiometry for the solvent–solute complex is questionable, as indeed is the notion of what constitutes an 'inert' solvent in this respect.

Where more than one polar group is present in a given solute molecule, each group might reasonably act as a site for solvent association, and in such cases a 1:1 stoicheiometry is less likely to obtain. Comparison of the relative magnitudes of the benzene solvent shifts for certain episulphides such as (3)

$$
\begin{array}{c}
\text{H} \quad \text{H} \\
\diagdown \text{C} \text{---} \text{C} \diagup \quad \quad \text{H} \\
\diagup \quad \diagdown \diagup \quad \text{CH---C---CN} \\
\text{H} \quad \text{S} \quad \text{H} \quad | \quad | \\
\quad \quad \quad \quad \quad \text{OR} \quad \text{H}
\end{array}
$$

(3)

suggests that solvent molecules are associated both with the episulphide ring and with the cyano-group.[44] Likewise, benzene solvent shifts for the protons of 2,6- and 3,4-dinitro-1-t-butylbenzenes[45] are irreconcilable with a simple 1:1 complex model involving the Johnson–Bovey[39] mapping. This is argued as evidence against a 1:1 association, although there remains the possibility that no simple relationship exists between solvent shift and solvent anisotropy in these cases. Observations relating to methyl-substituted biphenyls[47] are more readily explained in terms of a 'general solvation'[46] of the solute rather than a specific 1:1 solvent–solute complex, methyl groups being regarded as polar sites as a result of hyperconjugative interactions.[47]

There seems to be a close similarity between systems of charge-transfer complexes[48] under conditions studied by n.m.r. (where one component is in large excess) and systems for which ASIS effects are observed. N.m.r. data for such systems of charge-transfer complexes have in the past been fitted to a theoretical expression derived for a 1:1 molecular association with remarkable exactitude.[49] Nonetheless, despite this 'goodness of fit', other n.m.r. studies[50] have appeared which challenge the inference that such systems are in fact composed of complexes with a 1:1 stoicheiometry. These observations seem to be particularly relevant to Feeney and Pauwels' work.[42]

For the case of charge-transfer complex systems, an alternative to the postulate of a 1:1 complex could reasonably be the simultaneous existence of

---

[44] K. D. Carlson, D. Weisleder, and M. E. Daxenbichler, *J. Amer. Chem. Soc.*, 1970, **92**, 6232.
[45] I. Leupold, H. Musso, and J. Vičar, *Chem. Ber.*, 1971 **104**, 40.
[46] Y. Ichikawa and T. Matsuo, *Bull. Chem. Soc. Japan*, 1967, **40**, 2030.
[47] Y. Nomura and Y. Takeuchi, *J. Chem. Soc. (B)*, 1970, 956.
[48] R. Foster, 'Organic Charge-Transfer Complexes', Academic Press, London and New York, 1969.
[49] R. Foster and C. A. Fyfe, *Progr. N.M.R. Spectroscopy*, 1969, **4**, 1.
[50] M. I. Foreman, R. Foster, and D. R. Twisleton, *Chem. Comm.*, 1969, 1318.

1:1 (AD) and 2:1 (AD$_2$) complexes, where component D is in large excess relative to A, and two equilibria are considered having constants K_1 and K_2, respectively. For such a situation an expression [equation (7)] has been derived[51] which relates the shift of a given resonance of the minor component (A) to the various system parameters:

$$\Delta = \frac{K_1\Delta_1[D]_0 + K_1K_2\Delta_2[D]_0^2}{1 + K_1[D]_0 + K_1K_2[D]_0^2} \qquad (7)$$

where Δ is the observed shift, Δ_1 and Δ_2 the hypothetical shifts in the pure 1:1 and pure 2:1 complex, respectively, measured relative to the shift of the same resonance in uncomplexed A. The subscript '0' denotes the total (*i.e.* free plus complexed) concentrations. Such an expression is clearly rather cumbersome; it was, however, possible to extract values of K_1 and K_2 from studies of the benzene resonance in the presence of varying excess amounts of aqueous silver nitrate,[51] a system known to be composed of 1:1 and 2:1 Ag$^+$–benzene aggregates.[52] The values so obtained agree reasonably well with other determinations,[52] which provides some support for the validity of the treatment.

Foster, Dodson, and co-workers[53] have reconsidered certain charge-transfer complex systems previously evaluated on the basis of a 1:1 association. The authors note that equilibrium constants for such 1:1 complexation obtained from n.m.r. and optical measurements frequently do not agree.[48] As an example, constants obtained by the two methods for the system trinitrobenzene(TNB)–hexamethylbenzene(HMB) in carbon tetrachloride were 3.2 ± 0.2 l mol^{-1} (n.m.r.) and 4.4 ± 0.2 l mol^{-1} (optical) for the case [HMB] $\gg$ [TNB]. Rather more extensive data, obtained by both methods and treated on the basis that 1:1 and 2:1 complexes are both present, yield identical values for the two equilibrium constants of 10 ± 2 l mol^{-1} for $K_{1:1}$ and 1 ± 0.2 l mol^{-1} for $K_{2:1}$ from both sets of data.[53]

Similarly good agreement was obtained for the fluoranil–hexamethylbenzene–carbon tetrachloride system; other cases studied also indicate deficiencies in the simple 1:1 complex approach. It seems very probable therefore that 1:1 and 2:1 complexes *at least* are present in charge-transfer complex systems where one component is in large excess, and the clear implication is that ASIS effects might also be amenable to such a treatment.

Bertrand, Compton, and Verkade[54] demonstrate rather well the failure of approaches based on 1:1 associations, and interpret their results rather on the basis of a 'cage' of ordered solvent molecules about the solute. For the two series of dipolar compounds having the general structures (4) and (5) (Y = P

[51] M. I. Foreman, R. Foster, and J. Gorton, *Trans Faraday Soc.*, 1970, **66**, 2120.
[52] L. J. Andrews and R. M. Keefer, *J. Amer. Chem. Soc.*, 1949, **71**, 3644.
[53] B. Dodson, R. Foster, A. A. S. Bright, M. I. Foreman, and J. Gorton, *J. Chem. Soc. (B)*, 1971, 1283.
[54] R. D. Bertrand, R. D. Compton, and J. G. Verkade, *J. Amer. Chem. Soc.*, 1970, **92**, 2702.

Medium Effects

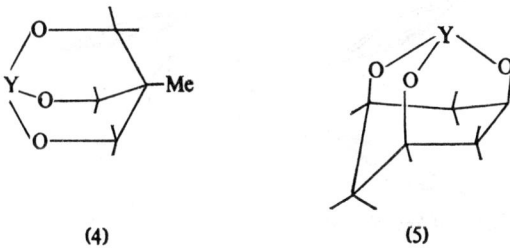

(4)   (5)

or As), the effect of benzene as solvent is to increase the shielding of protons at the positive end of the dipole, and to deshield those at the negative end. This could be explained in terms of a single benzene molecule associated at the positive end of the solute dipole, with its periphery so aligned as to deshield protons at the negative end.[55] It is argued that such an explanation is precluded in this case by the molecular geometries involved, and by the presence of the oxygen lone-pair electrons, which will tend to align a benzene molecule at the positive end of the dipole in such a way as to minimize the peripheral deshielding. It is proposed therefore that different benzene molecules must be associated with each end of the molecular dipole in order to explain the observed shifts. On the n.m.r. timescale, a 'cage' of ordered solvent molecules is envisaged as a time-averaged effect (Figure 1).

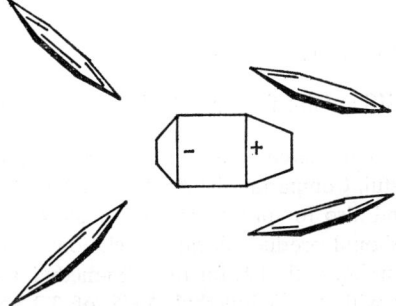

Figure 1 *Proposed orientation of benzene molecules about dipolar bicyclic solutes*[54]

Interestingly, in hexafluorobenzene the direction of the solvent shift is the reverse of that for benzene. The authors suppose that both solvents form a similar cage structure, but that the solute is orientated within the cage in a different sense in the two cases (Figure 2).

Engler and Laszlo[56] have produced an extremely lucid description of the 'solvent cluster' approach to ASIS effects. The time averaging due to the

---

[55] J. Ronayne and D. H. Williams, *Ann. Rev. N.M.R. Spectroscopy*, 1969, **2**, 83.
[56] E. M. Engler and P. Laszlo, *J. Amer. Chem. Soc.*, 1971, **93**, 1317.

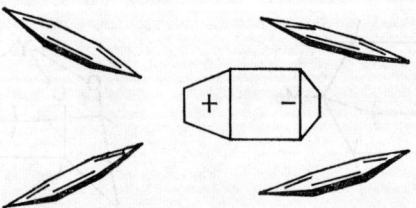

Figure 2 *Orientation of hexafluorobenzene molecules about dipolar bicyclic solutes*[54]

timescale of the n.m.r. experiment is stressed. It is supposed that, as a consequence of numerous solvent collisions with the solute, for which there is a multiplicity of solvent–solute orientations, the time-average effect is a 'solvent cloud' about the polar group(s). This will have the effect of modifying the *apparent* anisotropy of the polar group, which will alter the magnetic environment of a particular proton by an amount which depends on the position of the proton relative to the centre of the solvent 'cloud'. The ASIS for a given proton therefore has a 'site factor' associated with it which might reasonably be expressed as[57] in equation (8), where r and θ are the radius vector and inclination of the proton relative to a point dipole assumed to be located (for the ketones considered in this work) at the oxygen end of the carbonyl group.

$$\text{site factor} = \frac{3\cos^2\theta - 1}{r^3} \qquad (8)$$

Overall, it is proposed that:

$$\text{ASIS} = (\text{site factor}) \times (\text{solvent parameter}) \qquad (9)$$

For the ketones considered, particularly camphor, this description is remarkably successful. Comparing the ASIS for different proton groups of a single molecule, and also for proton groups of different molecules, over a series of solvents should produce linear correlations. Comparing the $C(9)$-methyl ASIS of camphor with that for the $C(8)$-methyl, and the $C(9)$-methyl ASIS of camphor with the 3,5-methyl ASIS of 3,3,5,5-tetramethylcyclohexanone produces just such correlations. That the expected proportionality between the ASIS for a particular proton and its calculated site factor for a given solvent also obtains may be seen from Table 4.

Accepting, then, the validity of the above description of ASIS, at least for the weakly associated camphor system, the rather telling point is made that proposals of a hypothetical orientation of solvent to solute in a specific complex are quite unnecessary, 'the geometry of the camphor solute in itself suffices to account for the solvent shifts observed'.

[57] H. M. McConnell, *J. Chem. Phys.*, 1957, **27**, 226.

Table 4[56] *Site factors for protons of camphor*

Proton Group	$\dfrac{10^3 \times (3\cos^2\theta - 1)}{r^3}$	Benzene ASIS $\times\ 0.5/\text{Hz}^a$
8-Me	7.0	6.4
9-Me	9.4	9.2
10-Me	−1.5	−1.4
exo-3-H	10.1	7.8[b]
endo-3-H	10.1	8.4[b]
4-H	15.4	13.2[b]
exo-5-H	11.0	10.5[b]
endo-5-H	13.4	13.4[b]
exo-6-H	8.9	9.9[b]
endo-6-H	5.8	4.2[b]

[a] 0.5 is the estimated proportionality constant (see the text); [b] P. V. Demarco, D. Doddrell, and E. Wenkert, *Chem. Comm.* 1969, 1418.

Considering next the solvent parameter, for aromatic solvents the ASIS for a given proton of the solute is shown to be proportional to the concentration of benzene rings in the solution (Figure 3); exactly the same proportionality

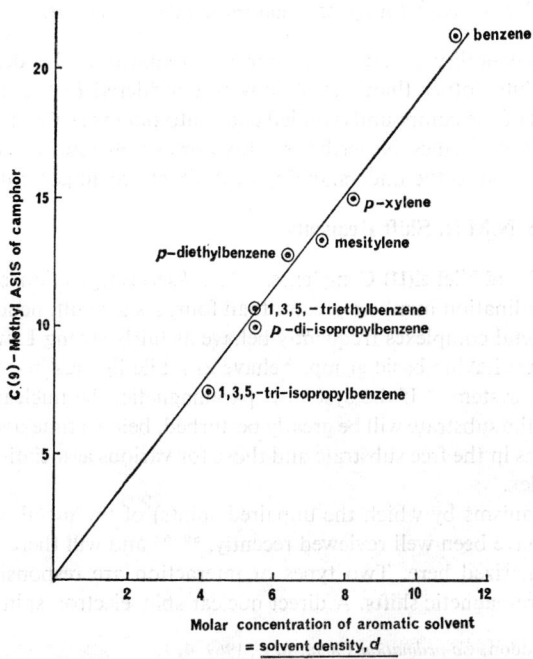

Figure 3 *Plot of the C(9)-methyl ASIS of camphor versus the molar concentration of the appropriate aromatic solvent*[56]

constant obtains if benzene itself is diluted with an inert solvent. The temperature dependence of the ASIS can be treated in a like manner, all of which argues against an interaction of the charge-transfer type.

For a given substituted benzene as solvent, Engler and Laszlo then rather neatly dissect the solvent parameter into a size contribution and an effect due to the polar character of the solvent, $(ASIS)_R$, using the correlation of Figure 3 to obtain the size factor and always referring to the $C(9)$-methyl ASIS for camphor, which effectively keeps the site factor constant. An example of such a breakdown for two solvents having similar solvent parameters, but for which the $(ASIS)_R$ terms are rather different, is shown in Table 5.

Table 5[56] *Contributions from 'electronic' and 'size' factors to the solvent parameters of benzaldehyde and benzotrichloride.*

	Benzene	Benzaldehyde	Benzotrichloride
$C(9)$-methyl ASIS of camphor/Hz^a	21·1	10·1	11·0
Molar concentration, d/M^b	11·3	9·9	7·0
Contribution to ASIS from molar volume/Hz^a	21·15	18·6	13·2
Perturbation due to solvent substituent, $(ASIS)_R$/Hz^a	0	−8·5	−2·2

^{a} at 60 MHz; ^{b} d = solvent density, M = molecular weight of solvent.

The full generality of such an approach remains to be demonstrated, although solutes other than camphor were considered in the above work. Certainly all of the compounds studied enter into only very weak associations with aromatic molecules. Nevertheless, this work surely represents a considerable contribution to the understanding of ASIS effects in general.

5 N.M.R. Shift Reagents

A. Cobalt(II) and Nickel(II) Complexes.—The d-series transition metals often exhibit co-ordination numbers greater than four. As a result, neutral four-co-ordinated metal complexes frequently behave as fairly strong Lewis acids, so that substrates having basic groups behave as labile ligands in a five- or six-co-ordinated system.[58] If the system is paramagnetic, the nuclear resonance positions of the substrate will be greatly perturbed, being a time-average of the shift positions in the free substrate and those for various associations with the metal complex.

The mechanisms by which the unpaired spin(s) of the metal affect ligand resonances have been well reviewed recently,[59-62] and will therefore be only briefly summarized here. Two types of interaction are responsible for the observed paramagnetic shifts. A direct nuclear spin–electron spin interaction

---

[58] D. P. Graddon, *Co-ordination Chem. Rev.*, 1969, **4**, 1.
[59] H. J. Keller and K. E. Schwarzhans, *Angew. Chem. Internat. Edn.*, 1970, **9**, 196.
[60] K. E. Schwarzhans, *Angew. Chem. Internat. Edn.*, 1970, **9**, 946.
[61] D. R. Eaton and W. D. Phillips, *Adv. Magn. Resonance*, 1965, **1**, 112.
[62] G. A. Webb, *Ann. Rep. N.M.R. Spectroscopy*, 1970, **3**, 211.

Medium Effects

is possible if there is a finite unpaired electron spin density at the appropriate nucleus, a situation which arises if the unpaired spin is in an orbital having some degree of s-character. This is the Fermi 'contact' interaction. Study of contact shifts therefore provides information about the nature of the metal–ligand bond, and often about the bonding within the ligand itself. The second possibility is the pseudo-contact term. This is a 'through-space' interaction between the magnetic dipole due to the unpaired spin(s) and the nuclear magnetic dipole, being a function of the anisotropy of the g-tensor and also of the geometry of the complex. The magnitude of the effect for the i-th nucleus, $(PC)_i$, is given by[63] equation (10) where $F(g)$ is a function of the g-tensor values; the form depends on the relative magnitudes of the electronic relaxation times and the rotational correlation times. C is a constant. For a given complex the term involving the principal values of the g-tensor will also be constant. The observed shift is therefore a function of the geometrical term $(3\cos^2\theta_i - 1)/r_i^3$. In these examples θ_i is the angle between the principal molecular axis and the vector of length r_i joining the i-th nucleus and the metal atom. The pseudo-contact term is therefore a potential source of information about the geometry of the complex.

$$(PC)_i = -C \times \frac{(3\cos^2\theta_i - 1)}{r_i^3} \times F(g) \qquad (10)$$

In general, complexes of Co^{II} and Ni^{II} might reasonably induce paramagnetic shifts by either or both of the above mechanisms. Clearly, therefore, the pseudo-contact and contact contributions to the overall shift must be evaluated with great care before eliciting further information from the observations.

Eaton has reported the behaviour of the proton resonances of aniline[64] and of substituted anilines[65] in the presence of Co^{II} tripyrazolylborate, $Co\{HB(pz)_3\}_2$. The large shifts observed are clearly due to the paramagnetism of the complex since the structurally similar, diamagnetic $Zn\{HB(pz)_3\}_2$ has little effect. The observed shifts are attributed entirely to the pseudo-contact term, although attempts to fit the observed shifts to equation (10) have been unsuccessful. Other geometrical factors are introduced to account for this, such as restricted tumbling of the aniline molecule.[64] From the relative shifts of the protons within a given aniline therefore, it is suggested that anilines approach perpendicularly to the three-fold symmetry axis of the Co^{II} complex (Figure 4) to form a loosely bound second co-ordination sphere. From studies of the temperature dependence of the shifts, the binding energy of this second sphere is found to be of the order of 2—3 kcal mol^{-1}. Electron-withdrawing substituents in the aniline ring are also reported to enhance the strength of the aniline–complex bond; for electron-donating substituents the converse is true.[65] It is thought, therefore, that the bonding is electrostatic in origin, the

---

[63] H. M. McConnell and R. E. Robertson, *J. Chem. Phys.*, 1958, **29**, 1361.
[64] D. R. Eaton, *Canad. J. Chem.*, 1969, **47**, 2645.
[65] D. R. Eaton, H. O. Ohorodnyk, and L. Seville, *Canad. J. Chem.*, 1971, **49**, 1218.

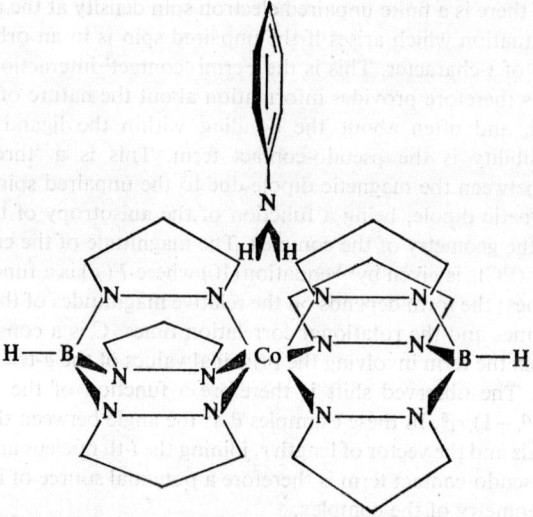

Figure 4 *Proposed geometry of the $C_6H_5NH_2$–$Co\{HB(pz)_3\}_2$ complex*[65]

nitrogen of the aniline behaving as a positively charged site attracted by the partial negative charge on the pyrazolyl nitrogens bonded to the cobalt atom.

Gillies, Szarek, and Baird[66,67] have studied the behaviour of simple aliphatic alcohols and amines with bis(acetylacetonato)cobalt(II), $Co(acac)_2$. No evaluation of the relative contribution of contact and pseudo-contact terms to the observed shift is given, although the authors refer to the shifts as 'contact' shifts. The α-CH_n protons are markedly perturbed by addition of $Co(acac)_2$, the effect being rapidly attenuated thereafter; β- and γ-CH_n protons are little affected. The magnitude of the effect on α-CH_n protons appears to be an indication of the structure of the alcohol; straight-chain alcohols are most shifted, those with methyl substitution on the α-carbon least, and structures where branching occurs at the β-carbon are intermediate between the two.

Earlier studies[68,69] of $Ni(acac)_2$ and $Co(acac)_2$ complexes suggest that basic ligands loosely bound to $Ni(acac)_2$ are subjected almost exclusively to a contact interaction, whilst with $Co(acac)_2$ the pseudo-contact term becomes important. From a study of cyclic saturated amines, Yonezawa and co-workers[70] concur with this view. For the $Ni(acac)_2$ case two processes are considered by which spin density is propagated throughout the ligand. Firstly, direct electron delocalization over the ligand π-orbitals will lead to

[66] W. A. Szarek, E. Dent, T. B. Grindley, and M. C. Baird, *Chem. Comm.*, 1969, 953.
[67] E. Gillies, W. A. Szarek, and M. C. Baird, *Canad. J. Chem.*, 1971, **49**, 211.
[68] R. W. Kluiber and W. DeW. Horrocks, jun., *J. Amer. Chem. Soc.*, 1965, **87**, 5350.
[69] G. N. LaMar, *Inorg. Chem.*, 1969, **8**, 581.
[70] T. Yonezawa, I. Morishima, and Y. Ohmori, *J. Amer. Chem. Soc.*, 1970, **92**, 1267.

Medium Effects

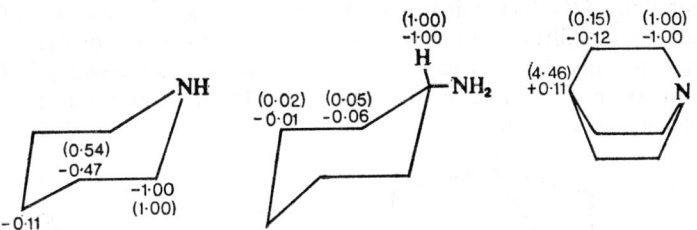

Figure 5 *Relative proton shifts of piperidine, cyclohexylamine, and quinuclidine affected by* Ni(acac)^2 *complexation. Calculated spin densities are marked in parentheses*[70]

deshielding of the protons. Secondly, spin polarization from unpaired electron density in the nitrogen lone pair will give rise to rapidly attenuating shifts alternating in sign along the ligand. The uniform direction of the observed shifts (Figure 5) implies that a σ-delocalization mechanism is predominant, although the increased shielding observed for the γ-proton of quinuclidine suggests a contribution from spin polarization. Spin densities calculated on this basis can be seen to parallel fairly closely the observed shift behaviour.

Rather different data are obtained from studies of the Co(acac)$_2$ complex;[70] this is attributed to a contribution from the pseudo-contact term. Assuming that the same distribution of spin densities obtains[69] for amines complexed to either Co(acac)$_2$ or Ni(acac)$_2$, the authors assessed the magnitude of the pseudo-contact interaction and deduced therefrom a Co—N bond length of about 2 Å.

The orientation of the nitrogen lone-pair electrons appears to affect the propagation of shifts throughout the amine.[70] From ^{13}C studies,[71] an axial

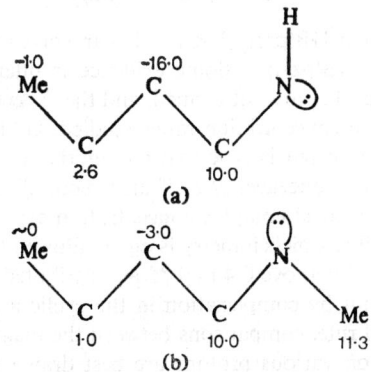

Figure 6 *Relative* ^{13}C *shifts of substituted piperidines affected by* Ni(acac)$_2$ *complexation:*
(a) N-*unsubstituted (lone pair equatorial)*
(b) N-*methyl (lone pair axial)*[71]

[71] I. Morishima, K. Okada, T. Yonezawa, and K. Goto, *Chem. Comm.*, 1970, 1535.

orientation of the lone pair was shown to lead to rapid attenuation of the paramagnetic shift; attentuation of the shift in the case of an equatorial lone pair, on the other hand, is very much less (Figure 6). Extended study[72] of the proton resonance behaviour at 220 MHz also illustrates this point: typical results are shown in Figure 7.

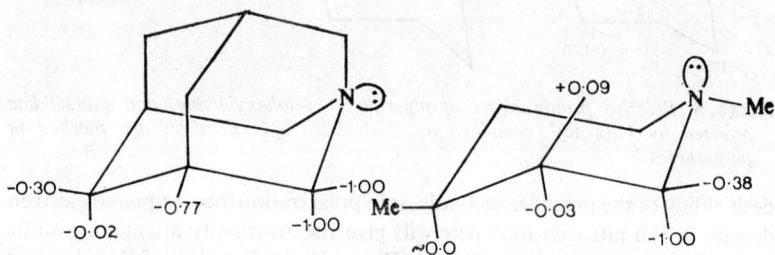

Figure 7 *Relative proton shifts for quinuclidine (equatorial lone pair) and an N-methylpiperidine (axial lone pair) on Ni(acac)$_2$ complexation from 220 MHz n.m.r. studies*[72]

It was remarked earlier that the substrate behaves as a labile ligand, exchanging rapidly on the n.m.r. timescale. The observed shift is therefore a time average of the shift of the uncomplexed ligand and the shift of the ligand in association with the metal complex. The situation in the systems discussed here may have been oversimplified. Eaton[64,65] discusses aniline–Co{HB(pz)$_3$}$_2$ systems in terms of a 1:1 association having an equilibrium constant K. The overall shift Δ_i for a given system is therefore:

$$\Delta_i = \frac{K[\text{Co}\{\text{HB}(\text{pz})_3\}_2]}{1 + K[\text{Co}\{\text{HB}(\text{pz})_3\}_2]} \tag{11}$$

For cases where $K[\text{Co}\{\text{HB}(\text{pz})_3\}_2] \ll 1$ a linear correlation should obtain between Δ_i and $[\text{Co}\{\text{HB}(\text{pz})_3\}_2]$. Some evidence is offered to suggest that higher complexes than 1:1 are not formed, and the expected correlation is in fact observed over the concentration range studied. Gillies *et al.*[67] report a similar linear correlation but believe that the limiting case is a 2:1 alcohol–Co(acac)$_2$ association. Yonezawa *et al.*[70] also report linear correlations, in this case for Δ_i *versus* metal complex concentration expressed on a mole % basis. Only one possible stoicheiometry is again allowed for the association; deviations from linearity above 0·4 mol % are attributed to conformational changes brought about by complexation in the cyclic amine systems being studied. As a general rule, comparisons between the magnitude of the effect of the shift reagent on various protons are best drawn by considering the slopes of such plots over ranges where the correlation is a linear one.

The behaviour of the anhydrous Co(acac)$_2$ system with pyridine in benzene solution provides a good indication of the possible complexity of the systems

---

[72] I. Morishima, K. Okada, M. Ohashi, and T. Yonezawa, *Chem. Comm.*, 1971, 33.

discussed above. The situation is described by the following equilibria;[73]

$$2Co_3(acac)_6 + 3py \underset{}{\overset{K_1}{\rightleftharpoons}} 3Co_2(acac)_4,py \quad K_1 \text{ ery large}$$

$$Co_2(acac)_4,py + py \underset{}{\overset{K_2}{\rightleftharpoons}} 2Co(acac)_2,py \quad K_2 = 2 \cdot 0 \, l^2 \, mol^{-2}$$

$$Co(acac)_2,py + py \underset{}{\overset{K_3}{\rightleftharpoons}} Co(acac)_2,py_2 \quad K_3 = 5 \cdot 8 \times 10^3 \, l \, mol^{-1}$$

The $Ni(acac)_2$ system in benzene is slightly less complex:[74]

$$2Ni_3(acac)_6 + 3py \underset{}{\overset{K_1}{\rightleftharpoons}} 3Ni_2(acac)_4,py \quad K_1 \approx 10^{10} \, l^2 \, mol^{-2}$$

$$Ni_2(acac)_4,py + 3py \underset{}{\overset{K_2}{\rightleftharpoons}} 2Ni(acac)_2,py_2 \quad K_2 = 2 \cdot 7 \times 10^5 \, l^2 \, mol^{-2}$$

Perry and Drago have discussed the possibility that other factors, such as hydrogen-bonding, ion-pair formation, *etc.*, may contribute to the observed paramagnetic shifts of ligands in cobalt and nickel complexes.[75]

It is perhaps therefore inappropriate to assume that only one $M(acac)_2$–substrate associated species contributes to the observed shift in any given system without more detailed study. Under such circumstances, deviations from linearity of plots of shift *versus* concentration of metal complex should be the norm.

B. Lanthanide Metal Complexes.—For paramagnetic systems, short electronic relaxation times tend to favour fairly narrow n.m.r. linewidths, which would otherwise be broadened due to the large fluctuating magnetic fields associated with paramagnetic species in thermal motion.[57,62] The electronic relaxation times of Co^{II} and Ni^{II} complexes are generally short enough that reasonably narrow n.m.r. lines are observed. However, line broadening is usually sufficient to degrade the fine structure of the resonance lines quite markedly, particularly for nuclei in close proximity to the metal atom.[67]

For the lanthanide series the situation is closely analogous to that for Co^{II} and Ni^{II} shift reagents. The lanthanides form six-co-ordinate metal complexes which may behave as Lewis acids, so that basic substrates become weakly bound as labile ligands; for such 'outer sphere' ligands, exchange is again rapid on the n.m.r. timescale. However, complexes of Eu^{III} and Pr^{III} are of especial interest in that the electronic relaxation times are generally sufficiently short that very narrow n.m.r. lines are observed, the fine structure frequently persisting to quite high relative concentrations of shift reagent to substrate.

The first reported use of a lanthanide shift reagent appears to have been by Hinckley,[76] who studied the downfield shift of protons of cholesterol which

---

[73] J. P. Fackler jun., *Inorg. Chem.*, 1963, **2**, 266.
[74] J. P. Fackler jun., *J. Amer. Chem. Soc.*, 1962, **84**, 24.
[75] W. D. Perry and R. S. Drago, *J. Amer. Chem. Soc.*, 1971, **93**, 2183.
[76] C. C. Hinckley, *J. Amer. Chem. Soc.*, 1969, **91**, 5160.

occurs on addition of the dipyridine adduct of tris-(2,2,6,6,-tetramethylheptane-3,5-dionato)europium(III) [the abbreviation Eu(tmhd)$_3$,py$_2$ is used[77] throughout this article, rather than the alternative Eu(DPM)$_3$,py$_2$]. Shifts of the cholesterol resonances were reported to occur without significant broadening of the lines, the magnitude of the shift being a function of the concentration of the Eu(tmhd)$_3$,py$_2$ complex. Hinckley believed that contributions to the observed shifts from the contact interaction would be negligible compared to those from the pseudo-contact term,[78] and that the angle-dependent term of equation (10) might also be neglected, in which case the paramagnetic shifts observed should be a function of r_i^{-3}.

Several other possibilities have since been considered as shift reagents. Williams and Sanders[79] obtained shifted spectra of benzyl alcohol and cyclohexanol which were effectively first-order spectra in carbon tetrachloride solution using the pyridine-free Eu(tmhd)$_3$ complex. The praseodymium analogue, Pr(tmhd)$_3$, generally induces upfield shifts of the proton resonances; the absolute shift is about three times that for the Eu^{III} complex.[77] Also reported are the directions and relative magnitudes of shifts induced in suitable substrates by several complexes of the form M[OP(NMe$_2$)$_3$]$_4$(ClO$_4$)$_3$ where M is a lanthanide metal.[77] Of the metal complexes studied, line-broadening effects are least for those of Pr^{III} and Eu^{III}.

Complexes of Pr^{III} and Eu^{III} with other ligands have also been considered.[80] By far the most successful work in this context is that of Rondeau and Sievers[81] who have reported the application of tris-(1,1,1,2,2,3,3-heptafluoro-7,7-dimethyl-4,6-octanedionato)europium(III) [Eu(fod)$_3$] as a shift reagent. This complex is a stronger Lewis acid than any hitherto considered, by virtue of the electron-withdrawing effect of the fluorine atoms, and also has the very considerable advantage of being highly soluble in solvents such as chloroform and carbon tetrachloride. Rondeau and Sievers report the successful use of this agent in cases where Eu(tmhd)$_3$ is totally ineffective.[81] A chiral ligand, 3-(t-butylhydroxymethylene)-d-camphor, complexed to Eu^{III}, has been successfully used to separate the resonances of (R)- and (S)-α-phenethylamine and also of (R)- and (S)-amphetamines,[82] thereby allowing a ready determination of enantiomeric purity. This technique should be a useful adjunct to other methods involving chiral solvents.[83,84]

Because of exchange averaging, it is not generally possible to observe the shifts of the wholly complexed substrate. Use of a paramagnetic shift term Δ_{SR} [equation (12)] has been proposed, where $\delta_{solvent}$ is the chemical shift of

---

[77] J. Briggs, G. H. Frost, F. A. Hart, G. P. Moss, and M. L. Staniforth, *Chem. Comm.*, 1970, 749.
[78] D. R. Eaton, *J. Amer. Chem. Soc.*, 1965, **87**, 3097.
[79] J. K. M. Sanders and D. H. Williams, *Chem. Comm.*, 1970, 422.
[80] J. K. M. Sanders and D. H. Williams, *J. Amer. Chem. Soc.*, 1971, **93**, 641.
[81] R. E. Rondeau and R. E. Sievers, *J. Amer. Chem. Soc.*, 1971, **93**, 1522.
[82] G. M. Whitesides and D. W. Lewis, *J. Amer. Chem. Soc.*, 1970, **92**, 6979.
[83] L. Mamlok, A. Marquet, and L. Lacombe, *Tetrahedron Letters*, 1971, 1039.
[84] W. H. Pirkle and S. D. Beare, *J. Amer. Chem. Soc.*, 1969, **91**, 5149.

$$\Delta_{SR} = \delta_{solvent} - \delta_{SR}^{n=1} \qquad (12)$$

the substrate resonance in a given solvent,[85] and $\delta_{SR}^{n=1}$ is the shift of the same resonance in the presence of a 1:1 mole ratio of substrate to shift reagent. This does not, of course, correspond to the shift in the pure complex. An alternative technique, which gives values for the relative paramagnetic shifts of the resonances of a given substrate, is to compare the slopes of the generally linear plots of shift vs. shift reagent concentration for the appropriate nuclei.[86,87] This approach avoids the need to extrapolate such plots, which is generally necessary for the former method.

Following Hinckley's suggestion,[76] a qualitative dependence of the paramagnetic shift, determined in one of the above ways, with distance from the metal atom has been reported by several authors.[86,88-90] Rather more detailed studies involving estimates of the r_i values for protons of a given substrate based on assumed metal–substrate bond lengths and angles have also been attempted. In a number of cases the expected dependence of the paramagnetic shift on r_i^{-3} was demonstrated.[79,80,91-93] On the other hand, similar approaches have led to observations of other than an r^{-3} dependence. Demarco et al., for example, plotted log (Δ_{SR}) against log r_i for cis- and trans-4-t-butylcyclohexanols, isoborneol, and borneol, which indicated an $r_i^{-2.2}$ dependence.[85] For adamantan-2-ol a dependence on $r_i^{-2.0}$ has been reported,[94] and on $r_i^{-2.3}$ for certain oxime systems.[95] An erroneous estimate of the metal–substrate bond lengths could account for these latter observations. Another possibility is neglect of the angle-dependent term. Certainly in cases where the angle dependence has been included very good agreement is often obtained between calculated and observed values. Briggs, Hart, and Moss[96] included the angle term in calculating proton shifts for borneol (Table 6). A similarly good correlation was observed for ^{13}C studies on borneol.[97]

Rather more striking examples of the dangers of neglecting the angle-dependent term exist. The acetamido and glycosidic methyl groups of β-methyl-N-acetylglucosamine show paramagnetic shifts in opposite directions on addition of Eu^{III}–lysozyme complex.[98] Likewise, Siddall[99] notes that

---

[85] P. V. Demarco, T. E. Elzey, R. B. Lewis, and E. Wenkert, *J. Amer. Chem. Soc.*, 1970, **92**, 5734.
[86] P. Bélanger, C. Freppel, D. Tizané, and J. C. Richer, *Chem. Comm.*, 1971, 266.
[87] D. R. Crump, J. K. M. Sanders, and D. H. Williams, *Tetrahedron Letters*, 1970, 4419.
[88] C. C. Hinckley, *J. Org. Chem.*, 1970, **35**, 2834.
[89] G. H. Wahl, jun., and M. R. Peterson, jun., *Chem. Comm.*, 1970, 1167.
[90] R. R. Fraser and Y. Y. Wigfield, *Chem. Comm.*, 1970, 1471.
[91] C. Beauté, Z. W. Wolkowski, and N. Thoai, *Tetrahedron Letters*, 1971, 817.
[92] W. Walter, R. F. Becker, and J. Thiem, *Tetrahedron Letters*, 1971, 1971.
[93] C. C. Hinckley, M. R. Klotz, and F. Patil, *J. Amer. Chem. Soc.*, 1971, **93**, 2417.
[94] A. F. Cockerill and D. M. Rackham, *Tetrahedron Letters*, 1970, 5149.
[95] Z. W. Wolkowski, *Tetrahedron Letters*, 1971, 825.
[96] J. Briggs, F. A. Hart, and G. P. Moss, *Chem. Comm.*, 1970, 1506.
[97] J. Briggs, F. A. Hart, G. P. Moss, and E. W. Randall, *Chem. Comm.*, 1971, 364.
[98] K. G. Morallee, E. Nieboer, F. J. C. Rossotti, R. J. P. Williams, A. V. Xavier, and R. A. Dwek, *Chem. Comm.*, 1970, 1132.
[99] T. H. Siddall, tert., *Chem. Comm.*, 1971, 452.

Table 6[96] ^{1}H n.m.r. shifts for borneol[a] in the presence of $Pr(tmhd)_3$

Proton	exo-2-H	exo-3-H	endo-3-H	4-H	exo-5-H	endo-5-H	exo-6-H	endo-6-H	8-Me	9-Me	10-Me
Paramag. Shift (Δ_{SR})[b]	45.0	16.6	34.2	10.1	10.5	16.0	15.7	35.6	7.78	8.01	17.8
$K(3\cos^2\theta_i - 1)/r_i^3$ [c] —(2-H) = 25;	49.0	15.8	30.2	9.84	11.0	16.1	16.0	38.7	7.84	7.88	16.5

[a] Numbering of borneol as shown on formula. Pr—O distance taken as 3.0 Å for optimum fit. Pr—O—(C-2) = 126°. Dihedral angle Pr—O—(C-2) —(2-H) = 25°; [b] $\Delta_{SR} = \delta_{CCl_4} - \delta_{Pr(tmhd)_3}$; [c] K is a scaling factor to allow direct comparison with the observed shifts.

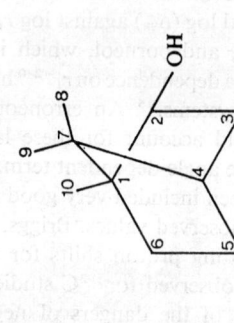

Medium Effects

the β-methyl doublet of *exo*-2,6-di-2-propylacetanilide is shifted upfield on addition of Eu(tmhd)$_3$; all other resonances shift downfield. Proton H(3) of compound (6) also exhibits an anomalous upfield shift with Eu(tmhd)$_3$,[100] as do the protons of the methoxy-group in (7).[101]

(6)

(7)

For a given complex the most likely source of a change in sign of the paramagnetic shift is the angle term $(3\cos^2\theta_i - 1)$. Positive values obtain for $0° < \theta_i < 54.7°$ and for $125.3° < \theta_i < 180°$, negative values arise for $54.7° < \theta_i < 125.3°$.[100] Whilst for most molecular geometries positive values of $(3\cos^2\theta_i - 1)$ are probably the rule, and in any case internal rotation about the substrate–metal bond will to some extent average out the angular differences, nevertheless the angle-dependent term should not as a rule be neglected.

Clearly, the pseudo-contact term is the major contributor to the observed shift in the lanthanide shift reagents. Some contribution from the contact interaction, particularly for protons close to the metal, cannot wholly be discounted, although the unpaired spin is in a low-lying metal *f*-orbital and unlikely therefore to be involved to any great extent in metal–ligand bonding.[90,102] Larger shifts than predicted by a pseudo-contact interaction have been observed[85,89,93,96] and some slight variation in coupling constants noted with increasing concentration of shift reagent.[80] Other indications of contributions from contact shifts have also been noted.[103,104]

[100] B. L. Shapiro, J. R. Hlubucek, G. R. Sullivan, and L. F. Johnson, *J. Amer. Chem. Soc.*, 1971, **93**, 3281.
[101] P. H. Mazzucchi, H. J. Tamburin, and G. R. Miller, *Tetrahedron Letters*, 1971, 1819.
[102] F. A. Hart, J. E. Newbery, and D. Shaw, *J. Inorg. Nuclear Chem.*, 1970, **32**, 3585.
[103] M. Hitanowski, L. Stefaniak, H. Januszewski, and Z. M. Wolkowski, *Tetrahedron Letters*, 1971, 1653.
[104] J. K. M. Sanders and D. H. Williams, *Tetrahedron Letters*, 1971, 2813.

Despite its widespread use, equation (10) was derived for a set of conditions which restrict its application, particularly to transition-metal complexes. Kurland and McGarvey[105] have discussed these limitations, and have derived a rather more sophisticated expression which avoids these difficulties. LaMar et al.[106] have demonstrated that this expression best represents the temperature behaviour of the proton resonances of $Ni\{HB(pz)_3\}_2$. These considerations have yet to be applied to shift reagent complexes.

There seems to have been no detailed n.m.r. study of the species present in solutions of basic organic molecules and lanthanide shift reagents, although many incidental observations reported are pertinent to such a discussion. The solubility of $Eu(tmhd)_3$ in chloroform, for example, is enhanced by the presence of basic solutes,[85] and the strength of the solute–solvent bond, judged from the size of the induced paramagnetic shifts, appears qualitatively to depend on the basicity of the substrate.[80] Addition of the shift reagent also appears to modify the equilibrium position of conformationally mobile systems; the *exo*:*endo* ratio of 2,6-di-propylacetanilide depends on the relative concentration of shift reagent,[99] as do the conformational possibilities of thian *S*-oxides.[90] The organic substrate is clearly therefore bonded in some way to the metal, albeit rather more weakly than the diketone ligands since there is no evidence that these are displaced. The situation in this respect is closely analogous to that obtained for the Co^{II} and Ni^{II} shift reagents.

Most authors have commented that there is a linear dependence of paramagnetic shift on the shift reagent:substrate ratio,[80,85-87,92-94,96,100] although deviations at low[80,85,100] and high[81,94] ratios have been noted. Apparently, deviations at low shift reagent concentrations are generally slight and are unlikely to greatly affect the accuracy of chemical shifts inferred for the substrate by backwards extrapolation.[85] Rondeau and Sievers[81] have studied the paramagnetic shifts over the widest shift reagent concentration range reported to date, up to effectively complete complexation of the substrate. These authors caution against the over-ready assumption that only one species in the solution is contributing to the overall shift. The linearity of the shift *vs.* shift reagent concentration plots over relatively large portions of the curve seems to argue for the predominance of one contributor; there does, however, seem to be room for more detailed study on this point, since it is a basic assumption of much of the work reported to date.

[105] R. J. Kurland and B. R. McGarvey, *J. Magn. Resonance*, 1970, **2**, 286.
[106] G. N. LaMar, J. P. Jesson, and P. Meakin, *J. Amer. Chem. Soc.*, 1971, **93**, 1286.

10
Oriented Molecules

BY P. DIEHL AND P. M. HENRICHS

1 Introduction

This article is designed to cover publications appearing between July 1969 and May 1971. Wide-line n.m.r. studies of solids and of adsorbed molecules are not included, and the majority of the papers deal with the n.m.r. spectra of compounds dissolved in liquid crystalline solvents.

Several review articles dealing with n.m.r. spectroscopy of compounds dissolved in liquid crystal solvents have appeared during the time limits concerned.[1] The present article only takes account of literature that is more recent than that discussed by Diehl and Khetrapal.[1e]

Whiffen has made a comparison of the geometric parameters obtained by n.m.r. and by other methods.[2]

2 Structure Determination

N.m.r. spectroscopy is a useful means of studying the structures of molecules dissolved in liquid crystal solvents in spite of problems presented by the effects of vibrations and, in some cases, anisotropy in the indirect spin–spin couplings.

One of the major limitations, however, has been that the spectra of systems containing more than seven or eight nuclei contain so many lines that an exact analysis is impossible due to peak overlap. Very recent work has indicated that spectral simplification may in the future be possible through hetero-nuclear decoupling of partially deuterated large molecules. In spite of the fact that the deuterium spectrum of oriented molecules is ordinarily several kilohertz wide, decoupling can apparently be achieved nevertheless through irradiation of the very weak double-quantum transitions directly in the spectral centre.[3] Calculations based on the spectra of several deuteriated molecules combined should be done with care, however, since there are some

---
[1] (a) L. C. Snyder and S. Meiboom, *Mol. Crystals Liquid Crystals*, 1969, **7**, 181; (b) P. J. Black, K. D. Lawson, and T. J. Flautt, *Mol. Crystals Liquid Crystals*, 1969, **7**, 201; (c) R. Briere, Commissariat A L'Energie Atomique Publication CEA-BIB, 167; (d) A. Saupe, in 'Magnetic Resonance', ed. C. K. Coogan, N. S. Ham, S. N. Stuart, J. R. Pilbrow, and G. V. H. Wilson, Plenum Press, New York, 1970, p. 339; (e) P. Diehl and C. L. Khetrapal, in 'NMR, Basic Principles and Progress', ed. P. Diehl, E. Fluck, and R. Kosfeld, Springer-Verlag, Heidelberg, 1969, p. 1.
[2] D. H. Whiffen, *Chem. in Britain* 1971, **7**, 57.
[3] S. Meiboom and L. C. Snyder, *Accounts Chem. Res.*, 1971, **4**, 81.

indications of a measurable isotope effect on orientation.[4] For those compounds which do have analysable spectra, the direct (dipolar) coupling constants are most easily interpreted in terms of structure for relatively rigid molecules such as benzocyclopropene,[5] ethylene,[6] phenylacetylene,[7] spiropentane,[8] and pyrimidine,[9] for which proton structures have been obtained from the H–H direct couplings alone. The structure of an organometallic complex has also been obtained.[10]

In another rigid molecule, p-dinitrobenzene, an interesting problem has arisen, though, in that the calculated proton structure appears to be temperature dependent. Such a dependence is possibly due to partial formation of charge-transfer complexes between the solute and the solvent (4,4'-di-n-hexyloxyazoxybenzene), so that the observed couplings are averages for both the complexed and non-complexed molecules. There does seem to be evidence for a special interaction since p-dinitrobenzene orients much more strongly than do p-dichloro-, p-dibromo-, and p-di-iodo-benzene.[11]

The ^{13}C–H direct couplings in allene have been used in conjunction with the H–H couplings for a complete structure determination. However, the C–C bond length appears to be 4% shorter than the corresponding distance as determined by electron diffraction.[12]

Direct couplings to nitrogen have been observed and used to calculate structural parameters for acetonitrile[13] and methyl isocyanide.[14] An interesting feature of one of the determinations of the structure of acetonitrile, however, was that the shape was found to be slightly different for three different solutions.[13a]

A combination of H–H, ^{13}C–H, and Hg–H direct couplings has been used to calculate an unusually small H–C–Hg bond angle of 106·2° in methyl mercuric chloride.[15] For this work it was merely assumed that the anisotropy of the indirect Hg–H coupling is negligible, but research by Englert to determine structural parameters for dimethyl mercury suggests this is reasonable.[16]

Spiesecke has calculated a H–P–H bond angle of 94° 41' ± 3' and a P–H bond length of 1·427 Å for phosphine using a combination of n.m.r. and microwave results.[17]

[4] (a) P. Diehl and C. L. Khetrapal, *Canad. J. Chem.*, 1969, **47**, 1411; (b) C. S. Yannoni, *J. Chem. Phys.*, 1969, **51**, 1682.
[5] J. B. Pawliczek and H. Günther, *J. Amer. Chem. Soc.*, 1971, **93**, 2050.
[6] W. Bovée, C. W. Hilbers, and C. MacLean, *Mol. Phys.*, 1969, **17**, 75.
[7] P. Diehl and C. L. Khetrapal, *Org. Magn. Resonance*, 1969, **1**, 467.
[8] A. D. Buckingham, E. E. Burnell, and C. A. De Lange, *Mol. Phys.*, 1969, **17**, 205.
[9] C. L. Khetrapal, A. V. Patankar, and P. Diehl, *Org. Magn. Resonance*, 1970, **2**, 405.
[10] C. L. Khetrapal, A. C. Kunwar, C. R. Kanekar, and P. Diehl, *Mol. Crystals Liquid Crystals*, 1971, **12**, 179.
[11] P. L. Barili and C. A. Veracini, *Chem. Phys. Letters*, 1971, **8**, 229.
[12] E. Sackmann, *J. Chem. Phys.*, 1969, **51**, 2984.
[13] (a) G. Englert and A. Saupe, *Mol. Crystals Liquid Crystals*, 1969, **8**, 233; (b) D. N. Silverman and B. P. Dailey, *J. Chem. Phys.*, 1969, **51**, 1679.
[14] C. S. Yannoni, *J. Chem. Phys.*, 1970, **52**, 2005; see also ref. 13(b).
[15] M. Ayres, K. A. McLauchlan, and J. Wilkinson, *Chem. Comm.*, 1969, 858.
[16] G. Englert, *Z. Naturforsch.*, 1969, **24a**, 1074.
[17] H. Spiesecke, *Z. Naturforsch.*, 1970, **25a**, 650.

Oriented Molecules

The largest amount of work with hetero nuclei has involved fluorine. It now appears that anisotropy in some fluorine indirect couplings, particularly between two fluorine nuclei, is important, and many of the structure determinations have been for the purpose of detecting deviations from structures found by other methods. For example, Buckingham and Dunn found that structural deviations for 1,2-difluoroethylene might be the result of anisotropies in H–F indirect coupling. Somewhat better agreement was obtained for vinyl fluoride.[18]

It appears that there is also a large anisotropy in the F–F indirect coupling of 1,1-difluoroethylene,[19] but Gerritsen and MacLean have nevertheless obtained structural information by using spectra taken at different concentrations and temperatures to allow them to calculate the J-value anisotropies.[20] These same authors were able to calculate a structure for 1,2-difluorobenzene based on evidence that J-value anisotropies are negligible for that compound. Long and Goldstein found structural parameters for tetrafluoro-1,3-dithietan which were consistent with what they calculated on the basis of a model compound.[21] More discussion of fluorine couplings is contained in the section on J-value anisotropies.

Recently, there have been a number of attempts to extend the n.m.r. method to molecules in which rather large internal motions, such as occur during internal rotations or ring inversions, are important. The basic situation is that in such cases the D-values must be expressed as weighted averages over the path of the motion. In principle, then, there is information in the D-values about the shape and size of the barrier to the motion. It has, however, been difficult to find cases where D-values to a methyl group show very great sensitivity to the size and shape of the potential barrier. Dimethyl mercury exemplifies one such insensitive case for which Englert assumed free rotation in his final calculations.[16]

Diehl and co-workers found that structural parameters obtained for 2,6-dichlorotoluene also depend very little on assumptions made about the methyl rotations, and again a zero barrier height was used.[22] For a related molecule, tetrachloro-p-xylene, the ratio between the methyl-proton radius and the distance between the protons in the two methyl groups was also found using a free-rotation model.[23] There was a sensitivity to the barrier height in 3,5-dichlorotoluene, but there were not enough D-values to allow the structure, the orientation parameters, and a barrier height all to be determined.[24]

In contrast to the above cases, the spectrum of ethane does not agree so well with a theoretical one based on free rotation as with one based on equilibrat-

[18] A. D. Buckingham and M. B. Dunn, *Mol. Phys.*, 1970, **19**, 721.
[19] H. Spiesecke and A. Saupe, *Mol. Crystals Liquid Crystals*, 1969, **6**, 287.
[20] J. Gerritsen and C. MacLean, *Mol. Crystals Liquid Crystals*, 1971, **12**, 97.
[21] R. C. Long, jun., and J. H. Goldstein, *J. Chem. Phys.*, 1971, **54**, 1563.
[22] P. Diehl, C. L. Khetrapal, W. Niederberger, and P. Partington, *J. Magn. Resonance*, 1970, **2**, 181.
[23] D. Canet and P. Granger, *Compt. rend.*, 1971, **272**, *C*, 1345.
[24] P. Diehl, H. P. Kellerhals, and W. Niederberger, *J. Magn. Resonance*, 1970, **3**, 230.

ing perfectly staggered conformations. Again, however, there are not enough D-values to permit an exact calculation of the barrier height.[25] So far, only for o-chlorotoluene have a structure and a methyl-rotation barrier height (1200 ± 600 cal mol^{-1}) both been obtained.[26]

Cyclobutane is an example of a molecule which has a different kind of internal motion, and Meiboom and Snyder have obtained structural information based on an average of two equivalent non-planar structures.[27] Since the height of the barrier to ring inversion in this compound has been found by i.r. spectroscopy to be only $1 \cdot 44$ kcal mol^{-1}, however, it would seem that an appreciable error could be introduced by a failure to take into account structures over the whole pathway to ring inversion.[28]

Buckingham et al. have studied cyclohexa-1,4-diene, a molecule for which ring inversion between non-planar boat conformations may also be important. Nevertheless, the planar form was thought to be more consistent with the data, although it is not clear from the paper how the necessary average between equivalent non-planar structures was made in the calculations. The rather improbable chair and twist-boat forms were also considered.[29]

An interesting aspect of the study of molecules with internal motion is that there is a possible rate competition among three different processes: one for the n.m.r. experiment, one for the internal motion, and one for the reorientation of the molecule in the liquid crystal. Benzaldehyde was studied for the purpose of determining whether or not the molecule had time to reorient itself after each turn of the substituent from one planar form to another, but the results did not permit a clear decision.[30]

Although it is general practice to ignore small vibrational effects in structure calculations, corrections can be made by considering them also to be small internal motions over which the related D-values are averaged. A rather detailed study of this type has been made for the methyl halides.[31]

3 Orientation Parameters

When molecular structures are determined from the n.m.r. spectra of compounds dissolved in liquid crystals, information is also obtained about the way the molecules are oriented by the solvent. In cases for which the number of D-values does not permit a complete analysis, at least orientation parameters can be obtained from an assumed structure.[32] In all instances, however, (even when a complete analysis is possible) it is ordinarily necessary to refer

[25] D. N. Silverman and B. P. Dailey, *J. Chem. Phys.*, 1969, **51**, 655.
[26] P. Diehl, P. M. Henrichs, and W. Niederberger, *Mol. Phys.*, 1971, **20**, 139.
[27] S. Meiboom and L. C. Snyder, *J. Chem. Phys.*, 1970, **52**, 3857.
[28] J. M. R. Stone and I. M. Mills, *Mol. Phys.*, 1970, **18**, 631.
[29] A. D. Buckingham, E. E. Burnell, and C. A. De Lange, *Mol. Phys.*, 1969, **6**, 521.
[30] P. Diehl, P. M. Henrichs, and W. Niederberger, *Org. Magn. Resonance*, 1971, **3**, 243.
[31] (a) J. Bulthuis and C. MacLean, *Chem. Phys. Letters*, 1970, **7**, 242; (b) J. Bulthuis and C. MacLean, *J. Magn. Resonance*, 1971, **4**, 148.
[32] (a) B. M. Fung and M. J. Gerace, *J. Chem. Phys.*, 1970, **53**, 117; (b) J. Courtieu and Y. Gounelle, *Bull. Soc. chim. France*, 1969, 2951; (c) B. M. Fung and I. Y. Wei, *J. Amer. Chem. Soc.*, 1970, **92**, 1497.

Oriented Molecules

to a known sign of a J-value in order to determine the signs of the orientation parameters. For example, since the signs of the indirect (scalar) coupling constants of acetone and dimethyl sulphoxide are known, it was possible to determine that the axis passing through the centres of the methyl groups are oriented differently for the two compounds in a lyotropic mesophase.[33] In another approach, the anisotropy in the dielectric constant has also been used in conjunction with n.m.r. data to determine the signs of orientation parameters.[34]

Correlations of orientations found for various different compounds have indicated that dispersion forces are usually the primary orienting factor,[35] and the primary axis of orientation is usually along the direction of the largest feature of the molecule. Even for monodeuteriomethane it has been found that there is a slight tendency for the plane of the three protons to orient parallel to the optic axis of the liquid crystal, since the C–D bond is slightly shorter than are the C–H bonds.[4b] There is a very recent report[36] which indicates that for substituted benzenes the direction which most resists rotational motion is most strongly oriented. No theoretical basis has been presented for this empirical correlation of the S-tensor with the inertia tensor.

There has been some indication recently that molecules which are known to form good charge-transfer complexes show a higher degree of orientation than expected. For example, the S-value of the axis perpendicular to the ring plane is -0.297 for 1,3,5-trinitrobenzene but only -0.161 and -0.172 for 1,3,5-trichlorobenzene and 1,3,5-tribromobenzene, respectively, at 40 °C. The nitro-group does not appear to be larger than the iodo-group so that a special interaction must be important.[37] Similarly, p-dinitrobenzene is reported to orient more strongly than do p-dichloro-, p-dibromo-, or p-di-iodo-benzene.[11]

An unusual temperature effect on the orientation of methyl iodide and methyl bromide was detected in a variety of solvents. In these molecules the orientation increased with temperature until close to the nematic–isotropic transition point when it finally decreased. Since most compounds show decreasing order with temperature, there may be a special interaction. Similar results were obtained for methylene halides.[38]

Two distinct orientations could be observed simultaneously for both *trans*-1,2-dichloro- and *trans*-1,2-dicyano-ethylene in 4,4′-di-n-hexyloxy-azoxybenzene. Changes in temperature caused one of the superimposed spectra to disappear. These results suggest that it may be possible to have two distinct areas of different orientation in the same sample.[39]

N.m.r. can be used to gain other information about the liquid crystal through studies of the orientation of the solvent itself as well as of the mole-

---

[33] J. Lindon and B. P. Dailey, *Mol. Phys.*, 1971, **20**, 937.
[34] G. E. Chapman, E. M. Long, and K. A. McLauchlan, *Mol. Phys.*, 1969, **17**, 189.
[35] J. Nehring and A. Saupe, *Mol. Crystals Liquid Crystals*, 1969, **8**, 403.
[36] J. M. Anderson, *J. Magn. Resonance*, 1971, **4**, 231.
[37] J. Courtier, Y. Gounelle, and J. Jullien, *Bull. Soc. chim. France*, 1969, 4184.
[38] I. Morishima, A. Mizuno, and T. Yonezawa, *J. Amer. Chem. Soc.*, 1971, **93**, 1520.
[39] T. Yonezawa, I. Morishima, K. Deguchi, and H. Kato, *J. Chem. Phys.*, 1969, **51**, 5731.

cules dissolved in it.[40] Yannoni has observed that if a smectic phase, for example, is allowed to form in the magnetic field, a satisfactory n.m.r. spectrum can be obtained. The orientation is independent of the external field once the liquid crystalline phase is formed.[41] There have also been studies on biological systems through investigation of the orientation of deuteriated collagen[42] and of deuterium oxide dissolved in rat phrenic nerve.[43]

Diehl *et al.* have found that by using an electric field they can affect the orientation of liquid crystal solutions.[44] The direct couplings of solute molecules consistently changed to half their initial magnitudes with reversed signs at field strengths larger than 4 kV cm^{-1} applied in a direction perpendicular to the magnetic field of 1·4 T.

4 Anisotropies in Indirect Couplings*

In principle, the anisotropic couplings observed in the spectra of compounds dissolved in liquid crystal solvents consist of two parts: one resulting from direct dipole–dipole interactions and one from an anisotropic part of the indirect coupling through the electrons. It is not generally possible to separate these terms without prior structural knowledge so that the existence of the second term represents a limitation on the value of the n.m.r. method in calculating structures. When couplings among protons only are involved, however, theoretical considerations suggest that there is a negligible indirect anisotropic contribution.[45] The close agreement between the geometries calculated from proton–proton coupling constants alone and from microwave and electron diffraction results are consistent with such a conclusion.[1e]

There is some indication that the anisotropy in the indirect coupling cannot always be ignored when nuclei other than protons are involved. For compounds containing fluorine there may be substantial disagreement between the n.m.r. structures and those of other methods. The ratio between the H–H and geminal H–F distances in 1,2-difluoroethylene, for example, was found

* There are a number of different nomenclatures used to refer to the interactions involved here. For the purpose of this article we will use the term '*D*-value' to refer to the total observed anisotropic coupling which consists of two parts, one of which will be called the direct or dipolar coupling and the other the anisotropic indirect coupling. We may write this mathematically as

$$D_{ij} = D_{ij}^{\text{dir}} + D_{ij}^{\text{ind}}$$

The isotropic indirect coupling (trace of the coupling tensor) is identical with the coupling commonly observed for compounds in ordinary isotropic solvents.

[40] P. Pincus, *J. Physique*, 1969, **30**–C4, 8.
[41] C. S. Yannoni, *J. Amer. Chem. Soc.*, 1969, **91**, 4611.
[42] G. E. Chapman, I. D. Campbell, and K. A. McLauchlan, *Nature*, 1970, **225**, 639.
[43] M. P. Klein and D. E. Phelps, *Nature*, 1969, **224**, 70.
[44] P. Diehl, C. L. Khetrapal, H. P. Kellerhals, U. Lienhard, and W. Niederberger, *J. Magn. Resonance*, 1969, **1**, 527.
[45] (a) M. Barfield, *Chem. Phys. Letters*, 1970, **4**, 518; (b) H. Nakatsuji, H. Kato, I. Morishima, and T. Yonezawa, *Chem. Phys. Letters*, 1970, **4**, 607; (c) H. Nakatsuji, K. Hirao, H. Kato, and T. Yonezawa, *Chem. Phys. Letters*, 1970, **6**, 541; (d) A. D. Buckingham and I. Love, *J. Magn. Resonance*, 1970, **2**, 338.

Oriented Molecules

to differ from that obtained by microwave spectroscopy by 2·4%, indicating a possible anisotropy in the indirect H–F couplings. However, the vibrational effects in this molecule have not been analysed and the ratio of the F–F and H–H distances is the same as that for the microwave structure within experimental error.[18] There was also disagreement with microwave results for trifluoroethane, but this is not a good case because of the difficulty in accounting properly for the hindered rotation.[25] Long and Goldstein concluded that the D-values they calculated for tetrafluoro-1,3-dithietan based on a model compound were consistent with those observed.[21]

Gerritsen and MacLean have adopted a somewhat different approach for 1,2-difluorobenzene. They assumed that the anisotropy in the H–H and H–F indirect couplings would be negligible and calculated a structure based on them alone. They then calculated a F–F D-value which was in satisfactory agreement with experiment and concluded that anisotropies of indirect couplings are negligible for this molecule.[20]

On the other hand, Krugh and Bernheim have concluded that there is an anisotropy in both the C–H and C–F indirect couplings in $^{13}CH_3F$. There was a negligible anisotropy in the H–F coupling.[46] A theoretical study has also been made of this molecule[45c] which indicates that a theoretical basis for the large suggested anisotropy is missing. Bulthuis and MacLean have calculated that vibrations cannot be used to explain the results for CH_3F. The authors conclude that solvent effects on the anisotropic scalar coupling must be the cause of the discrepancy.[31]

Spiesecke and Saupe have concluded that there is a strong anisotropy in the indirect F–F couplings in 1,1-difluoroethylene and tetrafluoroethylene.[19] Gerritsen and MacLean reached similar conclusions for 1,1-difluoroethylene and were actually able to calculate the separate elements of the $J(FF)$ tensor by making measurements at a variety of temperatures and concentrations.[20]

There is much less information about anisotropies in indirect couplings to nuclei other than fluorine. There may be an effect for some ^{13}C–H couplings, however.[45c] An unusually short C–C bond length calculated for allene might be explicable by such an effect,[12] but there is recent evidence that vibrational effects may also be particularly important for directly-bonded nuclei.[3]

A geometry could be found for methyl isocyanide using ^{14}N–H and H–H D-values which was in good agreement with microwave results, suggesting that anisotropies in ^{14}N–H indirect couplings are small.[13a,14]

In an interesting case, Englert found[16] that there is a negligible anisotropy in the indirect ^{199}Hg–H coupling in dimethyl mercury. Since this is a linear molecule, the relative position of the mercury is automatically known, independent of any external data.

5 Anisotropies of Chemical Shifts

Chemical shifts of compounds dissolved in liquid crystals may, in general, be

[46] T. R. Krugh and R. A. Bernheim, *J. Chem. Phys.*, 1970, **52**, 4942.

divided into an isotropic and an anisotropic term. Studies of the oriented spectra should, in principle, give information about the anisotropic term, which should then be relatable to the electronic structure of the molecule. Measurements of the anisotropies of proton shifts have been hampered, however, by a difficulty in finding a good reference compound. Slightly different anisotropies are found when methane is used as a reference instead of TMS, which may be attributable to an anisotropy in the TMS shift relative to methane.[47] The prevailing opinion now seems to be that methane is the more reliable reference.

The fact that the isotropic shift term for some compounds may be subject to a slight solvent effect has also been a problem, but methods have now been developed for molecules requiring only one S-value to describe the orientation, thus avoiding this difficulty. In one of these procedures the shifts are measured with the nematic phase at a variety of temperatures. The observed shift Δ_{obs} may then be expressed as a linear function of the S-value and the anisotropic shift $\Delta\sigma$, so that $\Delta\sigma$ can be obtained as the slope of a plot of Δ_{obs} vs. S.[48]

$$\Delta_{obs} = \Delta_{iso} + \frac{2}{3} S \Delta\sigma$$

Interestingly enough, in several cases studied so far, the isotropic shift Δ_{iso} found by this method is within 0·01 p.p.m. of the value found directly using isotropic solutions.[48] The method has recently been applied in a study of benzene[49] and in a determination of anisotropies in ^{13}C shifts in methyl iodide and methyl cyanide.[50]

Nevertheless, the temperature-change method has provoked some discussion of its validity because of possible errors due to differential solvent effects.[51] A more reliable variation may be to produce different degrees of orientation with spinning rather than temperature changes.[47]

A problem in interpreting anisotropic shifts is that there may be substantial solvent effects due to ring currents in the aromatic rings which exist in many nematic systems. Nevertheless, at least for tribromobenzene and trichlorobenzene, similar chemical shift anisotropies have been calculated by the temperature variation method for both 4,4'-di-n-hexyloxyazoxybenzene and a lyotropic mesophase, $D_2O-C_{10}H_{21}SO_4Na-C_{10}H_{21}OH-Na_2SO_4$, as solvents.[48b]

The anisotropies in the proton chemical shifts of methyl isocyanide and methyl cyanide have been measured by a two-phase method both without correction for the solvent shift between the two phases and with a correction

[47] C. S. Yannoni, *IBM, J. Res. Development*, 1971, **15**, 59.
[48] (a) K. Hayamizu and O. Yamamoto, *J. Chem. Phys.*, 1969, **51**, 1676; (b) K. Hayamizu and O. Yamamoto, *J. Magn. Resonance*, 1970, **2**, 377.
[49] J. Lindon and B. P. Dailey, *Mol. Phys.*, 1970, **19**, 285.
[50] I. Miroshima, A. Mizuno, and T. Yonezawa, *Chem. Phys. Letters*, 1970, **7**, 633.
[51] (a) A. D. Buckingham, E. E. Burnell, and C. A. De Lange, *J. Chem. Phys.*, 1971, **54**, 3242; (b) K. Hayamizu and O. Yamamoto, *J. Chem. Phys.*, 1971, **54**, 3243.

of 0·1 p.p.m. The results indicate that there is a more negative anisotropy for methyl cyanide than for methyl isocyanide, but a suitable explanation was not found.[13b] The same authors used a similar method to study ethane.[25]

Contrary to the situation for protons, anisotropies in fluorine chemical shifts are usually so large that errors of a few tenths of a p.p.m. caused by solvent effects should be negligible. Nehring and Saupe have measured anisotropies in the fluorine shifts of a number of fluorobenzenes.[52] They found an internal consistency for the compounds but could not correlate the data satisfactorily with the theory of Karplus and Das. Rather poor agreement with the theory was also obtained for *sym*-trifluorotribromobenzene, 1,1,1-trifluoro-2,2,2-trichloroethane, and trifluoromethyl iodide.[53] Studies have also been made of *p*-substituted fluorobenzenes[54] and of 1,1,1-trifluoroethane.[25]

6 Relaxation

The fact that thermal orientational fluctuations in nematic liquid crystals may cause nuclear relaxation through modulation of the direct coupling was pointed out early.[55] The linewidth contribution, T_1^{-1}, was predicted to depend upon the absolute temperature, the viscosity, and the Larmor frequency as $T\eta^{\frac{1}{2}}\omega_0^{-\frac{1}{2}}$. Furthermore, the above relaxation mechanism was estimated to be competitive with the normal translational diffusive one. This theory has been extended by Blinc *et al.* by inclusion of magnetic field effects on the thermal fluctuations.[56]

Experiments with *p*-azoxyanisole and *p*-azoxyphenetole agreed with the dependence $T\eta^{\frac{1}{2}}$; however, the variation with the Larmor frequency ω_0^{-n} was found with $\frac{1}{2}<n<1$.[57] In contrast, the temperature dependence of T_1^{-1} for *p*-azoxyanisole and *p*-azoxyphenetole was later found to disagree with the theory, whereas the $\omega_0^{-\frac{1}{2}}$ dependence was confirmed.[58] The authors concluded that the theoretical treatment of T_1 based on nearly perfectly ordered molecules with small fluctuations about the orientational axis should be considered an estimate, and the treatment cannot be used to infer properties of the liquid crystalline state. They suggested a modified theory recognizing not only the collective order fluctuations but also short-range molecular reorientation.[59] The theory has been tested for quadrupole as well as proton spin–lattice relaxation.[60] The frequency dependence is still not understood. Related

---

[52] J. Nehring and A. Saupe, *J. Chem. Phys.*, 1970, **52**, 1307.
[53] C. S. Yannoni, B. P. Dailey, and G. P. Caesar, *J. Chem. Phys.*, 1971, **54**, 4020.
[54] C. T. Yim and D. F. R. Gilson, *J. Amer. Chem. Soc.*, 1969, **91**, 4360.
[55] P. Pincus, *Solid State Comm.*, 1969, **7**, 415.
[56] R. Blinc, D. L. Hogenboom, D. F. O'Reilly, and F. M. Peterson, *Phys. Rev. Letters*, 1969, **23**, 969.
[57] M. Weger and B. Cabane, *J. Physique*, 1969, **30**–C4, 72.
[58] J. W. Doane and J. J. Visintainer, *Phys. Rev. Letters*, 1969, **23**, 1421.
[59] J. W. Doane and D. L. Johnson, *Chem. Phys. Letters*, 1970, **6**, 291.
[60] (*a*) J. J. Visintainer, J. W. Doane, and D. L. Fishel, *Mol. Crystals Liquid Crystals*, 1971, **13**, 69; (*b*) J. A. Murphy and J. W. Doane, *Mol. Crystals Liquid Crystals*, 1971, **13**, 93.

work confirmed that in this field conclusions are still contradictory.[61] Data of different authors do not agree very well.

The predicted anisotropy of T_1 was confirmed by experiments in which an electric field was applied to orient the liquid crystal optic axis with respect to the magnetic field.[62]

In the nematic phase of the n-propyl ester of 4-methoxybenzylidene-4-amino-α-methyl cinnamic acid a marked observed change in relaxation behaviour was attributed to a transition from a weakly to a strongly hindered molecular motion. Possible relations to a recently-suggested uniaxial-to-biaxial nematic phase transition were discussed.[63]

Studies of relaxation times in systems of dimethyl dodecylaminoxide in D_2O described by Hansen and Lawson showed that there are several relaxation mechanisms effectively leading to non-Lorentzian lineshapes.[64] They pointed out that diffusion of molecules through microscopic inhomogeneities may also result in line broadening. These and other authors[65] pointed out the complexity of relaxation processes in non-isotropic systems and discouraged interpretations based solely on linewidth measurements or incomplete relaxation data.

Further measurements of relaxation times were performed on C_6D_6 dissolved in 4,4′-n-hexyloxyazoxybenzene, and the results were interpreted in terms of the Redfield theory adapted to partially-oriented molecules. A correlation time of 0.5×10^{-9} s was found as compared with 0.5×10^{-10} s in the isotropic phase.[66] A correlation time of 10^{-11} s was derived for the liquid crystalline phase of poly-γ-benzyl-L-glutamate in CH_2Cl_2 and $CHCl=CHCl$ from the solvent chlorine spectra.[67] Nuclear quadrupole relaxation in ^{14}N has also been studied in order to derive information on the orientational fluctuations.[68]

Krüger and Spiesecke have demonstrated, by spin-echo experiments, that strongly-ordered solute molecules such as acetylene form a coherent system with the nematic solvent, whereas weakly-ordered ones like TMS do not. In the latter the rate of diffusion is reduced to that of the nematic solvent; there are, however, still rapid rotations leading to long relaxation times.[69]

7 Determination of Nuclear Quadrupole Coupling Constants

Nuclear quadrupole coupling constants may be derived from the spectra of partially-oriented molecules containing, simultaneously, nuclei with spin $> \frac{1}{2}$

[61] A. Farinha-Martins, Third Liquid Crystal Conference, Berlin, 1970.
[62] (a) R. Y. Dong and C. F. Schwerdtfeger, *Solid State Comm.*, 1970, **8**, 707; (b) C. E. Tarr, M. A. Nickerson, and C. W. Smith, *Appl. Phys. Letters*, 1970, **17**, 318.
[63] R. Y. Dong, M. Marušič, and C. F. Schwerdtfeger, *Solid State Comm.*, 1970, **8**, 1577.
[64] J. R. Hansen and K. D. Lawson, *Nature*, 1970, **225**, 542.
[65] (a) A. Farinha-Martins, *J. Physique*, 1969, **30**–C4, 83; (b) R. Y. Dong, W. F. Forbes, and M. M. Pintar, *Solid State Comm.*, 1971, **9**, 151.
[66] Y. Egozy, A. Loewenstein, and B. L. Silver, *Mol. Phys.*, 1970, **19**, 177.
[67] B. M. Fung, M. J. Gerace, and L. S. Gerace, *J. Phys. Chem.*, 1970, **74**, 83.
[68] B. Cabane and G. Clark, *Phys. Rev. Letters*, 1970, **25**, 91.
[69] G. J. Krüger and H. Spiesecke, *Ber. Bunsengesellschaft Phys. Chem.*, 1971, **75**, 272.

Oriented Molecules

and with spin = $\frac{1}{2}$. The latter are used to determine the orientation. The coupling constants determined depend upon an assumed absolute internuclear distance. Such measurements were performed on partially-deuteriated acetonitrile,[70] [$^2H_3$]phenylsilane, [$^2H_2$]phenylphosphine, and [$^2H_1$]benzylthiol.[32c] The deuterium quadrupole coupling constants for the last three molecules mentioned correlate well with force constants of the corresponding covalent hydrides and with electronegativities of Si, P, and S respectively. A decrease of the coupling with the number of substituted nitro-groups was detected in substituted nitrobenzenes.[71]

In MeNC, the ^{14}N-quadrupole coupling constant was found to be 57% of the gas-phase value.[14] The reason for this large deviation is not known. Further ^{14}N-couplings were determined in MeCN, MeNO$_2$, and PhNO$_2$.[72]

The quadrupole splitting of n.m.r. lines has been used by Hilbers and MacLean for determining the degree of orientation in samples partially oriented by an electric field. Contrary to the use of nematic liquid crystal solvents, this method provides a very low degree of orientation of the order of 10^{-4}. On the other hand, the internal electric field may be measured, as well as short-range interactions in the liquids studied. The internal field was found to be in accordance with the Onsager model.[73]

8 The Structure of the Liquid Crystal Phase

N.m.r. methods may provide important information on various structural aspects of liquid crystal phases. Measurement of relaxation times, self-diffusion coefficients and, particularly for lyotropic systems, studies of the quadrupole splitting of D_2O, have been undertaken.

In systems of n-octylamine–D_2O, n-octylammonium chloride–D_2O and $C_9H_{19}\cdot C_6H_4(OCH_2CH_2)_9OH$–$D_2O$, the D_2O orientation was measured as a function of temperature and of the concentration of added amphiphilic solutes.[74] Generally, the degree of orientation decreased with temperature, but there were exceptions to this rule. Linewidth data of ^{35}Cl, ^{81}Br, ^{23}Na, and ^{85}Rb resonance as well as quadrupole splitting of ^{35}Cl and ^{37}Cl, were discussed in terms of ion hydration and mesophase structure.[75]

Blinc et al. measured the self-diffusion coefficients of water by a proton spin-echo method, as well as the quadrupole splitting of D_2O in various phases of sodium palmitate.[76] Whereas no orientation was detected in the gel, the middle and neat soap phases displayed spectra of the powder type with the orientation going through a maximum in the mesophase range. At higher

---

[70] P. Diehl and C. L. Khetrapal, *J. Magn. Resonance* 1969, **1**, 525.
[71] I. Y. Wei and B. M. Fung, *J. Chem. Phys.*, 1970, **52**, 4917.
[72] M. J. Gerace and B. M. Fung, *J. Chem. Phys.*, 1970, **53**, 2984.
[73] (a) C. W. Hilbers and C. MacLean, *Mol. Phys.*, 1969, **17**, 433; (b) C. W. Hilbers and C. MacLean, *Mol. Phys.*, 1969, **17**, 517; (c) C. W. Hilbers and C. MacLean, *Ber. Bunsengesellschaft Phys. Chem.*, 1971, **75**, 277.
[74] A. Johansson and T. Drakenberg, Third Liquid Crystal Conference, Berlin, 1970.
[75] (a) G. Lindblom and B. Lindman, Third Liquid Crystal Conference, Berlin, 1970; (b) G. Lindblom, H. Wennerström, and B. Lindmann, *Chem. Phys. Letters*, 1971, **8**, 489.
[76] R. Blinc, R. Easwaran, J. Pios, M. Volpan, and I. Zupančič, *Phys. Rev. Letters*, 1970, **25**, 1327.

temperatures, an isotropic peak indicated defects and penetration of water into the palmitate framework. The coefficient of self-diffusion increases exponentially in the mesophase range and drops markedly in the transition to the isotropic liquid, where the diffusion rate must be controlled by the mobility of the Na-palmitate group.

Related studies on potassium laurate–D_2O mixtures revealed quadrupole splitting of the D-resonance in lamellar and hexagonal phases whereas, in the cubic phase, only a peak corresponding to isotropic D_2O was observed.[77] In the lamellar phase the splitting goes through a minimum as a function of water content. The results indicate differences in water mobility between close-to-polar-head and end-of-chain positions. A slow exchange is suggested between two types of water molecules, one bound to the barrier surfaces another free from the surfaces.

In a study of the water lineshape in lyotropic liquid crystals, the observed shape and its variations with sample spinning were understandable on the basis of an anisotropic magnetic susceptibility stemming from the parallel arrangement of the amphiphilic molecules in the lamellae.[78] This arrangement leads to a predictable magnetic field distribution experienced by the protons.

9 Miscellaneous

Anderson and Lee have extended the density-matrix treatment of chemical exchange for use with oriented compounds. They find agreement of their theory with the observed spectra of [$^2H_3$]dimethylacetamide,[79] but there are real problems for such studies in general, owing to the limited temperature ranges available, the changes in D-values produced by temperature changes, and difficulties in finding suitable compounds. For another exchanging molecule, bullvalene, the spectrum is characteristic of ten fully equivalent protons, but there is little information available.[80]

The oriented (nematic phase) spectra of several 1,2-disubstituted ethanes have been reported, but only a little information about the relative signs of coupling constants could be obtained.[81] A crude analysis of the spectrum of naphthoquinone has also been reported.[82] The spectrum of 3,3'-bisisoazole was found to be consistent with the molecule being virtually completely in a *trans* conformation.[83] There is a report that a spectrum can be observed for 1,3,5-trichlorobenzene oriented in a sulphur matrix.[84]

[77] J. Charvolin and P. Rigny, *J. Physique*, 1969, **30**–C4, 76.
[78] T. Drakenberg, A. Johansson, and S. Forsén, *J. Phys. Chem.*, 1970, **74**, 4528.
[79] (*a*) J. M. Anderson and C.-F. Lee, *J. Magn. Resonance*, 1971, **4**, 160; (*b*) J. M. Anderson, *J. Magn. Resonance*, 1970, **3**, 427.
[80] C. S. Yannoni, *J. Amer. Chem. Soc.*, 1970, **92**, 5237.
[81] P. J. Swinton and G. Gatti, *Spectroscopy Letters*, 1970, **3**, 259.
[82] J.-M. Dereppe, J. Degelaen, and M. Van Meersche, *J. chim. Phys., Physiochim., Biol.*, 1970, 1875.
[83] P. Bucci, P. F. Franchini, A. M. Serra, and C. A. Veracini, *Chem. Phys. Letters*, 1971, **8**, 421.
[84] M. El Moghazi, R. R. Ernst, and S. H. Günthard, *J. Magn. Resonance*, 1970, **3**, 480.

Author Index

Abragam, A., 115, 264
Abraham, R. J., 91, 200, 201, 205
Abramson, K. H., 188
Ackerman, P., 183
Adam, W., 59
Adamson, J., 95
Adcock, W., 35, 49
Addad, J. P. C., 129
Adlard, M. W., 45
Adler, R. G., 49
Afana'ev, V. A., 260
Afronzi, B., 159
Aguiar, A. M., 97
Ahmad, N., 238
Ahmad, T., 46
Akitt, J. W., 245
Aksnes, D. W., 200
Alberty, R. A., 217
Albrand, J. P., 70, 84, 98, 207
Albriktsen, P., 100, 101, 200
Alei, M. jun., 49, 68, 297, 298
Alexander, M. N., 153
Alexander, S., 220, 261
Alger, T. D., 31, 269
Alla, M., 251
Allen, F. H., 77
Allerhand, A., 112, 125, 139, 140, 189, 247, 268, 269, 282, 283
Allingham, Y., 100
Allred, A. L., 19, 84, 237
Alm, T., 213
Aminova, R. M., 38
Amith, C., 111, 265
Amos, A. T., 8
Anderson, J. E., 137, 154, 188, 245
Anderson, J. M., 112, 219, 220, 258, 261, 325, 332
Anderson, R., 304
Anderson, W. A., 182, 183, 262
Andersson, L. O., 19
Andrew, E. R., 162, 180, 288
Andrews, B. D., 41, 42
Andrews, L. J., 306
Anet, F. A. L., 33, 265
Angerer, J., 131
Anker, M. W., 236
Anteunis, M., 79, 100
Anton, A., 302
Anwar, M. N., 46
Aono, S., 58
Applequist, J., 280

Appleton, J. M., 41
ApSimon, J. W., 22
Arata, Y., 137
Archer, R. A., 47
Arlinger, L., 220
Armour, E. A. G., 26
Armstrong, R. L., 119, 152, 187
Arnaud, P., 93
Arnold, B., 286
Arramenko, G. I., 261
Arrighini, G. P., 13
Avogadro, A., 159
Assink, R. A., 127, 128
Aubke, F., 15
Austin, W. K., 93
Axenrod, T., 43, 68
Ayres, M., 322

Babb, A. L., 142
Bacon, J., 86, 247
Bacon, M., 61
Baici, A., 25
Bailey, D. S., 232
Baird, M. C., 112, 252, 312
Baker, E. B., 172, 184, 264
Baker, K. M., 304
Baker, M. R., 14
Balahura, R. J., 239
Baldeschweiler, J. D., 248, 257
Baldwin, J. E., 101
Ban, N. T., 143
Banney, P. J., 209, 265
Banwell, C. N., 113, 258
Baram, A., 135
Barber, B. H., 226
Barbier, C., 61
Barfield, M., 53, 64, 80, 90, 100, 102, 263, 326
Barili, P. L., 68, 322
Barkhash, V. A., 108
Barsukov, L. I., 277
Barthel, J. S., 126
Barthels, M. R., 85
Bartle, K. D., 29, 41, 46, 202, 203, 230
Barton, D. H. R., 265
Bartow, D. S., 38
Bartuska, V. J., 173
Batiz-Hernandez, H., 42
Batterham, T. J., 196
Battino, F., 29
Baughcum, S. L., 105
Bavin, P. M. G., 29, 41, 46, 202, 203, 230
Beachley, O. T., 19
Beall, H., 213
Beare, S. D., 316

Beauté, C., 317
Becker, E. D., 4, 43, 67, 68, 140, 170, 262, 263, 268, 299
Becker, R. F., 317
Bélanger, R., 317
Belikova, N., 47
Bell, R. A., 265, 266
Bennett, J., 77
Benson, R. C., 34
Bent, H. A., 65
Bentley, M. D., 29
Bergel'son, L. D., 277
Bergman, J. J., 29
Bernal, D., 296
Berneth, T., 202
Bernheim, R. A., 42, 63, 327
Bernstein, H. J., 39, 119, 169, 209, 252
Bersohn, R., 56
Berthier, G., 61
Berti, G., 37, 68
Bertrand, R. D., 49, 88, 98, 206, 207, 306
Bertrude, W. G., 206
Berus, E. I., 108
Betsuyaku, H., 148
Bettess, P. D., 35
Bhacca, N. S., 97, 238
Bierbeck, H., 22
Bigg, D. C. H., 98
Bilofsky, H. S., 213
Binsch, G., 68, 214, 218, 219, 261
Birkle, M., 177
Bissett, F. H., 226
Bissey, J. E., 84
Black, P. J., 321
Black, S., 80
Bladon, P., 86, 157
Blake, P. H., 25
Blicharski, J. S., 122, 123, 125, 151
Blier, J. E., 80
Blinc, R., 144, 146, 147, 150, 156, 159, 165, 291, 329, 331
Blizzard, A. C., 55
Bloch, F., 257, 258, 259, 262
Bloembergen, N., 135, 147, 151
Bloom, M., 116, 117, 120, 138
Bloor, J. E., 46, 93
Blout, E. R., 279
Bloxbridge, J., 46, 50, 112
Blum, H., 159

333

Author Index

Blume, R. J., 171
Bocchi, V., 92
Bock, B., 32
Bock, E., 122
Boden, N., 173
Bodner, G. M., 47
Boekelheide, V., 31
Boicelli, C. A., 99
Bond, M. R., 253
Bonera, G., 159
Bopp, T. T., 125
Borsa, F., 159
Borsdorf, R., 218
Bothner-By, A. A., 54, 98, 172, 196
Boublík, M., 280
Bourn, A. J. R., 68, 265
Bovée, W., 322
Bovey, F. A., 26, 279, 280, 304
Boylan, D. W., 45
Bradbury, E. M., 280
Bradbury, J. H., 281
Bradley, R. B., 43
Bramley, R., 196
Bramwell, M. R., 91
Brandts, J. F., 282
Brassfield, T. S., 291
Breitmaier, E., 47, 48
Breskman, D., 43
Brevard, C., 245
Brewster, A. I., 280
Brey, W. S., 4, 225
Bridges, F., 116
Briere, R., 321
Briggs, J., 47, 316, 317
Bright, A., 69, 206
Bright, A. A. S., 306
Briguet, A., 122, 139
Brinkman, A. W., 105
Broekart, P., 148
Brown, D. H., 86
Brown, R. C., 142
Brown, R. J. C., 122
Brown, T. H., 49
Brownlee, R. F. C., 15
Brownstein, S., 188, 266
Brune, H. A., 67, 232
Bucci, P., 254, 259, 332
Buckingham, A. D., 34, 39, 45, 53, 297, 322, 323, 324, 326, 328
Budenz, R., 106
Bugge, A., 84, 202
Bull, T. E., 126
Bulthius, J., 63, 324
Burngardner, C. L., 71
Burd, L. W., 172, 184
Burg, A. B., 87
Burke, T. E., 126, 132
Burnell, E. E., 45, 322, 324, 328
Burnett, L. J., 187
Burton, K., 265
Burton, L. L., 295
Burton, R., 47, 175
Bury, A. B., 205
Bushweller, C. H., 213, 226
Buslaev, Yu. A., 73, 86
Butler, J. N., 302

Buys, H. R., 93
Bystrov, V., 188, 266
Bystrov, V. F., 260, 277

Cabane, B., 144, 145, 329, 330
Caccamese, S., 29
Callaghan, D., 39
Calzolari, A., 228, 296
Campbell, I. D., 136, 326
Camus, A., 25
Canet, D., 289, 323
Cant, G. P., 166
Cantacuzène, J., 97, 218
Capert, J., 173
Cárdenas, C. G., 41
Cargioli, J. D., 42, 87
Carhart, R. E., 70
Carlson, E. R., 171
Carlson, K. D., 305
Carman, C. J., 46, 285
Carr, H. Y., 138, 141, 142
Carter, R. E., 216
Carver, J. P., 136, 279
Castellano, S. M., 202
Casu, B., 47
Cattolica, A. M. L., 118
Cavaliere, P., 254, 259
Cavalli, L., 48, 91, 95, 97, 204, 205
Cavelius, E., 159
Ceasar, G. P., 44, 329
Ceccarelli, G., 81, 37, 200
Chakrabarti, B., 100, 263
Chan, L. I., 300
Chan, S. I., 6, 14, 126, 132, 274
Chan, S. O., 141, 144
Chandler, W. D., 29
Chang, C. Y., 296
Chapman, D., 276, 277
Chapman, G. E., 325, 326
Charrier, C., 97, 226
Charton, M., 45
Charvolin, J., 147, 274, 332
Cheney, A. J., 47, 72
Cheradame, H., 284
Cheung, C. S., 59
Chew, K. F., 49, 247
Chezeau, J. M., 157
Chiellini, E., 81, 200
Chien, J. C. W., 282
Chien, M., 125
Chierici, L., 92
Chiglien, G. S., 253
Chiu, Y.-C., 157
Chivers, P. J., 79
Chinelnick, A. M., 240
Chow, Y. L., 80
Chuck, R. J., 176
Church, M. J., 70
Chuvylkin, N. D., 102
Ciganek, E., 226
Cinader, G., 156
Clark, G., 330
Clark, H. C., 49, 85
Clark, W. G., 145, 155
Claxton, W. E., 216, 284
Clementi, E., 68
Clementi, H., 68

Cochran, D. W., 139, 282, 283
Cockerill, A. F., 317
Coggon, P., 77
Cogley, D. R., 302
Cogne, A., 98, 207
Cohen, J. S., 281
Cohen, M., 136, 137
Cohen, M. H., 142
Cohen, S. M., 195
Cohn, M., 180, 292
Cole, T. C., 46, 93
Collins, S. W., 269
Colson, B. L., 214
Colter, A. K., 47
Colton, R., 236
Combrisson, S., 105
Compton, R. D., 306
Connick, R., 238
Connick, R. E., 213
Conti, F., 228, 296
Constable, J. H., 154
Cook, D. B., 42, 59
Cookson, R. C., 100
Cooley, J. W., 187
Cooper, M. A., 46, 49, 59, 71, 90, 173, 204, 263
Cooper, R. D. G., 47
Cotterrell, G. P., 31
Cotton, F. A., 261
Cotts, R. M., 142
Courtieu, J., 324, 325
Cowherd, L. C., 16
Cowley, A. H., 68, 76, 235, 236, 264
Cox, B. G., 216, 299
Cox, R. H., 46, 90, 93, 208, 300, 301
Crabb, T. A., 79, 100
Craig, L. C., 279, 282
Craig, W. G., 22
Crain, W. O., 47, 226, 285
Crane-Robinson, C., 280
Crapo, L. M., 117
Crecely, K. M., 82, 201
Crecely, R. W., 46, 82, 105, 201
Creswell, C. J., 213
Crosbie, K. D., 69, 86
Crowe, K. M., 43
Crump, D. R., 317
Cuniberti, C., 283
Cunliffe, A. V., 61, 101
Cuthbert, J. D., 155
Czubryt, J., 122

Dabrowski, J., 229
D'Agostino, J. T., 45, 230
Dahl, B. M., 99
Dahlqvist, K.-I, 213, 215, 216, 220
Dai-Ki Kang, 87
Dailey, B. P., 44, 322, 324, 325, 328, 329
Dale, A. J., 204
Danchin, A., 137
Daniel, E. S., 144
Daniels, F., 217
Das, T. P., 6, 56, 58
Davies, A. M., 42

Author Index

Davies, D. W., 15
Davis, D. D., 303
Davis, D. R., 68
Davis, F. A., 49
Davis, J. C. jun., 215
Davis, J. P., 266
Daxenbichler, M. E., 305
Dayan, E., 253
Daycock, J. T., 152
Dean, P. A. W., 73, 86, 245
Deane, J. W., 144
Deb, K. K., 46, 93, 215
Deber, C. M., 279
Debye, N. W. G., 84
Debye, P., 121
De-Clercq, M., 85
Degani, D., 156
Degelaen, J., 332
Deguchi, K., 325
de Haan, J. W., 91
Dehl, R. E., 290
De-Jeu, W. H., 56
de Kowalewski, D. G., 81, 201
De Lange, C. A., 45, 322, 324, 328
Della, E. W., 37
Delman, J., 122
Delpuech, J.-J., 225
Del Re. G., 66
Demarco, P. V., 22, 47, 317
Dence, J. B., 47
Dent, E., 312
Derbyshire, W., 49, 173, 247
Dereppe, J.-M., 332
Deutch, J. M., 115, 116
Devarajan, V., 67
Deverell, C., 125, 135, 189
Deville, G., 143
Deviller, J., 98
Dewar, M. J. S., 29, 31, 235
Dewkett, W. J., 213
Diaz, E., 100
Diehl, P., 209, 321, 322, 323, 324, 326, 331
Dierdof, D. S., 236
Diez, E., 201
DiGennaro, T. M., 125
Di Giacomo, A., 259
Dillon, K. B., 180
Dimic, V., 146
Dischler, V. B., 197
Ditchfield, R., 10, 55, 59
Dixon, J., 131
Dixon, W. T., 93
Doak, G. O., 73
Doane, J. W., 144, 145, 146, 329
Doddrell, D., 47, 97, 112, 139, 226, 269, 282, 283
Dodson, B., 306
Dombroski, J. R., 284
Dong, R. Y., 116, 117, 120, 144, 145, 330
Dorn, H. C., 47
Doskočilová, D., 289
Douglas, A. W., 79

Douglas, J. R., 109
Drabkin, G. M., 137
Drago, R. S., 19, 315
Drake, P. W., 130
Drakenberg, T., 215, 220, 275, 331, 332
Dreeskamp, H., 69, 70, 72, 76, 81, 85, 225, 263
Drozd, G. I., 48, 85
Du Bois, R., 148
Dubov, S. S., 48, 85
Dubrovnia, N. I., 277
Dufourcq, J., 157
Dugas, H., 46, 105
Dunell, B. A., 287
Dungan, C. H., 48
Dunmur, R. E., 73
Dunn, M. B., 323
Duplan, J.-C., 122, 139
Durand, A., 229
Durand, H., 229
Durette, P. L., 285
Dutta, C. M., 58
Dutta, N. C., 58
Dwek, R. A., 136, 241, 292, 317
Dyer, D. S., 86

Eastmond, G. C., 286
Easwaran, K., 147, 291
Easwaran, R., 331
Eaton, D. R., 310, 311, 316
Edzert-Maksić, M., 65
Egmond, G. J. N., 154, 288
Egozy, Y., 146, 188, 330
Ehrenberg, L., 165
Ehrlich, R. S., 138
Eisenberg, D., 127
El-Hanany, U., 156
Elleman, D. D., 44, 46, 90, 162
Ellett, J. D., 162
Ellis, A. F., 233
Ellis. P. D., 60
Ellner, J. J., 43
El Saffar, Z. M., 44, 154
Elvidge, J. A., 50, 112
Elzey, T. E., 317
Emanuel, R. V., 57, 59
Emerson, M. T., 195
Emsley, J. W., 35, 46, 93, 169, 203
Endo, K., 112
Engel, R., 112, 252
Englar, J. R., 286
Engler, E. M., 307
Englert, G., 192, 322
Erbeia, A., 139
Erlich, R. H., 301
Ernst, R. R., 171, 175, 184, 190, 258, 332
Evans, D. F., 73, 263
Evans, E. A., 50, 112
Evans, W. A. B., 161
Everett, G. W., 252
Ewing, D. F., 67, 81
Eyring, H., 142

Fackler, J. P. jun., 315
Falconer, W. E., 44
Farinha-Martins, A., 330
Farrar, T. C., 123, 140, 170, 268
Faucher, H., 61
Fedorov, L. I., 297
Feeney, J., 169, 261, 263, 286, 304
Feigenson, G. W., 274
Fenton, D. E., 84
Ferguson, R. C., 284
Ferretti, J. A., 83, 140, 268
Fessenden, R. W., 26
Fiat, D., 240
Ficarra, A., 24
Filleux-Blanchard, M. L., 229
Finer, E. G., 253, 277
Finocchiaro, P., 29
Fischer, R. D., 226, 246
Fishel, D. L., 329
Fixman, M., 123
Flatau, K., 32
Flautt, T. J., 321
Flint, W. L., 302
Florin, A. E., 49, 68, 297, 298
Flygare, W. H., 15, 34
Flynn, C. P., 180
Forbes, W. F., 145, 330
Ford, C. M., 43
Foreman, M. I., 208, 304, 305, 306
Forsén, S., 198, 213, 215, 216, 220, 257, 260, 275, 332
Foster, A. B., 95, 262
Foster, R., 305, 306
Fowler, R. H., 296
Franchini, P. F., 332
Franconi, C., 228, 296
Frankel, L. S., 238
Fraser, R. R., 30, 59, 79, 317
Fratiello, A., 303
Freeburger, M. E., 46, 67
Freedman, L. D., 49
Freeman, R., 139, 140, 141, 176, 184, 189, 190, 254, 255, 258, 264, 267, 268
Freppel, C., 317
Freude, D., 164, 287
Friebolin, H., 226, 286
Friedel, R. A., 31
Friedman, H. L., 302
Fritz, H., 110
Frost, G. H., 316
Fryer, P. A., 245
Fuhr, B. J., 217, 226
Fujieda, K., 47, 67
Fujii, Y., 33
Fujisawa, T., 46
Fujita, K., 25
Fukui, H., 103, 202
Fukumi, T., 137
Fukuta, K., 84
Fukuyama, M., 81
Fung, B. M., 49, 111, 255, 261, 289, 293, 301, 324, 330, 331

M

Author Index

Furtsch, T. A., 236
Fyfe, C. A., 305

Gabuda, S. P., 48
Gagarinskii, Y. U., 48
Gagnaire, D., 49, 70, 84, 98, 207, 262
Gaines, J. R., 154
Gambler, W., 106
Gansow, O. A., 46
Gardini, G. P., 92
Gardner, P. D., 31
Garrels, J. I., 228
Gatti, G., 332
Geils, R. H., 177
Gelan, J., 79
Geldard, J. F., 16
Gentzler, R. E., 240
Gerace, L. S., 330
Gerace, M. J., 324, 330, 331
Geraldes, C. F. G. C., 59
Gerig, J. T., 233, 294
Gerlack, D. H., 225
Gerritsen, J., 323
Gerritsma, C. J., 119
Gerritz, W. H., 137
Geschke, D., 156
Gestblom, B., 84, 176, 198, 202, 255, 258
Gestblom, G., 203
Giam, S., 48
Gibbons, W. A., 279, 282
Gielen, M., 84, 85
Gierer, A., 124
Giger, W., 67, 262
Gil, V. M. S., 59, 67
Gilbao, H., 246
Gill, D. F., 77
Gillen, K. T., 126, 127, 216, 247
Gillespie, R. J., 72, 86, 245, 247
Gillies, D. G., 43, 169, 174, 176, 214, 267
Gillies, E., 112, 252, 312
Gilson, D. F. R., 329
Ginsburg, D., 111, 188, 265.
Glasel, J. A., 189, 291
Gleicher, G. J., 31
Glickson, J. D., 280
Glushko, V., 139, 283
Glydes, H. R., 157
Goetz, H., 93
Goggin, P. L., 85
Goldammer, E. V., 131
Goldman, M., 115
Goldstein, J. H., 44, 46, 47, 64, 82, 93, 105, 201, 203, 285, 323
Goldstone, J., 58
Goldwhite, H., 84
González, M. P., 47
Goodfellow, R. J., 85
Goodisman, J., 15
Goodwin, B. W., 217, 266
Gordon, M., 11, 105
Gordon, R. G., 123
Gordon, S. L., 132, 255, 261, 269

Gorenstein, D., 234, 235
Gornastansky, S., 156
Gorton, J., 306
Gosnell. J. L. jun., 232
Goto, K., 70, 76, 313
Goto, T., 156
Gounelle, Y., 324, 325
Grace, M., 213
Graddon, D. P., 310
Gränicher, H., 156
Gragerov, I. P., 214
Graham, L. L., 296
Granger, P., 289, 323
Grannell, P. K., 165
Grant, C. W M., 262, 265
Grant, D. M., 31, 47, 53, 65, 80, 139, 197, 263, 269
Gray, G. A., 47, 67
Gray, K. W., 4, 112
Greeley, D. N., 49
Green, P. J., 49
Greifenstein, L. G., 232
Gribble, G. W., 109
Griffin, G. E., 95, 204
Griffin, R. G., 44, 162
Griffith, D. L., 214
Griffith, D. R., 66
Grimison, A., 59
Grindley, T. B., 312
Grinter, R., 49, 61
Grobet, P., 154
Grolini, J., 226
Gromb, S., 217
Gross, B., 143
Grover, T., 143
Gründemann, E., 218
Gründer, W., 164
Grunwald, E., 302
Grunwell, J. R., 15
Grutzner, J. B., 47
Gubaidullina, R. Z., 38
Günthard, W. H., 332
Günther, H., 107, 263, 322
Gueron, M., 137
Guggenberger, L. J., 225
Gurd, F. R. N., 139, 283
Gutowsky, H. S., 35, 54, 59, 122, 143, 173, 197, 224
Gylulai, H. G., 226

Hadamik, H., 93
Hadari, Z., 156
Haddon, R. C., 30
Haddon, V. R., 30
Haeberlin, U., 44, 138, 148, 161, 162, 164
Hägele, G., 98
Hafner, S., 180
Hague, J. F., 43
Hahn, E. L., 142, 165, 166
Haigh, C. W., 26, 90, 209
Halford, M. H., 226
Hall, C., 215, 247, 248
Hall, L. D., 47, 95, 175, 262, 265
Hallé, J.C., 107

Halliday, J. D., 177
Halls, P. J., 79
Haly, A. R., 290
Hamer, G. K., 90
Hammel, J. C., 65
Hammes, G. G., 293
Hampson, P., 49, 262
Hanabusa, M., 164
Hanebeck, H., 67
Hansen, E. A., 87
Hansen, J. R., 146, 173, 291, 330
Hansen, K. C., 97
Haque, F., 208, 304
Hardy, W. N., 116
Hargis, J. W., 206
Harmon, J. F., 130, 187
Harris, L., 226
Harris, R. K., 61, 74, 77, 83, 101, 112, 139, 191, 208, 209, 210, 212, 213, 214, 243, 253, 269
Harrison, L. W., 301
Harrison, P. G., 69
Harrison, W. G., 208
Hart, F. A., 47, 316, 317, 319
Hart, P. A., 266
Hartland, A., 165
Hartman, J. S., 72, 247
Hartman, S. R., 165
Hartmann, O., 84, 198, 202, 255, 258
Hartwell, G. E., 188
Harvey, R. G., 105
Haseda, T., 156
Haslam, E., 94
Haupt, J., 148
Hauser, H., 277
Hawkins, B. L., 226
Haworth, O., 183
Hawthorne, M. F., 49
Hayamizu, K., 45, 89, 328
Haynes, D. H., 293
Heidberg, J., 223
Hems, R., 262
Hengge, E., 69
Henold, K. L., 94
Henrichs, P. M., 324
Herbison-Evans, D., 172
Herring, F. G., 15
Hertz, H. G., 128, 131, 143
Hess, S., 117
Hetz, W., 232
Heuring, V. P., 225
Hewitt, R. C., 175
Hewson, M. J. C., 85
Hickmott, P. W., 46
Highberger, G., 302
Hilbers, C. W., 105, 322, 331
Hildenbrand, K., 72, 225
Hilgetag, G., 218
Hill, D. T., 226
Hill, H. D. W., 139, 140, 141, 177, 189, 190, 254, 267, 268
Hill, N. E., 131
Hinchliffe, A., 59

Author Index

Hinckley, C. C., 315, 317
Hindermann, D. K., 44
Hindmann, J. C., 128
Hinshaw, W. S., 119
Hinton, J. F., 296
Hirai, A., 156
Hirao, K., 55, 326
Hirst, R. C., 197
Hitanowski, M., 319
Hlubucek, J. R., 319
Hoarau, J., 56
Hobbs, C. W., 67
Hoefler, F., 69
Hoffman, E. G., 141
Hoffmann, R. A., 198, 215, 257, 260
Hogben, M. G., 72
Hogenboom, D. L., 124, 144, 153, 329
Hogen-Esch, T. E., 300
Holm, C. H., 224
Holm, R. H., 240
Homan, J. M., 85
Homer, J., 39
Horrocks, W. De W. jun., 240, 312
Horton, D., 285
Horvitz, E. P., 151
Howarth, O. W., 172
Howell, J. A. S., 72
Howery, D. G., 170
Huang, S., 24
Hubbard, P. S., 118, 119, 122
Huber, L. M., 44, 161
Huckerby, T. N., 105
Hudson, A. T., 46
Huestis, W. H., 283
Hüther, H., 67
Huitric, A. C., 25
Hultgren, G. O., 31
Hunt, B. I., 130
Hunt, E. R., 119, 138
Huntress, W. T., 125, 215
Hutton, H. M., 217, 226
Hynes, T. V., 153

Ichirawa, Y., 305
Ignatova, N. P., 97
Ihrig, A. M., 300
Iio, T., 278
Inamoto, N., 25
Isemura, T., 278
Ishii, Y., 76
Ivanov, E. N., 121, 149
Ivanova, T. M., 215
Ivin, S. Z., 48, 85
Iwata, Y., 48

Jackman, L. M., 30
Jackson, W. R., 235
Jaeckle, H., 138
Jaeschke, A., 226
Jaffé, H. H., 45, 230
Jakobsen, H. J., 71, 97, 199, 206, 262
Jameson, A. K., 35, 73
Jameson, C. J., 35, 59
Jantzen, R., 97, 218
Januszewski, H., 238, 319
Janzen, A. F., 104

M*

Janzen, E. G., 208
Jasinski, A., 288
Jautelat, M., 47
Jeener, J., 148
Jefford, C. W., 100
Jeffrey, K. J., 152
Jenkins, J. M., 88
Jenkins, P. N., 183, 265
Jenks, G. J., 128, 246
Jennings, W. B., 235
Jensen, R. K., 233
Jerome, F. R., 109
Jesson, J. P., 225, 320
Jhon, M. S., 142
Johannesen, R. B., 83
Johansson, A., 275, 331, 332
Johnson, A., 252
Johnson, C. E., 26, 304
Johnson, C. S., 120, 214, 223
Johnson, D. L., 145, 329
Johnson, H. W., 48
Johnson, L. F., 47, 282, 319
Johnson, M. D., 64
Johnson, P. A., 142
Johnson, R. N., 95
Jolley, K. W., 253
Jonas, J., 125, 126, 127, 128, 189
Jonassen, H. B., 99
Jones, A. J., 31
Jones, D. W., 29, 41, 46, 202, 203, 230
Jones, E. P., 155
Jones, E. W., 280
Jones, G. P., 152
Jones, J. R., 46, 50, 112
Jones, R. A. Y., 41, 79
Jones, R. G., 209
Jones, V. I. P., 232
Jordan, R. B., 238, 239
Joseph-Nathan, P., 47, 100
Joubert, F. J., 278
Juan, C., 54
Juds, H., 93
Jugie, G., 69
Jullien, J., 325
Jung, A., 112, 252
Jung, G., 47, 48
Junge, H., 32
Jurado, B., 237
Jurga, J., 140
Jurga, K., 131, 140

Kabuss, S., 226
Kainosho, M., 98
Kaiser, R., 190, 265
Kalck, P., 236
Kamezawa, N., 111, 224
Kamiénski, B., 16
Kamitani, T., 85
Kampa, L., 282
Kanekar, C. R., 322
Kanellakopulos, B., 226, 246
Kang, D. K., 205
Kanofsky, A., 43
Kaplan, J., 220
Kaplan, J. I., 151

Karim, A., 22
Karplus, M., 16, 54, 55
Karra, J. S., 121
Katô, H., 14, 55, 63, 325, 326
Kato, Y., 59
Katritzky, A. R., 41, 46, 79, 90, 284
Kauzmann, W., 127
Kavakoff, D. S., 260
Kawamoto, J. M., 85
Kawamura, K., 135
Kawasaki, K., 121, 138
Kearns, D. R., 286
Keat, R., 253
Keefer, R. M., 306
Keilich, G., 286
Keller, H. J., 214, 310
Kellerhals, H. P., 323, 326
Kellie, G. M., 47
Kelly, D. P., 175
Kemmerer, G. E., 121
Kemp, R. H., 91
Kern, C. W., 8
Kessemeier, H., 163
Kessenikh, A. V., 97
Kessler, H., 214, 260
Khanagov, A. A., 290
Kher, V. G., 132
Khetrapal, C. L., 321, 322, 323, 326, 331
Khodos, I. I., 156
Khodosov, E. F., 156
Khosla, A., 117
Khrapov, V. V., 46, 49
Kietrich, W., 143
Kimmich, R., 133
Kimura, B. Y., 46
King, N. L. R., 281
Kintzinger, J. P., 245
Kitching, W., 49
Kleier, D. A., 218, 219
Klein, H. F., 99
Klein, M. P., 139, 177, 326
Klemm, A., 142, 143
Klemperer, W., 72
Kleppner, D., 5
Klimova, A. I., 46, 49
Klotz, I. R., 278
Klotz, M. R., 317
Kluiber, R. W., 312
Knapp, P. S., 302
Knispel, R., 155
Koch, H. J., 47
Kohovec, J., 104
Koketsu, J., 76
Kosfeld, R., 143
Koster, D. F., 108
Kovar, R. A., 46, 248
Kowalewski, V. J., 264
Kowalsky, A., 180, 293
Kozerski, L., 229
Krishna, N. R., 244, 259
Krüger, D. P., 142, 330
Krugh, T. R., 63, 327
Krygowski, T. M., 16
Krynicki, K., 150
Kubo, R., 236
Kubo, V., 303

Kucheryaev, A. G., 241
Kugatova-Shernyakina, G. P., 215
Kuhlmann, K. F., 269
Kuhr, M., 32
Kumar, A., 244, 255, 259, 269
Kundla, E., 260
Kunitomo, M., 166
Kuntz, I. D., 291, 292, 301
Kunwar, A. C., 322
Kurland, R. J., 47, 320
Kuthan, J., 90
Kydon, D. W., 247, 248

Lacey, M. J., 66
Lack, R. E., 46
Lacombe, L., 316
Ladd, J. A., 214, 232
Lahajnar, G., 150, 156, 159
Laisaar, S., 270
Lallemand, J. Y., 202
La Mar, G. N., 237, 240, 251, 252, 270, 312, 320
Lambert, J. D., 68, 232
L'amie, R., 41, 230
Landau, M. A., 48, 85
Landesman, A., 143
Lang, D. V., 166
Langford, C. H., 239, 240
Lanir, A., 294
La Planche, L. A., 296
Lardon, M., 46, 99
Larsen, D. W., 245
Laszlo, P., 303, 307
Laurent, J. P., 69
Laussac, J. P., 69
Lauterbur, P. C., 282
Lavrencic, B., 152
Laws, E. A., 13
Lawson, K. D., 146, 321, 330
Lawson, P. J., 139, 283
Lazzeretti, P., 16, 66
Leach, J. B., 49, 246
Lebedev, V. A., 241
Lecourt, M.J., 208
Lee, A. C.-F., 112, 219, 220
Lee, A. G., 242
Lee, C.-F., 332
Lee, R. E., 303
Lee, S. S., 117
Leffler, A. J., 239
Lehman, P. G., 41
Lehn, J. M., 84, 231, 245
Leifson, O. S., 151
Leigh, J. S., jun., 180, 223
Lenk, R., 129
Lenzi, M., 49
Lequan, R.-M., 103
Le Quan Minh, 85, 208
Letcher, R., 265
Lett, R. G., 233
Letter, J. E., 238
Leupold, I., 305
Levy, G. C., 47, 87
Lewis, D. W., 316

Lewis, R. B., 317
Lichter, R. L., 82, 99, 111, 245, 269, 299
Lienhard, U., 326
Lincoln, S. F., 239
Lindblom, G., 135, 331
Lindman, B., 135, 248, 331
Lindon, J., 44, 325, 328
Lindqvist, I., 248
Lipman, A., 188
Lippert, E., 266
Lippi, G., 37
Lippmaa, E., 47, 184, 189, 246, 251, 270
Lipscomb, W. N., 8, 13, 14
Lipsicas, M., 120
Litchman, W. M., 31, 49, 65, 68, 298
Locker, D. R., 189
Lockhart, N. C., 157
Loewenstein, A., 146, 188, 246, 330
Logan, N., 49, 247
Long, E. M., 325
Long, G. J., 178
Long, K. R., 105
Long, R. A., 25
Long, R. C., jun., 44, 64, 323
Look, D. C., 189
Lotan, N., 278
Lotfullin, R. Sh., 149
Loudet, M., 217
Louick, D. J., 224
Loustalot, F., 217
Love, I., 53, 64, 326
Lubensky, T. C., 145
Luck, W., 296
Lütje, H., 140
Lunazzi, L., 66, 99
Lurie, F. M., 165
Lustig, E., 87
Lutz, R. E., 45
Luz, Z., 135, 241
Lyerla, J. R., 139
Lyle, J. L., 48
Lynch, L. J., 229
Lynden-Bell, R. M., 15, 55, 214

McArthur, D. A., 165
McCall, D. W., 224, 283
Macchia, B., 37
Macciantelli, D., 66
MacClement, W. D., 155
McClung, R. E. D., 123
McConnell, H. M., 55, 89, 223, 308, 311
McCourt, F. R., 117
Macdonald, C. G., 66
McDonald, M. P., 276
McEwen, G. K., 98, 206, 207
McFarland, W., 69
McFarlane, N. S., 45
McFarlane, W., 18, 65, 68, 72, 95, 109, 257, 262, 264
McGarvey, B. R., 320

Maciel, G. E., 47, 60, 61, 67, 173
McIver, J. W., 10, 54, 60, 67
Mackay, K. M., 49, 246
McKeever, L. D., 47
McKinnon, D. M., 82
McLachlan, L. A., 146, 275
McLauchlan, K. A., 322, 325, 326
MacLean, C., 63, 105, 322, 323, 324, 331
McNeel, M., 281
Macomber, R. S., 102
McPhail, A. T., 77
McQuilkin, R. M., 31
McWeeny, R., 8, 10, 26
McWilliam, D. C., 265
Maestro, M., 13
Maher, J. P., 263
Maier, W., 197
Majoral, J. P., 98
Maksić, Z. B., 65
Mali, M., 165
Malinowski, E. R., 67, 302
Malisch, W., 84
Mallion, R. B., 26, 90
Malmberg, M. S., 123
Mamlok, L., 316
Manatt, S. L., 46, 49, 59, 71, 76, 90, 173, 204, 263, 264
Mandell, L., 135
Mann, B. E., 47, 49, 69, 70, 72, 77, 191, 206
Manscher, O., 71
Mansfield, P., 161, 162, 165, 166
Maraviglia, B., 154
March, N. H., 142
Marianelli, R. S., 239
Maricic, S., 180
Mark, V., 228
Marks, R. E., 46
Marks, T. J., 261
Marquet, A., 316
Marsden, K., 150
Marshall, A. G., 245
Martin, G. J., 229
Martin, J., 98, 207
Martinelli, M., 254, 259
Martin-Smith, M., 46
Marton, J., 216
Marušič, M., 144, 330
Marynick, D., 32, 249
Maryott, A. A., 123
Mason, J., 19, 20, 49, 247
Masters, C., 49, 69, 70, 77, 206
Mathias, A., 49, 262
Mathieson, D. W., 22
Mathieu, R., 49
Matsubayashi, G., 24
Matsuo, T., 305
Matwiyoff, N., 19
Matzen, P. F., 218
Mays, M. J., 70
Mazzucchi, P. H., 319
Meakin, P., 225, 320
Mefed, A. E., 148
Mehring, M., 44, 162, 163, 287

Author Index

Meiboom, S., 321, 324
Meister, R., 130
Mel'nikov, N. N., 97
Metcalf, B. W., 31
Metras, F., 217
Meyer, G. H., 126
Meyer, H., 154
Meyers, S. M., 154
Michel, D., 156
Michelson, C. E., 49, 73, 86
Mildvan, A. S., 136
Miller, C. R., 132
Miller, D. P., 10
Miller, G. R., 319
Millett, F., 44
Mills, I. M., 324
Milovidova, N. D., 297
Miroshima, I., 328
Mitchell, G. H., 31
Mitsch, C. C., 49
Miyajima, G., 46
Mizuno, A., 70, 325, 328
Moccia, R., 13
Mochel, V. D., 216, 284
Mock, J. B., 178
Mock, W. L., 202
Modak, S. G., 132
Moghazi, M. El., 332
Mohanty, S., 119, 252
Moir, R. Y., 29
Moiseev, B. M., 297
Molin, Yu. N., 108
Mondelli, R., 92
Moniz, W., 131
Montavdo, G., 29
Montgomery, D. J., 143
Montgomery, L. P., 155
Moores, B. M., 187
Morallee, K. G., 292, 317
Moran, P. R., 166
Moreland, C. G., 49, 71
Morgan, G. L., 46, 85, 248
Morgan, L. O., 135, 239
Morgan, R. E., 124, 189
Moriarty, R. M., 224
Morishima, I., 63, 70, 84, 112, 312, 313, 314, 325, 326
Moritz, A. G., 178
Morrow, C. J., 97
Moss, G. P., 47, 316, 317
Moss, K. C., 72
Mountain, R. D., 123
Müllen, K., 104
Müller, D., 159, 287
Müller, K. P., 143
Müller-Warmuth, W., 142
Muetterties, E. L., 225, 245
Muir, A. R., 141
Mukherjee, R., 224
Muller, B. H., 130
Muller, N., 297
Munnings, R. H., 187
Muratova, A. A., 98
Murayama, D. R., 81
Murday, J. S., 142
Murphy, J. A., 146, 329
Murrell, J. N., 53, 55, 59, 113

Musher, J. I., 26
Musso, H., 32, 305
Myint, T., 5

Nachtrieb, N. H., 180
Nagel, M., 156
Nageswara Rao, B. D., 258, 259
Nakamura, A., 98
Nakamura, K., 135
Nakatsuji, H., 55, 63, 326
Nakazaki, M., 31
Nakushima, T. T., 173
Namanwirth, E., 260
Nasielski, J., 84, 85
Natusch, D. F. S., 146, 270, 275
Navech, J., 98
Navon, G., 188, 293, 294
Navon, S., 137
Neely, J., 238
Negrebetskii, V. V., 97
Nehring, J., 192, 325, 329
Nelson, J., 98
Nelson, J. H., 99
Némethy, G., 279
Nemorin, J., 46
Neranov, Yu. I., 137
Neu, J.-M., 289
Newbery, J. E., 319
Newman, R. H., 146, 275
Newmark, R. A., 74
Newton, R. F., 100
Ng, S., 108, 229
Nichols, D. I., 18
Nickerson, M. A., 146, 330
Nieboer, E., 292, 317
Niederberger, W., 323, 324, 326
Nielson, P. H., 99
Nishioka, A., 134
Nist, B. J., 25
Nixon, D. E., 136
Nixon, J. F., 68, 264
Noack, F., 130, 133
Noble, A. M., 18, 72, 86
Noble, J. D., 138, 151
Noggle, J. H., 126, 127, 216, 247, 266
Nomura, Y., 40, 305
Norbury, A. M., 49, 247
Norman, A. D., 77
Norris, M. O., 150, 159, 180
Nosel, W., 123
Nukada, K., 262, 299
Nummelin, A. J., 136

O'Brien, J. F., 49
Oda, R., 25
Odle, R. L., 180
Ogilvie, F. B., 88
Ogoshi, H., 41
Ohashi, M., 314
Ohmori, Y., 84, 312
Ohorodnyk, H. O., 311
Ohtsuru, M., 81, 84
Okada, K., 313, 314
Okamura, M., 76
Okazaki, R., 25

Olah, G. A., 69, 175
Oldfield, E., 276
Oliver, J. P., 85, 106
Olivson, A., 184
Olympia, P. L., jun., 111, 255, 261
Onak, T., 32, 49, 246, 249
Onaka, S., 50, 85
O'Neil, J. W., 226
Oosting, P. H., 119, 138
Oppenheim, I., 115, 116, 121
O'Reilly, D. E., 44, 124, 134. 144, 153, 154, 155, 329
Ormonoroyd, S., 303
Ortis, C. E., 233
Osredkar, R., 165
Ostlund, N. S., 10, 54, 55, 60, 67
Ostroff, E. D., 161
O'Sullivan, W. J., 137
Oth, J. F. M., 31
Ouchinnikov, I. M., 241
Overhauser, A., 264
Ozier, I., 4, 112, 117

Paasivirta, J., 47
Pachler, K. G. R., 59, 91
Packer, K. J., 133
Pajak, Z., 131, 137, 140
Palke, W. E., 14
Paolillo, L., 67, 68, 262, 299
Parfitt, R. T., 46
Parker, J., 214
Parker, R. G., 189
Parry, K., 91, 200, 201
Parry, K. A. W., 81
Partington, P., 323
Pascard-Billy, C., 105
Past, J., 184
Patankar, A. V., 322
Patel, D. J., 282, 286
Patil, F., 317
Pattison, V. A., 228
Pauwels, P. J. S., 263, 304
Paviot, J., 56
Pawliczek, J. B., 107, 322
Peak, S., 303
Peake, S. C., 85
Pearce, C. D., 46, 90
Pederson, B., 155
Peel, W. E., 276
Peeling, J., 217, 226
Peet, W. G., 225
Pehk, T., 47
Peller, S., 241
Pellizer, G., 25
Perlin, A. S., 47
Perraud, R., 93
Perry, W. D., 315
Pesek, J. J., 293
Petch, H. E., 155
Peterson, E. M., 44, 124, 134, 144, 153, 154, 155, 329
Peterson, J., 159

Peterson, M. R., jun., 17, 317
Petrakis, L., 108, 233
Petrii, O. P., 41, 108
Petrissans, J., 217
Petrosyan, V. S., 108
Petrosyants, S. P., 73, 86
Phelps, D. E., 139, 177, 326
Philipsborn, W. V., 265
Phillips, L., 35, 183
Phillips, W. D., 310
Phillpott, E. A., 303
Picard, M., 84
Pidcock, A., 77, 88
Pierre, J. L., 93
Pignolet, L. H., 240
Pilling, G. M., 31
Pince, R., 236
Pincus, P., 145, 326, 329
Pines, A., 44, 163, 287, 288
Pinkowitz, R. A., 250
Pinschmidt, R. K., 101
Pintar, M. M., 145, 150, 155, 156, 330
Pinto, A. J. L., 67
Pios, J., 331
Pirkle, W. H., 316
Pirs, J., 147, 291
Pivcovaá, H., 104
Platé, A., 47
Peoshe, W. H., 46
Pogorelyi, V. K., 214
Pälblanc, R., 49, 236
Pongs, O., 281
Pople, J. A., 9, 10, 11, 16, 25, 26, 53, 54, 60, 64, 67, 169, 209, 242
Popov, A. I., 301
Porter, R. D., 175
Pound, R. V., 147
Poupko, R., 246
Powles, J. G., 122, 130, 137, 150, 159
Praat, A. P., 260
Prausnitz, J. M., 304
Pregosin, P. S., 47
Preissing, G., 130
Prelesnik, A., 165
Pressman, B. C., 293
Price, R., 47
Primas, H., 258
Prins, K. O., 118
Pritchard, A. M., 131
Pross, A., 66
Pugmire, R. J., 47
Pujol, R., 98
Purcell, E. M., 141, 142, 147
Purcell, W. L., 239
Pyper, N. C., 77, 112, 214, 217, 243

Qureshi, A. M., 15

Rabenstein, D. L., 241
Rabideau, P. W., 105
Rackham, D. M., 317
Radley, K., 72
Rae, I. D., 41, 42

Raftery, M. A., 283
Ragsdale, R. O., 49, 73, 186
Rahkamma, E., 68
Ramasseul, R., 262
Ramaswamy, K., 67
Ramey, K. C., 224, 226
Ramsey, N. F., 5, 14, 51, 117
Randall, C. M., 143
Randall, E. W., 43, 47, 68, 91, 176, 214, 257, 317
Randic, M., 65
Rankin, D., 84
Rao, B. D. N., 215, 244
Rao, G. U., 226
Rao, U. R. K., 247
Rassat, R., 262
Ray, G. J., 47
Raynes, W. T., 17, 34, 39, 42, 296
Raza, M. A., 34, 296
Read, J. J. R., 99
Reddy, G. S., 73
Redfield, A. G., 147, 178, 214, 258
Reed, F. J. S., 85
Reed, G. H., 137
Reeves, L. W., 72, 133, 224
Reger, J. P., 73
Reich, H. J., 226
Reichert, B. E., 41
Reidlinger, A. A., 24
Reiter, R. C., 297
Remane, H., 218
Renaud, R. N., 30, 59, 79
Retcofsky, H. L., 31
Reuben, J., 136, 137, 292, 297
Reutov, O. A., 108
Reynolds, W. F., 39, 90, 91
Rhim, W. K., 44, 163, 287, 288
Richards, K. H. B., 161
Richards, R. E., 77, 131, 136, 170, 172, 177, 183, 214, 247, 248, 292
Richardson, J. W., 38
Riche, C., 105
Richer, J. C., 317
Rico, M., 201
Riddell, F. G., 47, 84, 216
Rider, K., 123
Ridley, A. B., 46
Rietz, R. R., 49
Rigamonti, A., 159
Rigny, P., 147, 150, 158, 274, 332
Rimerman, R. A., 294
Ripmeester, J. A., 287
Rix, C. J., 236
Rizvi, S. Q. A., 35
Roach, E., 301
Robert, J. B., 49, 84, 98, 207
Roberts, B. W., 68
Roberts, G. C. K., 261, 286
Roberts, H. G. Ff., 8

Roberts, J. D., 47, 68, 70, 71, 82, 97, 99, 111, 172, 179, 214, 226, 228, 245, 269, 285, 299
Robertson, R. E., 311
Robins, M. J., 47
Robins, R. K., 47
Robinson, H. G., 5
Robinson, L. G., 43
Rodak, M. I., 148
Rodmar, S., 203
Rogers, M. T., 296
Ronayne, J., 307
Rondeau, R. E., 316
Roomi, M. W., 46, 105
Root, G. N., 184
Roques, B., 105
Rose, P. I., 281, 297
Rossknecht, H., 69
Rossotti, F. J. C., 317
Rotaru, M., 216
Rothberg, J. E., 43
Roussel, J., 236
Rowbotham, J. B., 104
Rowsell, D. G., 84
Royden, V., 257
Rubenstein, M., 135
Ruddell, V. A., 25
Ruddick, J. D., 49, 85
Rudolph, R. W., 74
Rüterjans, H., 281
Rummens, F. H. A., 2, 91

Sack, R. S., 236
Sackmann, E., 85, 322
Sadlej, A. J., 7, 15
Saika, A., 16, 59
Saitô, H., 299
Salman, S. R., 46, 93, 203
Saluvere, T., 184, 189, 246, 270
Samitov, U. U., 90, 98
Sanders, J. K. M., 316, 317, 319
Sandhu, H. S., 142
Sano, H., 85
Santry, D. P., 53, 55
Santucci, S., 254, 259
Sardella, D. J., 90, 106
Sarko, A., 284
Sarney, S. G., 293
Sasaki, Y., 50, 85
Sato, K., 134
Sauer, J., 219
Saunders, J. K., 265, 266
Saunders, L., 22
Saunders, M., 225, 236
Saupe, A., 192, 321, 322, 323, 325, 329
Savitsky, G. B., 43
Schaefer, J., 285
Schaefer, T., 39, 82, 104, 105, 217, 226
Schaefer, T. S., 297
Schaeffer, C. D., 66
Schaeffer, R., 49
Schamp, N., 100
Schastnev, P. V., 102
Schaumburg, K., 199
Schechter, A. N., 281
Scheie, C. E., 44, 124, 153

Author Index

Schenck, A., 43
Schenck, G. E., 33
Scheraga, H. A., 278
Schick, H., 218
Schirmer, R. E., 266
Schlegel, J., 143
Schmid, H. D., 166
Schmid, G., 246
Schmid, H. G., 226
Schmidbaur, H., 84, 99
Schmidpeter, A., 69
Schmiedel, H., 164, 287
Schmutzler, R., 70, 73, 74, 85
Schneider, B., 289
Schneider, H., 123
Schneider, H. J., 48
Schneider, W. G., 19, 34, 169, 209, 297
Schröder, G., 31
Schuerch, C., 284
Schulz, G., 177
Schulz, G. W., 77, 214
Schultz, C. W., 74
Schumann, C., 69, 70, 76, 81, 263
Schumann, K., 69
Schuster, R. E., 303
Schwartz, M., 127
Schwarzhans, K. E., 214, 310
Schweiger, J. R., 68, 76, 264
Schweitzer, D., 138
Schwerdtfeger, C. F., 144, 330
Sciacovelli, O., 105, 111, 265
Sebastian, J. F., 15
Seddon, L. G., 49
Sederholm, C. H., 108
Seel, F., 106
Seifert, E., 286
Seiter, C. H. A., 274
Selbin, J., 238
Semin, G. K., 149
Sergeev, N. M., 41, 108
Sergeyev, N. M., 261, 262
Serra, A. M., 254, 259, 332
Servers, K. L., 205
Servis, K. L., 66, 87, 109
Seville, L., 311
Seymour, S. J., 125
Shannon, J. S., 66
Shapet'ko, N. N., 41, 108, 297
Shapiro, B. L., 106, 319
Sharp, A. R., 150
Sharp, D. W. A., 86
Sharp, R. R., 247, 248
Shaw, B. L., 47, 49, 69, 70 72, 77, 206
Shaw, C. F., 19, 84
Shaw, D., 169, 263, 267, 319
Shaw, K. N., 224
Shaw, R. A., 253
Sheldrick, G. M., 69, 242
Sheline, R. K., 44
Sheluchenko, V. V., 48, 85

Sheppard, N., 55
Sherer, S., 295
Sherman, E. O., 240
Sherwood, J. N., 157, 159
Shevelev, S. A., 247
Shigorin, D. W., 297
Shimizu, H., 125
Shimizu, T., 148
Shimokawa, S., 202
Shimomura, K., 122
Shimuzu, S., 76
Shinokawa, S., 103
Shiotoni, A., 99
Shodery, J. N., 46, 47, 49, 179
Shoup, R. R., 139, 267
Shporer, M., 241
Shrager, R. I., 281
Shulman, R. G., 282, 286
Shvetov-Shilovskii, N. I., 97
Sidall, T. H., tert., 45, 214, 317
Sidwell, W. T. L., 110
Siegel, M. M., 120
Siegert, A., 262
Sievers, R. E., 316
Sillescu, H., 251
Silver, B. L., 146, 188, 246, 330
Silverman, D. N., 322, 324
Simon, P., 297
Simon, W., 67, 262, 265
Simonnin, M.-P., 85, 103, 107, 208
Sinha, A. I. P., 49, 247
Sioto, H., 262
Skála, V., 90
Slade, R. M., 69, 206
Slichter, C. P., 16, 115, 165, 166, 224
Slusher, R. E., 166
Smail, G. A., 46
Smid, J., 300
Smith, A., 284
Smith, A. J., 49
Smith, C. W., 146, 330
Smith, D. W. G., 122
Smith, J. A. S., 65
Smith, R. A., 47
Smith, W. B., 300
Snarey, M., 79
Snowden, B. S., 125, 126, 157
Snyder, L. C., 321, 324
Socrates, G., 45
Sogn, J. A., 282
Sohma, J., 103, 202
Solomon, I., 129
Sondheimer, F., 31
Sone, T., 47, 67
Sono, H., 50
Spagnolo, P., 90
Speight, P. A., 119
Spencer, P. R., 166
Spialter, L., 46, 67
Spielman, J., 49
Spiesecke, H., 19, 322, 323, 330
Spiess, H. W., 44
Sprangle, P. A., 59

Spratt, R., 98
Springer, C. S., jun., 237
Stacey, L. M., 44, 162
Stainbank, R. E., 69, 206
Staniforth, M. L., 316
Stanko, V. I., 46, 49
Stec, W., 74
Steele, J., 188
Steele, W. A., 131, 149
Stefaniak, L., 238, 319
Steigel, A., 219
Steiner, P. R., 47, 265
Stejskal, E. O., 130, 142
Steltzer, O., 76
Stempfle, W., 141
Stengle, T. R., 240, 248
Stepisnik, J., 159
Stern, A., 279, 282
Sternhell, S., 66, 78
Stevens, R. M., 13
Stewart, R. P., 18
Stewart, W. E., 45, 214
Stilbs, P., 220
Stobart, S. R., 49, 246
Stock, L. M., 108
Stoddart, J. F., 200
Stodulski, L. P., 85
Stokes, J., 210, 212
Stone, J. M. R., 324
Storey, R. A., 46, 93, 203
Storhoff, B. N., 47
Stothers, J. B., 105
Strange, J. H., 124, 150, 157, 159, 180, 189
Stringer, J. R., 143
Strom, E. T., 125
Stynes, D. V., 237
Sudmeier, J. L., 81, 293
Sullivan, G.-R., 319
Summerhays, K. D., 60
Sutcliffe, L. H., 169, 253
Svanholm, U., 233
Svare, I., 156
Svirmickas, A., 128
Swaelens, G., 79
Swain, J. R., 68, 264
Swift, T. J., 213, 250
Swinton, P. J., 332
Sykes, B. D., 189
Sykora, S., 244, 250
Symons, M. C. R., 303
Szarek, W. A., 312
Szczesiak, E., 131
Sze, S. N., 77
Szerek, W. A., 200

Tabushi, I., 25
Taddei, F., 16, 66, 90, 99
Tänzer, C., 47
Taft, R. W., 15
Takahashi, K., 46, 47, 67
Takaya, Y., 24
Takeuchi, Y., 40, 46, 90, 305
Takumitsu, T., 41
Tallman, D. E., 293
Tam, J. W. O., 278
Tamburin, H. J., 319
Tamm, C., 110
Tanaka, S., 31
Tanaka, T., 24, 85

Author Index

Tanaka, Y., 299
Tangerman, A., 24
Tanner, J. E., 130, 142
Tarasov, U. P., 86
Tarpley, A. R., jun., 46, 47, 82, 93, 203, 285
Tarr, C. E., 146, 330
Taylor, R. P., 301
Tebbe, F. N., 225, 245
Terenzi, M., 159
Ternay, A. L., 105
Terrier, F., 107
Terry, H. W., jun., 301
Thiem, J., 317
Thijs, L., 24
Thoai, N., 317
Thomas, W. A., 91, 95, 201, 204, 215
Thomson, H. B., 296
Tiddy, G. J. T., 253, 273
Tiecco, M., 90, 99
Tison, J. K., 119, 138
Tizané, D., 317
Tobias, R. S., 67
Todd, D. K., 22
Todd, L. J., 47
Tomchuk, E., 122
Tomlinson, D. J., 133
Tonelli, A. E., 280
Tori, K., 81, 84
Torrey, H. C., 142, 188
Townsend, L. B., 25
Tracey, M. M., 80
Trappeniers, N. J., 119, 138
Trautmann, P., 289, 301
Trautwein, W.-P., 98
Travernier, T., 79
Treichel, P. M., 18
Trepanier, D. L., 79
Tsiang, H. G., 239
Tsuchihashi, G., 46
Tukey, J. W., 187
Tunstall, D. P., 156
Tupciauskas, A., 262
Turchi, I. J., 49
Turkel, P. M., 97
Turnbull, D., 142
Turner, A. B., 45
Turner, D. W., 141
Turner, M. J., 94
Turpin, M. A., 55, 59, 113
Twisleton, D. R., 305
Tzalmona, T., 152

Ulrich, S. E., 69, 84
Ustynyuk, Yu. A., 261, 262

Valic, M. I., 156
Van Artsdalen, E. R., 295
van Bronswijk, W., 19, 49
van Denason, F. W., 183
van den Berghe, E. V., 47
Vandendunghen, G., 84
Vanderhart, D. L., 139, 267
Van der Hart, W. J., 62
van der Kelen, G. P., 47

van Deursen, F. W., 265
Van Geet, A. L., 179
van Gerven, L., 154
van Hecke, G. R., 251, 252
van Hecke, P., 154
Van Leenwen, P. W. N. M., 260
Van Meersche, M., 332
van Putte, K., 154, 288
van Steenwinkel, R., 148
van Wazer, J. R., 48
Varga, J. A., 59
Vary, S., 100
Vasileff, T. P., 108
Vasyanina, L. K., 297
Vaughan, R. W., 44, 162
Velenek, A., 15
Venables, J. A., 157
Venanzi, L. M., 172
Veracini, C. A., 322
Verkade, J. G., 49, 88, 98, 206, 207, 306
Verrier, J., 49, 98, 207
Vičar, J., 305
Vidal, M., 93
Vilfar, M. J., 159
Vincens, M., 93
Vinogradov, L. I., 98
Virlet, J., 150, 158
Visintainer, J. J., 144, 329
Visser, H. D., 85, 106
Vladimiroff, T., 67
Voelter, W., 47, 48
Vogel, E., 151
Vogel, G., 90
Vogrin, F. J., 302
Vold, R. L., 44, 139, 141, 143, 144, 161
Volfan, M., 291, 331
Vrscaj, S., 150
von Ammon, R., 226, 246
von Bredow, K., 226
von Bronswijk, W., 247
von Jouanne, J., 223
von Philipsborn, W., 105, 111, 257

Waack, R., 47
Waddington, T. C., 180
Wade, C. G., 125
Waegell, B., 100
Wagner, J., 231
Wahl, G. H., jun., 17, 317
Waite, R. O., 302
Waldman, L., 117
Walker, B. J., 98
Wallach, D., 149, 150
Walstedt, R. E., 165
Walter, W., 317
Wander, J. D., 238
Wang, R., 134
Wang, C. H., 44, 161
Warashina, A., 278
Ware, D., 161
Warwick, A., 188
Wasielewski, M. R., 108
Wasylishen, R., 104, 105
Waterhouse, C. R., 77

Waugh, J. S., 26, 44, 118, 120, 139, 140, 161, 162, 163, 164, 287, 288
Webb, G. A., 215, 310
Weber, H. E., 49, 71, 173, 204, 263
Weber, W. P., 66
Weger, M., 144, 329
Wehrli, F. W., 67, 262, 265
Wei, S. C., 49
Wei, I. Y., 324, 331
Weigert, F. J., 71, 172, 228
Weinbaum, S., 57
Weinhaus, F., 154
Weisleder, D., 305
Wells, E. J., 143, 188
Wells, P. R., 209, 265
Wenkert, E., 139, 283, 331
Wennerström, H., 331
Wessels, P. L., 91
West, R. J., 239
Westerman, C. E., 296
Westhead, R., 49, 262
Westheimer, F. H., 234
Whalley, W. B., 22
Whiffen, D. H., 258, 321
White, D., 134
White, D. W., 49, 98, 206, 207
Whitehurst, P. W., 224
Whitesides, G. M., 316
Whyte, J. N. C., 286
Widdowson, D. A., 265
Wieder, M. J., 68
Wigfield, Y. Y., 317
Wildman, W. C., 47
Wilkinson, J., 322
Willard, A. K., 66
Williams, D. A. R., 216
Williams, D. H., 307, 316, 317, 319
Williams, D. L., 43
Williams, J. M., 155
Williams, R. J. P., 292, 317
Williams, R. W., 43
Williamson, M. P., 202
Wilson, R. G., 304
Wimett, T. F., 58
Winfield, J. M., 18, 72, 86
Wing, W. H., 171
Winstead, M. B., 228
Wirtz, K., 124
Wise, W. B., 224
Wiseman, M. N., 136
Witanowski, M., 238, 247
Wittekoek, S., 184
Woessner, D. E., 125, 126, 157
Wolfan, M., 147
Wolkowski, Z. W., 238, 317, 319
Wong, C. M., 217, 226
Wood, D. J., 90, 91
Wood, M., 128
Woplin, J. R., 74, 208
Wright, C. H., 97
Wulz, K., 232
Wyluda, B. J., 282

Author Index

Xavier, A. V., 292, 317

Yannoni, C. S., 332
Yalymova, S. V., 90
Yamada, F., 225
Yamada, H., 45
Yamamoto, H., 85
Yamamoto, K., 31
Yamamoto, O., 45, 89, 111, 224, 328
Yamane, T., 282
Yang, P. P., 255, 261
Yannoni, C. S., 44, 322, 326, 328, 329

Yarkova, E. G., 98
Yatsiv, S., 254
Yi, P., 117
Yim, C. T., 329
Yoder, C. H., 66
Yonemoto, T., 39, 266
Yonezawa, T., 55, 63, 70, 84, 112, 312, 313, 314, 325, 326, 328
Yoshida, Z., 41
Young, K. K., 43
Yue, C. P., 133

Zambelli, A., 285
Zamir, D., 156
Zeltmann, A. H., 239
Zhidomirov, G. M., 102
Zielen, A. J., 128
Ziessow, D., 111, 266
Zschunke, A., 218
Zuckerman, J. J., 43, 69, 84
Zürcher, R. F., 23.
Zumdahl, S. S., 59
Zumer, S., 159
Zupancic, I., 147, 156, 165, 291, 331
Zwanenberg, B., 24
Zwanig, R., 121